Microbial Biotechnology

Principles and Applications

Second Edition

Microbial Biotechnology

Principles and Applications

Second Edition

Editor

Lee Yuan Kun

National University of Singapore

NEW JERSEY • LONDON • SINGAPORE • BEIJING • SHANGHAI • HONG KONG • TAIPEI • CHENNAI

Published by

World Scientific Publishing Co. Pte. Ltd.

5 Toh Tuck Link, Singapore 596224

USA office: 27 Warren Street, Suite 401-402, Hackensack, NJ 07601

UK office: 57 Shelton Street, Covent Garden, London WC2H 9HE

British Library Cataloguing-in-Publication Data
A catalogue record for this book is available from the British Library.

MICROBIAL BIOTECHNOLOGY (2nd Edition)
Principles and Applications

ISBN 981-256-676-7
ISBN 981-256-677-5 (pbk)

Typeset by Stallion Press
Email: enquiries@stallionpress.com

Printed in Singapore by B & JO Enterprise

List of Contributors

Anwar, Azlinda, BSc (Hons), PhD
Division of Biomedical Sciences, Johns Hopkins Singapore
31, Biopolis Way, #02-01, The Nanos
Singapore 138669
Email: azlinda@jhs.com.sg

Candlish, John, PhD, LLM
Professor, Head of Department of Basic Medical Sciences
Faculty of Medicine and Health Sciences
University of Malaysia
Sarawak, Kuching, East Malaysia
Email: jkcandlish@dominomail.unimas.my

Chow, Vincent T. K., MBBS, MSC, DipMicrobiology, FRC Path, MD, PhD
Associate Professor, Head of Department
Department of Microbiology, National University of Singapore
5 Science Drive 2, Singapore 117597
Email: micctk@nus.edu.sg

Chu, Lee Man, BSc (Hons), PhD
Associate Professor
Department of Biology, The Chinese University of Hong Kong
Shatin, NT, Hong Kong SAR, China
Email: leemanchu@cuhk.edu.hk

Chua, Kim Lee, BSc (Hons), Dip Ed, PhD
Senior Lecturer
Department of Biochemistry, National University of Singapore
8 Medical Drive, Singapore 117597
Email: bchckl@nus.edu.sg

Inglis, Tim J. J., BM, DM, PhD, FRCPath, DTM&H, FRCPA
Medical Microbiologist
Division of Microbiology and Infectious Diseases
Western Australian Centre for Pathology and Medical Research
Locked Bag 2009, Nedlands WA 6009, Australia
Email: tim.inglis@health.wa.gov.au

Kwang, Jimmy D. V. M., PhD
Professor, Principal Investigator
Temasek Life Sciences Laboratory, National University of Singapore
1 Research Link, Singapore 117604
Email: kwang@tll.org.sg

Lee, Yuan Kun, BSc (Hons), PhD
Associate Professor
Department of Microbiology, National University of Singapore
5 Science Drive 2, Singapore 117597
Email: micleeyk@nus.edu.sg

Liu, Jian-Wei, BSc, PhD, MBA
Asian Biotechnology Licensing Manager
Syngenta Singapore Pte Ltd
6 Temasek Boulevard
#09-02 Suntec Tower 4, Singapore 038986
Email: jianwei.liu@syngenta.com

Loke, Paxton, BSc (Hons), PhD
Senior Research Officer
Ian Potter Hepatitis Research Laboratory
The Macfarlane Burnet Institute for Medical Research and Public Health
Commercial Road, Melbourne
Vic 3004, Australia

O'Toole, Desmond K., BSc, PhD
Associate Professor
Department of Biology and Chemistry, City University of Hong Kong
83 Tat Chee Avenue, Kowloon, Hong Kong, China
Email: bhsnuff@cityu.edu.hk

Sim, Tiow Suan
Associate Professor
Department of Microbiology, National University of Singapore
5 Science Drive 2, Singapore 117597

Song, Keang Peng, BDS, PhD
Assistant Professor
Department of Microbiology, National University of Singapore
5 Science Drive 2, Singapore 117597
Email: micskp@nus.edu.sg

Tan, Hai Meng, BSc, PhD
R & D Director
Kemin Industries (Asia) Pte Ltd
12 Senoko Drive, Singapore 758200
Email: hmtan@keminasia.com.sg

Tan, Theresa May Chin, BSc, PhD
Assistant Professor
Department of Biochemistry, National University of Singapore
8 Medical Drive, Singapore 117597
Email: bchtant@nus.edu.sg

Teo, Alex Y. L., BSc, PhD
Research Manager
Kemin Industries (Asia) Pte Ltd
12 Senoko Drive, Singapore 758200

Too, Heng-Phon, BSc, PhD
Associate Professor
Fellow of Singapore-Massachusetts Institute of Technology Alliance
Department of Biochemistry, National University of Singapore
8 Medical Drive, Singapore 117597
Email: bchtoohp@nus.edu.sg

Wong, P. K., BSc (Hons), PhD
Professor
Department of Biology, The Chinese University of Hong Kong
Shatin, NT, Hong Kong SAR, China
Email: pkwong@cuhk.edu.hk

Preface

Life science research and industry is developing rapidly all over the world. Microbial biotechnology is increasingly being regarded as a core subject in most university and polytechnic life science courses. There are already a number of excellent general textbooks on microbiology and biotechnology in the market that deal with the basic principles of the field. In order to complement this, this book aims to focus on the various applications of microbial biotechnological principles. A teaching-based format is adopted, whereby working problems, as well as answers to frequently asked questions, will supplement the main text. This textbook also includes real life examples on how the application of microbial biotechnological principles has achieved breakthroughs in both research and industrial production.

The first edition of the textbook was divided into seven parts:

Part I deals with the *Principles of Microbial Biotechnology*, which contains *Screening for Microbial Products, Bioprocess Technology, Enzymology, Manipulation of Genes, Applications of Bioinformatics and Biocomputing to Microbiological Research,* and *Real Time Polymerase Chain Reaction.*

Part II focuses on the *Food Production Involving Microorganisms and Their Products*, which includes the subheadings *Fermented Foods, Food Involving Yeast* and *Ethanol Fermentation, Fungal Solid State Cultivation, Food Ingredients, Enzyme Modified Food Products,* and *Regulation of Foods Involving Genetically Engineered Microorganisms.*

Part III deals with *Microbes in Agrobiotechnology*, and is subdivided into *Microbes and Livestock* and *Transgenic Plants*.

Part IV probes into the area of *Microbes in Medical Biotechnology*, under the subheadings of *Diagnostic Clinical Microbiology and Microbes: Friends or Foes*?

Part V elaborates on the roles of *Microbes in Environmental Biotechnology*. These include *Municipal Wastewater Treatment, Industrial Wastewater Treatment, Municipal and Industrial Solid Waste Treatment*, and *Regulatory Issues on Application of Natural and Genetic Engineered Microbes in Environmental Biotechnology*.

Part VI explores the potential of *Microbes in Alternative Energy*.

Part VII provides a glimpse of *Patenting Inventions in Microbiology*.

In this second edition of the textbook, most chapters were revised and two additional chapters were included. They are: Chapter 6 "Real Time Polymerase Chain Reaction (RTPCR)" and Chapter 9 "Fungal Solid State Cultivation". RTPCR is becoming a widely used molecular biology technique in biotechnology research and application, and it is our belief that a special chapter should be devoted to this subject. Growing fungi on wet substrate (solid state) is common in fermentation production of food. The inclusion of the chapter on "Fungal Solid State Cultivation" compliments the other chapters in the "Food" section of this textbook.

Although this book is written for university undergraduates and polytechnic students, it contains sufficient details to be used as a reference book for postgraduate students and lecturers. It may also serve as a source book for corporate planners, managers and applied research personnel.

Lee Yuan Kun
Editor

Contents

Part I

Principles of Microbial Biotechnology

The use of microorganisms for large-scale industrial purposes has a long history, which is long before the realization of the activities of the microorganisms. For centuries, beer, wine, vinegar, soy sauce and other fermented foods were produced through spontaneous fermentation of natural occurring microorganisms or the use of carry-over microbial seeds from the previous batch of production. The quality and productivity of these early products were very often inconsistent. The development of scientific screening and isolation methods allows the selection of desirable natural occurring or mutated microorganisms for specific purposes. These methods, coupled with the advancement of the technical know-how in large-scale sterilization of culture media, in provision of adequate oxygen supply and in mixing homogeneity of the culture systems, enable the exploitation of both anaerobic (yeast and some bacteria) and aerobic microorganisms (fungi and some bacteria). Common examples are: the development of large-scale processes for the production of citric acid, amino acids and antibiotics; in improved biotransformation of steroid hormones, as well as the mass production of many enzymes. The diverse catalytic activities of microorganisms are being used more and more widely to perform specific chemical reactions in the industrial production processes.

Microbial biotechnology was pushed to a new height in the eighties when continuous fermentation and airlift fermentation processes were developed for the production of food and feed grade microbial protein from industrial by-products, such as methanol and alkanes. These processes lead to considerable savings in capital, energy and labor costs.

Modern techniques of gene manipulation, and advanced bioinformatics and biocomputing are powerful tools for genomic and proteomic research. The scientific breakthroughs that ensued have made feasible the industrial manufacturing of non-microbial products, such as human growth hormone, interferon and viral vaccines.

It is the aim of Part I of this book to provide the readers with in-depth and comprehensive scientific knowledge in the areas as listed below, so as to facilitate the understanding of the various applications of micro-organisms and the production of their bioactive molecules in the biotechnological systems:

- Screening for microbial products
- Bioprocess technology
- Enzymology
- Manipulation of genes
- Application of bioinformatics and biocomputing.

Chapter 1

Screening for Microbial Products

Alex Y. L. Teo and Hai Meng Tan
Kemin Industries (Asia) Pte Ltd
12 Senoko Drive, Singapore 758200

1.1. Introduction

If the discovery of penicillin marked the beginning of the antibiotics era, then the development of its fermentation has ushered in what might be called the golden age of industrial microbiology. The onset of this has resulted in the production of a large number of plant or microbial primary and secondary metabolites that are of commercial importance. Whereas primary metabolism is universal among living systems, secondary metabolism is mainly carried out by plants and microbes and is usually strain specific. The primary metabolites that are important in the bio-industry are the organic acids, vitamins, amino acids and purine nucleotides. However, of all the traditional products made by bioprocess, it is the secondary metabolites that are of great importance and value to the human health. Secondary metabolites, particularly from microbial sources, are selective in their actions on pathogenic bacteria and fungi. The success rate was so impressive that the pharmaceutical industry screened secondary metabolites almost exclusively for antibacterial, antifungal and antitumor chemotherapy as well as against diseases not caused by bacteria, fungi or tumors.

1.2. Screening for New Antibiotics

The main approach in which new antibiotics have been discovered has been by screening of groups of microorganisms such as *Streptomyces*, *Penicillium*, and *Bacillus*. In the screening approach, a large number of possible antibiotic-producing microorganisms are obtained from nature and pure isolates tested for antibiotic production by observing for diffusible materials that are inhibitory to the test or indicator bacteria. The classical method for testing potential antibiotic-producing microbial isolates is the cross-streak method, used by Fleming in his studies on penicillin. Isolates that demonstrate evidence of antibiotic production are then subjected to further studies to determine if the antibiotics they produce are new. When an organism producing a new antibiotic is discovered, it is produced in large quantity, purified and tested for cytotoxicity and therapeutic activity in infected animals. Most new antibiotics will fail the *in vivo* testings but a few of these new antibiotics that prove useful medically are then produced commercially. Since antibiotic-producing strains isolated from nature rarely produce the desired antibiotic at sufficiently high concentration, it is necessary to isolate new high yielding strains. In the commercial production of penicillin, the yield of this antibiotic was increased by 50,000 times using strain selection and appropriate medium development. Strain selection involves mutagenesis of the wild type culture, screening for mutants and testing of these mutants for enhanced antibiotic production.

1.3. Screening for Beneficial Cultures

Crude lactic cultures are known to inhibit some psychrotrophs in milk and ground beef. A large number of lactic acid bacteria, singly or in combination, have been shown to display varying degrees of antimicrobial activity against pathogenic microorganisms. In addition, viable cultures or components of lactic acid bacteria are useful in the treatment of displaced endogenous intestinal microflora, which are characteristic of many intestinal disorders and enhanced gut permeability of the host. Such bacteria are able to survive gastric conditions to colonize the intestine, at least temporarily, by adhering to the epithelium. They have been reported to improve the growth rate and feed utilization in pigs, chicken and calves and to improve their feed conversion ratio. There is significant decrease in the occurrence of diarrhea observed in pigs and calves fed with these beneficial bacteria. Lactic cultures are also believed to neutralize the effect of enterotoxin from *E. coli*, which are pathogenic for pigs. The beneficial effects of these bacteria ranged from displacement of harmful bacteria such as *Clostridium perfringens* to reduction of bacterial urease activity to synthesis of vitamins, stimulatory effects on the immune system and contribution to digestion.

1.4. Screening for Antimicrobial Peptides

Microbial peptides with pronounced antimicrobial activity are commonly isolated from animals, plants, microbes and in fermented food. They are small and cationic with molecular masses of between 3,000 and 6,000 daltons. Post-translational modification of precursor peptides has been shown to introduce intramolecular thioether bridges to cationic peptides such as Pep 5, nisin, and subtilin. Although these peptides offer an important potential safety advantage over chemically synthesized preservatives when incorporated into food, many peptides are not suitable owing to the pathogenic nature of the producer-strains. Peptides such as colicins, epidermin and Pep 5 may be useful in topical application in creams and salves, but are unlikely to be approved for use in foods because of the nature of the producer-strains. Bovine lactoferrin, an antimicrobial component of colostrum and milk, helps in the protection of infants from gastrointestinal infections. Porcine pepsin cleavage of native lactoferrin produces low molecular weight peptides inhibitory to some Gram-negative and Gram-positive bacteria. In addition, hydrolysis of native lactoferrin at pH 2 and 120°C produces active peptides, which are bactericidal.

Methanol-acetone extracts of lyophilized fermented milk by lactic cultures have been reported to contain ninhydrin-positive materials. These proteinaceous materials do not lose antimicrobial activity when exposed to 100°C for 10 minutes, are active at pH 5.4 and not inactivated by pepsin treatment. Purified cationic, low molecular weight material isolated from *Streptococcus diacetylactis*, has been found to be heat stable and active towards several *Pseudomonas* species up to pH 6.0. Acidophilin, a low molecular weight (approximately 700 daltons) nitrogenous compound from *Streptococcus diacetylactis* and *Leuconostoc citrovorum* cultures, has potent antimicrobial activity. Acidophilin is also produced by *Lactobacillus acidophilus*. Antimicrobial compound(s) from *Streptococcus thermophilus* are likely to be amines of low molecular weights, which are heat stable. An antimicrobial substance from *Lactobacillus* sp. strain GG is inhibitory to Gram-positive as well as Gram-negative bacteria but not lactobacilli. It is active between pH 3 and 5, heat stable and resistant to proteinase K, α-chymotrypsin, trypsin, and carboxypeptidase A. It resembles a microcin from some of the characteristics studied. The antimicrobial substance from *Lb.* sp. strain GG has a molecular weight of less than 1,000 daltons and is soluble in acetone-water.

1.5. Screening for Low Molecular Weight Peptides

Microcins are low molecular weight amino-acid-oligopeptide antibiotics, which are resistant to some proteases, extreme pH values and heat, and soluble in methanol. In contrast, bacteriocins are larger proteins, which can be inactivated by proteases. Microcin A15 (< 500 daltons) is a bacteriostatic compound with a methionine moiety. Microcin B17, with a molecular weight of 4000 daltons, is sensitive to pronase, subtilisin, and thermolisin. Microcin C7, which is about 900 daltons, is sensitive to trypsin and subtilisin but loses little activity when exposed to 100°C for 30 minutes. Microcin C7 is a linear structure consisting of acetyl-methionine, arginine, threonine, glycine, asparagine, alanine and a supposed ethanolamine entity at the carboxyl-terminal end. Microcins D15, D93, and D140 are highly hydrophilic, basic and small peptides.

1.6. Screening for Potential Bacteriocins

Bacteriocins are an extremely heterogeneous group of substances. The original definition of bacteriocins referred to proteins of the colicin type produced by *Escherichia coli*. Bacteriocins are proteinaceous compounds usually with bactericidal activity against species closely related to the producer bacterium. These secondary metabolites also possess fungicidal, metal-chelating and immuno-modulating properties. Most bacteriocins are produced by Gram-positive bacteria such as *Bacillus*, lactic acid bacteria and *Streptomyces* against pathogenic and food spoilage microorganisms such as *Staphylococcus aureus* and *Listeria monocytogenes*. Produced by strains of Gram-positive and Gram-negative bacteria, they are characterized by lethal biosynthesis, intraspecific activity, and adsorption to specific receptors. Bacteriocins are active macromolecules possessing a narrow inhibitory spectrum of activity, protein in nature, plasmid encoded and without effect on producer cells. Reports have shown that most bacteriocins produced by Gram-negative bacteria act on closely related species. On the other hand, inhibition of Gram-negative bacteria has not been clearly demonstrated by purified bacteriocins from Gram-positive organisms. In this regard, the presence of a lipoteichoic acid receptor for pediocin in the host cell wall suggested that the bacteriocins from Gram-positive bacteria may not be inhibitory to Gram-negative organisms since the latter do not possess cell wall teichoic acid.

Many strains of Gram-positive fermentation starter culture bacteria produce bacteriocins. These bacteriocins possess narrow or wide spectra of activity. However, it is those with wider ranges of activity that are desirable as food biopreservatives. Characteristics that define a bacteriocin have been expanded to include their chemical nature, stability, mode of action, genetic determination, inhibitory spectrum, sensitivity to various proteases, mode of synthesis, and toxicity level to animals. The molecular weights of bacteriocins also vary greatly. They have been classified into small or large molecular-weight groups. There are presently three classes of bacteriocins: Class I bacteriocins which comprise lantibiotics that are small (< 5 kDa), proteins with intramolecular thioether rings and are further divided into type A (which are cationic and pore-forming proteins) and type B that are enzyme inhibitors and immunologically active peptides. Nisin, lactosin, subtilin and the heat stable lacticin are examples of Class I bacteriocins. Class II bacteriocins are small (< 10 kDa), heat stable non-lantibiotic peptides and comprise four subgroups: IIA to IID. Examples are pediocin, leucocin and entercin. Class III bacteriocins are large (> 30 kDa), heat labile proteins. In general, low molecular weight bacteriocins are sensitive to trypsin digestion but insensitive to heat inactivation.

Diplococcin and nisin are well-known bacteriocins produced by *Lactococcus cremoris* and *Lc. lactis*, respectively. Nisin contains a total of 34 amino acids and unusual lanthionines. Since 1969, nisin has been approved by the FDA as a preservative in processed cheese, cheese spread, pasteurized milk and dairy desserts. Ethylenediaminetetraacetic acid, Triton X-100 and Tween 20 have been used to potentiate nisin activity against Gram-negative organisms including *Salmonella typhimurium* and *Escherichia coli*. Other bacteriocins include pediocin PA-1 from *Pediococcus acidilactici*, which was found effective in preventing spoilage of salad dressing. Pediocin A from *Pediococcus pentosaceus* has a broader-range of activity against food spoilage pathogens.

Although numerous bacteriocins from Gram-positive bacteria have been isolated, characterized, and purified, very few from the propionibacteria are documented. *Propionibacterium acnes* from the cutaneous group, produces a bacteriocin that is bacteriostatic against both Gram-positive and Gram-negative anaerobes. The classical dairy propionibacteria also have been shown to produce bacteriocins.

Propionibacteria are Gram-positive rods, non spore-forming, pleomorphic, non-motile, mostly slow-growing aerobic to aerotolerant mesophiles that form small yellow, orange, cream, and dark red colonies. The optimum pH range for growth is between 6 and 7. However, there is no growth when the pH is less than 4.5. Their ability to produce inhibitory metabolites has been studied from as early as 1972. Food-grade bacteria such as propionibacteria can be grown in milk or cheese whey. The resultant medium, cells, acids and any other antimicrobial substances produced, may be dried and used as a natural preservative. For example, Microgard™, a grade A skim milk product fermented by *P. freudenreichii* ssp. *shermanii* and pasteurized is used as a biopreservative in cottage cheese. Approved by the Food and Drug Administration, it has been estimated that 30% of the cottage cheese produced in the United States has Microgard™ incorporated.

1.7. Screening for Non-Proteinaceous Fermented By-Products

Propionibacteria have a long history of use in the fermentation and dairy industries. In the dairy industry, propionibacteria play important roles in the development of the characteristic flavor, aroma, texture and eye formation in Swiss cheese. They are also well known for their ability to synthesize vitamin B_{12}. Propionibacteria have become increasingly important to the fermentation industries. Fermentation of glucose or lactate as carbon sources produces propionate, acetate and carbon dioxide. Typically, 3 moles of lactate result in the production of a 2:1:1 molar ratio of propionic acid, acetic acid, and carbon dioxide, respectively. Propionic and acetic acids are the two major products derived from the fermentation of lactose. The yield of propionic acid changes from approximately 33% (w/w) at pH 6.1–7.1 to about 63% at pH 4.5–5.0. However, acetic acid remains within the range of 9–12% at all pH values. The succinic acid yield decreases from 17% at pH 7.0 to 8% at pH 5.0. In contrast, pyruvic acid produced during fermentation of lactose is reconsumed. The production of acetic acid is closely associated with cell growth, whereas propionic acid can be both growth and non-growth associated.

In the production of Swiss cheese, the inhibitory effects of the undissociated forms of propionic and acetic acids are potentiated against Gram-negative bacteria by the low cheese pH. However, the extent of the fermentation is also restricted by the acidic pH. In addition, propionic acid is also known as an important mold inhibitor. As producers of propionic and acetic acids, the propionibacteria could serve as alternative sources of these organic acids when chemical synthesis becomes costly and environmentally prohibitive. In terms of the feed industries, organic acids have been added to the animal diets for many decades now. They are believed to promote animal growth by decreasing the gastric pH, reducing post-weaning disturbances, improving pepsin activity, decreasing bacterial growth and increasing nutrient digestibility in the gut. Organic acids are more effective on diets low in essential amino acids and minerals. Their preservative effects on the feeds are due to changing the physical conditions to prevent pathogenic growth and they are also lethal for some bacteria. Organic acids are widely used in Europe to inhibit pathogens such as *Salmonella* in both raw materials and finished feed. Organic acids in undissociated form are able to pass through the cell membrane where they dissociate into H^+ ions within the cell. The resultant pH causes the cell to expend energy to pump out the extra H^+ ions. The $RCOO^-$ anions produced can then disrupt DNA and protein synthesis. Organic acids have also been reported to enhance growth in pigs.

The concentration of diacetyl produced by strains of *Propionibacterium freudenreichii* ssp. *shermanii* in milk (10 μg/ml) was found to be non-inhibitory to yeast and non-lactic acid bacteria. Both propionic acid and diacetyl have attained GRAS (generally recognized as safe) status as food additives in the United States. Other volatile compounds such as sesquitrepenes are produced by propionibacteria.

1.8. Requirements of Screens

In order to ensure good screen performance, the screening system must work in the presence of the library to be tested. The library consists of a microbial collection that has undergone stringent selection and optimization to maximize the yield of secondary metabolites. This library is then screened to detect for the presence of active samples or "hits". Screens designed to test microbial metabolites must be sensitive, specific, selective and rapid. In terms of sensitivity, the use of appropriate technology to achieve a lower detection limit is important. Screens should be specific using appropriate selective or discriminatory tests, such as isotype specificity tests for molecular assays, or cytotoxicity measurement for cell-based assays to add value to the data obtained for screening hits. To enable rapid screening of samples, screens may be carried out with minimum numbers of steps using semi- or fully automated high throughput screening programs (Fig. 1.1). The data generated from each screen in which the samples have been tested should then be compared and analyzed so that selective "hits" can be identified at an early stage. Following hit selection, active metabolites will be purified and their structure elucidated using a range of biochemistry tools, which include chromatographic and spectroscopic techniques.

1.9. Screening Process

Several key areas are involved in the screening process of microorganisms and their metabolites. The first is the development and validation of assay(s), which upon further assessment for its reproducibility and compatibility

(a)

(b)

Fig. 1.1. An example of a High Throughput Screening System. (a) The Bioscreen C workstation, and (b) the output showing the growth curves of a test organism cultivated in the presence of different putative bioactive substances.

with the physicochemical characteristics of the samples, convert it to a primary or pilot screen. The primary screening process makes it possible for the estimation of hit rates and for large numbers of samples to be screened rapidly to obtain sufficient data to initiate a full screening program.

The second area is high throughput screening, which enables large numbers of potential samples to be screened with minimum numbers of steps using semi- or fully automated systems. The high throughput screening system will also perform retesting and confirmation of active metabolites from the initial screen. The third area is secondary screening of the active metabolites or "hits" from the primary screen to eliminate false positive results. At this stage, active fractions are further characterized based on heat stability, chemical classes, molecular weights, pH optimum and enzyme inhibition patterns. Following purification and structure elucidation of active metabolites, the potential molecules are then subjected to further *in vivo* testing.

1.10. Screening Methods

Prior to the development of preliminary screens, isolation procedures in search of useful secondary metabolites must be in place. Most of the examples given in this chapter concern the selective isolation of producer-bacteria and recognition of their secondary metabolites. Basically, selective isolation of producer-bacteria may involve: (i) selection and pre-treatment of the material containing bacterial strains of interest; and (ii) growth and isolation of producer-bacteria on selective laboratory media.

Actinomycetes are important producers of secondary metabolites including antibiotics, anti-tumor and cytotoxic compounds. Due to their growth requirements and characteristics, various pre-treatments can be used to increase the relative number of actinomycetes on isolation plates relative to bacteria. Drying and dry heating are some of the common methods to reduce the number of undesired Gram-negative bacteria which can cause swarming on the isolation plates since the aerial spores of most actinomycetes are relatively resistant to desiccation. Incubation conditions generally include temperatures of 25–30°C and periods of 7–14 days during which visible colonies of actinomycetes will be observed. Slower growing strains will require longer incubation periods. The judicious use of spore activating and bactericidal agents in isolation media such as sodium doedcyl sulphate, phenol, antibiotics and yeast extract in combination with dry heat can greatly increase the recovery of actinomycetes while suppressing the counts of bacteria. Some examples of antibiotics used in the isolation of specific actinomycetes include rifampicin, novobiocin and oxytetracylcine for *Acitnomadura, Micromonospora* and *Streptoverticillium*, respectively. Antifungal agents such as cycloheximide and nyastatin, which do not inhibit actinomycetes, are now routinely added to the isolation media to eliminate fungal contamination. Since very few actinomycetes can be identified based on colony morphology alone, the use of selective media with the incorporation of chromogenic or fluorogenic substrates for genus specific enzymes can be helpful for the isolation of specific taxa of actinomycetes.

Following isolation of producer-bacteria, the antimicrobial activities of secondary metabolites may be assessed by overlaying an isolation plate with a single test or indicator organism. A modification of this method, known as the agar-spot test has also been used for the detection of antagonistic activity between bacteria. However, the major limitation of both techniques is the potential contamination of selected colonies with the indicator organisms. In addition, they are also laborious screening methods as only one indicator organism at a time can be applied to each isolation plate. Alternatively, replica plating of selected colonies may be used to study a range of indicator organisms. However, the method has limitations due to the spread of motile bacteria on the surface of each isolation plate.

The sensitivity of an isolation medium can be influenced by its nutrient composition, pH, selective agents and incubation temperature. Typically, nonselective and/or selective enrichments are required to increase the sensitivity of detection of target bacteria. Generally, the stepwise enrichment process begins with pre-enrichment or primary enrichment, in which contaminated samples are incubated in a nutritious nonselective medium to allow growth of target bacteria. Pre-enriched samples are then transferred to a secondary selective enrichment medium, where the normal flora is suppressed but the target bacteria are allowed to grow. Finally, the producer bacteria is streaked on selective isolation agar and isolated as a pure culture. The pure producer bacterial culture is then transferred into a broth system to begin the fermentation and optimization process of enhancing the production of active secondary metabolites. Filtration of fermented broth through a membrane filter (0.20–0.45 μm pore size) is routinely used to separate cells from the aqueous phase. The filtrate is then subjected to screening against a spectrum of indicator organisms.

Fig. 1.2. A well diffusion assay used to screen for the presence of bioactive compounds against a lawn of indicator organism. The size of the zones of clearance around the wells is indicative of the relative potency of the different compounds.

The disk activity assay has been the method of choice used in the comparison of relative activities of active filtrates from various bacterial fermented broths. Sterile analytical paper disks (1.2 cm in diameter) are immersed in sample filtrate and placed onto the surface of solidified agar medium seeded with an indicator organism. Following overnight incubation at a specific temperature, the relative activities between different filtrates can be observed from size of the zone of inhibition on each medium seeded with indicator organisms. Alternatively, a modified agar-well diffusion method can be used to examine the activity of the active filtrates from various bacterial fermented broths (Fig. 1.2). In this method, wells are aseptically created by a hole-borer in a solidified medium seeded with indicator organisms. Similarly, the relative activities between different filtrates are determined based on the sizes of the zone of inhibition on the solidified medium. However, both techniques using the disk and well assays are laborious and time-consuming when screening the active filtrate against a wide range of bacteria. Therefore, to enable rapid screening of active samples, screens may be carried out using semi- or fully automated high throughput screening systems. The major advantage of using the high throughput screening process is in the quick turn-around time to screen numerous samples against a wider range of microorganisms. This screening method may also be extended to perform numerous bioassays to determine the minimum-inhibitory concentrations (MIC) of the active fractions against a range of indicator-microorganisms. In the minimum-inhibitory concentration assay, an active sample undergoes a two-fold dilution procedure using suitable broth and a portion from each dilution is transferred into the wells of a microtiter plate. Following incubation for 24–48 hours, the optical density of each culture broth in the wells is measured spectrophotometrically with a microtiter plate reader at a stipulated wavelength. Automated high throughput and target directed screening are now used together with structure-based design and combinatorial chemistry in drug discovery. High throughput screening is used to generate lead compounds in the initial part of the drug discovery process. Such lead compounds are then subjected to more vigorous biological and chemical analyses. The application of combinatorial chemistry generates a whole library of new variant compounds that can be tested for the elucidation of the structure-function relationship.

The task of identifying the main active component(s) in bacterial filtrates can be complicated by the presence of both organic acids and antimicrobial peptides or proteins. It has been established that the predominant organic compounds found in a fermented product of *P. freudenreichii* ssp. *shermanii* include diacetyl, acetic, propionic, and lactic acids as well as an antimicrobial peptide. A method to remove organic acids from fermented broth involves a lyophilization process at a vacuum of 150 millitorrs and a temperature of −40°C. Lyophilization under such conditions would cause the acids to volatilize and evaporate from crude fermented extract leaving behind the largely unaffected antimicrobial substances for further analysis. Alternatively, solvents may be used to extract organic acids from fermented products. The samples can be centrifuged to remove cell debris and growth medium components. Bioactive supernatants are then extracted with an equal volume of organic solvent (e.g. ethyl-acetate) for 1 hour. The fractions obtained are allowed to evaporate under vacuum to dryness using a rotary-evaporator set at 70°C to remove the organic solvent. Samples are rehydrated with distilled water and then analyzed for the presence of organic acids by gas-liquid chromatography. A simpler technique with hydroxyl-treated anion-exchanger has been found to be effective in binding organic acids and can be used to obtain acid-free active ingredient(s) from crude extract.

The proteinaceous nature of these antimicrobial substances is then determined using proteolytic enzymes. Generally, antimicrobial peptides are not sensitive to lipases, amylases, Rnases or Dnases. The effect of heat, pH, hydrostatic pressure, chemicals and ions may be included for assessment of the antimicrobial peptides present in the active filtrate or supernatant fractions.

The active peptides or proteins can be quantified using various molecular weight cut-off membranes at ultra-high pressure under nitrogen gas. All retentate and filtrate samples can be assayed using the disk activity or well-assay methods for antimicrobial activity against specific indicator organisms. Active molecular weight cut-off filtrate or retentate is then analyzed for total protein concentration using the Lowry Protein Assay. Isoelectric focusing is a sensitive technique for protein separation based on the principle that protein migration in an electrically charged field will cease at its isoelectric point. The isoelectric point of a protein is defined as the point of electrical neutrality (zero-charge). Individual proteins or peptides are separated into specific bands based on a pH gradient. The individual bands are then eluted and tested for antimicrobial activity against specific indicator-organisms. Other qualitative methods to determine molecular weights for partially purified active components include gel filtration and SDS-polyacrylamide gel electrophoresis. Further purification of the active compounds may involve multi-steps such as separation and concentration of the molecules by different chromatographic methods, including hydrophobic, ion exchange, and size exclusion chromatography. Purified antimicrobial peptides are then analyzed and sequenced to determine their amino acids composition using mass spectrometry (MS) and Edman degradation.

1.11. New Screening Technology

In recent years, more and more microbial genomes are being fully sequenced. With the development of new technologies in functional genomics and proteomics, the discovery of new drug molecules and targets is expected to gain added impetus and momentum. This will be further aided by more powerful bioinformatics and high throughput screening capabilities. The challenge lies in understanding the structure-function relationship of proteins, receptors and their ligands.

Increasingly, high throughput target screening and identification is making use of sequence-based functional genomics methods. Many proteins contain functional domains (e.g. ATP binding sites, protein-DNA interacting sites) that can be identified from sequence similarity with known genes. In the "gene-chip" technology, cDNA or oligonucleotides are deposited on silicon or glass substrates resulting in a dense array of DNA spots on solid supports that allow hybridization to fluorescent or radioactive RNA or DNA samples. The binding between the two sets of nucleic acids can then be captured digitally. Such a high throughput genomic screen has been used to compare the gene expression patterns between normal and disease tissues for example.

There are now a large number of biological receptors with predicted biological functions but no known ligands. Such receptors are called "orphans" and can be used to screen against a host of compounds or cell extracts to identify putative ligands in a process termed "ligand fishing". The orphan receptors can be matched to the ligands using high throughput functional screens such as looking for elevated Ca^{2+} levels in transfected mammalian cell lines. One particular type of technology is based on the integration of surface plasmon resonance (SPR)-based biosensor technology in conjunction with MALDI-TOF (matrix-assisted laser desorption/ionization time-of-flight) — a mass spectrometry technique. SPR technology allows the detection and quantification of the specific binding between orphan receptors and complex materials such as bacterial lysates or cell extracts. This forms the basis of a direct screen of such materials as potential sources of ligands. In practice, the receptors can be bound to a sensor chip, and the solution containing the test compounds is made to pass over the chip surface. In this way, the sensor chip is used for the affinity capture of orphan receptor ligands and other binding proteins. The non-destructive SPR detection system can also provide real-time information on the biomolecular interactions that are going on. Using MS, accurate molecular weight of the bound ligands can be determined. In many cases, the information obtained can be used to identify bound proteins through correlation with protein structure databases. Further proteolytic digestion of the bound ligands can be carried out to provide further sequence characterization and identification.

Further Readings

1. Al-Zoreky NJ, Ayres W and Sandine WE. Antimicrobial activity of Microgard™ against food spoilage and pathogenic microorganisms. *J. Dairy Sci.* **74**: 758–763, 1991.
2. Baird-Parker AC. Organic acids. In: *Microbial Ecology of Foods Vol. 1: Factors Affecting Life and Death of Microorganisms*, eds. JH Silliker *et al.* Academic Press, New York, 1980, pp. 126–135.
3. Basler AW, Der Hude V and Scheutwinkel M. Screening of food additive propionic acid for genotoxic properties. *Food Chem. Toxic* **25**(4): 287–290, 1987.
4. Branen AL, Go HC and Genske RP. Purification and properties of antimicrobial substances produced by *Streptococcus diacetilactis* and *Leuconostoc citrovorum*. *J. Food Sci.* **40**: 446–450, 1975.
5. Brock JH. Lactoferrin in human milk: Its role in iron absorption and protection against enteric infection in the newborn infant. *Arch. Dis. Child.* **55**: 417, 1980.
6. Charteris WP, Kelly PM, Morelli L and Collins JK. Selective detection, enumeration and identification of potentially probiotic *Lactobacillus* and *Bifidobacterium* species in mixed bacterial population. *Int. J. Food Microbiol.* **35**: 1–27, 1997.
7. Daeschel MA. Antimicrobial substances from lactic acid bacteria for use as food preservatives. *Food Technol.* **43**: 164–166, 1989.
8. Delves-Broughton J. Nisin and its uses as a food preservative. *Food Technol.* 100–117, 1990.
9. Demain AL and Fang A. The natural functions of secondary metabolites. In: *Advances in Biochemical Engineering Biotechnology Vol. 69*, eds. T Scheper and A Fiechter. Springer-Verlag, Berlin-Heidelberg, Germany, 2000, pp. 1–39.
10. Drake MC, Small L, Spence KD and Swanson BG. Rapid detection and identification of *Lactobacillus* spp. in dairy products by using polymerase chain reaction. *J. Food Prot.* **59**: 1031–1036, 1996.

11. Glatz BA. The classical propionibacteria: Their past, present, and future as industrial organisms. *ASM News.* **58**: 197–201, 1992.

12. Harrigan WF. In: *Laboratory Methods in Food Microbiology*, 3rd ed. Academic Press, San Diego, California, 1998, pp. 20–22.

13. Harris LJ, Daeschel MA, Stiles ME and Klaenhammer TR. Antimicrobial activity of lactic acid bacteria against *Listeria monocytogenes*. *J. Food Prot.* **52**: 384–387, 1989.

14. Herrero AA. End-product inhibition in anaerobic fermentation. *Trends Biotechnol.* **1**: 49, 1983.

15. Hettinga DH and Reinbold GW. The propionic-acid bacteria — A review, I, Growth. *J. Milk Food Technol.* **35**: 295–301, 1972a.

16. Hettinga DH and Reinbold GW. The propionic-acid bacteria — A review, II, Metabolism. *J. Milk Food Technol.* **35**: 358–372, 1972b.

17. Hettinga DH and Reinbold GW. The propionic-acid bacteria — A review, III, Miscellaneous metabolic activities. *J. Milk Food Technol.* **35**: 436–447, 1972c.

18. Hill DC. Advanced screening technology and informatics for natural products drug discovery. In: *Advances in Drug Discovery Techniques*, ed. AL Harvey. John Wiley & Sons Ltd, Chichester, UK, 1998, pp. 25–37.

19. Hurst A. Nisin. *Adv. Appl. Microbiol.* **27**: 85–123, 1981.

20. Jack WR, Tagg JR and Ray B. Bacteriocins of Gram-positive bacteria. *Microbiol. Rev.* **59**: 171–200, 1995.

21. Jay JM. Antimicrobial properties of diacetyl. *Appl. Environ. Microbiol.* **44**: 525–532, 1982.

22. Juffs HS and Babel FJ. Inhibition of psychrotrophic bacteria by lactic cultures in milk stored at low temperature. *J. Dairy Sci.* **58**: 1612–1619, 1975.

23. Klaenhammer TR. Bacteriocins of lactic acid bacteria. *J. Biochimie* **70**: 337–349, 1988.

24. Lechevalier HA and Corke CT. The replica plate method for screening antibiotic-producing organisms. *Appl. Microbiol.* **1**: 110–112, 1953.

25. Lee YK, Nomoto K, Salminen S and Gorbach SL. *Handbook of Probiotics*. John Wiley & Sons, NY, 1999.

26. Lewus CB, Kaiser A and Montville T. Inhibition of food-borne bacterial pathogens by bacteriocins from lactic acid bacteria isolated from meat. *Appl. Environ. Microbiol.* **57**: 1683, 1991.
27. Motlagh AM, Johnson MC and Ray B. Viability loss of food borne pathogens by starter culture metabolites. *J. Food Prot.* **54**: 873–878, 1991.
28. Muriana PM. Antimicrobial peptides and their relation to food quality. In: *Food Flavor and Safety.* American Chemical Society, Washington, D.C., 1993, pp. 303–321.
29. Nursey I. Control of *Salmonella. Kraftfutter.* **10**: 415–412, 1997.
30. Ray B. Methods to detect stressed microorganisms. *J. Food Prot.* **42**: 346–355, 1979.
31. Ray B and Daeschel M. Food biopreservatives of microbial origin. In: *Bacteriocins of Starter Culture Bacteria as Food Biopreservatives: An Overview.* CRC Press, Boca Raton, FL, 1992, pp. 177–205.
32. Ray B and Miller KW. Pediocin. In: *Natural Food Antimicrobial Systems*, ed. AS Naidu. CRC Press, Boca Raton, Florida, 2000, pp. 525–566.
33. Roller S. The biotechnological development of new food preservatives. *Biotechnol. Gen. Engin. Rev.* **9**: 183–206, 1991.
34. Jorgensen JA and Turnidge JD. Susceptibility test methods: Dilution and disk diffusion methods. In: *Manual of Clinical Microbiology*, 8th ed., eds. PR Murray, EJ Baron, JH Jorgensen, MA Pfaller and RH Yolken. Amercian Society for Microbiology, Washington, D.C., 2003, pp. 1108–1127.
35. Stevens KA, Sheldon BW, Klapes NA and Klaenhammer TR. Nisin treatment for inactivation of *Salmonella* species and other Gram-negative bacteria. *Appl. Environ. Microbiol.* **57**: 3613–3615, 1991.
36. Teo AY-L and Knabel SJ. Development of a simple recovery-enrichment system for enhanced detection of heat-injured *Listeria monocytogenes* in pasteurized milk. *J. Food Prot.* **63**: 462–472, 2000.
37. Williams ST and Wellington EMH. Principles and problems of selective isolation of microbes. In: *Bioactive Microbial Products: Search and Discovery*, eds. JD Bu'lock, LJ Nisbet and DJ Winstanley. Academic Press Inc. Ltd., London, UK, 1982, pp. 9–26.

38. Yarborough GG, Taylor DP, Rowlands RT, Crawford MS and Lasure LL. Screening microbial metabolites for new drugs — theoretical and practical issues. *J. Antibiotics* **46**: 535–544, 1993.

39. Zähner H, Drautz H and Weber W. Novel approaches to metabolite screening. In: *Bioactive Microbial Products: Search and Discovery*, eds. JD Bu'lock, LJ Nisbet and DJ Winstanley. Academic Press Inc. Ltd., London, UK, 1982, pp. 51–70.

Chapter 2

Bioprocess Technology

Yuan Kun Lee

Department of Microbiology, National University of Singapore
5 Science Drive 2, Singapore 117597
Email: micleeyk@nus.edu.sg

Bioprocess technology is the industrial application of biological processes involving living cells or their components to effect desired transformation of substrates. The major advantages of bioprocesses over traditional chemical processes are that they require mild reaction conditions, are more specific and efficient, and produce renewable by-products (biomass). The development of recombinant DNA technology has expanded and extended the potential of bioprocesses.

2.1. Microbial Growth

When all nutrients required for cell growth are present in non-growth limiting quantity, i.e. at sufficiently high concentrations (so that minor changes do not significantly affect the reaction rate) and the culture environment is favorable, most unicellular microorganisms reproduce asexually. The size and biomass of individual cell increase with time, resulting in **biomass growth**. Eventually, the DNA content doubles in quantity and cell division occurs upon complete division of a cell into two progenies of equal genome and approximately identical size, thereby increasing the population number. This increase in the population number of cells in a culture is known as **population growth** (Fig. 2.1).

Box 2.1

What is Biomass?

Biomass is defined as the materials essential for the structure and reproduction of a living organism. Thus, storage products such as starch and glycogen are not true biomass and should be subtracted from the weight of the biological material measured, to give the true biomass. The biomass is often expressed as a concentration, e.g. g-biomass/L-culture.

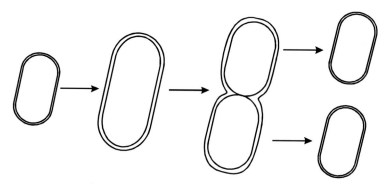

Biomass growth   Population growth

Fig. 2.1. Bacterial cell growth and division.

2.1.1. *Growth kinetics*

Growth kinetics deal with the rate of cell growth and how it is affected by various chemical and physical conditions. During the course of growth, the heterogeneous mixture of young and old cells is continuously changing and adapting itself in the media environment, which is also continuously changing in physical and chemical conditions.

There are distinct growth phases in the growth curve of a microbial culture. A typical growth curve includes the following phases: (1) lag phase, (2) growth phase (accelerating, exponential and decelerating growth phase), and (3) stationary phase and death phase as depicted in Fig. 2.2.

(1) Lag Phase
When an inoculum is introduced to a fresh culture medium, the culture may not grow immediately at the maximum rate, thus giving rise to a lag phase. This is essentially a period of adaptation of cells to a new environment (e.g. change in pH, increase in nutrient supply, reduced growth inhibitors). The cells increase in size and weight rather than increase in numbers. The duration of growth lag is defined in Box 2.4. Lag phase is considered a non-productive period of a fermentation process. Thus, it is often desirable

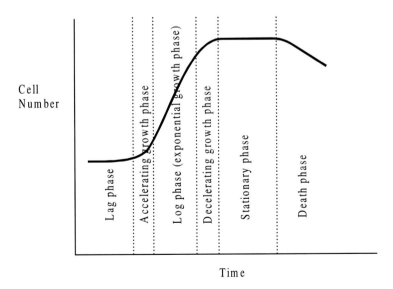

Fig. 2.2. Growth curve of a bacterial culture.

Box 2.2

Why is it Necessary to Understand Microbial Growth Kinetics?

Microbiologists should understand microbial growth kinetics for two basic reasons:

(1) For appropriate interpretation of growth data

Take for example the case of studying the pH profile of a newly isolated bacterium in a batch culture system. The new bacterial culture previously cultured at pH 7.0 was used as the inoculum for the same medium adjusted to pH 5.5 and 7.5. The cultures were incubated at the same temperature for a period of time. Thereafter either the final cell number was enumerated or biomass was measured and the results are as depicted in the following figure.

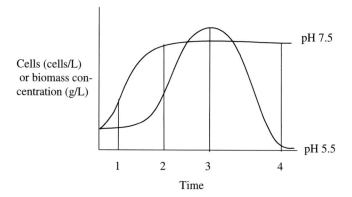

The interpretation of the outcome would depend largely on when the final sample was taken. If the cultures were sampled at time 1, it appears that at pH 5.5 growth of the culture was completely inhibited. In fact, the inoculum used was previously cultured at pH 7.0, was still in the lag phase at time 1, when the cells were adapting to the new pH condition.

If the cultures were sampled at time 2, it is easy to conclude that the culture could grow at both pH 5.5 and 7.5, and the bacterium grew faster at pH 7.5.

When the cultures were sampled at time 3, it may appear that pH 5.5 is a favorable pH condition for faster growth. The truth is that the mineral nutrients in the medium are more soluble in this acidic solution, rendering them more available for bacterial growth, hence a higher biomass concentration.

Box 2.2 (*Cont'd*)

If the cultures were sampled at time 4, the vast difference in the cell population of the two pH conditions may lead to the conclusion that the bacterial cells lysed at pH 5.5. However, it is the high maintenance energy requirement at pH 5.5 that caused rapid losses of biomass in the stationary growth phase.

In actual fact, the exponential growth rate of the bacterium was the same at pH 5.5 and 7.5.

(2) For optimization of growth and product formation

Knowledge of microbial growth kinetics is essential to determine when to harvest the culture for different purposes. For a growth-linked product, it is desirable to harvest the culture at the late exponential growth phase. On the other hand, for a non-growth-linked product, it would be desirable to harvest the culture at the stationary growth phase (see Sec. 2.2).

---

to minimize or control the duration of the lag phase. Base on its mechanism, growth lag could be differentiated into **apparent lag** and **true lag**.

Immediately after inoculation, a portion of the seed culture population may grow at the maximum rate while the rest fail to grow, thus resulted in an apparent **lag phase**. For example, if half of the cells in the inoculum are not viable, and the other half of the cells grow at the maximum rate, it may appear that the culture is growing at half of the maximum rate.

True lag occurs when the culture is not able to grow at its maximum rate initially due to one of the following factors:

(a) Change in nutrient

A change in nutrient supplied in the culture medium, e.g. concentration, component, will probably involve the induction of one or more new enzymes, which could take a few minutes to many hours. The lag is thus due to the induction of these new enzymes and the time required to synthesize the optimal amount of these enzymes. For some microbial cultures, the induction of enzymes to utilize a new carbon and energy substrate could only occur when a small amount of the original carbon and

Box 2.3

Why is the Inoculum Size Usually 10% v/v of the Culture Volume?

Inoculum is often taken from a culture in the late exponential or early stationary growth phase. The desired size is usually 10% of the culture volume. An inoculum size that is too large would dilute the rate-limiting substrate concentration in the medium, resulted in a lower specific growth rate. On the other hand, a low inoculum size would result in a long lag phase. In a culture of alga *Chlorella*, glycolate in the culture supernatant needs to be accumulated to a critical concentration before cell growth is possible (see also Inoculum Effect on True Lag, Sec. 2.1.1(1)).

Box 2.4

When does the Lag Phase in a Growth Curve End?

Let us plot the graph of ln X (X = biomass) versus incubation time, T, as shown in the figure below, and extrapolate the exponential growth line to meet the original biomass concentration. The time when the two lines intercept is defined as the end of the lag phase.

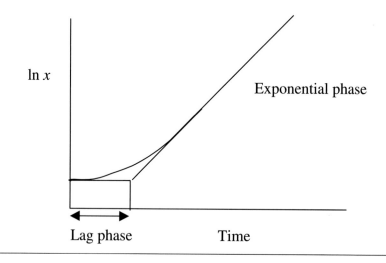

energy substrate is present. It is possible that the cells do not have sufficient carbon and energy storage to support synthesis of the new enzymes. For example, the penicillin producing fungus *Penicillium chrysogenum* could metabolize both glucose and lactose in a mixed substrate fermentation, but could not metabolize lactose if it is added into a glucose starved culture. It is always desirable to adapt an inoculum in the same medium as in the final culture system to minimize the duration of the lag phase.

At times, cells from a nutrient starved inoculum could experience **substrate-accelerated death**, when they are inoculated into a nutritionally rich medium (Fig. 2.3). This could be interpreted as a case where the rapid flux of the growth-limiting substrate (often the carbon-energy substrate) led to unbalanced enzymatic activities and disruption of cellular metabolic control. A rapid decrease in the intracellular level of some essential metabolites, such as cyclic nucleotide, will result in loss in viability. For example, a lactose-limited culture of *Klebsiella aerogenes* shows rapid death when starved in non-nutrient buffer containing lactose.

(b) Change in culture conditions
The change in physical environment of a culture often arises when cells are in the late exponential and stationary phases, where pH and other growth parameters have changed due to cellular activity, are inoculated into a fresh culture medium. The pH of a microbial culture may decrease due to

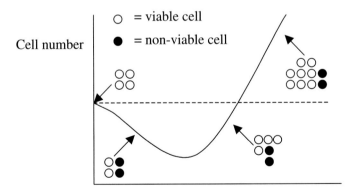

Fig. 2.3. Substrate accelerated cell death results from the transfer of nutritionally starved cells to a nutritionally rich medium.

the consumption of NH_4^+ in the medium or production of organic acids, e.g. lactic acid by *Lactobacillus* culture. The pH of culture medium may increase due to the consumption of NO_3^{++} in the culture medium. In these cases, growth lag represents the adaptation of cells to the new culture conditions.

In the studies of growth parameters in batch culture system, cultures are often inoculated with inoculum of the same source and then incubated at different conditions, e.g. various temperatures. Cells incubated at conditions that deviate from the original culture conditions of the inoculum may experience a growth lag. Some cultures may even fail to grow in the conditions that deviate too much from the original culture condition. Thus, the outcome of the study is cell history dependent and could lead to biased conclusions. It is therefore desirable to preculture the cells in some particular culture conditions before they are used as inoculum to study the effect on cell growth.

(c) Presence of an inhibitor

After the initial inoculation, a period of time may be required to reduce the amount of an inhibitory substrate (e.g. alcohol, phenol and hydrocarbon, or an inhibitory medium component like antibiotic and heavy metal) to a concentration which permits maximum growth rate.

In some cases, the product of inoculum may inhibit growth. For example, members of *Lactobacillaceae* when transferred from an anaerobic to an aerobic medium may accumulate H_2O_2 in the cells. This could happen when a heavy inoculum of *lactobacilli* (low oxygen partial pressure in the original culture) is inoculated into a freshly prepared medium (oxygen partial pressure at air saturation). Growth would only be possible after the synthesis of the enzyme peroxidase and removal of H_2O_2. As such, increasing the inoculum size would prolong the lag period, as the production of H_2O_2 is proportional to the cell concentration. On the contrary, for most of the other causes of growth lag, a bigger inoculum size would shorten the lag period (see below).

(d) Spore germination

Where the inoculum is made up of spores, vegetative growth is only possible after germination of spores.

(e) Inoculum effect

The physiological stage (age and size) of the inoculum is an important factor in determining the length of growth lag. The duration of lag period is prolonged when the inoculum added is at the initial stage of the exponential growth phase. Nevertheless, the lag period could be eliminated by the addition of culture filtrate taken from a culture in the late exponential growth phase, as the intermediate products in the filtrate are readily available by increasing the inoculum size, or by adding intermediates of citric acid cycle, such as asparagine.

Growth lag could represent "priming" of various essential metabolic pathways before substrates are metabolized at the maximum rates for conversion into biomass. Thus, a small inoculum may not be able to grow in a minimal medium containing only essential nutrient. An inoculum of *Klebsiella aerogenes* is not able to grow in a minimal glucose-ammonia medium when the initial cell concentration is less than 10^5/mL. The addition of amino acids such as asparagine, alanine or peptones (partially hydrolyzed proteins) helps to lower the minimal inoculum concentration.

The duration of the lag period is shortened when the growth of the inoculum added to the culture approaches the end of the exponential growth phase, as it consists mainly of young and active cells. The lag period increases when the inoculum is at the stationary growth phase. This could be attributed to the re-organization necessary in the cells to reverse the change caused by cessation of growth.

(2) Growth Phase

There are three stages at the growth phase, namely the accelerating growth phase, the exponential (or logarithmic) growth and the decelerating growth phase.

At the late lag period, the cells have adjusted to the new environment and begin to grow and multiply (**accelerating growth phase**), and eventually enter the **exponential** (or **logarithmic**) **growth phase**, where the cells grow and divide rapidly, at a relatively constant rate, as the exponential function of time. The time required to achieve a doubling of the number of viable cells is termed **doubling time** (t_d). It is also termed **generation time**, as this is the duration the culture takes to grow and produce a generation of cells. The number of cells in an exponentially growing microbial culture could be expressed mathematically as follows:

$$2^0 N_0 \rightarrow 2^1 N_0 \rightarrow 2^2 N_0 \rightarrow 2^3 N_0 \rightarrow 2^n N_0$$

where N_0 = Initial number of cells
            n = Number of doubling (generations)

It can be seen from Table 2.1 that the doubling time of bacterial cells measured at their optimal growth temperature varies widely. Some organisms take slightly longer than others to double their population. In general, the higher the optimal growth temperature, as in the thermophiles, the shorter is the doubling time. Furthermore, small, structurally simple bacteria such as the spherical and rod-shape bacteria double faster than large, structurally complex helical and filamentous bacteria.

The number of doubling (n) at a time interval t, is determined by the relation t/t_d. Thus, the number of cells (N_t) in an exponentially growing culture after incubated for some time, t, can be estimated as:

$$N_t = N_0 \cdot 2^n$$

$$= N_0 \cdot 2^{t/t_d}$$

$$N_t / N_0 = 2^{t/t_d}$$

$$\ln(N_t / N_0) = (\ln 2)\, t/t_d \tag{2.1}$$

Table 2.1. Doubling time of various bacteria measured at their optimal growth temperature.

Organism	Temperature (°C)	Doubling time (min)	Shape
Bacillus stearothermophilus	60	11	Single rod
Escherichia coli	37	20	Single rod
Staphylococcus aureus	37	28	Cluster coccus
Streptococcus faecalis	30	62	Chain coccus
Lactobacillus casei	30	96	Single rod
Nostoc japonicum	25	570	Filament
Anabaena cylindrica	25	840	Chain coccus with heterocyst
Treponema pallidum	37	1980	Helix

During the exponential growth phase, the growth rate of the cells is proportional to the biomass of cells. Since biomass generally can be measured more accurately than the number of cells, the basic microbial growth equations are very often expressed in terms of mass. A biomass component such as protein may be used as an alternative to direct weighing of biomass. Hence, Eq. (2.1) can be modified by assuming the biomass concentration at time 0 (initial) and time t as X_0 and X_t respectively:

$$\ln(X_t/X_0)/t = 0.693/t_d$$
$$d(\ln X)/dt = 0.693/t_d$$
$$d(\ln X)/dX \cdot dX/dt = 0.693/t_d$$
$$1/X \cdot dX/dt = 0.693/t_d$$
$$\mu = 0.693/t_d$$

μ represents the **specific growth rate** (g-biomass/g-biomass/h or h^{-1}) of the culture. It defines the fraction of increase in biomass over a unit time, i.e. an increase of certain g-biomass from every gram of existing biomass per hour. Specific growth rate represents the average growth rate of all cells present in a culture, but not necessary the maximum specific growth rate of the individual cells, as most microbial cultures divide asynchronously.

The expression of the rate of microbial growth as specific growth rate is crucial to avoid the effect of cell concentration. For a microorganism which has a doubling time of 1 h, the **output rate** of a culture with a concentration of 1 g/L is 1g-biomass/L/h, whereas at a biomass concentration of 10 g/L, the output rate is 10 g-biomass/L/h.

After the exponential growth phase is the **decelerating growth phase** where the culture is in a transient state. During this stage there are feed/back mechanisms that regulate the bacterial enzymes involved in key metabolic steps to enable the bacteria to withstand starvation. There is much turnover of protein for the culture to cope with this period of low substrate availability. Examples of the feedback mechanisms involved are Stringent Response (RNA content decreases), Ntr System and Pho System (activity of enzymes changes markedly).

Box 2.5

What is a Stringent Response?

Some bacteria (e.g. Enterobacteriaceae), on experiencing a depletion of nutrition, especially nitrogen source (amino acid starvation), can initiate a stringent response to reduce the rates of protein and other macromolecular synthesis, by decreasing the synthesis of ribosomal RNA. Using this mechanism of transcription control, the cells are able to shut down a number of energy-draining activities as a survival mechanism under poor growth conditions.

Box 2.6

What are Ntr and Pho Systems?

The activity of **Ntr system** is activated in response to nitrogen starvation, when ammonia is the growth limiting substrate. The bacteria are able to turn on transcription of genes for glutamine synthetase. The system enables the bacteria (e.g. Enterobacteriaceae) to assimilate very low levels of ammonia by catalyzing the assimilation of ammonia into glutamine in an ATP-dependent reaction. It also activates additional operons (e.g. *nac* operon) that are involved in the utilization of organic nitrogen sources. This permits the cells to use alternate sources of nitrogen.

A phosphate utilization network **(Pho system)** is activated on inorganic phosphate starvation (in Enterobacteriaceae), resulting in the production of a high concentration of alkaline phosphatase, so that phosphate can be obtained from organic sources.

(3) Stationary and Death Phase (or Decline Phase)

A growing bacterial culture eventually reaches a phase during which there is no further net increase in bacterial cell numbers. This is called the **stationary growth phase**. During the stationary phase, the growth rate is equal to the death rate. A bacterial population may reach stationary growth when a required nutrient is exhausted, or when inhibitory end products are accumulated or when physical conditions changed. The various metabolites formed in this stage are often of great biotechnological interest.

resulting onset of decline (death) phase. The onset of the decline phase is dependent on the type of organism and the culture conditions.

2.1.2. *Variations in growth kinetics*

Deviations from the norm in the growth kinetics described above were reported. These kinetic variations provide useful insights to the nutrient metabolism and physiological status of the cultures, which will be discussed in the following sections.

(1) Conserved Substrate
Some nitrogen and phosphorous substrates, even if they are growth limiting, may conserve in cells with or without simple modification, as the rate of uptake of the substrate is faster than the rate of metabolism of the substrate. In this case, the decelerating growth phase could be prolonged for a number of generations by drawing the substrate from the conserved substrate pool, with apparent depletion of the substrate in the culture medium, before the onset of stationary growth phase.

(2) Synchrony in Cell Division
Synchrony in cell division occurs when all cells divide at the same time. It could be induced by a marked change in the substrate level, culture

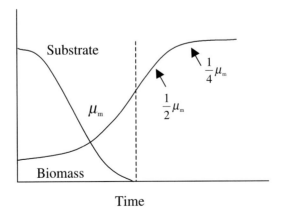

Time

Fig. 2.4. Bacterial growth curve with a rate-limiting substrate that is conserved within the cells.

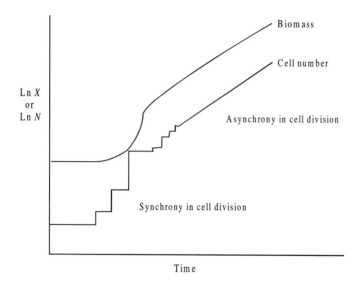

Fig. 2.5. Synchrony in cell division in a bacterial culture. N = cell number; X = biomass concentration.

temperature and light intensity, resulting in an initial rapid increase in the number of cells after the lag period. As the number of organisms increases, nutrients are consumed, metabolic wastes accumulate, living space may become limited and aerobes suffer from oxygen depletion. The culture normally becomes completely asynchronous after 2 to 3 cell divisions, due to the slight differences in the length of cell cycle of individual cells. The cells subsequently grow at a more or less constant exponential growth rate. The synchronous growth phase should not be included in the estimation of the exponential growth rate.

(3) Diauxic Growth

The presence of a preferential carbon substrate may inhibit the synthesis of the enzymatic system involved in the uptake and metabolism of a second carbon substrate. For example, *E. coli* culture may metabolize glucose preferentially in a mixed substrate medium containing both glucose and lactose. The lactose would be utilized as a carbon source only when the glucose is exhausted. Thus, in the growth curve of such a culture, an inflexion in the growth curve or even a decline in biomass occurs after

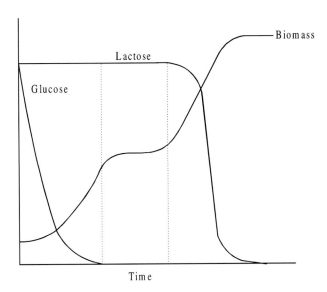

Fig. 2.6. Diauxic growth on two substrates, glucose and lactose in a *Escherichia coli* culture.

glucose is used up while lactose is being processed. This biphasic (two-phase) pattern of growth is known as diauxie or diauxic growth.

2.1.3. *Effect of growth-limiting substrate concentration*

The specific growth rate of a microbial culture is ultimately determined by the growth-limiting substrate concentration. This could be understood by regarding biomass as a component in a biochemical reaction:

$$X + s + \text{other nutrients} \leftrightarrow Xs \rightarrow \text{biomass} + \text{products}$$

where X = free biomass concentration
    s = concentration of growth-limiting substrate
    Xs = concentration of biomass-substrate complex

Therefore, the total biomass concentration can be expressed as:

$$X_T = X + Xs \tag{2.3}$$

The equilibrium concentration of the various components in the biochemical reaction described above is determined by the **Saturation Constant** (K_s, g/L):

$$X \cdot s/Xs = K_s \qquad (2.4)$$

K_s is defined as the concentration of the rate-limiting substrate to support the specific growth rate at half of the maximum value, i.e. $0.5\ \mu_m$. A higher K_s value would indicate that a higher concentration of the rate-limiting substrate is necessary to achieve a certain amount of biomass-substrate complex Xs in order to produce a certain amount of biomass. Thus, K_s is inversely proportional to the affinity of the cells for the substrate.

Since the **specific rate of substrate assimilation**, q (g-substrate/g-cell/h) is proportional to Xs formed,

$$q = q_m \cdot Xs/X_T \qquad (2.5)$$

where q_m is the value of q when substrate is saturated, i.e., the maximum specific rate of substrate assimilation.

By substituting Xs/X_T in Eq. (2.5) with $s/K_s + s$ derived from Eqs. (2.3) and (2.4), we obtain

$$q = q_m \cdot s/K_s + s. \qquad (2.6)$$

Assuming that all substrates assimilated are used for biomass production, the specific growth rate is directly proportional to the specific rate of substrate assimilation:

$$q = \mu/Y$$

where the yield constant Y (see Sec. 2.1.4) is the growth yield = dX/ds (g-cell/g-substrate).

Thus Eq. (2.6) can be re-written as:

$$\mu = \mu_m \cdot s/(K_s + s) \qquad (2.7)$$

This equation is often called the **Monod Relation**, since Jacques Monod in 1942 first showed empirically that the expression accorded well with the relation of bacterial growth rate to substrate concentration.

Box 2.7

How to Determine the Saturation Constant of a Growth-Limiting Substrate?

The saturation constant could be determined from the plots of specific growth rate, μ versus substrate concentration, s or $1/\mu$ versus $1/s$,

(a) According to Eq. (2.7): $\qquad \mu = \mu_m \cdot s/(K_s + s)$

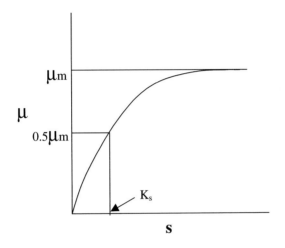

(b) According to the equation: $\qquad 1/\mu = (K_s + s)/\mu_m \cdot s$
$$= K_s/\mu_m \cdot s + 1/\mu_m$$

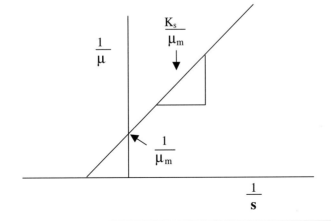

Monod Relation suggests that the specific growth rate of a microbial culture is determined by the highest possible growth rate of the culture (μ_m), the concentration of the growth-limiting substrate (s) and the affinity for the substrate (K_s). The value of μ_m is determined by growth parameters such as temperature, pH and medium osmolality.

Moreover, when

$$\mu = 0.5\ \mu_m$$
$$s = K_s$$

Hence, K_s could be determined by measuring the value of μ at various values of s (Box 2.7). Sample values of K_s for various carbon substrates and microorganisms are as shown in Table 2.2.

The K_s value for the same substrate (glucose) could be very different for two different microorganisms. In this case, *K. aerogenes* has a higher affinity for glucose than *S. serevisiae*. The microorganism *P. extorquens* has higher affinity for methanol (lower K_s value) than formate.

Knowledge of K_s values is of practical important for:

(1) Designing Culture Medium

In order to achieve the maximum specific growth rate, concentration of all essential substrates needs to be higher than their respective K_s value (see Sec. 2.3 for the design of growth medium).

(2) Predicting Growth Kinetics

Knowledge of K_s would allow us to estimate the specific growth rate of a microbial culture in respond to a certain concentration of the growth-limiting substrate according to Eq. (2.7). The theoretical maximum

Table 2.2. The saturation constant (K_s) of various carbon substrate and microorganisms.

Organism	Substrate	K_s ($\times 10^{-6}$M)
Klebsiella aerogenes	Glucose	20
Saccharomyces serevisiae	Glucose	140
Pseudomonas extorquens	Methanol	20
Pseudomonas extorquens	Formate	228

specific growth rate could be obtained from the plot of $1/\mu$ versus $1/s$ (Box 2.7).

(3) Interpreting Microbial Interactions in an Ecosystem

It is rare that a single species of microorganism is found in a natural habitat. The co-existence of different microorganisms competing for the same growth-limiting substrate could be understood from the following example.

Assuming that two bacteria, A and B, are present in a small volume of water competing for the same growth-limiting substrate. And μ_m of bacterium A is greater than that of bacterium B, while $K_{SA} > K_{SB}$. As such, we would expect a cross over at s_c in the μ versus s plots of the two bacteria (Fig. 2.7). When some amount of the substrate, which exceeds the s_c value, is added into the ecosystem, bacterium A (having a higher μ_m) would outgrow bacterium B in number, and the substrate decreases in concentration. However, when the residual concentration of the substrate drops below the s_c value, the specific growth rate of bacterium B would now be higher than that of bacterium A. Thus, the two bacteria take turns to increase in number, and no one bacterium would replace the other in the ecosystem.

The situation would be different if the concentration of the growth-limiting substrate does not decrease to below s_c or increase to above s_c. In this case, bacteria A and B would have same population size.

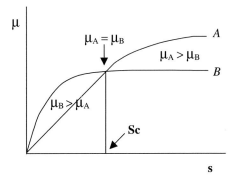

Fig. 2.7. Growth curves of two bacteria, A and B, in a mixed culture competing for the same rate-limiting substrate. μ = specific growth rate; s = rate-limiting substrate concentration.

2.1.4. *Growth yield*

It has been observed frequently that the amount of cell mass produced is proportional to the amount of a limiting substrate consumed.

Growth yield ($Y_{X/S}$, g-cells/g-substrate) is defined as the amount of biomass produced (dX) through the consumption of unit quantity of a substrate (ds), i.e.

$$Y_{X/S} = dX/ds.$$

It is an expression of the conversion efficiency of the substrate to biomass.

Knowledge of growth yield is of practical importance. In many industrial production processes, it is desirable to maintain high growth yield to achieve efficient conversion of the substrates to biomass and products. Biological waste treatment process represents the other extreme where it is desirable to achieve the lowest possible sludge (biomass) formation, with the maximum oxidation of carbon in the effluent to CO_2.

Growth yield also determines the cost of process operation. In the example of Microbial Protein (Single Cell Protein) production, the final product is biomass with a high protein content of about 50%. The yield value determines the biomass output rate as well as the cost of aeration and cooling can be illustrated as below:

(i) The cost of aeration (O_2 supply) in a fermentation process using methanol (a derivative of a by product of petroleum refinery industry) is determined by the O_2 demand of a fermentation system.

The mass balance of microbial growth could be expressed by the following equation:

$$a(CH_3OH) + (1.5a - 1.225c)O_2 \rightarrow c(CH_2O_{0.55}N_{0.25}) + (a - c)CO_2 + (2a - c)H_2O$$
methanol   biomass

For any value of growth yield on methanol (c/a) there is a corresponding O_2 requirement ($1.5a - 1.225c$). Thus, a low growth yield of biomass (c) would result in a higher demand ($1.5a - 1.225c$) of O_2 (air supply) for the oxidation of unit quantity of methanol (a) into CO_2. In essence, as it requires 2 moles of oxygen to produce 1 mole of carbon dioxide, while only 0.55 mole of oxygen is necessary to produce 1 C-mole of biomass, there is a higher O_2 demand if more carbon in methanol is oxidized to CO_2. Hence this results in higher cost of production due to the increase in the cost of aeration.

Insufficient O_2 supply would not allow the full oxidation of methanol, leading to even lower level of biomass production.

(ii) The energy input to sustain optimal culture temperature in a fermentation process using methanol.

Energy balance of microbial growth on methanol could be expressed using the following equation:

Thus, the lower the growth yield on the energy substrate, the larger amount of the chemical energy stored in the substrate would be released as heat through respiration process. In order to maintain the optimal culture temperature in the fermentation system, energy supply is needed for the circulation of cooling water through the system.

There are several biological factors and culture parameters that determine the growth yield of a microbial culture.

(1) Type of Microorganism
Among the examples cited in Table 2.3, the bacterium *Pseudomonas* has the lowest growth yield on glucose, whereas the fungus *Aspergillus* converted 75% of the carbon in glucose into biomass. The filamentous nature of the fungus facilitates separation of the cells from the culture medium. Therefore, *Aspergillus* would be a desirable candidate for Microbial Protein production. Other factors to be considered are the content of protein and the composition of amino acids. Digestibility of the biomass and safety aspect of the organism as food needs to be evaluated too.

Table 2.3. Growth yield of various microorganisms.

Microorganism	$Y_{X/S}{}^a$ (C eq^b biomass/C eq glucose)
Pseudomonas fluorescens	0.47
Penicillium chrysogenum	0.55
Candida utilis	0.68
Aspergillus nidulans	0.75

[a]Growth yield
[b]Carbon equivalent

(2) Substrates

(a) Carbon source

The C-substrate is also the energy source for heterotrophic bacterial culture. The energy content of different carbon substrates varies, which are as shown in Table 2.4.

The salt of organic acid oxalic acid has the lowest energy content (2.4 Kcal/g-carbon) among the C-substrates listed, and the maximum growth yield of microbial cultures reported in the literature is also the lowest (0.14 g-biomass/g-carbon). The maximum growth yield seems to increase with increasing energy content of the various C-substrates, down the list from oxalate to glycerol. However, higher energy content in methanol and methane does not support a higher growth yield.

The data imply that the microbial cultures were energy-limited in media containing the first four C-substrates (oxalate to glucose). Thus, higher growth yield was obtained with C-substrate of higher energy content.

In those C-substrates, which have an energy content of higher than 11.0 Kcal/g-carbon, the microbial cultures were probably carbon-limited. There is no difference in the growth yield with increasing energy content of the substrate.

This brings us to the question of whether it is possible to utilize the excess energy from an energy-rich substrate, such as alcohol and sugar alcohol, in the metabolism of the carbon in a carbon-rich substrate, such as organic acid and sugar. Such **co-metabolism** of the two substrates would achieve a growth yield that is higher than any one of the substrate

Table 2.4. Energy content of various carbon substrates.

C-substrate	Energy content (Kcal/g-carbon)	Maximum $Y_{X/S}$[a] (g-biomass/g-carbon)
Oxalate	2.4	0.14
Formate	5.1	0.53
Citrate	6.5	0.80
Glucose	9.2	1.03
Glycerol	11.0	1.42
Methanol	14.6	1.43
Methane	17.4	1.46

[a]Growth yield

Table 2.5. Growth yield of single substrates and substrate mixtures.

Microorganism	Substrate	Molar ratio	$Y_{X/S}$[a] (g-biomass/g-substrate)
Candida utilis	Sucrose		0.50
	Ethanol		0.68
	Sucrose/ethanol	13.8:1	0.68
Paracoccus	Formate		0.07
denitrificans	Mannitol		0.52
	Formate/mannitol	4:1	0.70

[a]Growth yield

alone. Table 2.5 presents the growth yield of single substrates and substrate mixtures.

In the case of *Candida utilis*, the energy from one part of ethanol was sufficient to bring up the growth yield from 13.8 parts of sucrose to 0.68 g-biomass/g-sucrose, which represents a 36% increase from 0.50 g-biomass/g-sucrose. When formate and mannitol were mixed in a molar ratio of 4 to 1, the growth yield from the mixture was 10 times that of formate and 1.3 times that of mannitol.

The enhancement of growth yield in **mixed substrate fermentation** can only be observed in microbial cultures where the substrates are metabolized simultaneously. Some C-substrates are assimilated preferentially over other C-substrates, and a diauxic growth kinetic is observed in medium containing more than one C-substrates (see Sec. 2.1.2(3)).

(b) Efficiency of ATP generation

The oxygen content and other culture parameters determine the pathways of which a carbon/energy substrate is metabolized to yield different amounts of biochemically available energy in the form of ATP. As can be seen in Table 2.6, 30 ATP moles are generated from a mole of glucose via oxidative phosphorylation, but only 1–3 moles of ATP are generated in anaerobic condition via the various pathways.

It is generally true that the growth yield of biomass per mole of carbon/ energy substrate is the highest under the condition of carbon/energy limitation. Under the condition of carbon excess, there is often production of overflow metabolites, e.g. acetate, α-ketoglutarate, which reduce yield of biomass on the C-substrate. Overflow metabolites are intermediate metabolites that are produced over and above the base line levels for the formation of biomass, resulting in the accumulation of metabolites within the cells or release of these into the culture medium.

(c) Nitrogen source

Inorganic N sources (NO_3, N_2) need to be reduced to the level of $-NH_2$ for incorporation into protein molecules, thus less reductant is available for

Table 2.6. The number of ATP molecules generated on the metabolism of glucose at different culture conditions and metabolic pathways.

Substrate	Condition	ATP (mole/mole substrate)
Glucose	Aerobic (P/O[a] = 3)	30
Glucose	Anaerobic (Formate:acetate = 1:1)	3[b]
Glucose	Anaerobic (Ethanol + CO_2)	2[b]
Glucose	Anaerobic (Ethanol + CO_2)	1[c]

[a]phosphorus oxygen ratio
[b]via Embden-Myerhof pathway
[c]via Entner-Duodoroff pathway

energy generation, with an up to 20% resultant decrease in growth yield on the energy substrate.

When organic N-substrates like amino acids are included in culture media, they may serve as pre-fabricated carbon skeleton and give a high apparent growth yield on glucose. If these organic N-substrates are catabolized, energy would be generated as an energy source.

(d) Environment

Oscillation or fluctuation in environmental conditions, such as pH, dissolved O_2 and supply of carbon substrate, may cause **uncoupling** of substrate oxidation from biosynthesis. The growth yield of *Pseudomonas methylotropha* on methanol was found to decrease with increasing interval of addition of methanol into the culture, and corresponding increase in the yield of carbon dioxide.

The phenomenon of uncoupling of substrate oxidation from bio-synthesis has been exploited in the design of bioreactor for wastewater treatment. In ICI's deep shaft process and Union Carbide's Unox process, oscillation in the temperature, oxygen concentration and substrate level are allowed in large volume (tower-type) bioreactors, giving rise to a high rate of substrate oxidation and corresponding low biomass growth yield. Thus, biodegradable carbon in wastewater is transformed into gaseous CO_2, and relatively little solid mass is left.

(e) Maintenance energy requirement

Microbes and cells require energy for both growth and other maintenance purpose, like turnover of cellular materials, maintaining concentration gradient across cell membrane and motility.

Under a particular set of environmental conditions, it is assumed that energy is consumed at a constant rate for maintenance, $(ds/dt)_m = mX$,

---

Box 2.8

Uncoupling

The separation **(uncoupling)** of the two processes: formation of the protonmotive force and its utilization to generate ATP, leading to dissipation of energy without ATP formation.

where m is **maintenance coefficient** (e.g. g-glucose/g-biomass/h), X is the biomass concentration (e.g. g-biomass/L). Hence, energy balance can be expressed as:

Total rate of   =   Rate of consumption   +   Rate of consumption
consumption   for growth   for maintenance

$\mu X/Y_E$ = $\mu X/Y_G$ + mX

where Y_E is overall growth yield and substrate is the energy source. Y_G is true growth yield, when m = 0. Hence,

$$1/Y_E = 1/Y_G + m/\mu. \qquad (2.8)$$

Thus, if m is a constant, the graph of $1/Y_E$ against $1/\mu$ will be a straight line, with slope m, and intercept $1/Y_G$ on the ordinate (Fig. 2.8). Alternatively

$$q_E \cdot X = \mu X/Y_G + m \cdot X$$

where q_E is specific rate of energy uptake = μ/Y_E. As shown in Fig. 2.9,

$$q_E = \mu/Y_G + m \qquad (2.9)$$

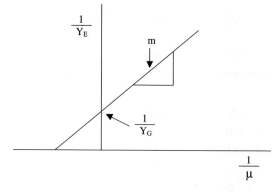

Fig. 2.8. A plot of $1/Y_E$ versus $1/\mu$ according to the equation $1/Y_E = 1/Y_G + m/\mu$, where Y_E is the overall growth yield, Y_G is the true growth yield; m, the maintenance coefficient; μ the specific growth rate.

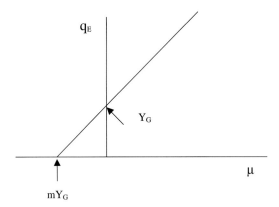

Fig. 2.9. Plot of q_E versus μ according to the equation $q_E = \mu/Y_G + m$, where q_E is the specific rate of energy uptake; μ the specific growth rate; Y_G, the true growth yield and m, the maintenance coefficient.

The fraction of energy that is channeled to maintenance would deprive the culture of the energy for growth, i.e.

$$\mu = \mu_m \cdot s/(s + K_s) - m \cdot Y_G. \tag{2.10}$$

Thus, if the maintenance energy requirement makes up a significant component in the energy balance equation, the plot of μ versus s would not pass through the origin (Fig. 2.10).

There are several factors that could affect the maintenance energy requirement of a culture.

• Environment
Environmental parameters, which fall outside the optimal range for growth, may have an effect on the maintenance energy requirement of microbial cultures. The energy presumably is consumed for maintaining a favorable intracellular environment for optimal metabolic activities and growth.

It appears that the ionic strength of media has an effect on maintenance energy requirement. For example, in the culture of *Saccharomyces cerevisiae* cited in Table 2.7, the incorporation of 1M salt resulted in a 10-fold increase in maintenance energy for maintaining an osmotic balance across the cell membrane. Whereas N_2-fixing *Azotobacter vinelandii* spent

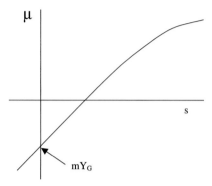

Fig. 2.10. The relationship between specific growth rate, μ and substrate concentration, s in a microbial culture with measurable maintenance energy requirement.

Table 2.7. Maintenance energy requirement of some protists (eukaryotes other than plants, animals or fungi) with glucose as the energy source.

Organism	Growth conditions	Maintenance energy (g-glucose/g-biomass/h)
Saccharomyces cerevisiae	Anaerobic	0.036
	Anaerobic, NaCl (1 M)	0.360
Azotobacter vinelandii	Fixing N_2, dissolved O_2 tension 0.2 atm	1.5
	Fixing N_2, dissolved O_2 tension 0.02 atm	0.15

energy to remove O_2 which could be inhibitory to the highly reduced nitrogenase system.

• Metabolite leakage
Microbial cells are known to release polysaccharides, glycolic acid, amines, amides and vitamins into the culture environment. These extracellular metabolites may play a role in their survival and reproduction, however, as they do not contribute directly to cell growth they are considered maintenance requirements.

● Intermittent energy supply

A decrease in growth yield is often observed if the energy source of a microbial culture is not supplied continuously (see (d) above). This could be due to the uncoupling of substrate oxidation to yield CO_2 and release of latent chemical energy as heat. The released heat may not serve any physiological function, but since it does not contribute to the growth and reproduction of a microbial culture, it is included in the category of maintenance energy.

2.2. Product Formation in Microbial Culture

Microbial cells produce a wide range of industrially useful biomolecules, which can be broadly grouped into three categories, based on their production kinetics (Table 2.8).

(1) Growth-Linked Products

These are products which formation (dp) is directly proportional to the amount of biomass formed (dX). That is:

$$dp = Y_{p/x} \, dX$$

where

$Y_{p/x}$ = product yield referred to biomass formed (g-product/g-biomass)

The rate of product formation could be expressed as follow:

$$dp/dt = Y_{p/x} \cdot dX/dt$$
$$= Y_{p/x} \cdot \mu \cdot X.$$

i.e.

$$dp/dt \cdot 1/X = Y_{p/x} \cdot \mu.$$

Thus, the specific rate of product formation is given by:

$$q_p = Y_{p/x} \cdot \mu. \qquad (2.11)$$

$Y_{p/x}$ could be a constant value, e.g. DNA production, or proportional to the specific growth rate, e.g. RNA production.

Table 2.8. Classification of microbial products.

Production kinetics	Type of microbial products	Examples
Growth-linked	Structural components of cells	Microbial protein, antigens
	Enzymes	
	extracellular	Amylases
	intracellular	β-Galactosidase
	Intermediary metabolites	Amino acids, citric acid
	Transformed substrates	Steroids
Partial growth-linked	End products of energy metabolism	Ethanol, methane, lactic acid
Non growth-linked	Energy storage compounds	Glycogen, starch
	Secondary metabolites	Antibiotics

In order to achieve high productivity ($R = qp \cdot X = Y_{p/x} \cdot \mu \cdot X$, g-product/L/h), the specific growth rate (μ) and the biomass concentration (X) of the culture should be maintained at the highest possible values.

(2) Partial Growth-Linked Product
When product formation is partly growth-linked and partly independent of specific growth rate, the specific rate of product formation could be expressed as follow:

$$q_p = Y_{p/x} \cdot \mu + \beta \tag{2.12}$$

The first term on the right hand side of the equation represents the growth-linked component, whereas β is a constant specific rate of product formation independent of the specific growth rate. Formation of end products of energy metabolism, e.g. lactic acid, follows this relation, where β includes lactic acid formation, which results from either the maintenance energy requirement or uncoupling of ATP.

(3) Non Growth-Linked Product
The product can either be, (i) q_p that is independent of specific growth rate, e.g. penicillin production by *Penicilium*, or (ii) q_p that varies with specific growth rate in a complex way, e.g. spore production by *Bacillus subtilis*:

$q_p = q_{pmax} - k\mu$, where q_{pmax}, the maximum specific rate of product formation, and k is a constant.

2.3. Culture Medium for Fermentation

Culture medium is an important factor in optimizing cell growth and product formation in fermentation processes. It provides not only the nutrients and culture environment necessary for optimal cell growth, but also induces production of desirable products, suppresses formation of undesirable products and represents a major component of the total cost of production. The following factors are to be considered in the design of a desirable culture medium.

(1) Major Nutrient Requirement

Bulk nutrient components include carbon, hydrogen, nitrogen, oxygen, phosphorus and sulfur that are required for the synthesis of biomass and target products. The chemical composition of fermentation products

Table 2.9. Chemical composition and formula of some fermentation products.

Product	Chemical Composition (% dry weight)						Chemical formula
	C	H	N	O	P	S	
Biomass							
Bacteria	53.0	7.3	12.0	19.0			$CH_{1.67}N_{0.20}O_{0.27}$
Yeast	44.7	6.2	8.5	31.2	1.1	0.6	$CH_{1.64}N_{0.16}O_{0.52}P_{0.01}S_{0.005}$
Fungus	43.0	6.9	8.0	35.0			$CH_{1.93}N_{0.16}O_{0.61}$
Antibiotics							
Erythromycin							$C_{37}H_{67}NO_{13}$
Penicillin G							$C_{16}H_{18}N_2O_4S$
Organic acids							
Citric acid							$C_6H_8O_7$
Lactic acid							$C_3H_6O_3$
Alcohol							
Butanol							$C_4H_{10}O$
Ethanol							$C_2H_6O_3$

listed in Table 2.9 provides some ideas on the minimum requirement of major nutrients.

A microbial culture would require more nitrogen for the synthesis of one mole of biomass than one mole of antibiotic. For example, for the production of one mole of bacteria, one mole carbon requires 0.20 mole nitrogen whereas for the production of one mole of erythromycin one mole carbon requires only 1/37 mole of nitrogen. Nutrient requirement is to a large extent determined by the product yield referred to per unit of biomass formed ($Y_{p/x}$, see Sec. 2.2). If the product yield for a particular growth substrate is not known, it is desirable to experimentally determine the requirement by carrying out batch fermentation in which the substrate concerned is supplied in growth-limiting concentrations while all other substrates are present in excess.

From such a study, the requirement for a particular substrate for the production of a unit biomass or product could be obtained from the slope of the plot (growth yield = biomass or product formed/substrate consumed) of cell concentration (or product concentration) versus substrate concentration. Often, in such a plot, the intercept is not at the origin. If there is carry-over of the substrate from the inoculum, the plot will intercept on the Y-axis, i.e. there is an increase in biomass even when no such substrate was added in the medium. An intercept of the plot on the x-axis indicates that a fraction of the substrate was not available for cell growth. The possible explanations are: (i) The energy substrate could have been consumed for maintenance purposes (see Sec. 2.1.4(2e)); (ii) the culture may have a high saturation constant (K_s, see Sec. 2.1.3) for the substrate; (iii) otherwise the substrate could have been "lost" by forming precipitation or complex with other non-cellular materials or destroyed during sterilization (glutamine, thiamine, riboflavin are heat labile) and storage (thiamine could be oxidized readily; riboflavin is light sensitive). It is common that magnesium, urea (or ammonium), sodium (or potassium) and phosphate form complexes and precipitated. For this reason, magnesium salt should be autoclaved separately from phosphate. In addition, trace elements may form less soluble salts with other cations in the medium during autoclaving. Chelating agents (e.g. organic ligands CDTA, EDTA and NTA) are often included in the medium to retain metal ions in solution, however they may compete with cells for the elements.

It should be kept in mind that nutrient requirement may depend on other available substrates or absence of certain substrates in the medium. Cells may metabolize organic substrates like amino acids when simple sugar is not available as energy and carbon sources (for energy substrate requirement see Sec. 2.1.4(2)). Sodium may replace some of the functions of potassium if the later is present in insufficient quantity.

(2) Essential Nutrients

Some strains of microorganisms (autotrophic strains) may have special nutrient requirement for growth and product formation, because of their inability to synthesize an essential amino acid, fatty acid, vitamin or nucleotide. Animal cells and tissue cultures require essential amino acids for growth (e.g. arginine, histidine, isoleucine, leucine, lysine, methionine, phenylalanine, threonine and valine).

(3) Trace Elements

These are elements which are essential for cell growth when present at low concentration, but are toxic at high concentration. The estimated requirements for some of the important trace elements are: Ca 0.100, Fe 0.015, Mn 0.005, Zn 0.005, Cu 0.001, Co 0.001, and Mo 0.001 (in g/100 g dry biomass). Their toxic concentrations are generally considered to be in the order of 10^{-4} M. Trace element deficiency usually slow down the growth rate (longer lag phase) rather than limiting the

Table 2.10. Typical composition of sugar cane molasses.

	% w/v		μg/100g dry wt
Dry matter	78–85	Thiamine	830
Sucrose	33	Riboflavin	250
Invert sugars	21	Pyridoxine	650
Other organic matters	20	Niacinamide	2100
Total nitrogen	0.4–1.5	Pantothenic acid	2140
P_2O_5	0.6–2.0	Folic acid	4
CaO	0.1–1.1	Biotin	120
MgO	0.03–0.1		
K_2O	2.6–5.0		
Ash	7–11		

Table 2.11. Typical composition of malt extract (often used for fungus, yeast and actinomycete cultivation).

	% w/w
Maltose	52
Hexoses (glucose, fructose)	19
Sucrose	2
Dextrin	15
Other carbohydrates	4
Total nitrogen	5
Ash	1.5
Water	2

Table 2.12. Typical composition of yeast extract produced by autolysis.

	% w/v		% w/w total amino acid
Dry matter	70	Alanine	3.4
Total nitrogen	9	Amino butyric acid	0.1
Protein	55	Arginine	2.1
NaCl	<1	Asparagine	3.8
		Cystine	0.3
		Glutamic acid	7.2
		Glycine	1.6
	ppm	Histidine	0.9
Thiamine	20–30	Isoleucine	2.0
Riboflavin	50–70	Leucine	2.9
Pyridoxine	25–35	Lysine	3.2
Niacinamide	600	Methionine	0.5
Pantothenic acid	200	Ornithine	0.3
		Phenylalanine	1.6
		Proline	1.6
		Serine	1.9
		Threonine	1.9
		Tyrosine	0.8
		Valine	2.3

Table 2.13. Typical composition of peptones.

	Casein peptone % w/w	Soy peptone % w/w
Total nitrogen	13	9
Amino nitrogen	7	2
Carbohydrate	—	29
NaCl	3	6
Ash	6	14
Water	5	4
Alanine	16.8	3.3
Arginine	30.2	1.1
Aspartic acid	8.7	2.4
Glutamic acid	38.6	5.6
Glycine	4.4	2.3
Histidine	13.3	1.1
Isoleucine	29.9	2.8
Leucine	71.9	9.4
Lysine	61.1	7.9
Methionine	22.7	3.3
Phenylanine	33.0	3.7
Proline	7.4	0.8
Serine	28.7	4.8
Threonine	21.5	1.9
Tryptophan	8.6	1.6
Tyrosine	14.0	0.7
Valine	36.4	3.8

Table 2.14. Typical composition of soy meal (often used in antibiotic fermentations).

	% w/w
Protein	50
Carbohydrate	30
(Sucrose, stachyose, raffinose, arabinoglucan, arabinan, acidic polysaccharides)	
Fat	1
Lecithin	1.8

biomass concentration. In a chemostat culture, one would observe a decreasing steady state biomass concentration with dilution rate, as in the cases of CO_2 (for photosynthetic culture) and O_2 limitation.

(4) Complex Substrates

Complex, sometimes indefinable substrates are often used in industrial fermentation for economic reasons. These are usually by-products of food industry, such as molasses from sugar refinery, malt extract from brewery, starchy waste from starch manufacturing factory, yeast extract (autolysis of baker's yeast at 50–55°C), peptones (acid or enzymatic hydrolysis of meat and animal organs or plant seeds) and soy meal from soybean oil factory (Tables 2.10, 2.11, 2.12, 2.13 and 2.14). Sulfite liquors and cellulose from paper industry are used. Even by-products of petroleum industry such as methane, methanol and alkanes (C_{12} to C_{18}) are suitable carbon substrates for some fermentation processes.

2.4. Microbial Cultivation Methods

In order to realize the production potential of a microbial culture, the provision of a culture system which would allow optimum expression of the desirable genetic traits is necessary. Culture systems could generally be categorized as closed systems and open systems. In a closed culture system, some essential components of the system (e.g. culture and products) cannot be supplied or withdrawn from the reactor during the process. Whereas in the open culture system, further addition of culture medium can be made and the microbial culture and products can be removed from the culture system during cultivation process.

2.4.1. Closed culture system

Based on their culture homogeneity, closed culture systems can be classified as homogeneously mixed batch culture and the plug-flow culture, where localized mixing of the whole culture is allowed.

(1) Batch Culture

A batch-wise culture system where culture medium and inoculum are added into a culture vessel at the beginning of the fermentation process and then incubated at suitable temperature and gaseous environment for an appropriate period of time is called a **batch culture**. The microbial culture in batch culture system goes through a lag phase, accelerating growth phase, exponential growth phase, decelerating growth phase, stationary phase and sometimes the decline phase depends on the end product desired (see Sec. 2.1.1). The substrate concentration in the culture medium and growth parameters, such as pH, changes correspondingly throughout the growth phases. Thus, the physiology of the microorganisms is always in a transient stage, subjected to a continually changing culture conditions. Consequently, product formation is confined to a certain period of cultivation, e.g. antibiotics would only be produced in the decelerating and stationary growth phases.

The batch culture system is still widely used in certain industrial processes, e.g. brewery industry, for its ease of operation; its less stringent sterilization; and its easy management of feed stocks. These advantages allow the use of unskilled labor and low risk of financial loss. Low level of microbial contamination in fermented products is at time tolerable, as long as the microbial contaminants are not pathogenic and do not alter the desired properties of the product, such as taste, color and texture.

(2) Plug-Flow Culture

In a plug-flow system, the culture vessel is usually in the form of a long cylindrical pipe (Fig. 2.11). Substrates and microbial inoculum are added at the inlet of the vessel, and the organisms carry out their biological activity as the liquid passes through the system. If there is minimum mixing in the culture such as in a pipe with large diameter, the culture will go through all the growth phases as in a batch culture, while being pushed along the pipe. That is, the composition of the nutrient, the cell population, gaseous content and products vary at different locations within the tubular reactor. By adjusting the rate that the culture is moving along the pipe and the length of the pipe, specific products for a particular growth phase could be obtained at the outlet of the vessel. Thus, this is a continuous flow culture process with the growth kinetics of a batch culture. Plug-flow culture systems have been used in the industry for continuous production of fermented milk and antibiotic production.

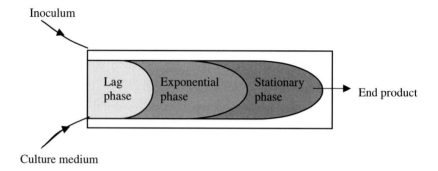

Fig. 2.11. A plug-flow culture system.

2.4.2. *Open culture system*

To maintain a microbial culture at a steady state, i.e. with its physiology fully adjusted to the culture conditions, concentration of substrates needs to be maintained and end product needs to be removed simultaneously. Thus, a fresh sterile culture medium is added to the bioreactor continuously to replenish the substrates that are consumed, and simultaneously diluting the end products that are accumulating in the system. This is often referred to as the continuous flow culture system.

Three main types of open system can be distinguished:

(1) Chemostat Culture
A chemostat is a special type of continuous flow culture system (Fig. 2.12), which consists of a perfectly mixed homogeneous suspension of microbial cells, in which the medium is fed at a constant rate (F) and culture is harvested at the same rate (F), so that the culture volume remains constant (V). The rate at which the culture is diluted through the addition of fresh medium is defined as:

Dilution rate, $D = F/V.$

Biomass balance is achieved when: Net increase = growth − harvest

$$V \cdot dX = V \cdot \mu X \cdot dt - FX \cdot dt$$
$$dX/dt = (\mu - D) \cdot X$$

Fig. 2.12. A Chemostat culture system.

If there is a **steady state** whereby the rate of growth equal the rate of cell removal, the biomass concentration remains constant, i.e. $dX/dt = 0$. Then, $(\mu - D) = 0$ or $\mu = D$.

Steady state in chemostat was experimentally established in bacterial, fungal and algal cultures. The theory indicates that it is possible to fix the μ of a culture at any value from zero to the maximum by adjusting the value of D.

A chemostat culture system can be used for the following purposes:

(i) To vary the specific growth rate (or metabolic rate) of a culture without changing the environmental condition, other than the concentration of growth-limiting substrate. This application is particularly useful in the understanding of the relationship between growth rate and product formation or a physiological function of the microbial culture.

(ii) To fix the specific growth rate while varying a culture condition, i.e. converse of (i). The application is useful in the study of the effect of the culture environment without interfering the growth rate.

(iii) To maintain substrate limited growth with a constant specific growth rate. This could not be achieved in a batch culture, as the growth rate is determined by the concentration of the substrate in the culture medium.

(iv) To obtain time-independent balanced growth (in steady state), with phenotype fully adjusted to environment, i.e. the culture is history independent, unlike cells in batch culture which performance is often dictated by the physiological state of the inoculum and culture conditions that the cells were exposed to prior to inoculation.

(v) To maintain maximum output rate of biomass or product. As is shown in Sec. 2.2, the rate of product formation is ultimately determined by the growth rate of the culture.

In the chemostat culture, the growth-limiting substrate balance can be expressed as:

$$\text{Net increase} = \text{Input} - \text{Output} - \text{Substrate used for growth}$$
$$V \cdot ds = F \cdot s_m \cdot dt - F \cdot s_r \cdot dt - V \cdot \mu \cdot X \cdot dt/Y_{X/S}$$

where

s_m = Substrate concentration in the fresh culture medium
s_r = Residual substrate concentration in the culture in steady state

$$ds/dt = F/V \cdot (s_m - s_r) - \mu \cdot X/Y_{X/S}$$
$$= D \cdot (s_m - s_r) - \mu \cdot X/Y_{X/S}$$

In steady state, $D = \mu$, $ds/dt = 0$,

$$X = Y_{X/S} \cdot (s_m - s_r) \tag{2.13}$$

Since $\mu = \mu_m \cdot s_r/(s_r + K_s)$,

$$D = D_c \cdot s_r/(s_r + K_s)$$

The critical dilution rate (D_c), which equals the maximum specific growth rate, is the highest possible dilution rate that a steady state could be established. Above which, complete washout (see (2)(i) below) of cells is observed (Fig. 2.13).

The primary effect of varying the dilution rate (D) is to change the residual concentration of the growth-limiting substrate, thereby affecting a change in μ, i.e. μ of a chemostat culture at steady state is determined by s_r only, independent of s_m.

The steady state biomass concentration can be defined and calculated by the following equation:

$$X = Y_{X/S} \cdot (s_m - K_s \cdot D/[\mu_m - D]).$$

Since μ_m, K_s and $Y_{X/S}$ are constants, the steady state concentration of biomass is effectively determined by s_m and D.

Chemostat culture system has been used in Microbial Protein production, e.g. ICI's Pruteen (*Methylophilus methylotrophus*) production from methanol; Rank Hovis McDougall's mycoprotein (*Fusarium graminearum*) production from carbohydrate containing waste from flour milling plant.

Box 2.9

Comparison of the Output Rate of Batch and that of Chemostat Culture

The **biomass output rate** ($R = \mu X$) can be compared quantitatively as follows:

Let X_m = Maximum biomass concentration in a batch culture

X_0 = Initial biomass concentration in a batch culture

μ_m = Specific growth rate in the batch culture until the rate limiting substrate is exhausted

t_c = Duration of the batch culture

t_a = Delay time which includes any lag period and the time required for recycling of the culture vessel between batches

t_d = Doubling time

Thus, in a batch culture:

$$t_c = 1/\mu_m \cdot \ln(X_m/X_0) + t_a$$
$$R_{batch} = Y_{x/s} \cdot s_m/t_c$$
$$= \mu_m \cdot Y_{x/s} \cdot s_m/(\ln[X_m/X_0] + \mu_m \cdot t_a)$$

assuming that in a chemostat, $D = \mu_m$.

Thus, in steady state:

$$R_{chemotat} = \mu_m \cdot Y_{x/s} \cdot s_m$$
$$R_{chemostat}/R_{batch} = \ln(X_m/X_0) + \mu_m \cdot t_a$$
$$= \ln(X_m/X_0) + 0.693 t_a/t_d$$

As a general practice, the ratio of $X_m : X_0 = 10:1$, which makes the maximum output rate of chemostat at least 2.3 times that of batch culture (when $t_a = 0$). Frequently, t_a is many times greater than t_d (doubling time of the culture).

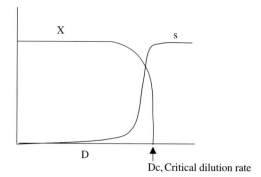

Fig. 2.13. The steady state biomass (X) and rate-limiting substrate concentrations (s) at various dilution rates (D) in a chemostat culture.

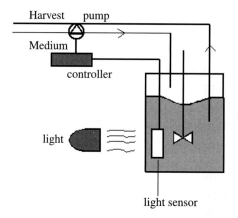

Fig. 2.14. Turbidostat culture system.

(2) Turbidostat

Turbidostat is a continuous flow culture in which the biomass concentration in the culture vessel is maintained at constant by monitoring the turbidity of the culture through the use of a photoelectric cell (an optical sensing device), and regulating the feeding rate of nutrient solution. Thus, the addition of fresh medium to the system is activated when the turbidity of the culture exceeds the set level. The specific growth rate of the culture is therefore allowed to adjust itself in respond to the growth-limiting parameter. Figure 2.14 is the schematic of a turbidostat.

The applications of turbidostat are as follows:

(i) To fix biomass concentration and allow the specific growth rate to adjust itself. This application is particularly useful for slow growing microorganisms and those with complex cell cycles. In the latter case, the μ_m of a culture may be very different at various stages in the cell cycle, resulting in complete **washout** (total removal of cells from the culture system through excessive dilution) of the culture in a fixed dilution rate chemostat.

(ii) To maintain a constant biomass concentration under unstable environment conditions, such as for the treatment of wastewater containing variable concentration of toxic compounds. Temporary suspension of specific growth rate by toxic compounds would result in washout in a simple chemostat with fixed dilution rate.

(iii) To select for fast growing microbial strains, e.g. resistance to an inhibitor. The dilution rate of a turbidostat is determined by the growth rate of the culture, thus evolution of a resistant strain would cause an increase in the dilution rate and washout of the slow growing cells (e.g. those could not tolerate the inhibitor) from the culture system.

(3) Fed-Batch Culture (FBC)

Unlike the conventional batch cultivation method where all the substrates are added at the beginning of the process, fed batch process is a batch culture with continuous feeding of nutrients while the effluent is removed discontinuously. The volume of the culture could be variable, by feeding the complete medium (**Variable Volume FBC**) or constant, by feeding the growth-limiting substrate in the form of solid, concentrated solution, gas or light (**Constant Volume FBC**). The critical elements are added in small concentrations at the beginning of the fermentation and continue to be added intermittently in small doses during the production.

In the case where at the end of a cultivation cycle, the culture is partially withdrawn and replaced with fresh complete medium, thereby initiating a new cycle. This is termed the **cyclic FBC** (Fig. 2.15).

Listed below are the various applications of FBC:

(i) Like the chemostat, FBC is a mean of achieving substrate-limited growth, but the specific growth rate of the culture is not a constant value at steady state. The specific growth rate of the culture would

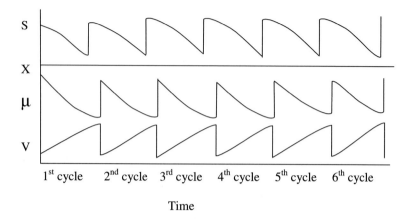

Fig. 2.15. Growth kinetics in a cyclic variable volume fed batch culture. V = volume of the culture, X = biomass concentration, μ = specific growth rate, S = rate-limiting substrate concentration.

decrease with time with increasing culture volume (in a variable volume FBC) or cell concentration (in a constant volume FBC).

(ii) FBC is also used to achieve periodic shift up and down in the specific growth rate with control over the frequency and magnitude of the shift. Such a periodic shift in metabolic rate may stimulate "overflow metabolism" and secondary metabolism, as in the case of cephalosporin C production.

(iii) To achieve a steady state in continuous flow culture of filamentous moulds. It is technically difficult to maintain a small volume chemostat of fungal culture, because of the difficulty in maintaining the homogeneity in the harvest stream of the culture system. Beside, FBC system requires only a mechanism to feed the substrate and maintenance of constant volume in the culture is not necessary.

An example of industrial FBC is the penicillin production process. Specific growth rate of *Penicillium chrysogenum* culture is maintained at below maximum values through appropriately adjusting the feed rate of carbohydrate, thus penicillin production rate is always maintained at the optimal level for a prolonged period.

(iv) The mechanism of FBC simulates some ecosystems, e.g. an infected bladder, is a case of variable volume FBC. The principle of continuous culture depicts that the pathogen in an infected bladder (cystitis) would eventually be washed-out if the dilution rate (i.e. rate of topping up with urine and discharge of urine from bladder) exceeds the specific growth rate of the pathogen.

Further Readings

1. Bull AT and Dalton H, eds. *Comprehensive Biotechnology Vol. 1: The Principles of Biotechnology*. Scientific Fundamentals, Pergamon Press, Oxford, 1985.
2. Crueger W and Crueger A. *Biotechnology: A Textbook of Industrial Microbiology*. Sinauer Associates, Sunderland Massachusetts, 1990.
3. Demain AL and Solomon NA, eds. *Biology of Industrial Microorganisms*. Butterworth, Stoneham Massachusetts, 1985.
4. Glazer AN and Nikaido H. *Microbial Biotechnology: Fundamentals of Applied Microbiology*. WH Freeman, New York, 1995.
5. Harrison DEF. Efficiency of Microbial Growth. In: *Companion to Microbiology*, eds. AT Bull and PM Meadow. Longman, London, 1978, pp. 155–180.
6. Kolter R, Siegele DA and Tormo A. The stationary phase of the bacteria life cycle. *Annu. Rev. Microbiol.* **47**: 855–874, 1993.
7. Mager WH and De Kruijff AJJ. Stress-induced transcriptional activation. *Microbiol. Rev.* **59**: 506–531, 1995.
8. Pirt SJ. *Principles of Microbe and Cell Cultivation*. Blackwell Scientific, Oxford, 1975.
9. Rehm HJ and Reed G, eds. *Biotechnology Vol.1: Microbial Fundamentals*. VCH Publishers, Weinheim, 1995.
10. Russell JB and Cook GM. Energetics of bacterial growth: Balance of anabolic and catabolic reactions. *Microbiol. Rev.* **59**: 48–62, 1995.
11. Slater JH. Microbial growth dynamics. In: *Comprehensive Biotechnology Vol. 1*, eds. AT Bull and H Dalton. Pergamon Press, Oxford, 1985, pp. 189–213.
12. Stouthamer AH and van Verseveld HW. Stoichiometry of microbial growth. In: *Comprehensive Biotechnology Vol. 1*, eds. AT Bull and H Dalton. Pergamon Press, Oxford, 1985, pp. 215–238.

13. Vanek Z and Hostalek Z, eds. *Overproduction of Microbial Metabolites: Strain Improvement and Process Control Strategies.* Butterworth, Stoneham Massachusetts, 1986.

Related World Wide Web Sites

- About Biotech
  http://outcast.gene.com/ae/AB/
- Access Excellence
  http://www.gene.com.ae
- Biotechnology Industry Organization
  http://www.accessexcellence.com/AB/BC/Bioprcess Technology.html
- Biotechnology Information Center
  http://www.inform.umed.edu:8080/EdRes/Topic/AgrEnv/Biotech
- Encyclopedia of Bioprocess Technology
  http://www.cbs.umn.edu/bpti/MCF/ebt.html
- Genetic Engineering News
  http://www.genengnews.com
- Microbiology Laboratory Simulation Programme
  http://www.leeds.ac.uk/bionet/compend/bnt05pst.html

Questions for Thought

1. Explain, with examples, how the effects of rate limiting substrate concentration on the growth of microbial culture could be quantified at both the theoretical and practical levels.
2. What are the conditions which allow the establishment of a stable mixed culture system? Use an example to describe the interactions among the microbial population in mixed culture fermentation.
3. How do you improve the productivity of a growth-linked product produced by a genetically engineered bacterium in a batch culture process? Give your reasons.
4. Two culture media each containing the same molar equivalent of carbon in the form of molasses or sucrose were used to culture yeast.

What could be the possible reasons that the yeast culture in the medium containing molasses was found to achieve a higher biomass growth yield?

5. Explain with examples, the importance of growth yield in fermentation processes.

6. Two batch cultures of yeast, containing an initial glucose concentration of 0.5 and 2 g/L respectively, were compared. It was observed that the specific growth rate and final biomass concentration of the culture containing 2 g/L glucose were lower than the one containing 0.5 g/L glucose. The final concentration of alcohol was, however, comparable. Give possible explanations.

7. Compare and contrast closed and open microbial culture methods.

8. If you are working for a dairy company, how would you convince your company that you could achieve a higher output rate for the production of fermented milk in a continuous flow fermentation process than the currently used batch culture system?

Chapter 3

Enzymology

Theresa May Chin Tan
Department of Biochemistry, National University of Singapore
8 Medical Drive, Singapore 117597
Email: bchtant@nus.edu.sg

3.1. Enzymes as Catalysts

Life depends on a series of coordinated chemical reactions. Many of these reactions however proceed too slowly to sustain life and hence proteins with catalytic functions have evolved. An enzyme is a protein that functions as a catalyst. The role of an enzyme (or a catalyst) is to speed up the rate of chemical reactions without undergoing any permanent changes itself. At the end of the reaction, the enzyme remains unchanged. The rate of a reaction in the presence of an enzyme is between 10^6 to 10^{14} fold higher than the uncatalyzed reaction. How is this achieved? In any chemical reaction, a reactant must contort into an unfavorable high-energy conformation before it changes into the product. The formation of this less favorable transition state represents an activation energy barrier that must be overcome before the reaction can occur. Enzymes act to lower the activation energy barrier and therefore facilitate the progress of the reaction (Fig. 3.1). Enzyme-catalyzed reactions can thus proceed at a greater rate.

Enzymes are protein molecules. Extreme conditions such as high temperature or extreme pH will affect the structure of an enzyme. This will in turn affect the catalytic activity. Most human enzymes have an optimal temperature around 37°C. At low temperatures, the rate of an enzymatic reaction is reduced. At high temperatures, denaturation of the enzyme occurs and loss of activity is observed. For most enzymes, an increase in reaction rate is usually observed as the pH goes from an acidic level toward physiological range and then declines as pH progresses towards the alkaline range. Hence, unlike chemical catalysts, enzymes catalyze reactions under relatively mild conditions: moderate temperature, atmospheric pressure and nearly neutral pH. In addition, the catalytic properties of many enzymes can be regulated within the living organism.

Each enzymatic reaction proceeds with great specificity. How do enzymes achieve specificity? An enzyme will bind to the substrate (reactant); form an enzyme-substrate complex and covert the substrate into the product. This occurs in a relative small portion of the enzyme called the active (or catalytic) site. The specificity of the enzymatic reaction results from the three-dimensional arrangement of specific amino-acids residues in the enzyme that forms the substrate-binding site. Two models have been used to describe enzyme-substrate interaction. In the lock and key model, the enzyme is viewed as the lock and the substrate as the key. Only one

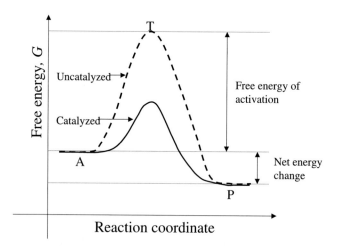

Fig. 3.1. Transition state diagram of an uncatalyzed and a catalyzed reaction. The net free energy change is the same for both but the free energy of activation is lowered in the presence of a catalyst. Hence, in the presence of a catalyst, it is easier for the reactant, A, to reach the transition state, T, and be converted to the product, P.

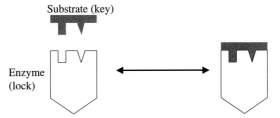

Fig. 3.2. Lock and key model for enzyme-substrate interaction. The enzyme is analogous to a lock while the substrate is viewed as the key that will fit specifically into the lock.

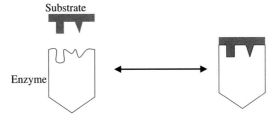

Fig. 3.3. Induced fit model of enzyme-substrate interaction. Substrate binding will alter the conformation of the active site such that the enzyme and substrate will fit each other more precisely.

key will fit each lock (Fig. 3.2). This model presents the active site as a rigid unchanging structure. In the induced fit model, the enzyme is not presented as a rigid molecule. Rather, it can change its conformation as the substrate binds and as the reaction proceeds (Fig. 3.3). Conformational changes have indeed been proven to occur by X-ray crystallographic studies on the yeast hexokinase enzyme.

3.2. Nomenclature

Before the Enzyme Commission (EC) recommended the present system of nomenclature, enzymes were given trivial names. The trivial names consisted of the suffix -ase added to the substrate of the enzyme (e.g. urease), or implied something about the nature of the reaction (e.g. alcohol dehydrogenase), or simply did not give any clue to what the enzyme does (e.g. trypsin).

The EC was set up by the International Union of Biochemistry (now known as the International Union of Biochemistry and Molecular Biology — IUBMB). Based on the EC recommendation, enzymes are classified and named according to the nature of the chemical reactions they catalyzed. There are six major classes of enzyme reactions (Table 3.1). The major classes are further subdivided into subclasses and sub-subclasses. Each enzyme has a unique EC number and a systematic name with the suffix -ase. The systematic name of an enzyme consists of the name of its substrate(s) followed by a word ending with "ase" specifying the type of reaction the enzyme catalyses. The enzyme, glucokinase, catalyzes the transfer of a phosphate from ATP to D-glucose. Its EC number is 2.7.1.2 and its systematic name is ATP:D-glucose 6-phosphotransferase. The number represents the class, subclass, sub-subclass and its arbitrarily assigned serial number in the sub-subclass.

The ENZYME database (http://tw.expasy.org/enzyme/) is a repository of information on the nomenclature of enzymes. It contains the following data for each enzyme: EC number, recommended name, alternative names, catalytic activity, cofactors, links to the SWISS-PROT protein sequence entries that correspond to the enzyme and links to human diseases associated with a deficiency of the enzyme.

Table 3.1. Classification of enzyme reactions.

EC number	Type of reaction	Class of enzyme
1	Oxidation-reduction reactions	Oxidoreductases
2	Transfer of functional groups	Transferases
3	Hydrolysis reactions	Hydrolases
4	Addition of groups to double bonds or formation of double bonds	Lyases
5	Isomerization reactions	Isomerases
6	Bond formation coupled with hydrolysis of high energy compounds such as ATP	Ligases

3.3. Cofactors

Enzymes have evolved to cope with the necessity of catalyzing chemical reactions within a living cell. To improve the repertoire of reactions which enzymes are able to catalyze, cofactors are often necessary. Without cofactors, the functional groups present on the side chains of amino acids on enzymes can only facilitate acid-base reactions, form certain types of transient covalent bonds and take part in charge-charge interactions. The presence of cofactors provides additional functional groups and allows enzymes to catalyze other types of reactions (e.g. oxidation-reduction reactions). Thus, enzymes can be divided into 2 groups: those that rely solely on their protein structure to carry out their catalytic function and those that require an additional cofactor to carry out the catalytic activity.

Cofactors may be metal ions (such as Zn^{++}, Fe^{++}) or organic molecules. Such organic molecules are referred to as coenzymes and are derived from vitamins. Flavin coenzymes (flavin adenine dinucleotide, FAD and flavin mononucleotide, FMN) are examples of coenzymes and the flavin component is derived from the vitamin riboflavin (also known as vitamin B_2).

Some cofactors are transiently associated with the given enzyme while others are permanently associated with their protein. The latter are called prosthetic groups. The catalytic active complex of enzyme-prosthetic group is called the holoenzyme. Without the prosthetic group, the protein portion is inactive and is referred to as the apoenzyme.

3.4. Regulation of Enzyme Activity

The catalytic activities of many enzymes are regulated in a living cell. Regulation can be achieved in many different ways. The different modes of regulation include: allosteric control, regulatory proteins, covalent modification of enzymes, synthesis of inactive precursors and control of enzyme synthesis.

(1) Control of Enzyme Synthesis
Bacteria can rapidly adapt to their environment by producing enzymes that can metabolize available nutrients. An example is the induction of lactose metabolizing enzymes, β-galactosidase and galactose permease.

Bacteria grown in the absence of lactose express little of these enzymes. In the presence of lactose, the synthesis of β-galactosidase and galactose permease is rapidly induced and the bacteria can now utilize lactose as a nutrient source. The enzymes are expressed only when required and regulation is achieved by controlling gene expression.

(2) Inactive Precursors

Regulation of enzyme activity can also be achieved by synthesizing enzymes as inactive precursors. The inactive precursors are known as zymogens or proenzymes. Digestive enzymes are usually synthesized as zymogens and proteolytic cleavage is needed to activate the enzyme. Chymotrypsin is secreted from the pancreas as chymotrypsinogen and it is activated in the digestive tract by the enzyme trypsin. Trypsin cleaves a small peptide from the N-terminal region and converts the inactive chymotrypsinogen to chymotrypsin.

(3) Covalent Modification

Protein phosphorylation is a major mechanism employed by hormones for intracellular signaling and for regulation of enzymatic activity. The transfer of a phosphate group from ATP to a specific serine or tyrosine residue on the target protein is mediated by protein kinases. Phosphorylation leads to a conformational change and can either activate or inactive the enzyme. Glycogen phosphorylase, which catalyzes the degradation of glycogen, is regulated by phosphorylation in response to the presence of hormones such as insulin and glucagon.

(4) Regulatory Proteins

Regulatory proteins can modulate the activity of an enzyme. These can either activate or inhibit enzyme activity. When the regulatory protein binds to its target enzyme, a conformation change occurs in the enzyme and this affects the function of the enzyme. Calmodulin, a 17 kD protein serves as a calcium sensor in eukaryotic cells. Ca^{++} can bind to multiple sites on calmodulin. The Ca^{++}-calmodulin complex in turn binds to many target enzymes and modulate their activity.

(5) Allosteric Control

The enzyme that catalyzes the first step of a biosynthetic pathway is usually regulated by the ultimate product (Fig. 3.4). This form of regulation is termed feedback regulation. Feedback regulation allows for quick adjustment to the cell's requirement for a particular product and prevents excessive synthesis. Feedback regulation makes use of the properties of allosteric enzymes.

Allosteric effectors (allosteric activators and inhibitors) affect the activity of allosteric enzymes. These effectors are molecules other than the substrates and bind at sites separate from the active site of the enzyme. The site where the allosteric effector binds is called the allosteric site (Fig. 3.5). Allosteric enzymes contain multiple subunits. They also exhibit positive cooperativity. Binding of the substrate to one subunit facilitates the binding of substrate to one or more of the remaining subunits. An allosteric activator serves either to activate the enzyme or to stabilize the active state of the enzyme, thus facilitating substrate binding in their own and other

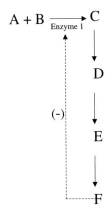

Fig. 3.4. Regulation of a pathway by feedback inhibition. C, D and E are intermediates in this hypothetical pathway. The final product, F exerts an inhibitory effect on enzyme 1 which catalyzes the first reaction in the pathway.

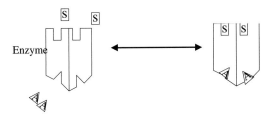

Fig. 3.5. Binding of substrate (S) molecules and effector (A) molecules to an allosteric enzyme. The allosteric site is distinct from the substrate-binding site.

subunits. Conversely, an allosteric inhibitor will inactive the enzyme and prevent substrate binding.

3.5. The Michaelis-Menten Equation

In 1913, Leonor Michaelis and Maud Menten proposed a general theory of enzyme action consistent with observed enzyme kinetics. The enzyme, E associates with its substrate S to form an ES complex with the rate constant of k_1. This association is reversible. The ES complex has two fates. It can dissociate to E and S with the rate constant of k_2 or the reaction can proceed and the product P is formed (with the rate constant of k_3).

$$E + S \rightleftharpoons ES \longrightarrow E + P$$

The Michaelis-Menten equation,

$$v = V_{max} [S]/K_m + [S]$$

relates the initial velocity (v) of an enzyme reaction to the concentration of the substrate, [S]. K_m is defined as the concentration of S at which the velocity of the reaction is at half the maximal. K_m is also equal to $(k_2 + k_3)/k_1$. V_{max} is the maximal velocity. A plot of v versus [S] would thus yield a rectangular hyperbola (Fig. 3.6). At infinite substrate concentration, the reaction proceeds under saturation conditions and the reaction velocity approaches V_{max}.

Rearrangement of the Michaelis-Menten equation can transform it into a linear equation. One such transformation is the Lineweaver-Burk transformation. By taking the reciprocal on both sides, the Michaelis-Menten equation is transformed into

$$1/v = (K_m/V_{max}) (1/[S]) + 1/V_{max}$$

A plot of 1/v versus 1/[S] would yield a straight line and the constants V_{max} and K_m can be easily determined (Fig. 3.7).

The value of K_m is dependent on the assay conditions (pH, temperature, presence of inhibitors or activators) but independent of the amount and purity of the enzyme. K_m is the substrate concentration at which half the

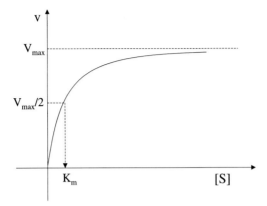

Fig. 3.6. The Michaelis-Menten plot of initial velocity (v) versus substrate concentration, [S]. V_{max} is the maximal velocity while K_m is the concentration of S at $V_{max}/2$.

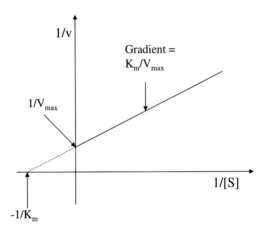

Fig. 3.7. The Lineweaver-Burk plot of 1/v versus 1/[S]. v is the initial velocity while [S] is the substrate concentration.

active sites of the enzyme are filled (i.e. $V_{max}/2$). When $k_2 \gg k_3$, then K_m provides a reasonable measure of the affinity of the enzyme for its substrate. A low K_m value would indicate that the enzyme has high affinity for the substrate. K_m values thus provide a means to compare the affinities of enzymes for their respective substrates.

The catalytic constant k_{cat} (also known as turnover number) of an enzyme is a measure of its maximal catalytic activity. It is defined as

$$k_{cat} = V_{max}/[E]_T$$

where $[E]_T$ is the total enzyme concentration. k_{cat} equals to k_3. It is the number of substrate molecules that is converted into the product by an enzyme molecule in a unit time when the enzyme is fully saturated with its substrate. However under physiological conditions, the substrate concentration is seldom at saturation and k_{cat} values cannot be interpreted meaningfully. The constant k_{cat}/K_m is used instead to provide a measure of the enzyme's catalytic efficiency under non-saturating substrate concentration ($[S] < K_m$).

The above kinetic analysis is for a single substrate reaction. Can it be applied to a multi-substrate reaction? It is beyond the scope of this book to describe such reactions but it suffices to say that K_m and V_{max} values can still be determined. The usual practice is to determine the velocity of reaction by varying just one substrate while keeping the concentration of all the other cosubstrates constant. In such situations, Michaelis-Menten kinetics will be observed.

Are there situations where a deviation from Michaelis-Menten kinetics is observed? A deviation from Michaelis-Menten behaviour is indicative

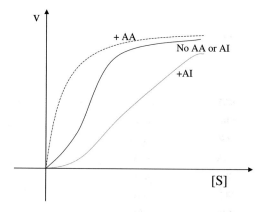

Fig. 3.8. Kinetic profile of an allosteric enzyme in the presence of an allosteric activator (AA) or an allosteric inhibitor (AI). In the absence of an activator or inhibitor, a sigmoidal curve is obtained.

of allosteric regulation (see Sec. 3.4). Hexokinase catalyzes the transfer of a phosphate from ATP to glucose. Hexokinase exists as a number of different tissue-specific isoenzymes. Isoenzymes are proteins with different amino acid sequences, which catalyze the same reaction. They also exhibit different kinetics and regulatory properties. The hexokinases found in most human tissues follow Michaelis-Menten kinetics and a plot of v versus [S] yield a rectangular hyperbola (Fig. 3.6). However, glucokinase (which is the hexokinase isoenzyme found in the liver) behaves differently and a sigmoidal curve is observed. Such sigmoidal curves are characteristic of enzymes that are allosterically regulated (Fig. 3.8).

3.6. Inhibition of Enzyme Activity

The activity of an enzyme can be regulated. The various modes of regulation have been described in Sec. 3.4. Besides these regulatory mechanisms, exogenous compounds can also affect the rate of an enzymatic activity. A compound that binds to the enzyme and decreases the velocity of the enzymatic reaction is an inhibitor. Irreversible inhibitors are compounds that form covalent bonds or other strong interactions with the enzyme. Examples of irreversible inhibitors with therapeutic uses include the antibiotic penicillin. Penicillin reacts covalently with bacterial glycopeptide transpeptidase to form the penicilloyl-enzyme complex; rendering the enzyme inactive. Glycopeptide transpeptidase is essential for cell wall biosynthesis. Thus, treatment with penicillin disrupts bacterial cell wall synthesis.

Reversible inhibitors, on the other hand, do not form covalent bonds with the enzyme and the inhibitor-enzyme complex can dissociate. There are three forms of reversible inhibition: competitive, uncompetitive and mix (or non-competitive) inhibition. All three types of reversible inhibition can easily be distinguished using the Lineweaver-Burk plot of 1/v versus 1/[S] (Fig. 3.9).

(1) Competitive Inhibition

A competitive inhibitor is a structural analog of the substrate. It resembles the structure of the substrate. It will thus compete with the substrate for binding to the active site of the enzyme (Fig. 3.10). Competitive inhibition can be overcome by increasing the substrate concentration. At high substrate concentration, all the active sites will be occupied by the substrate and the inhibitor will not be able to bind. Hence, competitive inhibition will lead to an increase in the K_m value but the maximal velocity (V_{max}) remains unchanged (Fig. 3.9). The mode of action of many therapeutic drugs is based on competitive inhibition. The therapeutic drug methotrexate is a structural analog of dihydrofolate. It competes with dihydrofolate for binding to the active site of the enzyme dihydrofolate reductase.

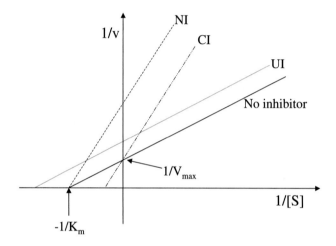

Fig. 3.9. Lineweaver-Burk plots in the presence of a competitive inhibitor (CI), a noncompetitive inhibitor (NI) and an uncompetitive inhibitor (UI).

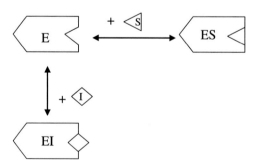

Fig. 3.10. Competitive inhibition. The inhibitor (I) competes with the substrate (S) for the active site of the enzyme (E).

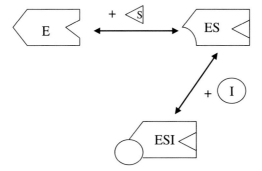

Fig. 3.11. Uncompetitive inhibition. The inhibitor (I) can only bind to the enzyme-substrate (ES) complex.

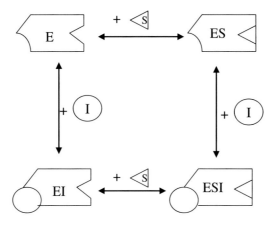

Fig. 3.12. Noncompetitive (mixed) inhibition. The inhibitor can bind to either the enzyme (E) or the enzyme-substrate (ES) complex.

(2) Uncompetitive Inhibition

An uncompetitive inhibitor behaves differently from a competitive inhibitor. It binds to the enzyme-substrate complex. Binding of the substrate to the enzyme will lead to a conformational change in the enzyme and allow the inhibitor to interact with the enzyme (Fig. 3.11). Formation of the substrate-enzyme-inhibitor complex prevents the reaction from proceeding. Uncompetitive inhibition changes both the K_m and V_{max} values (Fig. 3.9).

(3) Mixed Inhibition

Mixed inhibition is also known as noncompetitive inhibition. Noncompetitive inhibitors do not compete with the substrate for the active site. Noncompetitive inhibitors bind to the enzyme at a site that is distinct from the active site. Such inhibitors can bind either to the enzyme or the enzyme-substrate complex (Fig. 3.12). Mixed inhibition may or may not alter the K_m value of a reaction but a decrease in the V_{max} will be observed (Fig. 3.9).

3.7. Industrial Applications of Enzymes

Enzymes are useful catalysts. The advantages of enzyme compared to other catalyst include specificity (lack of side reactions) and catalysis under mild conditions (low temperature and atmospheric pressure). The catalytic ability of enzymes has been harnessed and used for centuries. The earliest uses include the use of enzymes present in microorganisms for brewing, bread making and cheese making. These had been carried out before there was any understanding of the biochemical reactions and the enzymes involved. Today, enzymes are used in the chemical, pharmaceutical and food industry. The sales of enzymes for industrial uses have doubled between the years 1983 and 1995. A further doubling is predicted by the year 2005. It is beyond the scope of this section to provide a detailed account of the many and varied uses of enzymes. A selection of examples is used to allow for the appreciation of the industrial and clinical application of enzymes.

(1) Brewing

An examination of the brewing process will illustrate how enzymes are utilized for the production of beer. Note that for brewing, intact yeast and barley seeds are used and isolation of the required enzymes is not necessary.

The starting material is starch, which is present in the barley seed. Starch is a polysaccharide and yeast is unable to ferment starch. This can be overcome by allowing the barley seeds to germinate. During germination, endo-1,3-β-D-glucanase, amylases and peptidases are released. Starch granules are found within the cells of the endosperm.

The walls of the endosperm are surrounded by polymers including 1,3-β-D-glucans. The enzyme endo-1,3-β-D-glucanase will degrade the glucans so that the amylases can come into contact with the starch granules. Amylases catalyze the hydrolysis of starch to simple sugars including glucose, maltose, and other oligosaccharides. In addition, peptidases are also released and these enzymes will catalyze the degradation of endogenous proteins within the seeds. The resulting peptides and amino acids are utilized by the yeast for growth. The primary role of yeast in brewing is to provide the necessary enzymes for anaerobic glycolysis. Ethanol is the by-product of anaerobic glycolysis. Hence, for brewing, both the germinating barley seed and the yeast provide the full complement of enzymes for converting the starting raw material starch to the final product ethanol.

(2) Antibiotic and Therapeutic Products
Microorganisms are also utilized for the production of many useful therapeutic products. These include the production of antibiotics such as erythromycin, streptomycin and penicillin by *Streptomyces spp* and *Penicillium spp* and the production of lovastatin by *Aspergillus terreus*.

(3) Clinical Applications
In clinical applications, enzymes can be used in one of the following manner: as diagnostic tools, as analytical reagents or as therapeutic agents.

Currently, more than 20 different enzymes are assayed routinely in clinical chemistry laboratories. Detection of abnormal levels of enzymes in body fluids can aid diagnosis of diseases. Elevated serum levels of aspartate amino transferase (AST), alanine amino transferease (ALT), alkaline phosphatase (ALP) and γ-glutamyltransferase (GGT) are observed in liver disease. How are these assayed?

The key feature when assaying for enzyme activity is the need for an end product that can be easily detected and measured. The product usually absorbs light in the UV or visible range. ALP is assayed by measuring the hydrolysis of 4-nitrophenyl phosphate. The products of this reaction are phosphate and 4-nitrophenol. The latter absorbs at 405 nm. However not all reactions produce at least a product which can be easily measured. An example is the assay for ALT activity.

Reaction 1:   alanine + 2-oxoglutarate $\longrightarrow$ pyruvate + glutamate
Reaction 2:   pyruvate + NADH + H$^+$ $\longrightarrow$ lactate + NAD$^+$

The products pyruvate and glutamate do not absorb light. Hence it is necessary to couple the ALT reaction (Reaction 1) to a second reaction (Reaction 2). The second reaction is catalyzed by lactate dehydrogenase (LDH). Pyruvate, the product from the ALT reaction, is used as a substrate for the subsequent LDH reaction. In the presence of the coenzyme NADH, LDH converts pyruvate to lactate. NADH absorbs at 340 nm while NAD does not. The change in A_{340} due to oxidation of NADH can be easily measured.

Enzymes are also used as analytical reagents in clinical chemistry applications. In the above example, the enzyme LDH is used as an analytical reagent for the assay for ALT. Another example is the use of the enzyme glucose oxidase to assay blood glucose levels.

Reaction 1:   glucose + H_2O + O_2 $\longrightarrow$ gluconic acid + H_2O_2
Reaction 2: H_2O_2 + reduced acceptor $\longrightarrow$ H_2O + oxidized acceptor

Glucose oxidase oxidizes glucose to gluconic acid and hydrogen peroxide (Reaction 1). The hydrogen peroxide produced then oxidizes an acceptor (Reaction 2). The acceptor commonly used is o-dianisidine. Oxidized o-dianisidine absorbs between 425 nm and 475 nm.

Enzyme-linked immunosorbant assay (ELISA) also utilizes enzymatic reactions for detection of antigen-antibody interaction (Fig. 3.13). The binding of an antibody to an antigen cannot be easily detected. To facilitate detection, a secondary antibody with an enzyme attached to it is used. Enzymes commonly used for ELISA assays include alkaline phosphatase and horse raddish peroxidase. Both enzymes can react with specific substrates to produce products that are easily detected.

Enzymes are also used for a variety of therapeutic purposes (Table 3.2). Streptokinase, urokinase and human tissue plasminogen activator (tPA) are effective activators of plasminogen. These are used as thrombolytic agents and are useful in removing blood clots. Intravenous administration of therapeutic enzyme molecules may elicit allergic, anaphylactic and immunological responses. In addition, the enzyme is usually rapidly removed from circulation and degraded. Much work had been carried out to overcome the antigenicity and short plasma half-lives of therapeutic proteins. An example is tPA. Human tPA has been cloned and the recombinant protein is used for myocardical infarction. tPA has poor solubility, is rapidly cleared by the liver and hence a large dose is necessary. However, the side effects of a large dose include bleeding and non-clotting. Recombinant DNA and protein engineering technology have been used to create an improved version of this protein. The engineered tPA is a smaller protein with increased solubility, better pharmacokinetics

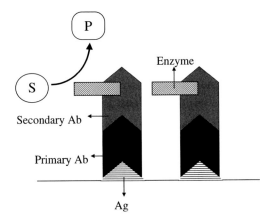

Fig. 3.13. Enzyme-linked immunosorbant (ELIZA) assay. The primary antibody (Ab) recognizes the immobilized antigen (Ag) and the secondary Ab will specifically interact with the primary Ab. These interactions can be detected using an appropriate substrate (S) for the enzyme attached to the secondary Ab. The product (P) formed can be easily measured.

Table 3.2. Therapeutic uses of enzymes.

Enzyme	Application
Collagenase	Skin ulcers
Trypsin	Anti-inflammatory agent
Uricase	Gout
Hyaluronidase	Drug administration
Tissue plasminogen activator	Thrombolytic agent
Streptokinase	Thrombolytic agent
Urokinase	Thrombolytic agent
Factor IX_a	Hemophilia B

and improved specificity and affinity at the clot with little effect on peripheral coagulation.

(4) Immobilized Enzymes

The attachment of an enzyme to a support will enable us to reuse the enzyme without a tedious recovery process. Enzymes may be immobilized on an inert support in a number of ways. These include adsorption, covalent linkages, entrapment within a matrix and encapsulation.

Immobilized β-galactosidase has been used to hydrolyze lactose to glucose and galactose. Immobilization is achieved by the adsorption of the enzyme onto silica particles. This immobilized enzyme has been used to produce a mixture of glucose and galactose, which is used as a sweetening agent. Immobilized β-galactosidase is also used to remove lactose from milk. Lactose intolerance is a common problem in many parts of the world and this has led to a demand for lactose-free milk.

Analytical reagents may also be immobilized. Glucose dehydrogenase can be immobilized onto nylon tubing and this can be used and reused for measuring blood glucose levels. As little as 0.18 mg of enzyme can be used for more than 1000 glucose determinations. The ability to immobilize enzymes and other analytical reagents has also led to a wide range of home testing kits for urine and blood samples.

The above examples illustrate how enzymes can be utilized for industrial and clinical applications. In some instances, the use of a suitable microorganism with the appropriate complement of enzymes is sufficient (e.g. brewing) but in other instances, purified enzymes are necessary (e.g. clinical applications). In addition, the progress in genetic engineering and recombinant DNA technology will lead to the production of genetically engineered microorganism and recombinant proteins which have improved features such as enhanced catalytic activity, increased solubility or improved thermal stability.

References

1. Steitz TA, Shoham M and Bennett WS Jr. Structural dynamics of yeast hexokinase during catalysis. *Philos. Trans. R. Soc. Lond. B. Biol. Sci.* **293**: 43–52, 1981.

2. Enzyme Nomenclature. *Recommendations of the Nomenclature Committee of the International Union of Biochemistry and Molecular Biology.* Academic Press, London, 1992.
3. Naqui A. Where are the asymptotes of Michaelis-Menten? *TiBS* **1**: 64–65, 1986.
4. Godfrey T and West S, eds. *Industrial Enzymology.* Macmillian, London, 1996.
5. Pollock J. *Brewing Science.* Academic Press, New York, 1987.
6. Buckel P. Recombinant proteins for therapy. *TiPS* **17**: 450–456, 1996.

Further Readings

1. Copeland RA. *Enzymes — A Practical Introduction to Structure, Mechanism and Data Analysis.* Wiley-VCH, Inc, USA, 1966.
2. Price NC and Stevens L. *Fundamentals of Enzymology — The Cell and Molecular Biology of Catalytic Proteins,* 3rd ed. Oxford University Press, 2000.

Relevant World Wide Web Sites

• About Enzymes

  http://tw.expasy.org/enzyme/

Questions for Thought

1. The table below shows the initial velocities (v) of an enzyme-catalyzed reaction with the corresponding substrate ([S]) concentrations

v (μmole/min)	[S] ($\times 10^{-5}$ mole/l)
200	100
180	35
130	12
100	6
50	2

(a) Determine the K_m and V_{max} values for the enzyme.

(b) In the presence of an inhibitor I, the K_m value increases but the V_{max} remains unchanged.

(i) What is the nature of this inhibition?

(ii) How can this inhibition be overcomed?

2. (a) Two different enzymes: A and B catalyze the same reaction. What is the term used to describe such enzymes?

(b) Compound X is a competitive inhibitor of Enzyme A. Another compound, Y interacts with the N-terminal domain of Enzyme A. The N-terminal domain of Enzyme B is different from that of Enzyme A. Kinetic analysis shows that Y is a non competitive inhibitor for Enzyme A. How will X and Y affect the K_m and V_{max} of Enzyme B?

3. A student was given 2 different preparations of enzyme E and told to determine the K_m and V_{max} values. Enzyme E was isolated from rat liver using the same protocol on two different occasions. He noticed that the K_m values were identical but the V_{max} values differ. How can he account for his observations?

4. (a) Explain the principle of coupled enzyme assays.

(b) You are interested in determining the concentration of ATP in your sample. With the following reagents

(i) glucose

(ii) glucokinase

(iii) glucose-6-phosphate dehydrogenase

(iv) $NADP^+$

outline how you would set up an assay to determine the concentration of ATP in your sample.

(c) In place of glucose, you are provided with lactose, galactosidase, ATP and reagents (ii), (iii) and (iv). Using these reagents, illustrate how you can determine the activity of galactosidase in your sample.

Chapter 4

Manipulation of Genes

Azlinda Anwar*, **Heng-Phon Too**[†] **and Kim Lee Chua**[†]
Division of Biomedical Sciences, Johns Hopkins Singapore
31, Biopolis Way, #02-01, The Nanos, Singapore 138669
[†]*Department of Biochemistry, National University of Singapore*
8 Medical Drive, Singapore 117597

This chapter focuses on the ways in which molecular genetic techniques can be used to manipulate genes in order to alter the expression and production of microbial products, including the expression of novel recombinant products. "Classical" (*in vivo*) genetic techniques are essentially limited by two factors. Firstly, they can only be applied to the existing genetic complement of an organism, i.e. they are restricted to naturally occurring genes or relatively minor modifications of these genes. It is not possible to get an organism to make a product totally foreign to that organism using these techniques. Secondly, with classical techniques, one can only work on the basis of phenotype and mutants are selected by their effect on the observable characteristics of the organism. This not only limits the changes that can be selected, but also means that it can be difficult to determine the nature of the mutation that has caused the alteration in the phenotype.

The advent of recombinant DNA technology (also referred to as gene cloning or *in vitro* genetic manipulation) has dramatically broadened the spectrum of microbial genetic manipulations. The basis of this technology is in the use of restriction endonucleases and DNA ligases as a means to specifically cut and paste fragments of DNA. In this way, foreign DNA fragments can be introduced into a vector molecule (a plasmid or a bacteriophage), which enables the DNA to be replicated after being introduced into a bacterial cell.

4.1. Construction of Recombinant DNA

Today many areas of biological research require the use of recombinant technology in one form or the other. The ability to modify and clone genes has become an essential tool in research and has accelerated the rate of discovery and the development of bioindustries.

The basic steps in DNA cloning involve the following steps (Fig. 4.1):

1. A fragment of DNA is inserted into a carrier DNA molecule, called a *vector*, to produce a *recombinant DNA*.
2. The recombinant DNA is then introduced into a host cell, where it can multiply and produce numerous copies of itself within the host. The most commonly used host is the bacteria, although other hosts can also be used to propagate the recombinant DNA.
3. Further amplification of the recombinant DNA is achieved when the host cell divides, carrying the recombinant DNA in their progenies, where further vector replication can occur.
4. After a large number of divisions and replications, a colony or clone of identical host cells is produced, carrying one or more copies of the recombinant DNA. The recombinant DNA is now said to be cloned.
5. The colony carrying the recombinant DNA of interest is then identified, isolated and analyzed.

From the schematic diagram, one can appreciate that DNA cloning is a relatively straightforward procedure. Essentially the basic principles in DNA cloning involve cutting a piece of DNA (restriction digestion) and pasting the restricted DNA into a vector (ligation). In order to produce more of this recombinant DNA for easy manipulations, the DNA is then introduced and propagated in host cells (transformation and amplification). Once the clone containing the desired DNA is selected, the recombinant plasmid can finally be isolated and further explored (analysis). Each of

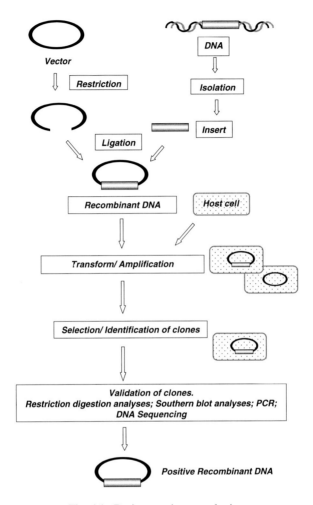

Fig. 4.1. Basic steps in gene cloning.

the steps in the schematic diagram will be described in detail in order to guide the students in the principles of DNA cloning.

4.1.1. *Enzymes used in gene manipulations*

Recombinant DNA technology will not be possible had it not been for the discovery and exploitation of the functions of restriction and modifying enzymes. Within a cell, these enzymes participate in essential processes such as DNA replication and degradation of undesirable DNA. A number of bacterial strains have been identified that can produce these enzymes in high quantities. After purification from the bacterial cell extracts, these enzymes are still able to carry out their natural reactions when the appropriate artificial condition is provided. Currently, a number of these enzymes are produced commercially by recombinant DNA technologies, and are supplied as highly purified and well-characterized enzyme preparations.

The enzymes that are used in gene manipulations can generally be classified into 5 classes based on their functions:

1. **Nucleases**, which cut or degrade DNA molecules.
2. **Polymerases** can copy and make new strands of DNA.
3. **Ligases**, which join pieces of DNA molecules together.
4. **Modifying enzymes**, which modify the DNA by adding or removing chemical groups.
5. **Topoisomerase**, which remove or introduce supercoils from covalently closed-circular DNA.

The functions of these enzymes and how they are used in DNA cloning will be described briefly. Three classes of enzymes that are more widely used in DNA cloning will however be described in detail. A point to note is that although most enzymes can be classified into a particular class of function, a few can display multiple functions. For example, polymerases have both the functions of making new strands of a DNA as well as degrading it.

(1) Nuclease

The function of this class of enzymes is to degrade or digest DNA by breaking the phosphodiester bonds that link one nucleotide to the other in a DNA strand. There are two types of nucleases:

(a) **Exonuclease**, which digests nucleic acids at either ends of the DNA.
(b) **Endonuclease**, which digest nucleic acids from within the DNA.

The main difference within different endonucleases or exonucleases is their ability to digest either single-stranded or double-stranded DNA.

For example, while the enzyme *Exonuclease* III (*Exo* III) is able to cleave double-stranded DNA, the enzyme *Exonuclease* VII (*Exo* VII) is only able to cleave single-stranded DNA. Similar distinctions can be applied to endonucleases. Whereas the endonuclease DNAase I, isolated from bovine pancreas, is able to digest double-stranded DNA, the mung bean nuclease (from the sprouts of mung bean) can only digest single-stranded DNA. Both DNAse I and mung bean nuclease are examples of endonucleases that cleave DNA at non-specific sites. There is yet another group of endonucleases that can only cleave DNA at restricted sites defined by specific DNA sequences. This group of enzyme is known as restriction endonucleases and is the most commonly utilized enzymes in recombinant DNA technology.

Restriction Endonuclease

This group of enzymes was first discovered when it was observed that some strains of bacteria were immune to bacteriophage infection. It is now known that these strains of bacteria had produced endonucleases to restrict the propagation of new phage particles by digesting the foreign DNA before it can replicate. During this process, the bacteria protect their genome by modifying their DNA through methylation to prevent the action of these degradative enzymes.

These restriction endonucleases have been given names consisting of 3 italicized letters (representing the bacteria from which the restriction enzymes isolated from), followed by a letter to identify the strain (if applicable), and end with a Roman numerical to distinguish the different enzymes isolated from the same bacteria. Some examples are given below to illustrate this nomenclature:

*Eco*R I is the first restriction/modification enzyme isolated from *Escherichia coli strain R*.

*Hin*d III refers to the third system isolated from *Haemophilus influenzae strain R_d*.

*Bam*H I was the first restriction/modification system isolated from *Bacillus amyloliquefaciens* H.

To date, more than 2900 restriction enzymes are known, and they have been generally classified into Type I, Type II or Type III specific classes.

Types I and III enzymes are complex, and are not commonly used in genetic engineering. These enzymes cleave the DNA at a substantial distance from their recognition sites. Type II restrictions enzymes, on the other hand, play a central role in recombinant technology.

The central feature of the Type II restriction enzymes is that each enzyme has a specific recognition sequence. The DNA is cleaved either *within* or *very near* to the recognition sequence. It is important to note that these enzymes are very specific and will cut the DNA only when their recognition sequences are present. The recognition sites of Type II enzymes are usually palindromic (two fold symmetry) and are 4, 5, 6 or more base pairs long. Some examples of the most commonly used restriction endonucleases are shown in Table 4.1.

Note that some restriction endonucleases, for example *Bam*H I, cut the DNA asymmetrically (as shown by the slash). This results in each fragment having four unpaired bases (in this case GATC) at both ends, referred to as "sticky" or cohesive ends. These single-stranded regions are complementary and will therefore tend to pair with DNA fragments with compatible sequences. For example, in the case of *Bam*H I-restricted DNA, the DNA can form complement pairing with either a *Bam*H I or *Bgl* II restricted DNA.

Some enzymes, such as *Pst* I, leave a four-base sticky end at the 3' end of the fragment, while others, such as *Sma* I and *Pvu* II, cleave in the center of the recognition sequence, generating blunt-ended fragments.

Table 4.1. Some examples of the frequently used restriction endonucleases. The sequences shown are that of the positive strand of DNA from the 5' to 3' direction.

Enzyme	Recognition site	Blunt or cohesive end
Hind III	A/AGCTT	cohesive
*Bam*H I	G/GATCC	cohesive
Bgl II	A/GATCT	cohesive
Pst I	CTGCA/G	cohesive
Sma I	CCC/GGG	blunt
Pvu II	CAG/CTG	blunt
Alu I	AG/CT	blunt
Not I	GC/GGCCGC	cohesive

(2) Polymerases

Polymerases are enzymes that can synthesize new strands of nucleic acids that are complementary to an existing piece of DNA or RNA. These enzymes can only do so when the template has a double-stranded region that can act as a primer to initiate nucleic acid polymerization.

There are 3 main types of polymerases that are frequently used in recombinant DNA technology (Fig. 4.2). The first, DNA polymerase I, attaches itself to the single-stranded portion of a largely double-stranded DNA and initiates synthesis from the 5' end of the DNA, replacing the

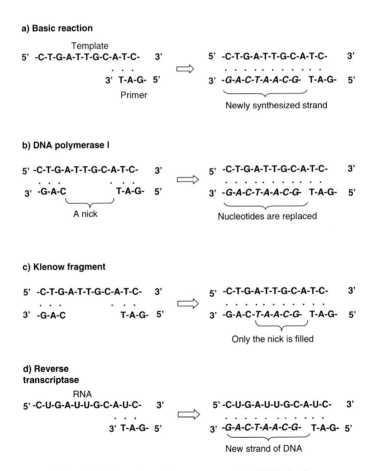

Fig. 4.2. The major polymerases used in DNA cloning.

old strand as it proceeds. One can therefore see that this enzyme has dual functions: DNA synthesis and degradative activities.

The second most commonly used polymerase enzyme is the Klenow fragment, which is essentially derived from DNA polymerase I. The DNA polymerase I enzyme has been modified to remove the DNA degradative function, retaining only the DNA polymerization function. This modified enzyme, called the Klenow fragment, is thus able to synthesize new DNA strands complementary to the single-stranded portion. It cannot, however, replace the preceding double-stranded DNA as it does not have the nuclease activity. The major application of the Klenow fragment is in DNA sequencing.

The third type of polymerase enzyme important in recombinant DNA technology is reverse transcriptase. Unlike the two other polymerases, the reverse transcriptase uses RNA as the starting template. The ability of this enzyme to synthesize a DNA strand complementary to an RNA template is crucial to a technique called cDNA cloning.

(3) Ligases

DNA ligases serve to link DNA strands together by forming phosphodiester bonds between the 5'-phosphate and the 3'-OH termini of the discontinuous strands. In the cell, these enzymes serve to repair single-stranded breaks (nicks) that arise during DNA replication. The enzymes can also join together 2 individual pieces of DNA together. The most commonly used DNA ligase is the T4 DNA ligase that is purified from *E. coli* infected with the T4 bacteriophage. The use of these enzymes in DNA cloning will be described in the later part of the chapter.

(4) Modifying Enzymes

These enzymes modify the DNA by either adding or removing a chemical group. There are numerous modifying enzymes, but the 3 most commonly used ones are:

(a) **Alkaline phosphatase** removes a phosphate group from the 5' end of the DNA. This is usually done to prevent vector religation.
(b) **Polynucleotide kinase** acts in reverse of the alkaline phosphatase by adding a phosphate group to the 5-terminus of a DNA.
(c) **Terminal transferase** adds on one or more nucleotides on the 3' end of a DNA.

(5) Topoisomerases

This set of enzymes is able to change the conformation of a closed circular DNA by adding or removing supercoils. These enzymes are used mainly in study of DNA replication, and are not too widely used in DNA cloning.

4.1.2. Use of restriction endonucleases in recombinant DNA technology

Restriction digestion of DNA is one of the most fundamental techniques in recombinant DNA technology. A typical DNA restriction digestion reaction is illustrated in Fig. 4.3. The amount of DNA to be digested will depend on the nature of the experiment. For example, if a routine cloning experiment is to be performed, typically 1–2 μg of DNA is sufficient. However if one is constructing a genetic library, more starting DNA ($\sim$10–20 μg) is needed. The second point to consider is the choice of the restriction endonuclease to be used. There are hundreds of commercially available restriction enzymes, supplied as purified enzymes with known concentrations. By convention, 1 unit of enzyme is defined as the amount required to digest 1 μg of DNA in 1 hour at 37°C. Before adding the

Fig. 4.3. Restriction of the λZAP DNA using *Hin*d III and *Sma* I. The sizes of each fragment are indicated in base pairs.

enzyme, the DNA solution must be adjusted to provide a suitable reaction condition to ensure optimal enzyme activity. Most enzymes function in pH 7.4, but require different Mg^{2+} concentration and ionic strength (usually provided by NaCl). All commercial enzymes are supplied with their respective buffer in a 10X working concentration. This concentrated buffer will have to be diluted to a 1X concentration in a digestion reaction. The optimal temperature for most restriction endonucleases to function is at 37°C, although there are some enzymes that require a lower temperature to function optimally (e.g. *Sma* I is used at 25°C).

To illustrate this technique, let us consider digesting a sample of λZAP DNA (of concentration 0.5 μg/μL) with *Hin*d III (Fig. 4.4).

We would start with 2 μg of DNA for restriction digestion. Hence 4 μL of the DNA sample is needed for the reaction, which is performed in a typical 1.5 mL microfuge tube. Next, we would need to provide the proper environment for the enzyme to work in. A typical restriction digestion reaction is carried out in a 20 μL volume. Therefore, we would need 2 μL of a 10X *Hin*d III buffer solution. The restriction endonuclease can now be added. A commercial *Hin*d III enzyme is typically supplied at a concentration of 10 U/μL. Since we are using 2 μg of DNA, a 0.5 μL of the enzyme (5 U total) is more than sufficient to completely digest the

Fig. 4.4. Restriction digest of λZAP DNA with *Hin*d III.

DNA. Since the total reaction volume required is 20 μL, 9.5 μL of water is finally added to the reaction mix. The digestion reaction is then carried out at 37°C for 1 hour to allow for complete digestion of the DNA. At the end of the incubation period, the enzyme must be destroyed to prevent accidental digestion of other DNA molecules that may be added at a later stage. This can be achieved by simply denaturing the enzyme by heating the reaction mix at 37°C for 10 min, or adding EDTA to chelate the Mg^{2+}, thus rendering the enzyme inactive.

4.1.3. Analysis of the result of restriction endonuclease digestion

As can be predicted from Fig. 4.3, restriction digestion of the DNA will yield multiple DNA fragments, the sizes of which will depend on the distance between the specific restriction sites for *Hin*d III. Clearly, a method is needed to separate the DNA fragments in order to determine the number and sizes of the different fragments. One of the easiest methods to achieve this is by using the technique of gel electrophoresis. Eventually we would also like to be able to acquire more information about the DNA in terms of the relative positions of different restriction sites within the DNA. This is called a restriction map.

(1) Separation of DNA Molecules by Gel Electrophoresis
Nucleic acids have net negatively charged phosphates with pKa of ~2. Hence in neutral conditions (pH ~7), the molecules will have an overall negative charge. Thus when the DNA molecules are subjected to an electric field in a buffer, the molecules will migrate to the positive end. The DNA molecules, to a great extend, possess a constant charge to size ratio, irrespective of their sizes. Hence nucleic acids of all sizes will migrate at a similar rate in liquids and it is therefore not possible to separate them in liquids. One way to separate the DNA fragments according to their sizes is by carrying out electrophoresis in gels that act as molecular sieves. Gels are semi-solid matrices of polymers entrapping a buffer. Since the DNA molecules have to weave through this matrix mesh to get to the positive end, the sizes of the molecules become a factor; the smaller the DNA fragment, the faster it can migrate through the gel. Two of the most commonly used gel matrices are agarose and polyacrylamide, or a mixture

Table 4.2. Guide to resolution of linear DNA versus gel concentrations.

Agarose (%)	Resolution for linear DNA (kb)
0.5	1.0–40
0.8	0.5–30
1.0	0.3–20
1.5	0.2–5
2.0	0.1–2

of both. In practice, the composition of the gel matrix will determine the sizes of the DNA to be separated. For example, a 1% agarose gel is used to separate DNA fragments of sizes 500–10,000 bp (Table 4.2).

If separation of smaller pieces of DNA fragments is required, differing by just bases, then a polyacrylamide gel can be used. The resolution of double-stranded DNA fragments using nondenaturing polyacrylamide gel is shown in Table 4.3:

Table 4.3. Resolution of DNA in polyacrylamide gels.

Polyacrylamide (%)	Double-stranded DNA (bp)
3.5	100–1000
5.0	100–500
8.0	60–400
12	50–200
20	5–100

(2) Visualization of DNA Molecules in a Gel

The easiest way to visualize DNA fragments in a gel is by using fluorescent dyes. A number of fluorescent dyes can be used for this purpose. One of the most commonly used is ethidium bromide, EtBr. EtBr molecule intercalates (inserts) between the stacked bases of double-stranded nucleic acid or nucleic acid stacked into some secondary structures, as in the case of RNA. The intercalated EtBr will emit a fluorescent light when excited at the UV range (200–400 nm), indicating the position of the DNA fragment. It is important to note that prolonged exposure of the DNA to UV will induce strand breakage within the DNA. Other fluorescent dyes, for example SYBR Gold and SYBR Green, are also available for DNA visualization.

(3) Estimation of the Sizes of DNA Molecules

In gel electrophoresis, the DNA fragments will be separated according to their sizes. The most common way to estimate the sizes of DNA is by including a DNA marker in the electrophoretic run. DNA markers are standard restriction digests, usually of λ DNA, comprising fragments of known sizes. As the sizes of the DNA markers are known, the fragment sizes in the experimental digest can be estimated by comparing the positions of the DNA bands in the two tracks.

There are a number of DNA markers available commercially catering to various ranges of DNA sizes. The choice of markers used in electrophoresis is dependent on the size range of the experimental DNA fragments. For example, if the size of a PCR product, less than a kilobase long, is to be estimated, then it would be more appropriate to use a 100 bp (size range of 100–500 bp) DNA marker rather than a λ DNA/ *Hin*d III marker (size range of 125–3,000 bp).

(4) Mapping the Positions of Different Restriction Sites

The next step in a restriction analysis is to determine the relative position of different restriction sites within the DNA, i.e. to build a restriction map of the DNA. This is necessary if one wants to locate and cut a particular region of the DNA.

To construct a restriction map, a series of restriction digests need to be performed. As a guide in choosing the types of restriction enzymes to use, one should first identify one or more unique restriction sites in the vector (i.e. 5' or 3' to the DNA insert) to anchor a relative map position. A set of digestion experiments is then performed, using one restriction enzyme per tube. The number and sizes of the fragments produced by each restriction endonuclease must then be determined by gel electrophoresis. Next, a series of double digestions, where the DNA is restricted with two restriction endonucleases, needs to be done. If the enzymes share similar conditions required for optimal enzymatic activity, then the double

Enzyme	Number of fragments	Sizes (kb)
EcoR I	3	2.0, 4.0, 5.0
Hind III	2	3.0, 8.0
EcoR I + Hind III	5	1.0, 1.0, 2.0, 3.0, 4.0

Conclusion:
1. The number of restriction sites for each of the enzymes is: EcoR I   3
                                                              Hind III   2
2. The Hind III and EcoR I sites can be mapped as follow:

Fig. 4.5. Construction of a restriction map.

digestion can be done in a single-step reaction (i.e. the enzymes are added at the same time). However, if the enzymes do not share similar enzymatic buffer conditions, then the digestion has to be done one after the other. Comparing the results of the two types of digestion will allow many of the restriction sites to be mapped within a piece of DNA (Fig. 4.5).

4.1.4. DNA ligase and its use in recombinant DNA technology

The final step in the construction of a recombinant DNA molecule is the joining of the DNA fragment and the vector in a process known as ligation. As mentioned in Sec. 4.1.1, this reaction is mediated by DNA ligases.

Figure 4.6 shows two examples of a ligation reaction, one between blunt-ended DNA fragments and the other between DNA with cohesive

a) Ligating blunt ends

b) Ligating cohesive ends ligation

c) A typical linker

G-G-C-A-T-*G-G-A-T-C-C*-T-T-A-G-C-T
| | | | | | | | | | | | | | | | |
C-C-G-T-A-*C-C-T-A-G-G*-A-A-T-C-G-A

*Bam*H I restriction site

d) Ligation using linkers

Linker   Blunt-ended DNA

DNA ligase

Bam HI

Fig. 4.6. DNA ligation.

ends. In general, ligation reactions between cohesive ends are more efficient compared to blunt end ligations. This is generally governed by the kinetics of the two ends of the DNA molecules meeting each other before a ligation reaction can proceed. In the case of cohesive end ligations, the single-stranded sequences from both strands of the DNA form a complement with each other, thus forming a stable albeit transient structure for the ligase to work on.

In contrast, without the aid of complementary sequence formation, blunt-ended DNA molecules rely on chance association to bring the two ends together. In order to increase the efficiency of ligation, blunt-end ligations usually have to be done in the presence of higher amounts of DNA molecules. Alternatively, we can modify the blunt ends of DNA fragments to form cohesive ends by adding short nucleic acids in the form of linkers or adaptors.

(1) Linkers

Linkers are short, blunt-ended pieces of double-stranded DNA of known sequences, with one or more restriction sequences inserted within it (Fig. 4.6). The linkers, which can be synthesized in large quantities, are attached to blunt-ended DNA molecules by DNA ligase. Although this is a blunt end ligation, this reaction can be performed with high efficiency as the linkers can be added in high concentrations. It is important to realize that the use of high linker concentrations would inevitably cause multiple linkers to be attached to either side of the DNA molecule. However, restriction digestion with the appropriate enzymes will create cohesive ends within the linkers. This modified DNA-linker molecule will now be ready to be ligated to a cloning vector restricted with the appropriate restriction enzymes.

(2) Adaptors

Adaptors are similar to linkers in that they are also double-stranded small pieces of nucleic acids that are synthesized *in vitro*. They are however different from linkers in that they do not have blunt ends, but are made with cohesive ends instead (Fig. 4.7). To prevent the formation of adaptor

a) An example of a typical adaptor and a 5'-modified adaptor

b) Typical adaptors could ligate to one another

c) Ligation with the modified adaptors

Fig. 4.7. Use of adaptors in the cloning of DNA.

concatamers during ligation, the 5' cohesive ends of the adaptors could be modified to incorporate a 5'-OH group instead. Hence, the enzyme DNA ligase will not be able to make phosphodiester bonds between the adaptors (recall that DNA ligase only makes a phosphodiester bond between a 5'-phosphate terminus and a 3'-OH terminus; see Sec. 4.1.1). Hence, the adaptors can be ligated to the DNA molecules but not to themselves. After the adaptors have been attached, the 5'-OH cohesive ends are then modified to 5'-phosphate by adding a phosphate group using the enzyme polynucleotide kinase (see Sec. 4.1.1). The modified DNA-adaptor molecules are now ready to be inserted into an appropriate cloning vector.

4.2. Host Organisms

The use of living cells as hosts to propagate and amplify recombinant DNA is one of the cornerstones in molecular biology. Among the many hosts used in biotechnology, bacteria have been and still are extensively used to produce genetic vectors as well as in the production of recombinant proteins. Other hosts including yeast, plant, insect and mammalian cells, are also exploited for the production of recombinant proteins for specialized purposes. Every host has its own unique properties that can be harnessed for different uses.

One of the objectives of molecular biotechnology is to produce therapeutic proteins. Today, many of the commonly used bacterial strains have been selected or engineered to make them suitable for specific purposes. As an example, the host restriction or modification systems of many laboratory strains have been rendered deficient in these systems and hence allow the propagation of unmodified DNA (e.g. DNA fragments from ligation or PCR products).

E. coli is the abbreviated name for the bacterium in the Family Enterobacteriaceae called *Escherichia* (Genus) *coli* (Species). *E. coli* represents approximately 0.1% of the total bacteria within an adult intestine. This bacterium is a small rod-shaped organism with a circular genome of about 4500 kb in length. The whole genome has now been sequenced. Its genetic property has been the subject of intensive research over the last 30 years and a wide variety of mutants have been isolated and characterized. With the recent availability of their genomic sequences, it is now possible to carry out targeted mutations to produce novel engineered strains for recombinant DNA studies and protein production. The rapid growth of *E. coli* both in minimal and rich media makes it an ideal host for the propagation of recombinant DNA and proteins *in vivo*. Almost all laboratory strains are derivatives of the wild type isolates K-12 or B strains and generally do not carry *Eco*R I or other type II restriction systems.

4.2.1. *Growth*

There are two common ways to grow *E. coli* in the laboratory: on solid or in liquid media. These media are designed to attain both high biomass and yield of recombinant DNA. This is especially useful for low yield vectors (e.g. BACs) where the number of copies in one cell can be very low in commonly used media, e.g. Luria broth (LB). The development of rich media has been successful for both fed-batch fermentations and small-scale applications. Some of these media go by the labels of "Superbroth", "Terrific Broth" and H15. By obtaining a high yield of vectors, there is therefore no need to use large centrifuges for subsequent manipulations of the host cells. It is also possible to affect the solubility of recombinant proteins by manipulating the growth parameters and media contents.

4.2.2. *Handling, culturing and storage*

It is very important, when working with bacteria, that the cultures are kept pure and properly stored. Note that air-borne bacteria can often confound proper storage. Hence, it is advisable to work in the flow hood. All glassware and media should be sterilized either by autoclaving or filtering through 0.2 or 0.4 μm filters.

Pure cultures can be obtained by propagating bacterial cultures from single, isolated colonies on solid agar plates. This procedure is a powerful and simple method to obtain single, pure species. If the stock is contaminated with another strain and there is a mean to verify the correct strain, then it is possible to plate out and select for the correct strain. For example, XL-1 blue cells (Stratagene) harbor a transposon that confers tetracycline resistance. If this cell culture is contaminated with another strain without this genetic marker, then by simply plating out into individual clones on a selection media (e.g. tetracycline), the contaminant can be eliminated.

The bacterial cultures can be stored for long periods in stab vials, as frozen glycerol cultures or as lyophilized cultures. Stabs should be transferred every 2 years whereas frozen glycerol cultures at $-80°C$ or lower can be stored indefinitely. Storage at $-20°C$ in a common-use freezer will result in lose of viability due to fluctuations in temperature with frequent opening and closing of the freezer.

For frequent laboratory use, bacteria can be streaked out on solid agar plates and stored at 4°C. It is advisable to replate these cultures every 1–2 month to maintain viability.

4.2.3. *Introduction of genes into bacterial host cells*

The ability to transfer recombinant DNA into a living host is a necessity in molecular biology. In nature, genetic transfers are common occurrences. There are 3 mechanisms of genetic transfer in prokaryotes:

- **Conjugation**: transfer via "conjugative plasmids" requiring cell-cell contact.
- **Transduction**: transfer by bacteriophage. Plasmids (usually smaller than the phage genome) can be encapsidated into phage proheads to produce plasmid-transducing agents. This form of gene transfer is highly efficient.
- **Transformation**: introduction of "naked" DNA into bacteria cells. This is one of the most commonly used methods to introduce recombinant DNA into host cells.

(1) Transformation

This method of transferring genetic material (as free DNA) to recipient cells does not require the contribution from intact donor cell (e.g. need of pili, i.e. the "sex organs" of bacteria). This method is not restricted to plasmids alone but can also be used to transform large DNA molecules. Figure 4.8 illustrates the two general methods for transforming bacteria:

(a) A chemical method utilizing salt and heat shock to promote DNA uptake.
(b) Electroporation utilizing electric shock to facilitate DNA entry.

Fig. 4.8. Transformation of bacterial cells using plasmid DNA.

To assess the success of a transformation, there are three related steps that need to be considered: the entry of the DNA (uptake), production of a selectable phenotype (e.g. antibiotic resistance) and the propagation of these transformed cells to form visible colonies.

(2) Competence State

In most host cells, including *E. coli*, competent cells do not occur naturally. For these cells, a competent state has to be induced artificially. Competence is usually a consequence of a nutritional shift-down (decrease of the availability of one or several nutrients, or of available energy) and usually result in decrease in the rate or a blockage of DNA synthesis. The exact mechanism is not well understood.

Several growth media and conditions have been developed empirically to induce competence. The kinetics of competence varies greatly from one strain to another (i.e. some strains may show competence for several hours and drops off with time). A number of metabolites can stimulate the establishment of competence in some strains. Although the biochemical mechanisms are not clearly understood, important modifications of the cell envelopes take place during development of competence.

(a) Chemical induced competence

This method of inducing competence is used extensively in many laboratories. There are a number of available procedures using various chemicals including calcium chloride, cobalt chloride, rubidium chloride and others. The transformation efficiency ranges from 10^6 colony forming units (cfu)/μg to greater than 10^9 cfu/μg of DNA, rivaling electroporation. As with other means of transformation, it is highly dependent on the strains used.

The strategy is as follows: cells are grown overnight to act as the starting inoculum. A small portion of this will be used to seed a fresh batch of media, and are then grown with high aeration. When the cell growth is in the logarithmic phase, the cells are collected and processed. Here, the cells are chemically treated to induce competence. All along the process, the cells are treated gently and in the cold to prevent lysis. It is thought that these chemical treatments facilitate the recruitment of channels where DNA molecules can enter. In the process of DNA uptake, the molecules are incubated with the cells at 4°C for a period of half to one hour. During this time, DNA molecules, being predominantly negatively charged, complex with cations (calcium, rubidium or cobalt) and precipitate onto the cell surface. A brief heat-shock treatment enhances the uptake of these complexes through the channels. There is still no concrete evidence that this actually happens although all the observations are consistent with this model of DNA uptake.

The cells can be prepared and kept frozen in glycerol and stored at −70°C for at least 6 months with little loss in efficiency. Batches of the cells prepared in exactly the same manner will vary. Hence it is essential to determine the transformation efficiency of every batch made. Several factors will affect the transformation efficiency, including the density of the cells at harvest, the absolute requirement for maintaining the cells at 0–4°C throughout the processing and the period of time at 4°C before freezing down for storage.

Competent cells of high efficiency for transformation can be prepared in the laboratory. However, this method is rather involved and tedious, requiring experience in treating the cells. It is no longer necessary to prepare competent cells if the usage is low. There are now commercially available competent cells that come in the form of supercompetent or ultracompetent cell preparations.

Since some portions of the host cells will not have taken up the DNA, it is necessary to be able to select for the recombinant colonies. This selection marker, usually in the form of an antibiotic resistance gene, is carried by the transforming DNA. Plating the transformed bacteria onto selective media would thus enable the transformants to be distinguished from the non-transformants. The next challenge is to distinguish colonies containing the recombinant DNA from the ones that contain self-ligated vectors. With most cloning vectors, insertion of the DNA fragment would destroy the integrity of a marker gene present in the vector. Recombinant colonies can then be identified by the characteristics that are not displayed by the host cells due to the inactivated gene.

(b) Electroporation
In general, this method produces better transformation efficiency and is more successful with many different cell types including plants, yeast, insects and mammalian cells that are resistant to chemical methods. Large molecules are also electroporated with high efficiency. Thus, this method can also be used to transfer drugs or even whole genome into cells of all types. The maximal efficiency occurs when the pulse is strong enough to yield 40–80% cell death. Therefore for electroporation, one needs a higher number of cells to start off in order to obtain a reasonable number of transformants. Like chemical methods, each cell type requires an optimal set of parameters to produce a maximum number of transformants. These are usually determined empirically and many protocols have been published.

Electroporation (electrically induced pore formation) is currently viewed as a bilayer membrane phenomenon, where short (μs to ms) electrical field pulses cause the transmembrane voltage to rise to about 0.5–1.0 V, thus causing pore formation. The membrane is then reversibly permeabilized, thereby enhancing the transport of molecules across the membrane.

The electroporators generally have two components: a generator to provide voltage pulses of electrical fields and a chamber in which cells are subjected to the electrical fields (Fig. 4.9). The transient power required by the generators to deliver the electrical energy to the chambers can easily far exceed the electrical power available in the laboratory outlets. To circumvent this limitation, electrical energy is first stored in capacitors. The accumulated voltage is then discharged, at high power, into the chamber.

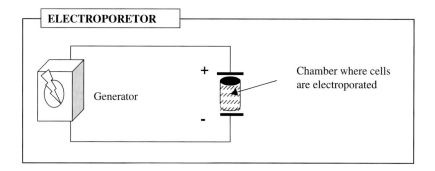

Fig. 4.9. Delivery of DNA into cells by electroporation.

Bacteria require much higher field strengths (6–25 kv) than most eukaryotic cells. For electroporation, the density of cells to be used should be approximately $1–2 \times 10^6$ cells/mL. However, this may give rise to a significant proportion of fused cells. This is particularly true with cells that do not possess cell walls (e.g. mammalian cells). This process is called "*electrofusion*".

The electrocompetent cells are prepared in the same way as the cells prepared by chemical competence, except that only water or a 10% glycerol solution is used throughout the preparation. As with chemical methods, the efficiency will vary between different batches of cells, and between different strains. In the preparation of electrocompetent cells, all centrifugations must be gentle and all manipulations must be carried in the cold (0–4°C).

(3) Transfection of Bacterial Cells

There are two ways by which recombinant DNA in a phage vector can be introduced into bacteria: transfection and *in vitro* packaging. Transfection is a process equivalent to transformation, except that the DNA to be transferred is a phage DNA rather than a plasmid. As in the transformation process, the phage DNA is incubated with the competent bacterial cells and the DNA uptake induced by heat-shock. Alternatively the phage recombinant DNA can also be packaged *in vitro*, using the necessary purified phage proteins, to form mature phage particles, which can now infect host cells. This method of genetic transfer is more efficient than transfection with the recombinant phage DNA. When the infected cells are spread onto a solid agarose media, cell lysis caused by the propagated phage particles, can be visualized as plaques on a lawn of bacteria.

4.2.4. Other hosts

Beside *E. coli*, the other commonly used hosts include yeast, plant, insect and mammalian cells in culture. These hosts are eukaryotes and are generally used for the expression of specific proteins rather than for propagating recombinant DNA. Protein expression and the controls involved will be discussed in the latter part of the chapter.

Unlike prokaryotes, eukaryotes can carry out a number of post-translational modifications of expressed proteins. These include:

1. Proteolytic processing of precursor proteins.
2. Specific glycosylations.
3. Post-translational modifications of amino acids.
4. Correct disulfide bond linkages.

Precisely how an expressed protein is modified post-translationally is highly dictated by the type of eukaryotic host used. Hence, with certain proteins, the correct choice of a host is essential.

(1) Yeast as Expression Systems

The common baker's yeast (*Saccharomyces cerevisiae*) has been used extensively as a host to express cloned eukaryotic genes. There are a number of reasons for this. This is a single cell and is extremely well known genetically (the whole genome has been completely sequenced) and physiologically. It is easier to manipulate and can be grown in both small culture vessels and large-scale bioreactors. There are several strong promoters and naturally occurring plasmids that have been isolated and well characterized. This organism is capable of carrying out some post-translational modifications. The most important factor is that this strain of yeast has been in use in baking and brewing industries, and has been listed by the U.S. Food and Drug Administration as a "generally recognized as safe" (GRAS) organism. Thus, the use of this organism for the production of human therapeutics (e.g. drugs or pharmaceuticals) does not require as much verification as compared to unapproved host cells. A number of proteins have been produced in this host including, Hepatitis B and HIV vaccines, insulin, blood coagulation factor XIIIa and others.

Other strains of yeast are also used for the purposes of recombinant protein production. They include *Kluyveromyces lactis* (used commercially to produce lactase), *Schizosaccharomyces pombe*, *Pichia pastoris* and *Hansenula polymorpha*.

(2) Cultured Insect Cells as Expression Systems

Fruit fly (*Drosophila melanogaster*) and baculoviruses are currently used to produce high amounts of recombinant proteins. In the former case, a number of cell-lines (e.g. S2) have been developed and have attained immortalized non-tumorgenic growth. These cells in culture can then be transfected with vectors with the appropriate promoters.

Baculoviruses infect invertebrates, including many insect cells. This virus does not infect humans and hence, is explored as a potential host to produce therapeutic proteins. As this virus harbor a large genome, a number of steps is involved in generating recombinant baculovirus which will then be used to infect insect cells. The infected cells in culture will then produce the protein of interest.

(3) Plants as Expression Systems

There are good reasons to develop plants that express foreign substances. The addition of certain gene(s) often improves the agricultural, horticultural or ornamental value of a crop plant. These "transgenic" plants can act as living bioreactors for the inexpensive production of many economically important metabolites or proteins. There are a number of effective DNA-delivery systems and expression vectors that can be used on a range of plant cells.

(4) Mammalian Cells as Expression Systems

Immortalized mammalian cells grown *in vitro*, has been invaluable in the studies of gene expressions and mechanisms underlying many diseases (e.g. cancers). Many mammalian cells from a variety of body tissues have been successfully grown in plastic wares bathed in specialized liquid media in controlled environment (37°C; 5% CO_2/95% O_2), often supplemented with bovine (cow) serum. After a defined number of generations, these cells die in a programmed fashion. Occasionally, some cells escape death and they thrive in the environment and begin to develop immortality. This does not necessary mean that the cells are "cancerous". A cell-line can stay immortal and not be a tumor. Tumorigenic cells can stay immortal in culture and when transplanted into immune deficient mice, develop into full-blown cancers. There are a large number of mammalian cells used for molecular genetic applications. Some of these are also used to produce recombinant proteins.

4.3. Vectors

Plasmids are widely distributed throughout prokaryotes. These are self-replicating entities (replicons), independent of the chromosome. They can be stably inherited in an extrachromosomal state, but are generally dispensable. However, not all extrachromosomal nucleic acids are necessarily plasmids. Some of the heterogeneous circular molecules found in *Bacillus megaterium* and the replicative forms of the filamentous phages (M13, Fd, f1) are not plasmids. However, parts of these replicative forms have been incorporated into the backbones of many vectors.

Most plasmids exist as *double-stranded* circular DNA. The replication of all plasmids is, to an extent, dependent on their host. Plasmids also exhibit selective replicative capabilities in specific hosts. This is referred to as the host-range of the plasmids. Therefore not all hosts can harbor the same plasmids.

The development of recombinant DNA technology has enabled the modifications of existing plasmids to improve their use as cloning vectors. There are many types of cloning vectors including plasmids, fosmids, bacteriophages and others. They all have common features that make them suitable for use as vehicles for genetic engineering. These vectors can also be easily modified to enable them to express heterologous proteins efficiently.

4.3.1. *Some general requirements for a cloning vector*

1. All vectors must be able to replicate within a host cell if these are to be used as "shuttle vectors". Some episomal vectors (vectors that are not integrated into the chromosomes) do not seem to replicate

in mammalian cells (inappropriate origin of replication), but can be effectively used to express recombinant genes. Some mammalian expression vectors, however, do replicate under intense selection pressures.

2. There must be efficient ways of introducing vector DNA into a host cell.
3. There must be ways to isolate the vector DNA from the host DNA.
4. The presence of specific restriction sites in the vector that can be used for cloning an insert DNA.
5. The vector should have at least a selectable marker in the vector to indicate the presence of the vector in host cell. This can be antibiotic or other xenobiotic markers, or merely reporter groups, for e.g. luciferase, green fluorescent protein (GFP) and β-galactosidase.
6. It is desirable to have unique restriction enzyme sites within a selectable or screenable marker so that the presence of a DNA insert in the cloning vector can be selected or screened.

A number of these vectors are available commercially. They can be used for cloning in prokaryotes, yeast, baculovirus, mammalian and other

Table 4.4. Some examples of vectors and their maximum insert sizes.

Vectors	Size of inserts (kb)
Plasmids (pBR322)/Phagemids (pGEM, pBSK)	0–10
Bacteriophage λ (insertion vectors; e.g. λgt10, λgt11, Lambda Zap, Unizap XR, Zap-express)	0–10
Bacteriophage λ (replacement vectors; Lambda Fix II, Lambda DASH II, EMBL, Charon)	9–23
Cosmid vectors (SuperCos 1, pWE15, COS)	25–45
Bacteriophage P1	60–100
PAC (P1 artificial chromosomes)	100–150
BAC (bacterial artificial chromosomes)	100–300
YAC (yeast artificial chromosomes)	200–2000
HAECs	
MACS	

cells (e.g. plants). These vectors come in many different names. They are designed to have specific functions (host specific, expression, recombination) and can encode inserts of different sizes (Table 4.4).

4.3.2. Vector systems

Vectors can be classified as "shuttle vectors" and "expression vectors". Shuttle vectors are used generally for the propagation of DNA. The prokaryotic vectors for carrying DNA inserts, which are of high copy numbers, are regarded as shuttle vectors. By simple manipulations, these vectors can be made to express proteins as well. The distinctions between these two types are gradually eroding, as new vectors that can be used for both propagation and expression have been designed. Some of these vectors are packaged into viruses for high efficiency gene transfer (infection as opposed to transduction or transformation).

There are basic features and principles that all vectors rest on. The vectors available today are rather versatile, with many variations of a common theme. The designs of eukaryotic vectors have evolved from those found in prokaryotic systems, often borrowing and wedding specific functions. The earlier plasmids are derivatives of naturally occurring sequences that are host compatible, e.g. pBR322. Nowadays, many commercial companies, including Promega, Invitrogen, Clontech, Stratagene and others, have spliced together sequences from phage DNA and plasmids, giving rise to a modified type of plasmid called *phagemid*. Phagemids are therefore plas*mid*s that have been synthetically manipulated to contain a small segment of sequences from a filamentous *phage* (M13, fd or f1).

4.3.3. Prokaryotic cloning vectors

A generic prokaryotic cloning vector has some of the following common features:

1. An origin of replication: *ori*.
2. An antibiotic resistance gene: e.g. ampicillin, kanamycin, tetracycline, etc.

3. A selectable marker: e.g. color selection with β-galactosidase (*lacZ*).
4. An available multiple cloning sites (MCS).
5. Other accessories:

   (i) Promoters for *in vitro* syntheses of RNA: SP6/T7/T3.
   (ii) f1 origin to allow production of single-stranded DNA.
   (iii) MCS sequences that allow unidirectional deletion of inserts.

Let us examine the structures of two commonly used vectors in the laboratory that are commercially available, Bluescript (pBSK; Stratagene) and pGEM (Promega) (Fig. 4.10). They are known as phagemids, and are designated as **pBSK(+/−)** and **pGEM-XZf(+/−)**, where **X** denotes vector number, **f** the presence of phage origin of replication, and **(+/−)** the orientation of the single-stranded DNA produced.

(a) **pGEM series**. This series comprises the vectors pGEM-3Z, pGEM-4Z, pGEM-3Zf(+/−), pGEM-5Zf(+/−), pGEM-7Zf(+/−), pGEM-9Zf(−), pGEM-11Zf(+/−) and pGEM-13Zf(+). The parental plasmid for this series is the pUC vector. These vectors contain the pMB1 origin of replication. The early series of pGEM vectors (e.g. pGEM-3Z and -4Z) do not contain an f1 origin for the production of single-stranded DNA. This feature, plus the addition of new MCS sequences, is incorporated into the later generation of pGEM vectors to allow for increased versatility and convenience in gene manipulations.

(b) **pBIISK(+/−)/pBIIKS(+/−)**. Both are the same vectors, except that the orientation of the MCS site has been reversed between the T7 and T3 promoters. These phagemids contain the ColE1 of origin of replication. The pBIISK vectors are derivatives of pUC19 (which is a derivative of pBR322) and, like the pGEM vectors, replicate in high numbers because of the lack of the pBR322 *rom* gene.

(c) Both pGEM and pBIISK/KS vectors use ampicillin resistance as a selection marker. Therefore transformants will acquire resistance to ampillicin.

(d) Both vectors have phage promoters; T7/SP6 in the pGEM series, and T7/T3 in pBIIKS/SK. The promoter sequences flank the insert and contain all the information required for the initiation of RNA transcription by T7, T3 or SP6 RNA polymerase. These polymerases

Fig. 4.10. Vector Maps of pBluescript SK (Genbank #X52528 [SK(+)] and #X52330 [SK(−)]) and pGEM-7Zf(+) (Genbank X65310). Adapted from Stratagene and Promega, 2001.

do not cross-initiate and hence, can be used to produce highly specific RNA transcripts. The RNA transcripts produced can be used for a number of studies. For example, the transcripts can be used as riboprobes for nucleic acid hybridizations, as a template for *in vitro* translation of proteins, or in gene expression studies.

(e) Both vectors have two origins of DNA replication, one which is specific for *E. coli* (ColE1 for BSK and pMB1 for pGem), and the other from the single-stranded DNA phage f1. The ColE1 or pMB1 origins are used for plasmid replication. However, following phage f1 infection, the other origin is used and single-stranded DNA is produced. This is useful for single-stranded DNA sequencing and some mutagenesis strategies.

Early cloning studies were rather cumbersome. The most common strategy was to subclone the DNA fragment from a plasmid or lambda vector, into filamentous bacteriophage M13 vector for the production of a single-stranded DNA template that can then be used for sequencing. Once sequencing has been completed, the template had to be recloned into another specialized vector, such as a vector designed for efficient *in vitro* transcription, or an expression vector for protein expression. If site-directed mutagenesis experiments were to be carried out, the mutant forms of interest had to be then subcloned at each step. These rather time-consuming steps have now been eliminated though the development of vectors that possess, in a single construct many features required for specialized DNA manipulations.

(f) **Genetic markers**. Both vectors have convenient ways to select for the presence of DNA insert colorimetrically. This selection method is based on the biochemical properties of the *lac* operon (see Sec. 4.3.4).

(g) **Protein expression**. Foreign DNA encoding protein sequences can be expressed in these two vectors, as long as the sequences are in-frame to the expression of β-galactosidase gene, which is under the control of the *lac* promoter. An adaptation of this strategy is the use of pGEM-T vector where random sequences of a protein can be incorporated and expressed as a fusion protein of β-galactosidase. It is advisable to use a host with the *lac*I^q present in order to increase the yield of proteins produced.

The T7 promoter can also be exploited for protein expression. By inserting the foreign DNA encoding protein sequences under the control of the T7 promoter and adding a ribosome-binding site (RBS) plus an ATG start site, it is possible to keep the expression tightly turned off until a T7 RNA polymerase is expressed in certain host cells (e.g. BL21(DE3)).

(h) **Gene mapping**. The unique *BssH* II restriction sites flanking the T3/T7 promoters of pBIISK offer an easy and rapid method to map large and small DNA inserts. To do this, complete *BssH* II digestion of the clone is first carried out, followed by partial restriction digestion with the restriction enzymes of choice. The partial fragments from the restriction digestion will then be separated by agarose gel electrophoresis, transferred to a nylon membrane, probed with labeled T3/T7 to determine the lengths of the partial fragments (see Sec. 4.4.1). The different lengths of the digested fragments identify the location of the restriction enzyme sites relative to the T3/T7 positions.

(i) **MCS unidirectional deletion of insert**. An alternative to the above partial restriction map analysis is simply to exploit the use of the MCS. If the insert DNA has restriction sites similar to that in the MCS, complete restriction will result in the excision of some vector and insert sequences. If sequencing is carried out from one direction, it will then be possible to gain internal sequence information.

(j) **Unidirectional deletions with *Exonuclease* III**. The restrictions sites have been arranged to allow unidirectional deletions of the cloned insert with *Exo* III. This enzyme is a 3–5' single-stranded exonuclease and require double-stranded DNA template. Hence, *Exo* III does not digest 3' end single-stranded overhangs, but exclusively digest the 3' ends of blunt ends or 5' overhangs. Therefore, double digestion with a 3', and a 5' or blunt end producing enzyme, followed by *Exo* III digestion, and the subsequent removal of the remaining single-stranded DNA by mung bean nuclease will result in unidirectional deletions of the insert DNA. This method allows the serial deletion of large inserts for sequencing from one direction.

4.3.4. *Lac operon: A model for the control of gene expression*

A number of cloning and expression vectors utilize the *lacZ* system as either a selection marker or as a reporter gene. It is therefore important to have a basic understanding of this system.

The *lac* operon found in *E. coli* is one of the best-known examples of a bacterial operon (Fig. 4.11). Structurally it consists of 3 genes: *lacZ* (coding for β-galactosidase), *lacY* (coding for a permease which is

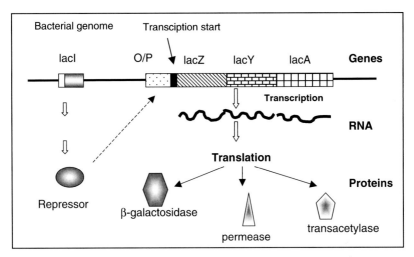

O/P = operator and promoter

Fig. 4.11. The *lac* operon.

needed for uptake of lactose) and *lacA* (coding for an enzyme called thiogalactosidase transacetylase). A regulatory region containing the promoter site is found at the 5' end, overlapping with a sequence known as the operator.

The *lacI*^{q} repressor protein binds to the operator region that physically overlaps with the promoter (which drives the *lac* operon) and the first 20 or so bases to be transcribed, thus preventing the initiation of transcription. The *lacI*^{q} gene is not part of the *lac* operon. The bound *lac* repressor protein has an affinity for a molecule called allolactose, which is a derivative of lactose. This binding to allolactose is allosteric and will prevent the repressor from binding to the operator region. Therefore, in the absence of lactose, the repressor is active and binds to operator, preventing transcription. However in the presence of exogenous lactose, the repressor is inactivated and the operon is then expressed. The cell then produces the enzymes that are necessary for the metabolism of lactose. As lactose will be metabolized as soon as induction begins, thus diminishing the inductive effect, it is not a good inducer of the *lac* operon. In practice, a synthetic analogue of lactose that will not be broken down easily by *β*-galactosidase is used instead. An analogue that is commonly used in the laboratory is isopropyl-1-thio-β-galactosidase (IPTG). Not all substances that are substrates for breakdown by β-galactosidase, can act as inducers as they are not recognized by the repressor. One example of such a chemical is 5-bromo-4-chloro-3-indolyl-β-D-galactopyranoside, commonly known as X-gal.

Some *E. coli* have a modified *lacZ* gene, lacking a segment referred to as *lacZ'* which codes for the α-peptide portion of β-galactosidase. These bacteria can only synthesize the full active enzyme when they harbor a plasmid that carries the missing *lacZ'* segment. This phenomenon is called α-complementation and is exploited in a technique used to distinguish a recombinant colony within a background of non-recombinants. In this technique, the transformed bacterial cells are plated onto solid media containing X-gal and IPTG. The active β-galactosidase enzyme breaks X-gal down to form a product that is deep blue in color. Since the MCS site resides within the *lacZ'* gene in most plasmids, insertion of a DNA fragment into the vector will inactivate the β-galactosidase enzyme, resulting in the formation of white colonies instead. This selection technique is also applicable to distinguish bacterial cells infected with recombinant bacteriophage DNA that has the *lacZ'* gene in its genome (e.g. λZAPII vector).

4.3.5. λ bacteriophage vectors

Two problems hinder the use of plasmids for all genetic manipulations: the severe size limitation of the insert DNA, and the efficiency of gene transfer. Although the efficiency of transformation has dramatically increased by the use of electroporation and specific chemical treatments, it remains a hindrance. As an alternative, infection as a way of introducing DNA into cells show efficiency close to unity (100%). Attention was therefore directed at bacteriophage λ as it possesses a number of desirable qualities required for cloning, a highly efficiency method of gene transfer as it infects the bacterial cells, and the ability to insert large DNA fragments ($\sim$20 kb) by manipulating the genome.

To design vectors based on this phage, it is necessary to insert the DNA of interest into the λ replicating sequences (replicon) *in vitro*. The resulting recombinant DNA must be able to infect *E. coli* at high efficiency. The latter requirement is met by developing an *in vitro* packaging system, which mimics the way in which wild-type λ DNA is packaged in a phage coat protein, resulting in high infection efficiency.

Two major strategies are used in generating λ vectors: replacement vectors and insertion vectors. As the name implies, replacement vectors are those which have some segments of its sequences deleted to accommodate the sequences of interest. It is observed that DNA molecules between the sizes of 37 to 52 kb can be stably packaged into the phage head. Exploiting this finding, removing the region of λ genome that is nonessential for growth (central portion where the genes are necessary for integration and recombination) and inserting the DNA of interest to make up the required size, foreign DNA of up to 23 kb in length can be easily incorporated. Hence, these vectors are commonly used for genomic cloning (Fig. 4.12).

For cDNA (complement DNA generated from mRNA) cloning, where the sizes of DNA fragments are generally not as large (<10 kb), these sequences can be cloned into modified λ vectors (insertion vectors).

With an insertion vector, the non-essential segment has been deleted, and the two arms ligated together. Therefore this vector has at least one restriction site into which a DNA fragment can be inserted. The newer versions of λ insertion vectors (e.g. Lambda ZAP II vector) allow easy *in vivo* excision of insert DNA in a phagemid form. This is enormously

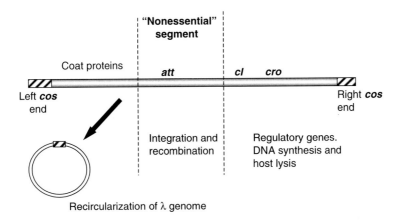

Fig. 4.12. The bacteriophage λ as a cloning vector.

convenient as it is easier to isolate and purify plasmid DNA as opposed to the purification of phage DNA.

The *in vivo* excision depends on the DNA sequences of the phagemid (BSK) and the presence of a variety of proteins, including filamentous phage (M13)-derived proteins. The M13 phage proteins recognize the f1 bacteriophage "origin of replication" found within the pBluescript sequences. This f1 origin of replication can be divided into 2 overlying parts: (a) the site of initiation (I), and (b) the site of termination (T) for DNA synthesis. These two regions have been subcloned separately into the lambda ZAP II vector. The lambda phage vector is made accessible to the M13-derived proteins by simultaneously infecting a particular strain of *E. coli* (the SOLR strain) with the lambda phage vector and the bacteriophage. The "helper" proteins (from M13) then recognize the initiator DNA that is within the lambda vector. One of these proteins nicks one of the two strands of DNA and a new DNA synthesis begins, duplicating the DNA sequences downstream of the nick site. This synthesis continues through the cloned insert until the termination signal is encountered. The single-stranded DNA molecule is then circularized by the gene II product of the M13 phage. This results in a circular DNA molecule containing the DNA sequences between the initiator and terminator sites.

Fig. 4.13. The newer versions of λ insertion vectors (e.g. Lambda ZAP II vector) allow easy *in vivo* excision of insert DNA in a phagemid form.

In this case, this includes all sequences of the pBluescript and the insert DNA (Fig. 4.13).

4.3.6. *Other vectors*

(1) Cosmid Vectors

Effective packaging of the λDNA molecules can only occur if the *cos* sites are present. It is now known that only a small region around the *cos* site is required for recognition by the packaging system. Plasmids have been constructed containing a fragment of the λDNA including the *cos* site. These are called **cosmids** and are used in conjunction with the *in vitro* packaging system. The packaging into the phage head requires a large amount of DNA. If the vector is only 5 kb in length, then to be incorporated into the phage head, an insertion of between 32 to 47 kb of DNA (gene of interest) is required. This large amount of DNA cannot be inserted into any of the known λ vectors. Note that after packaging *in vitro*, the phage can infect the appropriate host. The recombinant cosmid DNA is injected and circularizes like a phage DNA, but replicates as a normal plasmid without expressing phage proteins. Selection of transformants is done on the basis of a vector-drug-resistance marker carried on the plasmid portion.

(2) Bacterial Artificial Chromosome (BAC) Vectors

Most of the commonly used vectors for propagating small inserts have high copy numbers (discussed in other sections). One disadvantage of possessing high copy numbers is that these vectors often show structural instability of inserts (especially for large inserts and those with repeated sequences common in higher eukaryotes). Thus, this limits the use of these high copy number bacterial vectors.

   To overcome these limitations, attention has recently been focussed on vectors that exist in low copy number replicons, e.g. *E. coli* fertility plasmid, the F-factor. This plasmid occurs naturally in many strains and is maintained at a copy number of 1–2 per cell. Vectors based on the F-factor system can accept large foreign DNA fragments (>300 kb). Using electroporation, it is possible to introduce these recombinants into *E. coli* efficiently. However, it must be noted that the copy number will be low within the cells. The consequence of this is that, to prepare these vectors, efficient isolation procedures must be used.

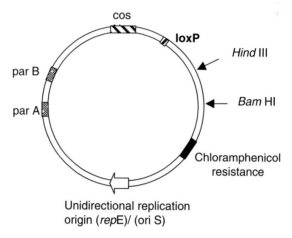

Fig. 4.14. A BAC vector.

An example of a BAC vector derived from a mini-F plasmid is shown in Fig. 4.14.

(3) Yeast Artificial Chromosomes (YACS)

As the upper size limit of cosmid cloning is about 35,045 kb, to cover the whole human genome in a theoretical contiguous chain would require about 70, 000 clones. What is required is the development of vectors that can accommodate insert sizes even bigger than in the cosmids. This will reduce the number of clones to be handled. The minimum structural element for a linear chromosome is the presence of an origin of replication (autonomous replicating sequence, *ars*), telomeres and a centromere.

A linear molecule that behaves like a chromosome in yeast has been constructed by combining an *ars* and a centromere (both from yeast) with telomeres from Tetrahymena (Fig. 4.15). These YACS allow the insertion of large pieces of foreign DNA (~1 Mb) into this linear artificial chromosome.

There are a number of operational problems with YACS. The existence of chimeric DNA segments from different regions of the genome into a single clone is one of such problems. Another problem is that of instability, which tends to delete internal regions in the inserts. Other problems include the difficulty of removing the yeast host chromosome background from the

Fig. 4.15. Yeast artificial chromosome as a cloning vector.

YACS and low yield of DNA from YACS. A number of these concerns are currently being addressed.

(4) Shuttle Vectors

Gene cloning and associated techniques are also extremely valuable for investigating all aspects of the behavior of a wide range of bacteria. *E. coli* genes are often expressed satisfactorily in other Gram-negative organisms, but the reverse is usually not true. Genes from *Pseudomonads*, for example, are only poorly expressed in *E. coli* and have to be analyzed in their original hosts. In many cases, the techniques available for these organisms are limited. For example, transformation (or electroporation) frequencies may be relatively low, and the methods for screening gene libraries may be difficult. It is therefore advantageous to carry out the initial cloning of a gene using *E. coli* as the host organism. After the desired DNA construct has been obtained, it can be recovered and then inserted into the target host. This procedure requires a vector, which can replicate in either organism. Certain plasmids are capable of existing in a variety of hosts and can function as shuttle vectors. A shuttle vector carries replication origins for two different plasmids on the same vector (Fig. 4.16). This will facilitate preliminary manipulations in a well-characterized system, such as *E. coli*, and then transfer the clone to the organism of interest. If the clone is to be expressed in both hosts, then it may also be necessary to provide two promoters as well.

(5) Broad Host Range Vectors

Broad host range plasmids can be used as an alternative to dual-origin shuttle vectors. These plasmids belong to two incompatibility groups, P and Q/P4. P group plasmids are usually derived from *Pseudomonads* and the prototype of this group is RP4 (also known as RP1 or RK2). RP4 is a broad host range plasmid and is able to replicate in some Gram-positive

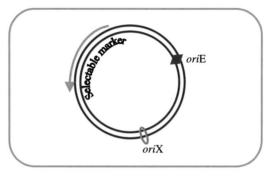

In *E. coli*, plasmid replication is
initiated using *ori*E.

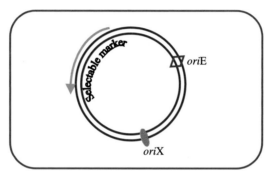

In the alternative host, the plasmid
replication is initiated using the second
origin of replication, *ori*X.

Fig. 4.16. Shuttle vectors with dual origins of replication.

bacteria and in most Gram-negative bacteria. The P group of plasmids
contain replication and transfer functions which are widely dispersed on
the bacterial chromosome, thus making it extremely difficult to construct
lower molecular weight derivatives which could be used as cloning
vehicles. One example is pRK2501, which is 11 kb in length, but lacks all
transfer and mobilization functions. Since this vector is transfer-deficient
and non-mobilizable, it satisfies important requirements for the biological
safety of vectors with broad host-range. A system, which employs two
plasmids, has been developed for cloning in *Rhizobia* (Fig. 4.17). The first
plasmid, pRK290, is the cloning vehicle which is transfer-deficient but still

The donor *E. coli* cells contains plasmid pRK2013 with RK2 transfer functions and ColE1 replicon. Replication of pRK2013 is restricted to *E. coli.*

pRK2013 is transferred into recipient *E. coli* containnng recombinant pRK290 which carries the RK2 replicon and confers broad host specificity.

pRK2013 mobilizes pRK290 and allows it to be transferrred into the target recipient.

Target recipient containing recombinant pRK290 plasmid.

Fig. 4.17. Use of broad host specificity plasmids for gene transfer into other host organisms.

possess all mobilization functions (Tra⁻ Mob⁺) and can replicate equally well in *E. coli* and *Rhizobium* because it contains the RK2 replicon. This vector can be mobilized by a second plasmid residing in the donor cells, and hence can be transferred to suitable recipient cells, e.g. *Rhizobium*. The mobilizing plasmid in the donor is pRK2013, which possesses all transfer functions and also contains the neomycin resistance region of RK2 and the ColE1 replicon, which restricts its replication in *E. coli* but not in *Rhizobium*. When this plasmid enters *E. coli* cells, the transfer functions of pRK2013 are expressed and mobilize plasmid pRK290, which can replicate in *Rhizobium* cells.

4.3.7. *Prokaryotic expression system*

In general, expression vectors should have the following properties:

1. **A promoter**, which marks the start of gene transcription. This sequence is recognised by the RNA polymerase.
2. **Inducible elements**, which provides the ability to control the level of protein expression.
3. **Accessory elements for stability** allows for the maintenance of a stable transcript.
4. **Transcriptional termination**, which marks the end of a gene where transcription should stop.
5. **Affinity tag** that allows the protein to be easily isolated and purified (optional).

The promoter is the most important factor of an expression vector. In all expression vectors, the cloning site is located just downstream from a promoter and ribosome binding site (RBS), allowing heterologous regulation of gene expression. The important sequences for *E. coli* RNA polymerase are the TTGACA and TATAAT (Pribnow box) sequences found at position -35 and -10, respectively. The most commonly used promoters are those from the *lac* or *trp* operons, a combination of the two called the *tac* promoter (*trp* and *lac*), and the λ_{PL} promoter from phage λ. These promoters are inducible and, in the case of the *lac* promoter, transcription or derepression of the gene is brought about by the addition of IPTG. The ability to control levels of expression is important: continuous high levels of expression may affect plasmid stability or the protein product may be toxic to the host organism.

Efficient translation of mRNA in prokaryotic cells cannot occur without a ribosomal binding site (RBS). This includes an initiation codon, AUG, together with a consensus sequence, AGGAGGU, which typically lie 3–12 bases upstream of the AUG. This is the Shine-Dalgarno sequence (SD). The sequence of the RBS, together with the secondary structure of the mRNA, has an effect on translation efficiency of mRNA. To obtain expression of foreign genes in *E. coli*, it is necessary to incoporate a RBS into the recombinant DNA construct. Furthermore, the SD sequence must be located at the optimal distance from the translation start codon.

More sophisticated expression vectors, with more efficient expression signals, are used when it is necessary to obtain very high levels of formation of the product of the cloned gene, for example in commercial processes. Extremely high levels of specific protein, representing perhaps 50% of total cell protein, can be produced using such a vector. However, diverting so much of the cell's resources into a product that is useless as far as the cell is concerned will slow down the growth rate (which will also exacerbate plasmid instability). The usual strategy for overcoming this problem is to use a promoter that is subject to repression, so that the bacteria can be grown to high cell density before product formation is induced by altering the growth conditions.

The choice of a promoter will depend on the requirements of the expression. For example, if it is desired that the cloned DNA be regulated to prevent inhibition of host cell growth, the *trp*, *lac* or λ promoters might do well. The appropriate repressor molecule is supplied as a regulator. The *tac* promoter, a combination of the -35 region of the *trp* promoter and the -10 region of the *lac* promoter, also has a high efficiency of expression. Many phage promoters are also very efficient for in protein expression. For example, the phage T5 promoter out-competes all *E. coli* promoters, thus allowing the expression of large amounts of the recombinant protein in *E. coli*.

An example of a prokaryotic expression vector that is designed for high protein expression and easy purification is the pQE-9, which belongs to the pQE series of vector (Fig. 4.18).

This vector uses the phage T5 promoter, which can be recognized by the *E. coli* RNA polymerase. The incorporation of a double *lac* operator repressor system provides a tightly regulated, high-level expression of recombinant proteins in *E. coli*. In addition, the vector contains a 6xHis tag segment that will allow easy purification of the expressed recombinant products *in vitro*.

Phages such as T3, T7 and the *Salmonella* phage SP6, code for their own RNA polymerases. Each of these polymerases specifically recognizes

Fig. 4.18. An example of a prokaryotic expression vector that is designed for high protein expression and easy purification. Adapted from Qiagen 2001.

its own particular promoter and ignores those of other phages or the host bacterium. Thus, by using such a phage promoter to drive the expression of a foreign gene, the expression of recombinant protein will be absolutely dependent on the availability of the phage RNA polymerase. By regulating the availability of the phage RNA polymerase using an inducible promoter, such as the λ_{PL} promoter, which is repressed by providing the cI repressor *in trans*, allows the expression of the recombinant protein to be tightly regulated.

Sometimes, a particular mRNA transcript made from cloned DNA is poorly translated in the host cell. This lack of efficiency may be due to an increased susceptibility to degradative enzymes or to the use of rare codons in the mRNA. The only solution for the latter problem is to change the host cells to one that has more suitable pattern of codon usage or else to use site-specific mutagenesis to change the codons themselves. In the case of the former problem, changing the structure of the mRNA molecule itself can often prevent excessive degradation. In particular, mRNA stability can be enhanced by adding sequences such as REP onto the end of the cloned DNA so that a suitable hairpin loop is formed. The loop prevents access of the exonuclease to the end of the mRNA.

4.3.8. *Isolation and purification of plasmids*

An obvious prerequisite for cloning in plasmids is the ability to purify the plasmid DNA. The strategy is to release the plasmid DNA from the cell and then it purify away from the genomic DNA and cell debris. The step that demands the most attention is the initial steps of cell lysis and neutralization of the reaction. The incomplete lysis of the cells will reduce the recovery and purity of the plasmid DNA. It is important to note that cell lysis must be carried out with minimal physical force. After neutralization, the chromosomal DNA of high molecular weight can be removed together with the cell debris, by high-speed centrifugation, to yield a clear lysate where the plasmid resides. There are a number of ways of isolating plasmids and many of these rely on the alkaline lysis procedure that is both simple and effective.

(1) Alkaline Lysis
This is largely based on the observation that there is a narrow range of pH (12–12.5) within which linear DNA, but not supercoiled DNA, denatures (Fig. 4.19). However, prolonged exposure to this pH will eventually denature even the supercoiled DNA. The method is rather simple but highly effective, even with large vectors like BACs. Plasmid containing cells are first lysed in high pH and SDS. The lysate is then neutralized. The chromosomal DNA renatures and aggregates to form an insoluble network. The high concentration of acetate also causes the protein-SDS complexes (and excess dodecylsuphate) to precipitate. Provided the pH of the alkaline lysis step is carefully controlled, super-coiled plasmids will remain in the native state. The insoluble precipitates are then removed by centrifugation. The plasmid can be purified by extracting any remaining bacterial proteins using a mixture of phenol and chloroform (organic phase). The purified plasmid, which is retained in the aqueous phase, is then recovered by precipitation using salt and ethanol. This process is known as ethanol precipitation.

(2) Boiling Method
This is based on the principle that the supercoiled DNA is not denatured as much as the chromosomal DNA. When the extract cools down, the chromosomal DNA will reanneal but due to its complexity will form

Fig. 4.19. Isolation of plasmid DNA using alkali lysis.

insoluble "tangles" and hence precipitate. As in alkaline lysis, this can be removed by centrifugation. Although this method is similar to the above method, it is not highly reliable and is dependent on host strain.

4.4. Identification of Clones

The success or failure of a cloning experiment relies mainly on the ability to identify and select for the particular clone of interest. This is especially true if a large collection of recombinant DNA representing the entire genome of an organism (a genetic library) has been cloned. Once this hurdle is overcome and a clone of interest is obtained, the scientist is then able to use a wide number of techniques to explore and extract information from the gene.

There are many ways of identifying and analyzing DNA often with the ultimate objective of gaining an insight into its sequence and function. This section describes some of the methods commonly used to isolate, identify and confirm the clone carrying the gene of interest. One of the simplest but powerful methods to identify the correct recombinant DNA directly is by a technique known as **hybridization probing**. This technique forms the basis for methods such as colony and plaque lifts, as well as Southern blot hybridization. The piece of unknown DNA can also be identified by PCR using specific primers. The DNA can eventually be confirmed by a technique called DNA sequencing.

4.4.1. *Hybridization probing: Its use in identifying specific clones*

Hybridization of nucleic acids immobilized on filters occupies a central role in molecular biology. This method has been used extensively in the detection of gene expression and identification of genes. This is based on the properties of strand separations (denaturation), the ability to immobilize nucleic acids on solid matrices and the ability to **probe** for the gene of interest by the formation of a stable duplex between complementary nucleic acid sequences.

In filter hybridization, nucleic acids are bound to a special filter (membrane) in such a way that intramolecular reannealing is prevented, but yet allows the formation of a hybrid with a single-stranded probe added in solution. The probes are incubated with the filter-bound nucleic acids (targets) under conditions favoring the formation of duplexes. Unbound probes are then washed away and the probe-target hybrids are visualized. We shall illustrate the use of this technique in the examples below.

(1) Colony Blotting and Plaque Lift Assays
Nucleic acid hybridization can be used to identify the DNA of interest contained in either a bacterial colony or a phage plaque. This method has been extensively used to screen cDNA and genomic libraries, constructed

Fig. 4.20. Identification of positive clones by colony transfer onto a membrane and hybridization probing.

in either bacteria (colony blot) or bacteriophages (plaque lifts), for clones carrying the desired genes of interest.

The first step in trying to identify the clones of interest within a library is to transfer the colonies or plaques onto either nitrocellulose or nylon filter membranes (Fig. 4.20). The filter membranes are placed on the bacterial colonies or plaques and marked with Indian ink on the plates for orientation purposes. After a few minutes, colonies/plaques would have been blotted onto the membranes. The membranes are then lifted off the plates, denatured and treated so that the DNA released from the colonies or plaques are immobilized onto the membranes. The single-stranded DNA can then be tightly bound to the membrane by baking it at 80°C if a nitrocellulose membrane is used, or irradiating the membrane with UV if a nylon membrane is used instead.

Next, to locate the DNA of interest within a library of DNA, the membranes are then probed using a labeled piece of complementary DNA, in a solution of chemicals that will promote nucleic acid hybridization. Positive colonies or plaques containing related sequences to the probe used are then identified and subsequently rescued from the agar plates.

In this process, unrelated colonies or plaques in close proximity to the positive clone will inevitably be rescued simultaneously. Further serial dilutions and screenings of the colonies or plaques are therefore necessary to isolate pure clones. These clones are then amplified for the isolation and purification of the insert DNA. Finally the DNA fragments can then subcloned into other vectors of choice for further characterizations, for example restriction mapping and sequencing.

(2) *In vitro* Labeling of Probes

The probes used to identify target sequences can be in the form of double-stranded DNA, single-stranded DNA (synthetic oligonucleotides) or as RNA (see Sec. 4.3.3(iv)). These probes can be labeled either internally or at the terminus end. Two competing methods can be used to generate these probes: radioactive and non-radioactive. The principles underlying both methods are the same except for the differences in the tags and detection methods used.

In general, long probes are usually labeled internally to achieve high specific activities. These probes generally provide the strongest hybridization signals. A major disadvantage of internally radiolabeled probes is that it suffers from radioautolysis. Non-radioactive labeling detection methods are becoming more acceptable and in some cases, can replace radioactive methods. However, this is not without limitations. More steps are generally required for non-radioactive methods and this may introduce more errors. In addition, more optimization of the experimental conditions is required. A major advantage of using non-radioactive methods is that it is not hazardous and the probes generated can be stored indefinitely.

Below are two examples of the commonly used technique for specific probe labeling. The principles are essentially the same for both radioactive and non-radioactive formats. All methods listed here are commercially available as kits.

(a) Nick translation

The term "translation" in this method must not be confused with the translation of mRNA into proteins. Translation here refers to the progressive extension of the newly made strand of DNA by the polymerase.

Single-stranded nicks (internal cleavage of one strand of the duplex DNA) are first generated at random in the double-stranded DNA using low levels of DNAase I. DNA polymerase I is then added. At the nicked sites, the 5' to 3' *exonuclease* activity of DNA polymerase I remove nucleotides from the 5' end of the nick, while the 5' to 3' *polymerase* activity adds dNTP and labeled dNTP to the 3' end (see Sec. 4.1). Hence, the nick is translated downstream, with the labeled nucleotides replacing the nucleotides located between the original and the new position of the nick. This results in a nick-translated probe that is labeled on both strands with interspersed sequences of the newly synthesized labels and the original (unlabeled) DNA. This method works best with probes of 500 base pair or more. The incorporated dNTP can either be radiolabeled nucleotides or chemically modified non-radioactive nucleotides (Fig. 4.21).

(b) Random priming

This uses a large excess of random hexamers (or longer oligomers) as nonspecific primers. Both single-stranded or double-stranded DNA is first denatured. Random hexamers (NNNNNN, where N is any nucleotides) are then incubated with the template where they will hybridize to specific sequences by chance. Next, the Klenow fragment of T7 DNA polymerase will extend the primer-template. The template strand remains unlabeled, whereas the newly synthesized strand(s) are fully labeled by Klenow. As little as 1–5 ng of template DNA is sufficient for synthesis

Fig. 4.21. Labeling of probe DNA by nick translation.

RANDOM LABELING METHOD

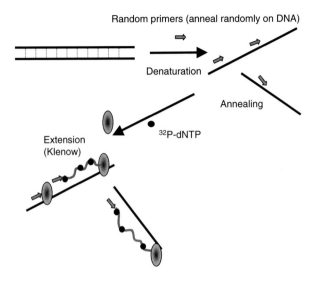

Fig. 4.22. Labeling of probe DNA by random priming.

of high-specific probes. It appears that probes less than 500 bp are more efficiently labeled than longer probes. There are commercially available Klenow fragments that do not have the 3'–5' proofreading exonuclease domain. Thus, these enzymes have higher processitivity (Fig. 4.22).

(3) Southern Blot: Its Use in Confirming the Identity of a Clone
Once positive colonies have been identified by colony of plaque hybridization, and the DNA isolated and purified (Sec. 4.3.8), it is often a good practice to confirm the identity of a given piece of DNA by restriction map (see Sec. 4.1.3) followed by Southern blot analyzes. Southern blot is a method developed for the transfer of single-stranded and denatured double-stranded DNA from an agarose gel to a membrane. The blotted DNA is immobilized on the membrane but is still able to hybridize to labeled probes.

The method involves is as shown in Fig. 4.23.

The aim of treating the gels before the blotting step is to aid the transfer of large fragments of DNA onto the membrane. The rate of DNA transfer

Fig. 4.23. Steps in Southern blotting.

A. Upward transfer of DNA from gel to membrane.

B.   Assembly of an "upward" transfer southern blot

Fig. 4.24. Southern blotting using the "upward transfer" method.

from the gel to the membrane depends on the size of the DNA. The larger the DNA, the less efficient is the transfer to the filter. Using this procedure, the DNA in the gel is depurinated, denatured and neutralized using a series of solutions.

Capillary blotting is the most commonly used transfer method in many laboratories, as it does not require expensive instrumentation. There are two ways to carry this out. Traditionally, Southern blotting is carried out in the "upward transfer" fashion (Fig. 4.24). Quite often, the use of too heavy a weight at the top of the assembly will result in the gel getting squashed. This can result in the smearing of the DNA during transfer. An alternative to this mode of transfer is the "downward transfer". Basically, it shares the same principle, except that the arrangements of the assembly is opposite to that of the "upward transfer". The advantage is that it does not deform the gel during transfer. It is recommended as a rapid method for transfer, using minimal amounts of transfer solution. However, the transfer for a defined short period of time may not result in complete transfer.

After the restricted DNA fragments have been immobilized onto the membrane, the position and identity of the desired piece of DNA can be detected and visualized by nucleic acid hybridization, using its complementary labeled probes.

4.4.2. *Polymerase chain reaction*

As mentioned earlier, a successful cloning experiment relies mainly on the ability to identify and select the particular clone of interest. In this respect, the polymerase chain reaction (PCR) has proven to be an invaluable technique for the amplification of a specific copy of a gene of interest, and thus enhancing its successful cloning and subsequent selection and identification. Using a pair of gene-specific primers (GSPs), it is possible to selectively amplify millions of copies of a gene of interest from the microbial genome. In addition, by the incorporation of restriction endonuclease sites into these GSPs, PCR cloning will also enhance the efficiency of ligation between vector and insert DNA, or to allow in-frame insertion into an expression vector for the production of recombinant proteins.

Fig. 4.25. Steps in the polymerase chain reaction.

(1) Basic Principles and Optimizing PCR

In PCR, short oligonucleotides act as specific primers hybridizing to a target DNA molecule. Under optimal conditions, two oligonucleotide primers complementary to the opposite strands of the template, can then be used to initiate DNA synthesis *in vitro* using thermostable DNA polymerises (Fig. 4.25). The hybridization conditions are carefully controlled to limit *in vitro* DNA synthesis due primarily to template sequences located between the two oligonucleotide primers. After annealing, the DNA polymerase will then extend the primers, generating an additional new copy of the template sequences located between the two primers. On completion of the primer extension, the sample is heated to denature the DNA duplex, regenerating approximately twice as much single-stranded template for primer annealing in the next cycle. A series of cycles is then performed, where the DNA template is denatured, oligonucleotide primers annealed specifically, and the primers extended by the thermostable DNA polymerase in each cycle. The repeated cycling results in a geometric accumulation of the products flanked by the two primers.

In slightly more detail, the initial extension products resulting from the original template do not have a distinct length because the polymerase will continue to synthesize new DNA until it either stops or is interrupted by the start of the next cycle. In the second cycle, the extension of products also result in variable lengths. However, by the third cycle, the segment between the primers is synthesized with a defined length. From the fourth cycle onwards the target sequence is amplified exponentially. Potentially, after 20 cycles of PCR, the products should be amplified 2^{20}-fold, assuming 100% efficiency during each cycle. However, in practice, this is seldom achieved, largely because the efficiency is highly dependent on the template and primers used, as well as the cycling conditions and reactants involved.

In later cycles, the accumulation of amplified products does not expand exponentially indefinitely but levels off ("plateau effect"). There are many reasons to account for this. Some of these factors include the gradual loss of polymerase activity and hence becoming a rate-limiting step. Another factor is the reannealing of target strands as their concentration increases. The reannealing of target strands then competes with primer annealing. Practically, the process levels out after microgram quantities of products are obtained and this is sufficient for many downstream procedures, e.g. subcloning into vectors, sequencing etc.

(2) Cloning of PCR Products

PCR has allowed the isolation and amplification of any region within a complex DNA in a few hours. Cloning the PCR products into appropriate vectors allows the generation of the amplified products without having to repeat the reaction every time the products are needed. Therefore it is more convenient to amplify the gene of interest from its recombinant clone than from its original source, which could range from RNA to a DNA genome. In addition, if the PCR products are to be use in further studies, e.g. expression or the generation of single-stranded RNA probes, then it is necessary to clone the amplicons. In this section, some strategies to clone PCR products will be discussed. It is important to note that a well-planned cloning strategy is needed before starting the actual PCR reaction, as many of these strategies involve modification of the primers that are used in the PCR reaction.

(a) Blunt-end and T-A cloning

The type of DNA polymerase used in PCR will determine the type of vectors that can be used to clone the PCR products. For example, if Taq, Amplitaq or Tth DNA polymerase had been used in the reaction, the resulting amplified products will have a single A residue overhang at the 3' end. This overhang nucleotide will facilitate ligation if the vectors used to clone these PCR fragments contain a complementary T nucleotide at the 3' ends of the vector. This is called T-A cloning and is illustrated in Fig. 4.26. An example of this type of cloning vector is pGem-T.

The orientation of the DNA insert cannot be controlled, as the overhang nucleotides on the inserts and vectors are bilateral. Precautions have to be taken to avoid non-recombinants due to religation of vector molecules, which have lost the T-overhang nucleotides due to a contaminating exonuclease activity. This problem is easily overcome with cloning vectors that has the multiple cloning sites within the *lacZ'* gene, as in pGem-T. As described in Sec. 1.4.3.5, the presence of the DNA insert will result in the formation of a white colony, in contrast to a non-recombinant blue colony.

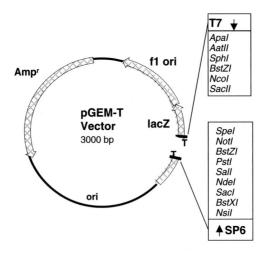

Fig. 4.26. A plasmid vector for TA cloning of PCR products.

The use of other DNA polymerases, for example *Pfu* or *Ultm*a, will however result in the formation of blunt-ended amplified products. As in the case of blunt-end ligations (Sec. 4.1.4), the ligation of blunt-ended PCR product to a vector is not as efficient as T-A cloning. However, this ligation reaction can be improved by using high amounts of starting inserts and vector. The choice of vectors that can be used for blunt-end ligation is broad, as long that the vector contains restriction sites that will result in blunt-ends, e.g. *Sma*I, making it terminally compatible with the amplified products. Alternatively, there are commercial vectors available for blunt-end cloning, for example, pCRScript and pCRScript™Direct (Stratagene).

(b) Introduction of restriction sites for cloning

To aid in efficient and unidirectional PCR cloning, restriction sites can be added to the 5' ends of one or both nucleotide primers that are to be used for PCR (Fig. 4.27). The sequences of these modified primers will be

Fig. 4.27. Restriction sites can be added to the 5' ends of one or both nucleotide primers that are to be used for PCR.

incorporated into the amplified PCR products. These amplified DNA can then be digested with the appropriate restriction endonucleases to generate either blunt or cohesive ends, as required by the scientists. By the use of different restriction sites at either ends of the DNA, it is thus possible to clone a DNA insert unidirectionally into a vector.

The addition of a few extra sequences to the 5' end will not significantly affect the primer specificity, since this is determined mainly by the sequences at its 3' end. There are, however, a few points that one needs to keep in mind when using this method:

1. The restriction site(s) used in the primer sequences should not have recognition sites within the DNA of interest.
2. The restriction sites that are added to the final product via the primers often need extra bases at the 5' end to allow for efficient DNA digestion.
3. It may be essential that sequential digestions be performed on the amplified products if the enzymes chosen do not have optimal activity in a common buffer.

The techniques described above are widely used in recombinant DNA technology and provide one of the most reliable and simplest methods for PCR cloning.

4.4.3. *DNA sequencing*

Probably the most important technique available in recombinant technology is DNA sequencing, which enables the scientist to decode the precise order of the nucleotides within a piece of DNA. The two of the most commonly used sequencing techniques are:

- Dideoxy chain termination method (Sanger), and
- Chemical degradation method (Maxam-Gilbert).

The former is the method of choice for the high throughput sequencing. Although the two methods differ in their strategies (Sanger method is enzymatic and Maxam-Gilbert method is purely chemical-based), the fundamental principle is the same. The principle is that two single-stranded DNA molecules differing in length by just one base can be resolved into distinct bands by some form of separations. The usual way of separation is the use of denaturing polyacrylamide gel electrophoresis.

(1) Dideoxy Chain Termination Sequencing (Sanger Sequencing)

This method is conceptually rather simple. It uses polymerases to synthesize a new template by extending the primer and occasionally incorporates dideoxynucleotides (ddNTPs) instead of the usual deoxynucleotides (dNTP) during chain elongation at the 3' end (Fig. 4.28). When the dideoxynucleotide is added to the 3' end of an elongating chain, no further elongation is possible as the 3' end nucleotide lacks a 3'OH that is essential for forming the phosphodiester bond between the primer and additional dNTPs for elongation. Thus, the chain is terminated.

In manual chain-termination sequencing, each of the 4 dNTPs (dATP, dCTP, dGTP and dTTP) is added separately to one of the 4 primer extension reactions. In addition, a radiolabeled ddNTP is included in the reaction mix. The reaction with ddGTP is called the "G" reaction; the one with the ddATP is called "A" reaction, etc. Chain termination will occur in a base-specific fashion depending on which ddNTP is present during the extension reaction. For example, termination will occur only at the "G" position (complementary base of C) when ddGTP is present, or exclusively

Structures of adenosine analogues

Adenosine

Deoxyadenosine

Dideoxyadenosine

Fig. 4.28. Dideoxynucleotides are used as chain terminators during DNA synthesis.

Termination with ddATP

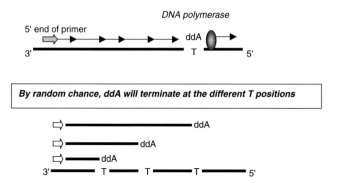

Fig. 4.29.   Incorporation of a dideoxynucleotide results in chain termination.

at the A when ddATP is present (Fig. 4.29). The concentration of ddNTP is kept low relative to the corresponding dNTP. Hence, base-specific chain termination will only occur in a small percentage of the time at a given position in the sequence. As the reaction proceeds, a mixture of all possible termination products for G, A, T, or C will be generated in the 4 primer-extension reaction tubes.

The radiolabeled products from the 4 base-specific termination reactions are then resolved on a high-resolution polyacrylamide gel (Sec. 4.1.3). Up to 300–500 bases in length can be separated. Each terminated species will be visualized as a separate band on the gel with the shortest primer-extended products running fastest down the gel. The base sequence of the primer-extended products can then be read from the 5' to 3' ends by identifying which reaction mix (G, A, T, C) contains the next longest species, beginning with the shortest fastest migrating species as seen on the autoradiogram of the gel (Figs. 4.30 and 4.31).

(2) DNA Polymerases
The properties of these polymerases have been discussed in other chapters. It suffices to say that the majority of those used in manual and automated sequencing are genetically modified for longer extension (high processivity) and diminished exonuclease activities. For manual

Add: α-^{32}P-ATP/dNTP/Polymerase

| **G** lane
+ ddG | **A** lane
+ ddA | **T** lane
+ ddT | **C** lane
+ ddC |

After reaction, run on PAGE

The sequence is read from the bottom up: **5'-GAATC**

Fig. 4.30. Separation of DNA fragments during DNA sequencing.

Fig. 4.31. An example of a DNA sequencing gel after Sanger dideoxy chain termination sequencing. Each lane represents a chain termination reaction with a specific dideoxynucleotide.

sequencing, some examples of the commonly used polymerases include Sequenase, *Bst* and Klenow *exo⁻*.

Sequenase is a genetically modified version of the bacteriophage T7 DNA polymerase. The unmodified polymerase (wild type) is highly processive, with an average of 2000–3000 nucleotides polymer formed (before dissociating from the template) with an elongation rate (extension) of about 300 nucleotide/second.

Before the availability of Sequenase, Klenow polymerase was the enzyme of choice for manual sequencing (Sec. 4.1). Klenow polymerase is a modified enzyme derived from *E. coli* DNA polymerase I. The unmodified version is unsuitable for sequencing because of its 5' to 3' exonuclease. Klenow polymerase was originally prepared by proteolytic cleavage of the parent enzyme, but now as genetically modified enzyme.

(3) Alternative Labeling/Detection Strategies

There are other strategies available besides incorporating radiolabeled nucleotides:

(a) End labeling of primer

Here, the primer is labeled instead of incorporating labeled nucleotides into the strands. T4 polynucleotide kinase is used, which catalyze the transfer of a ^{32}P or ^{35}S (in the γ-position) to the dephosphorylated 5' end (not the 3' end as it is where extension occurs). This method of labeling result in a chain-terminated molecule carrying just a single-labeled atom rather than the multiple labeled molecule as described above.

(b) Labeling with tags for chemiluminescence

This is an alternative to radiolabeled studies. Chemiluminescence can be as sensitive as radiolabeled approach, however, this involve more manipulations and can result in low sensitivity if mishandled (see discussion on chemiluminescence). Here the 5' end of the primer can be modified to contain a tag (e.g. biotin). Extension and termination is carried out as usual. At the end of the gel run, Southern transfer is carried out and the membrane is then processed for chemiluminescence.

(c) Fluorescent *labels*

This is extensively used in automated sequencing. Special imaging is required including a laser for monochromatic excitation and a detector with different wavelength discrimination. Fluorescent-labeled fragments with 4 different colors can be analyzed in the same lane. As the different sizes pass the detector, one of the four colors will be recorded and deciphered using a software program.

Two main types of automated sequencers are commercially available. The first type uses single-label, 4-lane separations, similar to the above manual sequencing, whilst the second type uses 4 labels, single-lane separation (Fig. 4.32). The latter type is often the instrument of choice for large-scale genome projects.

Sequencing is carried out with dye-terminators (similar to the use of radiolabeled ddNTPs) and the samples electrophoresed on a denaturing polyacrylamide gel. Fluorescent signals are detected after excitation by a laser positioned directly in front of the detector. Automated DNA sequencing has now become a routine in many laboratories because of its ease of use and the greater amount of sequencing data that can be obtained. An example of a sequencing result using the automated fluorescent sequencing is shown in Fig. 4.33.

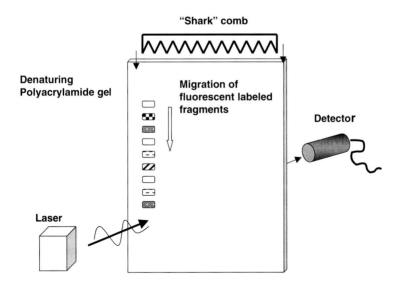

Fig. 4.32. An automated DNA sequencer which uses 4 labels, single-lane separation.

Fig. 4.33. An example of a sequencing result using the automated fluorescent sequencing is shown.

(4) Sequencing Strategies

In sequencing a large piece of DNA (10 kb), the amount of sequences that can be read from a single template is often in the 500–700 bases region. Thus, various methods have been devised to obtain the rest of the sequences. These methods can be broadly divided into two types: random and ordered sequencing.

Random sequencing is a simple method to obtain large portion of the template quickly. The template is first fragmented physically by shearing the DNA into smaller pieces, subcloned into vectors and the recombinants sequenced all at one go. This is an example of a "shot-gun" method. Another is the use of transposon-mediated sequencing method where a transposon randomly integrates into and excise parts of the DNA. Sequencing of a collection of these recombinant transposons will also yield large amounts of data at a go. An advantage is that some difficult regions (G-C rich) may be broken up during the fragmentation step, thus, facilitating the ease of sequencing. However, both methods suffer from one major drawback. They both can obtain most of the sequences quickly but the last 10% of the sequence will pose some difficulty with the presence of similar clones. Usually, to obtain the last bit of the template sequence, ordered approaches are used.

Ordered sequencing comes in many forms. The simplest but most tedious and expensive method is to "walk" the whole template. Here, the sequencing data is used to synthesize new primers that can then be used to sequence further down. Moving the template closer to the primer progressively is another approach. One of these methods is to generate "collapse clones" where internal restriction sites and those on the multiple cloning site of the vector can be used to delete off sequences. The unidirectional deletion (nested deletion) and religation of template sequences is yet another approach which is easy to carry out. This uses enzymes, e.g. *Exo* III and *Bal*31, which can delete the template in a specific direction.

Obviously, directional sequencing is more convenient and once a fragment can be localized, the rest of the fragments produced can be zoomed in and overlapping clones used for the rest of the project. Furthermore, the deleted (shorter clones) can be used for expression or mutagenesis studies.

4.5. Mutagenesis

Recombinant DNA and gene-cloning technologies allow scientists to isolate and characterize any given gene from any living organism. In addition, these genes may also be modified *in vitro* in order to produce new restriction enzyme cleavage sites, to change a particular codon, or to introduce site-specific mutations by insertion and deletion of nucleotides. This procedure has allowed scientists to dissect genes at the nucleotide level by changing one nucleotide at a time and examining the effect of each change on the function of the gene or its product. *In vivo* mutagenesis protocols, on the other hand, either makes use of bacterial transposons to either mediate mutation via random insertion of the transposon into the host genome, or the replacement of the normal gene with an altered gene sequence via homologous recombination.

4.5.1. *Random insertional mutagenesis*

A gene that has been marked with a transposon is relatively easy to map, either by genetic crosses or by physical mapping using restriction

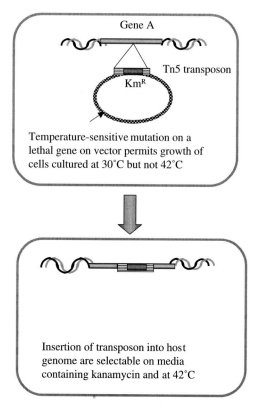

Fig. 4.34. Gene disruption using a transposon carrying a kanamycin resistance marker.

endonucleases. A typical strategy is to construct a plasmid carrying the transposon, using temperature-sensitive vector, i.e. a plasmid that is unable to replicate at an elevated temperature. When this plasmid is introduced into bacterial cells, and the cells are grown at a temperature permissive for plasmid replication (say 30°C), all the cells will carry the plasmid. When the culture is then shifted to the non-permissive temperature (say 42°C) at which the plasmid is unable to replicate, the only cells which retain the transposon (usually selectable by an antibiotic resistance marker) are those that have incorporated the transposon in its chromosome. Thus, using a transposon, such as Tn5, which carries the kanamycin resistance gene, will allow selection of transposon insertions by plating the culture on agar containing kanamycin at 42°C (Fig. 4.34).

This generates a library of transposon mutants, with the transposon inserted at a variety of chromosomal positions, and therefore causing a range of mutations. After the desired mutants have been selected using an appropriate phenotype screen, e.g. the loss of bacterial motility, the genes into which the transposon has inserted may be identified. This is achieved by cutting the total genomic DNA of the mutant with a restriction enzyme, which does not cut the transposon itself, and then cloning the collection of fragments to produce a gene library. If the library is then plated on

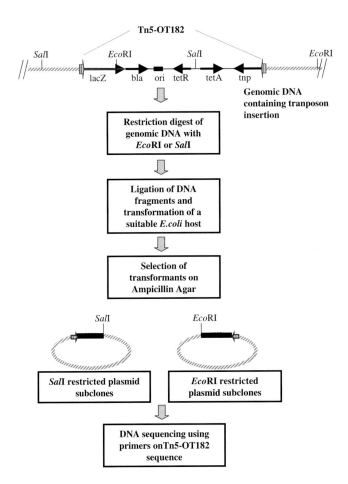

Fig. 4.35. Strategy for the identification of DNA sequences flanking the Tn5-OT182 transposon insertion site.

kanamycin agar, only those clones containing Tn5 will be able to grow. These clones will contain not only Tn5 but also some of the host sequence flanking the transposon. The cloning of DNA sequences flanking the transposon insertion is made easier when a transposon with self-cloning capabilities, e.g. Tn5-pOT182, is used. In this example, the genomic DNA is digested with a restriction enzyme that cuts once within the transposon (e.g. *Eco*RI or *Sal*I), leaving its plasmid origin of replication and an antibiotic selection marker (ampicillin resistance) intact. The other *Eco*RI or *Sal*I restriction site would be in the host genomic DNA sequence flanking the transposon insertion. Self-ligation of the digested DNA fragments results in recircularization of the digested genomic DNA fragments. Upon transformation of *E. coli* host cells and selection on ampicillin-containing agar, only the recircularlized DNA containing origin of replication on the transposon and the antibiotic reisistance marker would result in ampicillin-resistant transformants. The disrupted host gene is then identified by isolating the plasmid DNA from such clones and sequencing the DNA flanking either ends of the known transposon sequence (Fig. 4.35).

4.5.2. *Gene replacement mutagenesis*

Having identified a gene of potential interest through DNA sequence analysis, one of the best ways of determining its function and significance is to deactivate it in the host organism. This may not be easy to achieve by the classical procedures of mutagenesis and selection, especially if the phenotype associated with that gene is not known, or not readily selectable. However, if the cloned gene is inactivated *in vitro*, it can be used to replace the wild-type chromosomal gene by homologous recombination, a process also known as gene knock-out (Fig. 4.36). In this case, the central part of the cloned gene has been removed and replaced by an antibiotic resistance gene, such as kanamycin resistance. The inactivated gene is constructed on a suicide vector, which cannot replicate in the host cell. In this example, the mobilizable suicide vector, pJQ200mp18, contained the *sacB* gene. Cells carrying the intact *sacB* gene, i.e. those that had not undergone a gene replacement event, will be lethal when plated on media containing 5% sucrose. The mutated DNA is introduced into recipient cells by conjugation. After conjugation, selection on kanamycin and streptomycin

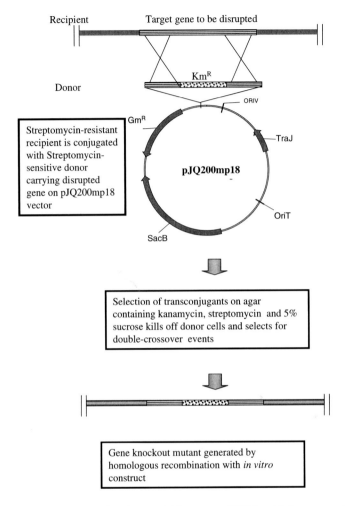

Fig. 4.36. Targeted mutagenesis using a suicide vector, pJQ200mp18, for gene replacement by homologous recombination.

agar containing 5% sucrose will identify cells in which the kanamycin resistance gene has been inserted into the target gene by homologous recombination, involving a double crossover. The addition of streptomycin selects against donor cells, which are streptomycin-sensitive, and the kanamycin selects against non-recombinant recipient cells. The target gene is thus deactivated. The phenotype of the knock-out mutant can then be studied using appropriate assays. A confirmation that the changed phenotype is actually due to inactivation of the gene done by introduction of the wild-type gene on a plasmid (complementation). Complementation is an important step, as the phenotypic changes could be due to polar effects on other genes, rather than directly caused by inactivation of the target gene. This not only provides a useful way of assessing the function of a cloned gene, but can also be used for producing attenuated strains of a pathogenic bacterium by inactivating a gene coding for toxin production or another virulence factor.

4.5.3. Site-directed mutagenesis

It is sometimes desirable to replace a gene with one with specific alterations. This requires a combination of the techniques of gene replacement and site-directed mutagenesis. The knocked-out gene can be complemented with a plasmid carrying a specifically mutated gene copy. This enables us to test, for example, the significance of a specific amino acid in the activity of an enzyme, by replacing it with a series of different residues and determining the activity of the product. Or we could look at the precise requirements for the function of a regulatory sequence on the DNA such as a promoter site, by introducing defined mutations into that region.

A number of protocols have been established for site-directed mutagenesis using hybridization of a mismatched oligonucleotide to a DNA template, followed by second-strand synthesis by a DNA polymerase. One of these is a site-directed mutagenesis procedure developed by Deng and Nickoloff that eliminated the need for subcloning and generating single-stranded DNA templates by employing double-stranded plasmid DNA (Fig. 4.37). The procedure involves simultaneously annealing two oligonucleotide primers to the same strand of heat-denatured double-stranded plasmid DNA. One of the primers (the selection primer) anneals to a region that contains a unique, nonessential restriction site. The other primer (the mutagenic primer) introduces a chosen mutation into the plasmid. Extension of these primers by T4 DNA polymerase followed by ligation of the resulting molecules with T4 DNA ligase results in a population of plasmid molecules, some of which contain the desired mutation but no longer contain the unique restriction site. The plasmid is

Plasmid contains a
unique non-essential
restriction site

Site to be
mutated

**Plasmid is denatured
and primers added**

Selection primer
anneals to the
restriction site

Mutagenic primer
anneals to the
target site

**Annealed primers are
extended using T4
DNA polymerase**

**Restriction digestion with
selection restriction
enzyme linearizes parental
homoduplex**

**Transformation into
repair-deficient *E.coli*
cells (XL*mut*S)**

**Plasmid DNA is prepared
and a second round of
restriction digestion with
selection restriction
enzyme enriches circular
mutant plasmid**

**Transformation into *E.coli*
XL1-Blue cells and screen
colonies for the desired
mutation**

Fig. 4.37. Protocol for double-stranded, site mutagenesis.

treated with a restriction enzyme that will linearize the parental plasmid but
not the heteroduplex. Since transformation of circular DNA is many folds
more efficient than that of linearized DNA, the circular heteroduplex would
be preferentially transformed. Using a mismatch repair-defective strain of
E. coli (e.g. XL*mut*S) that cannot distinguish the original unmutated
strand of DNA from the newly created strand containing the desired

mutation, both strands of the heteroduplex would be equally present in the newly replicated DNA. The *mut*S-deficient strain randomly selects one of these strands as "correct" and changes the other strand to be complementary to the chosen strand. DNA is isolated from the plasmids and subjected to a second round of restriction digest, which will enrich the population of mutant plasmids after transformation into an *E. coli* strain (e.g. XL1-Blue).

4.5.4. *PCR-based mutagenesis*

Various mutagenesis procedures based on polymerase chain reaction (PCR) are available. A versatile and relatively easy method is the "megaprimer" procedure (Fig. 4.38). This method uses three oligonucleotide primers and two rounds of PCR performed on a DNA template containing the cloned gene to be mutated. Primers A and B represent the "flanking" primers that can map either within or outside the cloned gene. M represents the internal "mutant" primer containing the desired mutation, which may be a point mutation, a deletion or an insertion. In the first round of PCR, primer pair A and M1 is used. The PCR product, A-M1, is purified and used as one of the primers for the second round PCR together with the flanking primer B, hence the term "megaprimer". Prior separation of the two strands in the PCR product A-M1 is unnecessary since this will be achieved during the denaturation step in the second round of PCR. The wild-type cloned gene is used as the template for both PCR reactions. The final PCR product (A-M1-B) containing the mutation can be used in a variety of standard applications, such as sequencing and cloning in expression vectors. The inclusion of restriction sites to primers A and B would facilitate the cloning of the final mutant PCR product. This is especially important when sequences upstream of A or downstream of B are essential for gene expression or regulation, and therefore, must remain unaltered.

This protocol may also be applied to the mutagenesis of protein-coding genes and the study of structure-function relationship. In this case, primers A and B can be kept constant and a variety of mutant primers (M1, M2, M3...) can be used to produce the various mutants.

Fig. 4.38. Site-directed mutagenesis using megaprimer PCR.

4.5.5. *Ordered deletions using exonuclease III*

Ordered deletions can be used in delineating sequences that are important for the function of a gene, such as those required for transcription. Deletions are generated by limiting, unidirectional digestion of DNA with *E. coli* exonuclease III (*Exo* III). *Exo* III preferentially digests one strand of a double-stranded DNA by removing nucleotides from 3' ends if the end is blunt or has a 5' protrusion, but not if the end has a 3' protrusion of 4 or more bases (Fig. 4.39). The plasmid clone is first digested with restriction endonuclease A, which generates a 5' protrusion or blunt end next to the target sequence, and restriction endonuclease B that generates

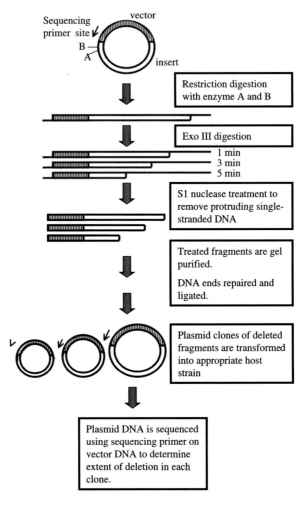

Fig. 4.39. Generation of deletion mutants using *Exo* III.

a 3' protrusion next to the sequencing primer site. The linearized plasmid DNA is then digested with *Exo* III, and aliquots of the digest are taken at a number of time points so as to yield a set of deletions of the desired lengths. These DNA fragments are then subjected to S1 nuclease digestion to remove the single-stranded DNA overhangs. The deleted DNA fragments may then be ligated to a vector with a promoterless reporter gene, such as the green fluorescent protein (GFP) or *lacZ*, to analyze the promoter activity. The cloned fragments can also be sequenced to determine the extent of the deletion.

Further Readings

1. Ausubel FM *et al. Current Protocols in Molecular Biology.* John Wiley and Sons, Inc., 2001.
2. Balik S. Site-directed mutagenesis *in vitro* by megaprimer PCR. In: *In Vitro Mutagenesis Protocols Vol. 57*, ed. M Trower, Humana Press, New Jersey, 1996, pp. 203–215.
3. Barth PT and Grinter NJ. Map of plasmid RP4 derived by insertion of transposon C. *J. Mol. Biol.* **113**: 455–474, 1977.
4. Clark D and Henikoff S. Ordered deletions using exonuclease III. In: *In Vitro Mutagenesis Protocols Vol. 57*, ed. M. Trower, Humana Press, New Jersey, 1996, pp. 139–147.
5. Deng W and Nickoloff J. Site-directed mutagenesis of virtually any plasmid by eliminating a unique site. *Anal. Biochem.* **200**: 81–88, 1992.
6. Ditta G, Stanfield S, Corbin D and Helinski DR. Broad host range cloning system for Gram-negative bacteria: Construction of a gene bank of *Rhizobium meliloti. Proc. Natl. Acad. Sci. USA* **76**: 7347–7351, 1980.
7. Figurski DH and Helinski DR. Replication of an origin-containing derivative of plasmid RK2 dependent on a plasmid function provided *in trans. Proc. Natl. Acad. Sci. USA* **76**: 1648–1652, 1979.
8. Kahn M, Kolter R, Thomas C, Figurski D, Meyer R, Remaut E and Helinski DR. Plasmid cloning vehicles derived from plasmids ColE1, F, R6K and RK2. *Methods Enzymol.* **68**: 268–280, 1979.
9. Sambrook J and Russell DW. *Molecular Biology: A Laboratory Manual.* Cold Spring Harbor Laboratory Press, 2001.

Chapter 5

Applications of Bioinformatics and Biocomputing to Microbiological Research

Paxton Loke* and **Tiow Suan Sim**[†]
Ian Potter Hepatitis Research Laboratory
The Macfarlane Burnet Institute
for Medical Research and Public Health
Commercial Road, Melbourne, Vic 3004, Australia
[†]*Department of Microbiology, National University of Singapore*
5 Science Drive 2, Singapore 117597

5.1. Introduction

This is the decade of life sciences. Indeed, the advent of recombinant DNA technology has fueled the road ahead for research and development in various disciplines. With computer-based technologies in parallel and in support of life science research, it is conceivable that much progress can be made in a shorter period of time. Bioinformatics is the application of computing to the analysis of biological data and this is in the form of manipulation of biological sequence data of DNA and proteins, as well as their structures.

Research in the field of Microbiology is no longer dependant on traditional techniques of microscopy and whole cell macro-analysis and manipulation. Since the early days of sequencing genomes of viruses and bacteriophages, microbiologists have now progressed to sequence more complex and larger genomes. With molecular biology techniques available for cloning and sequencing microbial genes, the biochemistry and physiology of a particular microbe can now be studied at the genomic level. Exploitation and improvement of genes, which may have industrial and commercial applications, are now possible in addition to the analysis of complex gene interactions.

In general, the analysis of gene sequences is easily achievable with fruitful results. On the contrary, the elucidation of structure and function is harder. As most analyzes are based on inferences, the limited information available on the biological characteristics of sequenced genes and their annotations may not be entirely helpful. In some cases, sequences that are closely related or homologous may share a common structure but not function. For example, lysozyme, which hydrolyzes bacterial cell wall polysaccharides and α-lactalbumin, which is a non-catalytic regulatory milk protein required for lactate synthase to transfer a galactose molecule from UDP-galactose to D-glucose, shares an identity of 50% (70% similarity) although both perform and are required in different functions or reactions. There are also proteins that share similar structures but have dissimilar functions, such as the serine proteases and integral membrane porins where structures like β-barrels are found. It is also interesting to note that in nature, the three-dimensional (3D) structure of proteins is more conserved than the underlying sequences. With that notion, different sequences can adopt similar folds.

Sequence analysis is a valuable tool in molecular biology as it assists in the extraction of information and biological knowledge, thus allowing us to infer biological information from sequence data. Due to the inference nature of sequence analysis, it becomes a "chicken and egg" predicament. In order to properly and accurately infer function from sequence, we must have a large starting pool of sequences that were previously obtained and closely related with known functions. This is to ensure a significant match to the unknown gene sequence, thus allowing us to make a conjecture of the function of the unknown sequence. On the other hand, we must first elucidate the function of the gene cloned. Thus, does sequence begets function or vice-versa? Nevertheless, earlier research efforts have ensured a starting pool of information, where gene function was revealed by complementation of arrays of defective mutants.

Moreover, our focus should not only be concentrated on DNA sequences and thus neglect protein sequences but both. As proteins are the main component of living cells that carry out functional tasks, their diversity in structure and function cannot be ignored. Each protein has a unique amino acid sequence that determines its 3D shape and biological activity. It is coded by a corresponding DNA sequence, which can be deciphered by the genetic code. With today's recombinant DNA technologies, it is relatively straightforward to obtain a DNA sequence and subsequently, the primary sequence of the encoded protein as the order of the amino acid residues is consequential.

However, in the early days, it took years to complete an entire protein sequence for the initial breaking of the genetic code, for example, the peptide hormone insulin in 1955 using chromatographic and labeling techniques and the ribonuclease in 1960. Nowadays, the typical approach is to obtain the coding DNA sequence at the outset followed by the usage of new conceptual translation methods to interpret or deduce the resultant protein sequence. While the primary sequence of the protein can be predicted with ease and precision, the predictions for secondary and tertiary (not to mentioned quaternary structures) structures are less accurate.

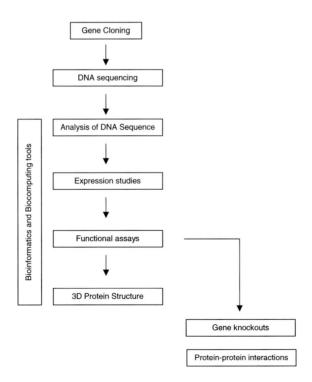

Fig. 5.1. Cloning strategies for a functional gene.

With the huge amount of microbial DNA sequence data currently available in databases, the reliability and accuracy of such data are crucial. Inaccuracies in DNA sequence information may lead to errors in translations due to frameshifts, thus generating subsequent protein sequences which may not be the actual encoded protein. In this chapter, sequence analysis using current computational programs in databases would be explored together with some of the more common tools in bioinformatics that will aid microbiological research. In essence, the standard route from cloning to elucidating the function and structure of a protein and ultimately, its natural role in the cell is shown in the flowchart (Fig. 5.1).

5.2. Gene Cloning

DNA sequences can be obtained from the partial or full cloning of the gene of interest. Traditionally, "shotgun" cloning has been performed but the process of screening to obtain positive clones of interest has been extremely tedious. Another method of cloning known as "reverse genetics" necessitates the purification of a small amount of protein (which is coded by the gene of interest). Protein sequencing of the purified protein would then reveal the order of a short fragment of about 10–20 amino acids. Thus, due to the degeneracy of the DNA code, a number of possible degenerate probes (a mixture of oligonucleotides, depending on the number of amino acids chosen initially) can then be designed and used for the hybridization of a genomic library. As prokaryotic microorganisms such as bacteria would not have introns in their genomic sequences, such methods would allow for the isolation of the complete gene of interest, and subsequently, the elucidation of the DNA sequence. For eukaryotic microorganisms, the generation of a different type of library, called the complementary DNA (cDNA) library is required.

cDNA libraries are generated from the copying of the mRNAs present in a particular cell or tissue. In this aspect, only the genes that are transcribed (exons) will be copied into the cDNA library. This will avoid the problem of introns and thus, when cloning is performed, an uninterrupted coding sequence will be obtained. One interesting usage of cDNA libraries is that the pattern of gene expression can be obtained from the various stages in the development of a cell or when a microorganism is grown under different conditions. This is because the genes contained in a cDNA library are dependant on the expression of the mRNAs, which in turn are responsive to various physiological or environmental signals. Thus, certain genes could either be turned on or off or if the genes were more frequently transcribed in response to certain signals, the corresponding amounts of mRNA would then increase.

An efficient and rapid method now being used for cloning is the polymerase chain reaction (PCR) (Box 5.1). Essentially, this is a copying method which is able to generate large amounts of DNA, depending on the number of cycles used in the replication process. Primers or short oligonucleotide sequences that are designed based on known sequences would facilitate the easy copying of the gene of interest. Even with primers designed based on related gene sequences from closely related microbial species, the amplification of the gene can also be performed. Other relevant uses of PCR today include applications in the field of molecular diagnostics, environmental monitoring and forensic medicine. The availability of sequence information from microbial genomes (completed

---

Box 5.1

How is PCR Performed?

PCR allows for the amplification of a gene in order to make numerous copies of itself. There are three major steps in a PCR reaction, namely the denaturation, the annealing and the polymerization steps. These are performed in a cyclical fashion, normally for 30–40 cycles. The first step is a denaturation step that is essential for melting the double-stranded DNA into single-stranded DNA (template) and this is performed at temperatures of 94–96°C. Following that, the annealing of a short oligonucleotide sequence (primer) to the single-stranded DNA would occur at temperatures ranging from about 50–60°C. Once the formation of the primer-template has taken place, a thermostable DNA polymerase would then begin copying the template. The third step, which is the extension step, allows for the optimal copying of the template by the polymerase at 72°C. The cycle continues with an exponential increase (2^n) in the number of gene copies generated.

Box 5.1 *(Cont'd)*

Box 5.2

Microbial Genome Sequencing

The sequencing of microbial genomes has provided a large information base for the understanding of the microorganism at the genomic level in relation to its living environment, for the exploitation of genes for commercial and industrial purposes and for the understanding of microbial evolution. The sequence of the 4.6 Mb genome of *Escherichia coli*, the workhorse of microbiology, was completed in 1998. Other interesting genomes which have been completed include *Mycoplasma genitalium*, which has the smallest bacterial genome (0.6 Mb) and archaea bacteria such as *Methanococcus jannaschii, Methanobacterium thermoautotrophicum* and *Pyrococcus furiosus*. With the minimalist genome of *M. genitalium* available, comparisons of genes obligatory for free living existence as compared to other species of bacteria can be performed. Interestingly, when the genome of *M. jannaschii* was first examined, 65% of the putative genes sequenced had no known homology to previously discovered genes. Generally, researchers have found that approximately 35% of newly predicted genes are also not related to any other genes in the databases during the analysis and annotation of microbial genome sequences. Thus, there is a vast potential for discovering and studying new genes for basic research and utilization of these biological resources.

Box 5.2 *(Cont'd)*

Currently, 88 microbial genomes have been completed with another 20 genomes at the annotation phase and 114 more sequence projects ongoing at the various sequencing centers around the globe. With the increasing number of microbial genomes available, it is now possible to obtain similar genes across many species and this will assist not only in the phylogenetic analysis of evolutionary relatedness between species but also in the PCR cloning of closely-related gene homologs. In the analysis of microbial genomes, it is also not uncommon to encounter cases of "lateral gene transfers" between bacterial species. The most well known phenomena are the transfer of antibiotic resistance genes via mobile autonomous DNA vehicles such as plasmids or "jumping elements" known as transposons. Other areas of wide-ranging interest include the analysis of genomic organization which may shed light on gene regulation, microbial pathogenesis especially opportunistic pathogens and the interactions of different microbial species in their natural habitats. Thus, the use of bioinformatics would be relevant in light of the enormous amount of sequence data that a researcher may need to plough through before making any specific conclusions. Although genes can be identified by the means of computational analysis, ultimately "wet lab" experiments would need to be performed prior to any specific assessments. To their credit, the generation of DNA sequence data by non-profit institutions is now made available in the public domain as expeditiously as possible, thus signaling heady times for researchers in microbiology.

and unfinished) (Box 5.2) in the databases would also allow for the ease of primer design and homology searches across various microbial genomes for related genes.

5.3. DNA and Protein Sequencing

The deoxyribonucleic acid (DNA) sequencing methodology can be credited to the work of Frederick Sanger in 1974. His technique involved using a DNA polymerase to copy DNA strands with the incorporation of deoxyribonucleoside triphosphates (dNTPs) and terminating the elongation with dideoxyribonucleoside triphosphates (ddNTPs). These ddNTPs would be tagged with a radioactive label, thus allowing the identification of the various sizes of DNA fragments by X-Ray autoradiography. Four different and separate tubes, each containing the termination base of adenosine (A), cytosine (C), thymidine (T) and guanine (G) were required to generate different DNA sizes, which were then lined up and read in the exact order. This method is popularly known as the "chain termination" or "Sanger's Dideoxy" method. Another method which arose about the same time and championed by the American scientists Alan Maxam and Walter Gilbert, called the "cleavage" method, was not taken up as favorably by the scientific community.

Although this technique was initially very slow and tedious, the process was sped up using fluorescent dye-labels and thus permitting the reaction to be carried out in a single tube. This process was further improved with the inclusion of a thermostable DNA polymerase for the extension. Hence, the entire reaction can now be carried out using a PCR thermocycler. With the use of a thermostable polymerase, the formation of DNA secondary structures can be reduced as higher temperatures (just like in PCR) could be used to denature the template. The resultant cycle sequencing products (different lengths of DNA fragments terminated in a fluorescent dye-labeled dideoxynucleotide) would be resolved by size on a denaturing polyacrylamide gel. The emissions of each fluorescent dye would then be scanned and read by lasers, which are part of the automated gel readers. An electropherogram would be generated, where the different bases indicated by four different colors line up to give the ensuing sequence order (Box 5.3).

Box 5.3

Automated Fluorescent DNA Cycle Sequencing

1. A mixture of normal dNTPs (dATP, dTTP, dCTP, dGTP) together with a small amount of the relevant ddNTPs, thermostable DNA polymerase, primer and template would be subjected to thermal cycling conditions.
2. Incorporation of a particular ddNTP (for example, a ddCTP) would terminate the chain elongation reaction.
3. Different sizes of DNA fragments (which terminate in a relevant ddNTP) would be subsequently generated.

Box 5.3 (*Cont'd*)

4. The different fragments would be separated by polyacrylamide gel electrophoresis.
5. The automated gel reader would then scan through the fluorescent dye-labeled fragments and identify the final base of each fragment (base calling).
6. An electropherogram (an example only) would be generated with four colors depicting each base. The original DNA sequence would then be recreated. Depending on factors such as the quality of the template and the G+C content of the gene, reads of about 400 to 700 bases are commonly obtainable.

In protein sequencing, the general strategy would be to cleave the protein into smaller fragments and then sequence each fragment and determine their respective overlaps (Box 5.4). Firstly, the amino acid composition of a particular protein would be determined, followed by sequencing at the N-terminal and C-terminal portion. From then on, the

Box 5.4

Protein Sequencing Strategy

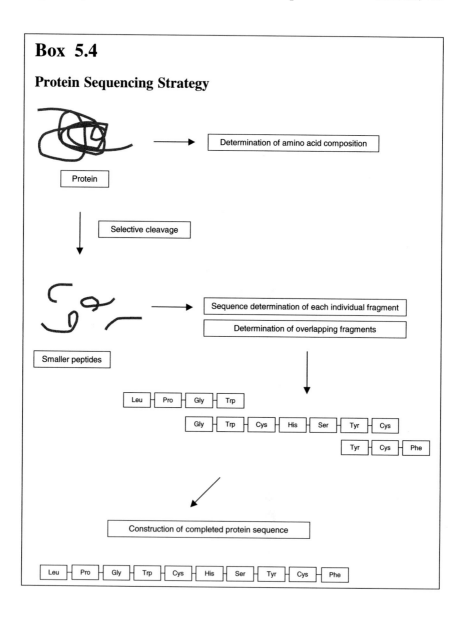

protein would be cleaved into smaller peptide fragments. The sequence of these peptide fragments would then be determined individually until an overlapping sequence is obtained. This is performed using different proteases that cleave at different locations of the peptide such as chymotrypsin, pepsin and trypsin or a chemical agent such as cyanogen bromide. With the overlapping fragments determined, the entire protein sequence is then reconstructed.

However, it has become the norm to obtain the primary sequence of a protein from the conceptual translation of the DNA sequence itself because the progress of DNA sequencing technology has exceeded that of the protein sequencing techniques. The protein product is more commonly obtained by expression of the gene sequence (a complete open reading frame) in heterologous systems and subsequently purified to obtain large amounts for analysis. On the other hand, protein sequencing is still useful and needed as direct translation methods do not tell us very much about the actual role of the protein in a microorganism. Whether the actual protein would be produced in the microorganism under certain physiological or environmental conditions or whether it is processed proteolytically or modified post-translationally, the direct cloning and sequencing of a gene and the subsequent translation to the protein product would not provide us with such information. As such, although the magnitude of DNA sequence data, due mainly to the fervor of the microbial genome sequencing projects, far exceeds that of direct protein sequencing, scientific information that could be gleaned from direct protein sequencing is still pertinent.

5.4. Outline of a Search Strategy

Before embarking on a search, it would be prudent to devise a search strategy. This will assist the researcher in determining the steps that need to be taken and the various options available to determine or to find out some clues concerning the sequence of interest (Fig. 5.2). Most of us would like to know whether a newly obtained sequence codes for any known protein or carries a complete or partial open reading frame (ORF). If it does code for some known protein homolog, we would then proceed to find out the function of the protein or whether it belongs to a certain class of tertiary structures. Thus, using biocomputing tools to help us address

Fig. 5.2. Outline of a search strategy for a useful gene based on computational methods.

some of these queries would hopefully point us in the right direction before we begin to conduct "wet lab" experiments to confirm the findings predicted or suggested by computational analysis.

The initial step before conducting a search would be to confirm the accuracy of the DNA sequence obtained. This would require sequencing the DNA fragment of interest in both directions in order to remove any possible ambiguities as errors in the sequence may give rise to frameshifts or changes in the codon, which would then give us an incorrect protein sequence. Once the DNA sequence has been properly determined, the first step in a search strategy would be to obtain the translated protein product in six frames and if possible, ascertain the correct ORF. In the case of partial genes, only a partial ORF may be identified. After that, a search of the primary protein database would be performed, followed by searches through the secondary databases such as BLOCKS, Pfam, PRINTS and PROSITE. The secondary searches would most likely provide information on crucial motifs or functional features within the protein molecule itself. A search of the tertiary structure database would then be performed to see if any homologous structures are available. Sometimes, the information obtained from the database of secondary or tertiary protein folds might provide some clues in relation to the function of the protein. When all the relevant hits or matches have been obtained from the different databases, they would need to be compared to see whether all the returned query results are in agreement. In an ideal situation, all of the hits should point to the same protein or its homolog. Otherwise, one would need to make the best possible judgment from the results obtained. Thus, it would be wiser to use more methods or databases for the searches as no one method is infallible and no one database is absolute.

The return of a perfect match, especially from the primary protein database, would be excellent as it tells exactly the nature and the function of the protein. However, it also means that the queried sequence is not novel but has already been discovered. Thus, sometimes it is more valuable to obtain a significant hit which is not exactly matching as this would then be considered a novel discovery. Another consideration regarding sequence searches is the twilight zone, which is a zone of sequence identity less than 20% as this zone represents alignments or matches which may still appear credible or reasonable but are no longer statistically significant. Whenever this happens, further analyzes at the secondary or tertiary levels might be more revealing of its function.

The definitive test of function still needs to be determined experimentally although the ultimate aim of biocomputational analysis is to predict the function of a given protein from the primary sequence itself. Until that becomes a universal foolproof method, proper laboratory experiments are still very much required. We should not, however, discount the power of computing but instead make the most out of the applications available to us in our research.

5.5. Data Mining for Gene Hunting

Data mining involves the searching of databases for complementary or related sequences resembling novel DNA or protein sequences. It is important not to rely on specific methods but to explore different databases with different tools and then establish a consensus view or opinion from various approaches. This is because no database is complete. Using the Internet, a global network of computer networks, one can access different sites especially those in the United States of America, which are by far the most popular sites although other European and Asia-Pacific sites are available. The databases and computational programs that are freely available can be accessed using common web browsers such as Internet Explorer or Netscape Navigator. The leading bio-information provider in the United States of America is the National Center for Biotechnology Information (NCBI), which is part of the National Institutes of Health (NIH). GenBank, the DNA sequence database maintained by NIH is one of the three international providers of sequence information, together with the European Molecular Biology Laboratory (EMBL) database and the DNA Database of Japan (DBBJ). Across the Atlantic, a consortium of bio-information sites, which include specialist providers such as the Sanger Centre, the UK-MRC Human Genome Mapping Project Resource Centre and the European Bioinformatics Institute, form the European Molecular Biology network (EMBnet).

Since sequence information is exchanged on a daily basis among the GenBank, EMBL and DBBJ databases, a search within one of these databases would normally suffice. For example, sieving through GenBank necessitates the use of the Basic Local Alignment Search Tool (BLAST®) to assist us in going through a huge collection of sequences, whether we are starting with a DNA or protein sequence. BLAST® comes in different forms but is fundamentally a set of similarity search programs intended for the exploration of a particular database. With BLAST®, the search for a similar gene would be easier than finding a needle in a haystack although sometimes, one must still validate and confirm the accuracy of the sequence data through subsequent experiments. This is not to say that the sequence data is inaccurate as it is widely held that almost all of the sequences deposited in the databases are of excellent quality but nevertheless, the verification of the deposited sequences is the responsibility of the depositor, not the database provider itself.

If a search through the public databases results in a null return, it may be necessary to search through the raw or unannotated microbial genomic sequence data in the databases of the genome-sequencing center that is involved in the sequencing project. This kind of search is sometimes useful as full-length genes which are yet to be annotated may be discovered. In normal practice, the raw sequence data generated by these sequencing centers would first be annotated by their own computer biologists before depositing into either one of the 3 public sequence databases. The sequence data from most of the sequencing centers that are publicly funded or related to academic institutions would be easily accessible without charge whereas a fee is customarily required for accessing the sequence databases of commercial companies which are sequencing and exploiting genes discovered for profit.

At times it may be prudent to join an email subscription list where up-to-date scientific information can be received either on a daily or weekly basis. This is a good method to keep track of the latest information, which sometimes may be released unannounced except through email by database providers or sequencing centers. Scientific bulletin boards may also be useful for posing questions or asking for suggestions online, depending on how sensitive the ideas or findings may be. On the contrary, one may not get a response immediately if the chosen bulletin board is not particularly active. If scientific sites and databases fail in coming up with the relevant answers, one may decide, as a last resort, to search the entire web of information using popular search engines such as Google, Lycos or Altavista. There are plenty of such engines available and the choice of a particular engine is subjective as each of them relies on a different algorithm or search method. Nevertheless, as a vast number of sites and any databases therein are easily accessible within a "click" away, depending on the speed of the Internet line and the current Internet traffic, it is handy for the scientist to take advantage of the ever-increasing information available on the Internet.

5.6. Sequence Comparisons and Alignment

It is, *a priori*, important to obtain and ensure that the sequence data is correct and of high quality before embarking on any methods of comparison. An incorrect sequence would predict an incorrect "mutation" and it is prudent to ensure that an actual mutation has taken place rather than a fortuitous error in the sequence. Understanding mutations affecting a protein's function is significant, for example, in the hemoglobin A chain, a codon for glutamic acid (GAA), when altered to GUA, results in fatal sickle cell anemia in a homozygous individual. Likewise, the accuracy of sequence data is critical in phylogenetic analysis as "silent" mutations can only be detected at the DNA level but not at the protein level. Differences at the DNA level are likely to reflect more precisely the relationship distance between species when phylogenetic trees are drawn.

As mentioned earlier, DNA sequencing can now be performed with ease owing to the high-throughput application of automated fluorescent cycle sequencing. Although the problem or question of redundancy exists, the analysis of DNA sequences are necessary, to mention a few of the applications involved, for the inference of intron/exons, the locations of open reading frames and phylogenetic analysis. From the DNA sequence, it is imperative that the location of a possible coding region (an ORF) for a protein be determined (Box 5.5). With the predicted primary protein sequence, we can then compare with other known proteins in the databases and possibly formulate some functional and structural suggestions or even study the molecular phylogeny of this putative protein (either similarity or divergence).

There are numerous tools for sequence analysis freely available in the public domain and these are easily accessible online from different websites. Typically, text-based queries are used to identify relationships between a newly determined sequence and a known gene or protein

Box 5.5

ORF Analysis (Introns, Exons and ESTs)

ORF analysis attempts to determine the correct gene coding sequence for a putative protein product. This requires the identification of the right start codon, usually a methionine (ATG), and an uninterrupted DNA sequence in the correct frame (out of six possible choices, ending in a stop codon of either TAA, TAG or TGA). The central dogma of molecular biology states, in simple terms, that DNA begets mRNA by transcription, which is then translated into protein. In prokaryotic systems (like bacteria), the mRNA is directly translated into protein whereas in eukaryotes, the introns would be spliced out and the exons are then joined together before the mRNA gets translated. Sometimes the gene products could be in different lengths due to the alternative splicing of mRNA.

It is also possible to watch out for Kozak sequences (CCGCC<u>AUG</u>G) in the mRNA transcript, which may hint towards a possible starting coding sequence. Alternatively, one may compare the codon usage or detect any possible promoter or ribosome-binding sites. If a homologous protein sequence is available in the database, the likelihood of finding the correct start codon increases when a pairwise alignment or comparison is made.

Expressed sequence tags (ESTs) are short segments of DNA, which are generated from a cDNA library or mRNA from tissues or eukaryotic organisms by single-pass sequencing from either the 5' or 3' ends. They could represent coding or non-coding regions and normally consist of exon-type material. Although they are generated randomly, ESTs could be a rich source of information for finding novel genes or for identifying coding regions of a known gene that could be expressed under certain conditions. Similar to standard sequence databases, EST databases are available for searching and comparing against known ESTs generated from different organisms.

family to assess the extent of shared similarity. Some of these methods or approaches are described below.

(1) Pairwise Database Searching

In database searching, queries can either be protein, DNA or RNA. The underlying algorithm is mathematical in nature and seeks to provide the closest match between a query and the subjects available in the database. Two of the more popular programs available are FastA and BLAST®. Efficiency is necessary as databases increase rapidly and most of these searches are carried out in real-time mode. Speed is of essence although it is largely dependent on the length of the query sequence and the ever-increasing database size. Both of these programs can be run in default or "automatic" mode although more advance users may choose to modify the parameters involved. A typical output from BLAST® would show the extent of sequence identity, similarity and a statistical value (Box 5.6). Newer versions of BLAST® are able to generate gapped alignments and a hybrid of BLAST®, called Position-Specific Iterated (PSI)-BLAST®, is able to derive motifs from an initial database search.

(2) Multiple Sequence Alignment

When it is necessary to compare two or more sequences, be it protein or DNA, the use of multiple sequence alignment algorithms come into play. These algorithms are powerful tools for finding similar or even dissimilar biologically significant features. Similarities (or conventionally known as "consensus") that are extracted from the sequences are normally useful for predicting motifs and consequently, any functional characteristics and for studying the evolutionary history of these sequences. As such, a multiple sequence alignment is a basic prerequisite in order to carry out phylogenetic analysis among a group of homologous sequences.

Clustal is one of the more popular tools available and is arguably the most well known among others, and can be used with many different input formats such as FastA/Pearson, GCG/MSF and SWISS-PROT. This program can be downloaded for usage in the DOS environment but a web-based version is also available. Clustal approaches multiple sequence alignment in an evolutionary manner and the scoring of sequences are relative to each other. The output formats are normally in Clustal format (default unless specific) but the program is flexible enough to accommodate output formats such as Phylip or GCG.

Box 5.6

Example of a BLAST® Output

BLASTP 2.2.1
Reference: Altschul, Stephen F., Thomas L. Madden, Alejandro A. Schäffer, Jinghui Zhang, Zheng Zhang, Webb Miller, and David J. Lipman (1997), "Gapped BLAST and PSI-BLAST: a new generation of protein database search programs", Nucleic Acids Res. 25:3389-3402.

RID: 997544736-18339-23564

Query= (120 letters)
Database: Non-redundant SwissProt sequences 99,924 sequences; 36,332,182 total letters

```
                                                              Score   E
Sequences producing significant alignments:                  (bits)  Value
gi|121529|sp|P13006|GOX_ASPNG  GLUCOSE OXIDASE PRECURSOR (GL...229    7e-61
gi|3023881|sp|P81156|GOX_PENAG GLUCOSE OXIDASE (GLUCOSE OXY...142     6e-35
gi|3287841|sp|Q92452|GOX_TALFL GLUCOSE OXIDASE PRECURSOR (G... 130    4e-31
```

Alignments

```
>gi|121529|sp|P13006|GOX_ASPNG GLUCOSE OXIDASE PRECURSOR (GLUCOSE OXYHYDRASE)
(GOD)
BETA-D-GLUCOSE: OXYGEN 1-OXIDO-REDUCTASE)
Length = 605

Score = 229 bits (583), Expect = 7e-61
Identities = 114/120 (95%), Positives = 114/120 (95%)

Query: 1    TENPNISALVIESGSYESDRGPIIEDLNAYGDIFGSSVDHAYETVELATNNQTALIRSGN 60
            TENPNIS LVIESGSYESDRGPIIEDLNAYGDIFGSSVDHAYETVELATNNQTALIRSGN
Sbjct: 61   TENPNISVLVIESGSYESDRGPIIEDLNAYGDIFGSSVDHAYETVELATNNQTALIRSGN 120

Query: 61   GLGGSTLVNGGTWTRQHKAQVDSWETEFGNEGGNWDNVAAYSLQAERHRAPNAKQILAGH 120
            GLGGSTLVNGGTWTR HKAQVDSWET FGNEG NWDNVAAYSLQAER RAPNAKQI AGH
Sbjct: 121  GLGGSTLVNGGTWTRPHKAQVDSWETVFGNEGWNWDNVAAYSLQAERARAPNAKQIAAGH 180
```

Using a partial protein fragment (120 amino acids) which is highly homologous to the glucose oxidase from *Aspergillus niger* (a microbial source), the following output was obtained from BLASTP (protein database search). For the purpose of this example, only the relevant parts are shown. The following terms are explained as follows:

(i)   Identity — percentage of exact sequence match (either amino acid or DNA base).

(ii) Positives — percentage of identities and similarities between the query sequence and subject sequence.

(iii) Expect (E-value or p-value) — probability or expected frequency value. Denotes statistical significance. Low values indicate greater statistical significance (less by chance), which is when the E-value approaches zero, the less likelihood that the match would have occurred by chance.

Box 5.6   (*Cont'd*)

(iv) Query — query sequence entered by the user (input).
 (v) Subject — sequence in the database which is being compared to.
(vi) Length — length of the entire subject sequence in the database.

In the DOS-based Clustal program, a text file is first generated with all the sequences (typically in FastA format) needing to be analyzed. This file is the input file and it has to be entered into the program. Following that, the user can define some of the parameters involved by selecting the 'options' portion of the program. These include the type of output format, the arrangement of the sequences aligned (either according to input order or by a default output order) and whether or not the sequences are numbered. The consensus sequences (in Clustal format) obtained which are exactly identical (whether protein or DNA) will be denoted by an asterisk — "*" (Box 5.7). If shading or coloring of the sequences according to sequence identity or similarity is required, the output file (without numbering) can be entered into the BOXSHADE program, which is available online or as a DOS version that is downloadable. One drawback however, is that manual adjustment or realignment of the sequences cannot be done unless other programs are used to manipulate the outputs. Clustal is also able to generate a phylogenetic tree using the Phylip output format and provide bootstrapping values for the tree generated.

(3) Secondary Databases
Searching through secondary databases is an approach which is slanted towards recognizing motifs and patterns within a group of related sequences. Information from secondary databases is derived from primary databases and this is to ensure that we do not miss out regions which are repetitious or are part of a multi-domain sequence. Furthermore, the initial level of noise found in primary databases can be avoided. There are different secondary databases based on different methods of analysis and the diverse biological features therein.

- PROSITE is a database which consists of protein families and domains and it taps into the resources of SWISS-PROT. It primarily identifies

Box 5.7

Example of a Clustal Output

```
CLUSTAL W (1.74) multiple sequence alignment


C_ACREMONIUM       MGSVPVPVANVPRIDVSP-LFGDDKEKKLEVARAIDAASRDTGFFYAVNHGVDLPWLSRE 59
S_CLAVULIGERUS     -MPVLMPSAHVPTIDISP-LFGTDAAAKKRVAEEIHGACRGSGFFYATNHGVDVQQLQDV 58
S_LIPMANII         -MPVLMPSADVPTIDISP-LFGTDPDAKAHVARQINEACRGSGFFYASHHGIDVRRLQDV 58
S_GRISEUS          -MPIPMLPAHVPTIDISP-LSGGDADDKKRVAQEINKACRESGFFYASHHGIDVQLLKDV 58
S_JUMMONJINENSIS   -MPILMPSAEVPTIDISP-LSGDDAKAKQRVAQEINKAARGSGFFYASNHGVDVQLLQDV 58
N_LACTAMDURANS     ---MKMPSAEVPTIDVSP-LFGDDAQEKVRVGQEINKACRGSGFFYAANHGVDVQRLQDV 56
S_CATTLEYA         -MPVLMPSADVPTIDISPQLFGTDPTPRRTSRGRSTRPARGSGFFYASHHGIDVRRLQTW 59
FLAVBACTERIUM      ----MNRHADVPVIDISG-LSGNDMDVKKDIAARIDRACRGSGFFYAANHGVDLAALQKF 55
L_LACTAMGENUS      ----MNRHADVPVIDISG-LSGNDMDVKKDIAARIDRACRGSGFFYAANHGVDLAALQKF 55
A_NIDULANS         MGSVSK--ANVPKIDVSP-LFGDDQAAKMRVAQQIDAASRDTGFFYAVNHGINVQRLSQK 57
A_CHRYSOGENUM      MGSVPVPVANVPRIDVSP-LFGDDKEKKLEVARAIDAASRDTGFFYAVNHGVDLPWLSRE 59
P_CHRYSOGENUM_1    MASTPK--ANVPKIDVSP-LFGDNMEEKMKVARAIDAASRDTGFFYAVNHGVDVKRLSNK 57
P_CHRYSOGENUM_2    MASTPK--ANVPKIDVSP-LFGDNMEEKMKVARAIDAASRDTGFFYAVNHGVDVKRLSNK 57
                    *.** **:*  * *  :      :       ..* :*****.:**::: *.


C_ACREMONIUM       TNKFHMSITDEEKWQLAIRAYNKEHESQIRAGYYLPIPGKKAVESFCYLNPSFSPDHPRI 119
S_CLAVULIGERUS     VNEFHGAMTDQEKHDLAIHAYNPDNP-HVRNGYYKAVPGRKAVESFCYLNPDFGEDHPMI 117
S_LIPMANII         VNEFHRTMTDQEKHDLAIHAYNENNS-HVRNGYYMARPGRKTVESWCYLNPSFGEDHPMI 117
S_GRISEUS          VNEFHRTMTDEEKYDLAIHAYNKNNP-RTRNGYYMAVKGKKAVESWCYLNPSFSEDHPQI 117
S_JUMMONJINENSIS   VNEFHRNMSDQEKHDLAIHAYNKDNP-HVRNGYYKAIKGKKAVESFCYLNPSFSDDHPMI 117
N_LACTAMDURANS     VNEFHRTMSPQEKYDLAIHAYNKNNS-HVRNGYYMAIEGKKAVESFCYLNPSFSEDHPEI 115
S_CATTLEYA         SNES-TTMTDQRSTTWRSTRYNENNS-HVRNGYYMAPGRETVESWCYLNPSFGEDHPMM 117
FLAVBACTERIUM      TTDWHMAMSAEEKWELAIRAYNPANP-RNRNGYYMAVEGKKANESFCYLNPSFDADHATI 114
L_LACTAMGENUS      TTDWHMAMSPEEKWELAIRAYNPANP-RNRNGYYMAVEGKKANESFCYLNPSFDADHATI 114
A_NIDULANS         TKEFHMSITPEEKWDLAIRAYNKEHQDVRAGYYLSIPGKKAVESFCYLNPNFTPDHPRI 117
A_CHRYSOGENUM      TNKFHMSITDEEKWQLAIRAYNKEHESQIRAGYYLPIPGKKAVESFCYLNPSFSPDHPRI 119
P_CHRYSOGENUM_1    TREFHFSITDEEKWDLAIRAYNKEHQDQIRAGYYLSIPEKKAVESFCYLNPNFKPDHPLI 117
P_CHRYSOGENUM_2    TREFHFSITDEEKWDLAIRAYNKEHQDQIRAGYYLSIPEKKAVESFCYLNPNFKPDHPLI 117
                     .  :: :..   **  :   : * ***  .   ::: **:*****.*  **. :
```

This is the output (partially shown only) from the alignment of 13 isopenicillin N synthase (IPNS) isozymes from different microbial species obtained from GenBank. IPNS is a secondary metabolism enzyme that is involved in the penicillin and cephalosporin antibiotic biosynthetic pathway. The version of Clustal used is ClustalW 1.74 and the meaning of the symbols are as follows:

(i) "*" — positions are fully conserved.

(ii) ":" — one of the following "strong" groups is fully conserved: STA, NEQK, NHQK, NDEQ, QHRK, MILV, MILF, HY, FYW.

(iii) "." — one of the following "weaker" groups is fully conserved: CSA, ATV, SAG, STNK, STPA, SGND, SNDEQK, NDEQHK, NEQHRK, FVLIH, FYM.

biological signatures by locating a single most conserved motif or pattern. The motifs are shown as regular expression patterns.

- BLOCKS classify highly conserved regions of proteins and arrange them in a "block" fashion for comparison. These are multiple-aligned ungapped segments of sequences.
- PRINTS is a protein fingerprints' database. It consists of a group of conserved motifs and the fingerprints are determined by iterative scanning of the primary protein databases. In an analysis, more than one motif may be identified along a certain sequence, thus allowing a more thorough assessment.
- IDENTIFY is derived from BLOCKS and PRINTS. The motif obtained is displayed as a fuzzy expression pattern, which tolerates amino acid residues within a certain functional grouping.
- InterPro is another protein resource database that combines the various features of Pfam, PRINTS, PROSITE and so on. It is a useful source of information for proteomic analysis.
- ProDom is a protein domain database where the homologous domains are automatically compiled.
- Pfam runs on Hidden Markov Models and consists of multiple sequence alignments of many common domains. Families of aligned domains would then be created for analysis.
- Profile analyzes distance sequence relationships and takes into consideration sequences with very few conserved residues. The algorithm involved tolerates insertions and deletions in the sequences.

5.7. Protein Structure Prediction

One of the ultimate goals in protein structure prediction is to be able to predict the tertiary conformation from the primary sequence alone. However, due to the complexity of protein folding, one should not be over-enthusiastic about the accuracy of available programs and software packages in terms of predicting the structure of proteins. It is difficult to predict how a protein will eventually fold, if given a primary sequence, due to the numerous factors involved. Thus, structure and function cannot simply be inferred from sequence alone. Typically, there are four levels of protein structure, from the primary to the quaternary (Box 5.8) and

Box 5.8

The Four Levels of Protein Structure

- Primary structure — the linear polypeptide sequence of amino acids.

- Secondary structure — local regular regions such as α-helices, β-strands.

- Tertiary structure — packing of secondary structures, giving the overall fold.

- Quaternary structure — arrangement of several polypeptide chains.

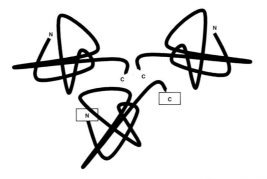

computational methods are now able to predict secondary and tertiary structures and hopefully, provide some insights into the function. Experimental methods such as NMR or 3D X-Ray crystallography are still very much the pillars of elucidating protein structures and they would be the final validation point of any predicted tertiary structure. Nevertheless, it has been possible to predict to a certain extent and to a fair degree of accuracy beyond merely a polypeptide backbone if a homologous structure is available in the structure database and is used as a starting template.

An approach to protein structure prediction should start with the primary protein sequence (Fig. 5.3). After which, it may be informative to obtain as much data on the primary structure as possible, such as disulfide bonds, active site residues and any available spectroscopic data. Searching the databases may also uncover known homologs or any functional domains. If a closely related structural homolog was available in the protein structure database, i.e. the PDB, then tertiary structure modeling would be relatively straightforward. The primary sequence would then be entered into the modeling program to obtain a predicted structure by homology. Modeling programs include the SWISS-MODEL and Modeller. The SWISS-MODEL

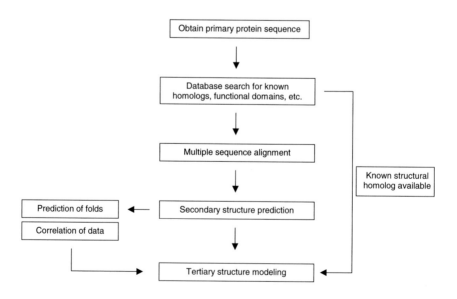

Fig. 5.3. An approach to Protein Structure Prediction.

is an automated protein-modeling program that is reasonably easy to handle and would generate a full set of atomic coordinates from a homology-modeling run.

If there are no known homologs, secondary structure predictions can then be performed although the accuracy of most programs is only approximately 50–60%. Links to most of these secondary prediction sites, which include SOPM, JPRED, PREDATOR, PHD, nnPredict and so on, are available from the ExPASy server. As an example, the PHD program, which is based on neural networks, is able to provide secondary structure predictions of a protein with locations of the α-helices and β-sheets, solvent accessibility and transmembrane helix predictions. It might also be more useful if a collection of programs is used to generate a consensus representation of all the secondary structures generated instead of relying on a single program. Analysis of folds is also another method for predicting whether or not a certain amino acid sequence can adopt a particular fold. These are carried out by fold recognition methods which will attempt to predict how known folds (available in the databases) would fit into the sequence being analyzed. Although not highly accurate and *ab initio* predictions are not possible, the analysis of folds would contribute towards predicting the overall structure. Classification of protein structures can be found in the SCOP and CATH databases.

After all the secondary structure and fold analyses have been performed, it would then be necessary to create a predicted tertiary structure. This structure would need to be refined as more information is available. Basically, homology modeling and structure prediction would be improved and made much easier if more protein structures are available for comparison. This is to ensure that the entire spectrum of structure types is covered. This would not be an easy task as comparing to the millions of sequences in GenBank, the number of protein structures in PDB is currently only approaching 20,000.

5.8. Useful Tools and Websites on the Internet for Microbiological Research

Some of the more common tools and free programs available on the Internet are listed (Table 5.1). As the information on the Internet constantly

Table 5.1. List of useful tools and websites on the Internet.

Description of sites	Name of website	Website address
Resource site for Molecular Biology information, e.g. Genbank, PubMed	National Center for Biotechnology Information	http://www.ncbi.nlm.nih.gov
International repository for the processing and distribution of 3-D macromolecular structure data	Protein Databank (PDB)	http://www.rcsb.org/pdb/ index.html
European Molecular Biology Network	EMBnet	http://www.embnet.org
European Molecular Biology Laboratories	EMBL	http://www.embl-heidelberg. de
DNA Database of Japan	DBBJ	http://www.ddbj.nig.ac.jp
European Bioinformatics Institute	EBI	http://www.ebi.ac.uk/ ebi_home.html
Genome Research Centre founded by the Wellcome Trust and MRC	Sanger Centre	http://www.sanger.ac.uk
Protein Information Resource	PIR	http://pir.georgetown.edu
Annotated protein sequence database	SWISS-PROT	http://www.expasy.ch/sprot/ sprot-top.html
Homologous Structure Alignment Database	HOMSTRAD	http://www-cryst.bioc.cam. ac.uk~homstrad/
Composite database of SWISS-PROT, PIR, GenBank and NRL-3D.	OWL	http://www.bioinf.man.ac.uk/ dbbrowser/OWL
Database of protein families and domains	PROSITE	http://www.expasy.ch/prosite
Protein fingerprints database	PRINTS	http://www.bioinf.man.ac.uk/ dbbrowser/PRINTS
Database for aligned ungapped segments of the most highly conserved regions of proteins	BLOCKS	http://www.blocks.fhcrc.org
Database of multiple sequence alignments and HMM of common protein domains	Pfam	http://www.sanger.ac.uk/ Software/Pfam
Searching both BLOCKS and PRINTS	IDENTIFY	http://dna.Stanford.edu/ identify
Profile scan server for searching against a protein database	Profiles	http://www.isrec.isb-sib.ch/ software/FSCAN_form. html

Table 5.1 (*Continued*)

Description of sites	Name of website	Website address
Structural classifications of proteins database	SCOP	http://scop.mrc-lmb.cam.ac.uk/scop
Protein structure classification	CATH	http://www.biochem.ucl.ac.uk/bsm/cath_new/
Motif search utility of GenomeNet, Kyoto University, Japan	Protein and Nucleic Acid Sequence Motifs search	http://motif.genome.ad.jp/
Restriction enzyme analysis of DNA sequences	Webcutter	http://www.firstmarket.com/cutter/cut2.html
Restriction enzyme database	Rebase	http://rebase.neb.com/rebase/rebase.html
Analysis of DNA and protein sequences and structures (many tools and links available)	ExPASy Molecular Biology Server	http://www.expasy.ch/
Analysis of protein sequences	ExPASy Proteomics tools	http://tw.expasy.org/tools/
BLAST® server (all types available) at the National Center for Biotechnology Information	NCBI BLAST®	http://www.ncbi.nlm.nih.gov/BLAST/
Sequence search analysis	FastA server	http://www.ebi.ac.uk/fasta33/
Frame analysis of protein-coding region of high G+C content	FramePlot 2.3	http://www.nih.go.jp/~jun/cgi-bin/frameplot.pl
Promoter Prediction	Neural Network Promoter Prediction	http://www.fruitfly.org/seq_tools/promoter.html
Molecular Analysis Tools at the University of Adelaide	A Pack of Molecular Analysis Tools	http://www.microbiology.adelaide.edu.au/learn/index.htm
Pretty Printing and Shading of Multiple-Alignment files	BOXSHADE 3.21	http://www.ch.embnet.org/software/BOX_form.html
Online version of Clustal, a multiple sequence alignment program	Clustal	http://www.ebi.ac.uk/clustalw
Automated comparative protein modeling server	SWISS-MODEL	http://www.expasy.org/swissmod/SWISS-MODEL.html
Sequence utilities	BCM Search Launcher: sequence utilities	http://searchlauncher.bcm.tmc.edu/seq-util/seq-util.html
Color Interactive Editor for Multiple Alignments	CINEMA	http://www.bioinf.man.ac.uk/dbbrowser/CINEMA2.1

Table 5.1 (*Continued*)

Description of sites	Name of website	Website address
A collection of links to information and services useful to Molecular Biologists	Pedro's BioMolecular Research Tools	http://www.public.iastate.edu/~pedro/research_tools.html
ProDom with BLAST® (graphical output)	ProDom	http://prodes.toulouse.inra.fr/prodom/doc/blast_form.html
Codon usage database	Countcodon	http://www.kazusa.or.jp/codon/countcodon.html
Collection of protein secondary structure analysis and information sites	GenomeWeb Protein Secondary Structure	http://www.hgmp.mrc.ac.uk/GenomeWeb/prot-2-struct.html
Resource site for biologists	BioMedNet	http://www.bmn.com/
Resources for the microbiologist	The Microbiology Network	http://microbiol.org/
Culture Collection	American Type Culture Collection	http://www.atcc.org/
Culture Collection	Agricultural Research Service Culture Collection	http://nrrl.ncaur.usda.gov/default.htm
Comprehensive Science-Specific Search Engine	Scirus	http://www.scirus.com/
Google	Internet Search Engine	http://www.google.com
Lycos	Internet Search Engine	http://www.lycos.com
Altavista	Internet Search Engine	http://www.altavista.com

changes and web pages are frequently replaced, removed or updated, one needs to check whether these sites are still functional. One would not believe the "speed" of the Internet as a lot of changes could take place even within a few weeks of an undergraduate year, let alone the entire course. It would be useful to maintain a list of "Favorites" or "Bookmarks" (depending on your default browser) where sites which are most frequently used can be easily accessible and monitored.

5.9. Summary

From the success of sequencing the first-ever genome, the bacteriophage ϕX174, microbial genome sequencing efforts never looked back. The scientific endeavors of cloning genes, currently in bounteous amounts and the completion of many more microbial genomes have created new fields of research such as comparative genome analysis, functional genomics, proteomics and structural genomics. These are now becoming the main players in the post-genomic age. With heightening demands of increased biological information, bioinformatics will continue to play an essential role in assimilating this massive and ever-increasing amount of information. As bioinformatics is a rapidly developing field and computing processing power can only increase (according to Moore's Law), so will the sophistication and complexity of the biocomputational software programs. Both biological research and computing are intertwined and will continue to be mutually reliant as science continues to advance.

Further Readings

1. Burge CG and Karlin S. Finding the genes in genomic DNA. *Curr. Opin. Struct. Biol.* **8**: 346–354, 1998.
2. Glick BR and Pasternak JJ, eds. *Molecular Biotechnology: Principles and Applications of Recombinant DNA*. ASM Press, USA, 2003.
3. Higgins D and Taylor W. Multiple sequence alignment. *Methods Mol. Biol.* **143**: 1–18, 2000.
4. Higgins D and Taylor W, eds., *Bioinformatics: Sequence, Structure and Databanks*. Oxford University Press, UK, 2002.
5. Holm L and Sander C. Mapping the protein universe. *Science* **273**: 595–603, 1996.
6. Koel P. Protein structure similarities. *Curr. Opin. Struct. Biol.* **11**: 348–353, 2001.
7. Orengo CA, Jones DT and Thornton JM, eds., *Bioinformatics: Genes, Proteins and Computers*. BIOS Scientific Publishers Ltd, UK, 2003.
8. Westhead DR, Parish JH and Twyman RM, eds. *Instant Notes Bioinformatics*. BIOS Scientific Publishers Ltd, UK, 2002.

Chapter 6

Real Time Polymerase Chain Reaction

Heng-Phon Too[*,†] **and Azlinda Anwar**[‡]
*Department of Biochemistry, National University of Singapore
8 Medical Drive, Singapore 117597
†MEBCS, Singapore-Massachusetts Institute of
Technology Alliance
‡Division of Biomedical Sciences, Johns Hopkins Singapore
31, Biopolis Way, #02-01, The Nanos, Singapore 138669

Polymerase Chain Reaction (PCR) is increasingly becoming an essential tool for quantifying nucleic acids in many clinical and biological fields of studies. A recent innovation in PCR is the development of a state-of-the art real time detection method. Since the first report of real time detection of a PCR process by Higuchi and co-workers, several thousand publications have appeared, attesting to the acceptance of this technology into the mainstream of many research and clinical disciplines (Fig. 6.1). Real time PCR (also known as kinetic PCR or quantitative real time PCR) has been increasingly employed in diagnostic microbiology, virology and parasitology. It is also used in the field of hematology, oncology and immunology to aid, among others, in the detection of aberrant gene expressions, changes in chemokines levels and genotyping. In the field of biotechnology, real time PCR has been used to genotype genetically modified organisms, monitor gene expression in bioprocesses and for evaluating the safety of biologicals. As a research tool, this technology has been employed in validating data obtained from high-throughput transcriptional profiling using DNA microarray.

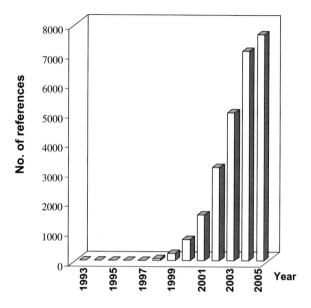

Fig. 6.1. Number of publications (including patents) with the keywords "Real time PCR" as reported using SciFinder Scholar 2004.

The importance and wide use of real time PCR as a quantitative tool in the life sciences is due to the inherent wide dynamic range of quantification (10^7–10^9 fold), high assay sensitivity (detection of less than 10 copies), high precision (<2% intra-assay variation, <3% inter-assay), with no post-PCR manipulations, thus minimizing the risk of cross-contaminations. In addition, a distinct advantage of this technology is the ease in obtaining reliable quantitative definitions of specificity, sensitivity and efficiency of an assay as compared to the conventional end-point PCR-based assays. Thus in view of the high accuracy of measurement and reliability of real time PCR, this technology will continue to be an essential tool for the quantification of gene expressions and genotyping.

6.1. What is Real Time PCR?

The PCR process can be visualized as a cycle of three stages: denaturation, annealing and extension (see Chapter 4). During this process, sections of a double-stranded (ds) DNA template are first separated at high temperatures (>90°C). On cooling (typically between 50–70°C), the primers anneal onto the desired sections of the template, and extended by a DNA polymerase at optimal temperatures (typically between 68–75°C). Traditionally, on completion of a PCR experiment, the products are fractionated based on their sizes and detected either using DNA binding dyes (usually ethidium bromide) or by more elaborate methods, for example Southern blot-based methodologies. This end-point, conventional approach however suffers from many technical limitations, including the need for post-PCR manipulations, which increases handling time and may potentially introduce experimental errors. Several attempts have been made to develop quantitative conventional-based PCR including the use of serial dilutions of cDNA, altering the amplification cycle for each gene, using exogenous competitor templates, performing limiting dilution assays or using PCR mimics with similar primer sequences that would co-amplify with the products. All these methods however require extensive post-PCR handling and analyses. In addition, quantification of the amounts of the genes of interests is neither accurate nor reliable when determined by measuring the amounts of products made at the end point. In the initial cycles, PCR generates copies of a DNA template exponentially but this process will however not continue indefinitely. Inhibitions due to substrate depletion, loss of enzyme activity and accumulation of amplicons which compete for primer binding and other products (pyrophosphates) may be responsible for the attenuation of the exponential increase of the amplified products, resulting in the plateau phase of PCR (Fig. 6.2a). Thus, regardless of the varied amounts of starting templates, similar amounts of products may be observed at the plateau phase.

In real time PCR, the accumulation of amplicons is detected and quantified periodically by fluorescence at every cycle. Since there is no requirement to isolate the probe-target hybrids from the non-hybrid probes for analysis, these assays are often referred to as "closed" or homogenous systems. A typical amplification profile of relative fluorescence unit (RFU) against the cycle number seen in real time PCR is shown in Fig. 6.2a. In the initial cycles of PCR, the fluorescent signal emitting from the sample is low and is considered to be background signal, defining the baseline for the amplification of the sample. As more amplicons accumulate in the later cycles, the signal gradually increases above the baseline. Since the

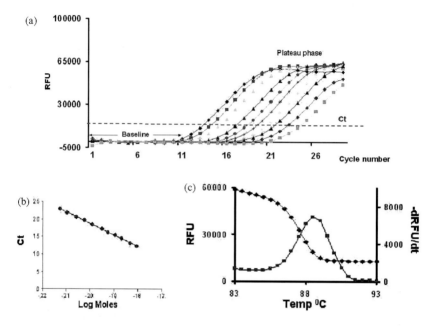

Fig. 6.2. (a) A SYBR Green I real-time PCR profile of background subtracted relative fluorescence unit (RFU) against the number of amplification cycles of two-fold serially diluted DNA template. (b) A plot of the Ct against the amount of initial template yields a straight line, with the amplification efficiency calculated from the slope. (c) Post-PCR melt curve analysis.

intensity of the fluorescence signal is directly correlated to the concentration of amplicons, a fixed fluorescence threshold set above the baseline would therefore indicate an equal amount of amplicons in different samples. This threshold parameter is often referred to as Ct (threshold cycle) or Cp (cross-over point), and is calculated by the software of the instrument as the cycle number at which the fluorescence is significantly above a defined threshold. Most instruments calculate the average background fluorescence gathered from the first few cycles and use the readings to define statistically significant threshold signals. A linear relationship exists between the amounts of initial templates and the Ct values. Thus, a plot of the log of initial template amounts for a set of dilutions against Ct is a straight line (Fig. 6.2b). The quantification of an unknown sample is therefore carried out by comparing its Ct value against the Ct values of known amounts of serially diluted standards by interpolation.

6.2. Instrumentation

A number of real time PCR instruments with varying capabilities and costs are available commercially. With the recent surge in interest for the quantification of gene expressions and the availability of economically affordable instruments, it is now not uncommon to find these instruments in individual laboratories. The demand for such instruments will continue and differences in costs, performances and features between instruments will gradually diminish.

Currently, real time PCR detection is by fluorescence hence all instruments share some common features: a module dedicated to the amplification process where the heating and cooling processes occur; a light source to excite the fluorophores; a detector to capture the emission of the fluorophores and a data analysis module where the signal is processed and displayed. The instruments either have stationary detectors with stationary sample modules (e.g. iCycler iQ, ABI 7500), rotating samples with stationary detectors (e.g. Lightcycler® 2, Rotor-gene 3000™) or mobile detectors with stationary samples (e.g. Chromo4 Opticon, MX 3005P™). Many instruments control the heating and cooling process by employing either a solid state-Peltier or a hybrid of Peltier, resistive and convective technologies, or resistive heating plates. The reaction vessels used are either polypropylene or polyethylene polymer molded into microtubes or multiwell plates, or thin glass capillaries. Most of the solid-state Peltier instruments use a 96-well plate format, while the higher throughput, automated instruments using a 384-well plate format (e.g. ABI 7900HT, Lightcycler® 480). A useful feature of some (e.g. iCycler iQ), but not all real time PCR instruments, is the ability to perform temperature gradient real time detection, which allows for the rapid optimization of annealing temperatures of many reactions simultaneously. In addition, many instruments have between two to six detection channels to enable multiplex studies. However, for many studies (singleplex reaction studies), the need of only one or two channels is sufficient, hence newer and more economical instruments are being produced to meet this demand (e.g. My iQ, MiniOpticon).

Another useful feature of most real time PCR instruments is the ability to perform melt curve analysis post-PCR (Fig. 6.2c). This feature not only allows the determination of possible contributions of extraneous DNA fragments to the observed signals, but also allows for the discrimination between multiple amplicons. In well-optimized PCR conditions, a melt curve can also be used to quantify and distinguish between highly related amplicons, and to detect single nucleotide polymorphisms.

6.3. How DNA is Detected in Real Time PCR

Currently, a number of different experimental formats are available for the detection of amplicons by real time PCR (Fig. 6.3). For simplicity, these can be classified into either "sequence-independent" or "sequence-dependent" methods.

Fig. 6.3. Overview of the quantification strategies in real time PCR.

6.3.1. *Sequence-independent method*

In the sequence-independent method, this approach uses fluorescent molecules that bind to DNA and exhibit little or no sequence preferences (Fig. 6.4a). Hence, detection is independent of the amplicon sequences. The binding of multiple DNA-binding fluorophores to a single amplicon will increase the sensitivity of detection. Hence, the intensity of the signal is dependent on the mass of the DNA. This widely used method is cost-effective, simple to design and easily adapted to established conventional PCR protocols. However a major disadvantage of this method is the inability of these DNA-binding fluorophores to discriminate between specific and non-specific amplicons, including primer-dimers, which may be generated during the reaction. However, in a well-designed and optimized PCR, specific amplicons will be produced in early cycles and primer-dimers, which are generated in later cycles, can be distinguished by using

Fig. 6.4. Sequence-independent DNA-binding dyes. (a) Binding of dyes (⊙) onto ds DNA results in an enhancement of fluorescence (✳) at the appropriate excitation wavelength (Ex) and emits energy at a lower wavelength (Em). (b–d) Structures of some commonly used DNA-binding dyes.

proper controls and by using melt-curve analysis. Although it is possible to reduce the contribution of signal from primer-dimers to the overall signal by detecting the fluorescence at temperatures higher than the melting temperature of the primer-dimer, the formation of the amplicons will be reduced. Another limitation of this sequence-independent method is its inability to be used in multiplex PCR reactions, which require simultaneous discrimination of multiple amplicons.

Currently, the most commonly used DNA-binding fluorophore is SYBR Green I, an asymmetric cyanine dye (Fig. 6.4b). It has several advantages over ethidium bromide (Fig. 6.4c) in that SYBR Green I is less toxic and exhibits a low intrinsic fluorescence, which is increased by 800–1000 fold on binding to ds DNA. A number of other DNA-binding fluorophores, including TOTO-1, YOYO-1, SYBR Gold, Picogreen, LC Green and BEBO (Fig. 6.4d), have been successfully used in real time PCR. Although these dyes are generally thought to be non-sequences specific, recent studies however have shown that some of these dyes may exhibit some sequence specificity, which can therefore affect the accuracy of discriminating between amplicons using melt curve analysis.

6.3.2. *Sequence-dependent method*

In this approach, the identity of the desired amplicon is interrogated with a probe, which is a short specific complementary oligonucleotide labeled with one or more fluorescent dye(s). Unlike the sequence-independent fluorophores, the signal generated here is independent of the size of the amplicon. Only one molecule of the probe is hybridized to every single amplicon. With this method, it is therefore possible to perform multiplex real time PCR to detect multiple targeted amplicons using distinct probes labeled with different dyes. An advantage of multiplexing is the significant reduction in the requirement for samples, labor and time. However, multiplexing is not as straightforward as singleplex assays as substrate limitations, spectra overlaps of probes, probe-probe and probe-primer interactions, and product inhibitions can pose significant challenges.

The recent progress in oligonucleotide synthesis and the availability of numerous polyaromatic quenchers and fluorescent labeled nucleobases has resulted in the development of a plethora of linear and conformationally constrained fluorescent probes for sequence-dependent methods. With some of these probes (e.g. the Molecular Beacon probes; Figs. 6.7a and 6.7b), the closed (quenched) form has the reporter (donor) and quencher (acceptor) fluorophores in close proximity to each other in space, while the open form has these fluorophores spatially separated. The fluorescence enhancement in the open form directly measures the interaction of the probe with its target. In another format (e.g. hybridization probes; Fig. 6.5), the donor and acceptor are brought into close proximity on hybridization to the target and the emission of the acceptor measures the interaction of the probe with the target. The change of fluorescence wavelength during the close spatial proximity of these fluorophores is due to the transfer of energy between the reporter and the quencher, a process commonly described as fluorescence resonance energy transfer (FRET). In addition, depending on the mode of interaction between these dyes, part of a fluorophore labeled-DNA probe can also participate in quenching of the fluorescence.

The quenching of fluorescence can result from a number of molecular interactions including excited-state reactions, molecular rearrangements, energy transfer, static (contact) and collisional (dynamic) quenching.

Fig. 6.5. Hybridization probe. In the adjacent probe format, excitation (Ex) of the hybridized probe enables the donor (D) to transfer energy to the acceptor (A), with the emitted energy (Em) detected at the appropriate wavelength. The donor/acceptor can either be on the same strand (a) or on the opposite strands (b).

A dynamic quenching is the result of the collisional encounters between the quencher and donor dyes in the excited state. The donor returns to the non-excited ground state without the emission of a photon. FRET (also known as Förster mechanism) is a dynamic quenching process that is used to explain the singlet-singlet transitions involved in the vast majority of donor-quencher pairs. The reporter and quencher fluorophores are physically in close proximity within a range of 20–100 Å (the Förster radius). With this mechanism of quenching, the reporter and quencher can be matched according to their spectral characteristics, i.e. they are chosen so that the emission spectra of the reporter and absorption spectra of the quencher overlap. In static (contact) quenching, a non-fluorescent dimer complex is formed between the reporter and the quencher at the ground-state. Here, the reporter and acceptor fluorophores are brought into close contact at distances less than the Förster radius. Such static quenching can be a dominant mechanism with certain reporter-quencher pairs in dual-labeled oligonucleotides, producing non-fluorescent heterodimers with their own distinct absorption spectra. In this case, pairs of donor-acceptor can also be chosen such that their absorption bands overlap in order to promote a resonance dipole-dipole interaction mechanism within a ground-state complex. Unlike FRET, fluorescence-quenched probes based on static quenching mechanisms, is less restricted to the requirements of the donor and acceptor to have an extensive emission-absorption spectral overlap. This therefore reduces the constraints on fluorescence-quenched probe design and expands on the ease of construction and the breadth of their applications. Existing assays employing FRET-based probes might significantly be improved by the use of these fluorescence-quenched probes.

Dynamic quenching is generally applicable to linear probe-based assays (e.g. adjacent hybridization probe, TaqMan), while static quenching appears to be more relevant to conformational probe-based (e.g. Molecular Beacons) assays. A number of organic molecules have successfully been utilized as quenchers, including fluorescent (e.g. TAMRA) and non-fluorescent compounds (e.g. Dabcyl, methyl red, BHQ series, QSY series, Disperse Blue, Iowa Black, Epoch Eclipse Quenchers). New organic molecules with wider absorbance spectrum will continue to be explored as the next generation of quenchers.

6.3.2.1. *Linear probes*

These probes are thought to adopt random coil conformations in solution. On hybridization to their targets, changes in fluorescence of these probes are detected as signals. Depending on the format, the probes can be reused at every cycle (e.g. Biprobe, Magiprobe, Induced FRET [iFRET] probe, Hybeacon probe, Light-up probe, Eclipse-MGB probe, Adjacent hybridization probe, LNA probes), degraded by DNA polymerase during PCR cycling (e.g. Double-Flap probe, Taqman probe, Taqman-MGB probe), or incorporated into the amplicons (e.g. G-quenching linear probe).

An example of a reusable linear probe is the adjacent hybridization probe which is the format often used in the Light Cycler instrument (Roche Applied Science) for sequence-dependent method of probing (Fig. 6.5). The principle underlying the adjacent hybridization probe format is first described by Heller & Morrison in 1985. These probes are easier to design and synthesize when compared to some conformationally restricted probes (e.g. Molecular Beacons). The assay uses two labeled probes hybridizing adjacent to each other on the target sequence (Fig. 6.5a). One probe is 5'-labeled with an acceptor fluorophore and the other at the 3'-end with a donor fluorophore. In solution, the two probes are spatially apart and the acceptor does not fluoresce. However, FRET occurs when the two probes are adjacent to one another when hybridized to the target sequence. Typically, the two probes are separated by one or more intervening nucleotides on the target sequence. FRET can still occur even when the two probes are separated by as many as 17 nucleotide bases, however with diminished fluorescence intensities. The probes can be designed to hybridize to either strands of the target (Fig. 6.5b), with the inverted position of a donor-acceptor pair (3' acceptor and a 5' donor) being equally efficient in the FRET process. In order to prevent the probes from acting as primers, the 3' end of the oligonucleotide can be modified with a phosphate, amine or biotin moiety to prevent oligonucleotide extension. An advantage of this probe format is that the concentration of the probe does not decrease, enabling the use of post-PCR melt curve for amplicon analysis.

In contrast, the commonly known Taqman probe assay uses the principle of probe degradation to generate fluorescent signal, and hence are not reusable in every PCR cycling (Fig. 6.6a). This format is an adaptation of

Fig. 6.6. Taqman probe format. (a) The reporter (R) is quenched by the rquencher (Q) in solution, and the eventual hydrolysis of R by the 5' nuclease activity of the polymerase results in fluorescence enhancement. (b) Structure of a 5' FAM-TCATTACAATA-TAMRA-3' Taqman probe.

the initial work of Holland and co-workers, where the probe is 5' terminally or internally labeled with a reporter fluorophore and the 3' terminus labeled with a quencher (Fig. 6.6b). The unbound probe adopts a random coil structure with the reporter and quencher thought to diffuse into a Förster radius by thermal motion, resulting in the quenching of the reporter fluorescence. Strand displacement of the probe occurs and the probe, which has annealed to the target, and the reporter fluorophore are subsequently cleaved off by the 5' nuclease activity of the DNA polymerase. The excitation of the cleaved reporter fluorophore thus results in fluorescense enhancement, which is directly proportional to the amount of specific amplified products. Hence, this probe format is known as either the 5' Nuclease, Hydrolysis or Taqman probe format. The 3' terminus of the probe is modified to prevent the probe from extension during PCR. As the probe is degraded with every cycle, the effective concentration of the probe decreases gradually and hence is not available for post-PCR melt curve analysis.

The Taqman probe is designed to have an annealing temperature 5–10°C higher than the amplification primers to ensure probe hybridization and hydrolysis at the extension phase. The optimal distance between reporter and quencher fluorophores appears to be 6–10 bp apart and they can be placed internally when long oligonucleotide probes are used. The extent of hydrolysis and hence the accuracy of quantification is dependent on a number of factors, including the position of the amplification primer with respect to the probe, the sequence of the 5' nucleobase on which the reporter is attached to, and the 5' nuclease activity of the DNA polymerase.

In instances where there is a need to detect similar targets of interest amplified using the same amplification primers, instead of producing individual probes for each target with different reporter-quencher pairs, a Universal Taqman probe could be designed to hybridize to a universal non-target sequence which is tagged onto the 5' end of one of the primers. Hydrolysis of the probe hybridizing onto the tag sequence thus reports the presence of these amplicons. The specificity of this approach however lies solely on the primer as the probe only detects the tag, which is incorporated into the amplicon. Hence, mispriming will result in the generation of false signals.

A recent extension of the Taqman probe assay is the development of shorter but more specific minor groove binding (MGB) probes. These probes have a 5' reporter (fluorescein) and 3' quencher (TAMRA) fluorophores conjugated to the minor groove binding moiety, dihydrocyclopyrroloindole tripeptide. Unlike Taqman probes, the MGB probes are shorter (10–15 bases) and have been shown to fold into the minor groove of ds DNA at the terminal 5–6 base pairs. A modification of this is used by the MGB-Eclipse probe, where the quencher (Eclipse dark quencher) is conjugated to the 5' end of an oligonucleotide, and the reporter at the 3' end. Similarly, LNA (Locked Nucleic Acid) probes are shorter than Taqman probes and can achieve high specificity of detection. These MGB Eclipse and LNA probes are effectively insensitive to the 5' nuclease activity of the DNA polymerase.

The underlying design of another linear probe format, the G-quenching probe, is based on previous observations that the fluorescence of a linear single fluorophore-labeled oligonucleotide can be quenched on hybridization to an unmodified complementary strand containing guanine (G) nucleobases at particular positions. These linear dequenching probes show a decrease in fluorescence on hybridization to the targets. A recent interesting modification of this approach is the incorporation of modified G (isoG) for the quenching of a fluorescent probe during PCR (Plextor^{TM} Technology).

6.3.2.2. Conformationally-restricted probes

These types of probes have some degree of secondary structure in solution and, on hybridizing to the target, changes in the structures result in fluorescence enhancements. Some of these probes can either be incorporated into the amplicons (e.g. Scorpion primer-probe, Amplifluor probe, dsDNA hybridization primer-probes, LUX primer-probe, Q-PNA primerprobes), or reused in subsequent PCR cycles (e.g. Molecular Beacon probe).

An example of a conformationally-restricted probe is the Molecular Beacon probe, which is a dual-labeled oligonucleotide with a fluorescent reporter at one end and a quencher at the other (Fig. 6.7a). The loop structure contains a sequence complementary to the target of interest and the ends of the probe contain a short self-complementary sequence (5–7 bases). In the absence of the target, the short self-complementary sequence form a stem-loop hairpin structure in a unimolecular reaction that serves to position the reporter in close proximity to the quencher, resulting in the quenching of the reporter. However, in the presence of the target, the sequence in the loop structure hybridizes with the complementary target sequences in a bimolecular reaction, enabling the molecule to unfold into an open form. The reporter and quencher are then physically separated and the reporter fluoresces on excitation. The fluorescence quenching in the Molecular Beacon probes is not strictly FRET-based when dark quenchers are used (e.g. Dabcyl or BHQ), as the reporter-quencher are physically closer than the Föster radius. The increase in fluorescence intensity upon hybridization to the target can vary widely, ranging from less than 10 fold to greater than 200 fold. A modification of the Molecular Beacon probe, called the wavelength-shifting Molecular Beacon probes, has been reported where a third fluorophore (a harvester) is conjugated

Fig. 6.7. (a) Molecular Beacons, (b) Wavelength-shifted Molecular Beacons with a harvester dye (H) and (c) Amplifluor primer-probes.

onto the reporter to give a substantially brighter fluorescence compared to the conventional probe (Fig. 6.7b). Yet another modification is the development of a Taqman-Beacon probe, where the 5' nuclease activity of the DNA polymerase is used to degrade the Molecular Beacon moiety. A variation of the Molecular Beacon probe format is the Cyclicon, which is a complex fluorescent-tagged pseudocyclic oligonucleotide.

The Amplifluor probe (previously known as the Sunrise probe) is another form of a conformationally-constrained probe (Fig. 6.7c). In this format, the amplification primer contains a hairpin structure with the reporter and quencher fluorophores conjugated on the stem. When this primer-probe is incorporated into the PCR product, the hairpin structure is unfolded and a fluorescent signal can be detected. Hence, this is a variation of the Molecular Beacon with an extended sequence acting as a primer. A disadvantage of this format is that mispriming of the primer-probe will result in the generation of non-specific signals. In order to increase the flexibility of this format, a universal sequence can be incorporated into the amplification primer and used for detection using Amplifuor-Uniprimer probes.

6.4. Quantification in Real Time PCR

Depending on the objective of a study, real time PCR quantification can be carried out by two approaches: a "relative quantification method" and an "absolute quantification method" (Fig. 6.3). The latter method implies that the absolute copy number of the unknown is determined. However, this is a misnomer as the quantity of the unknown sample is determined by interpolation from a set of predefined standards. Hence, the latter approach should appropriately be termed as "comparative standard curve method" or "standard curve method." The standards used in real time PCR can be in the form of a cDNA, genomic DNA or plasmid DNA, and are assumed to have identical properties to the unknown samples. The accuracy of the results when using this method depends solely on the accuracy of the measurements of the standards. Thus, a highly reproducible quantification of the unknown implies a precise but not necessarily an accurate "absolute" result.

The relative quantification method is often used to measure changes of a target gene when examined in two different conditions. Quantification is carried out relative to the control gene(s) by subtracting the difference in the Ct values of the control and target genes. The resulting difference in cycle number (ΔCt) is the exponent of the base 2 (due to the doubling function of PCR), which represents the fold difference between these two genes. There are some assumptions underlying this method, which includes the choice of a control gene that is assumed to have a constant expression level under the experimental condition, that the amplification efficiencies of both the control and target genes are assumed to be similar, and that the expression levels of the control and the target genes are assumed to be similar. If all these assumptions can be satisfied, an advantage of this method is that a large number of genes can be surveyed simultaneously and, if required, further characterization can be conducted with the standard curve method.

6.5. Data Analyses

The analytical sensitivity, efficiency and specificity of a real time PCR assay can be accurately determined. The analytical sensitivity or the detection limit of an assay is defined by the amplification and detection of the highest dilution of the target. It is possible to detect a single copy of a target but the quantitative accuracy decreases as the amount of template copy number approach unity. This is partly due to technical and methodological inaccuracy during sample preparation, amplification and detection, and to sampling errors arising from stochastic distribution of low-copy-number template molecules.

The efficiency of the amplification process can be calculated from a plot of template dilution versus their respective Ct values. At the exponential phase of PCR, the accumulation of product is related to the initial amount of starting target by the following equation,

$$N_n = N_0 (1 + \varepsilon)^n \qquad (6.1)$$

where N_n is the number of amplicons at cycle n, N_0 is the initial number of target molecules, ε is the efficiency of amplification ($\varepsilon \leq 1$) and n is the number of cycles needed to produce N_n. Thus, a 100% efficient amplification process will produce a doubling in the number of amplicons (N_n). Note that the amount of N_n present at any specific number of cycle (n) is dependent on N_0.

Taking the log of equation (6.1):

$$\log N_n = \log (N_0) + n \log (1 + \varepsilon). \qquad (6.2)$$

Rearranging equation (6.2):

$$n = 1/\log (1 + \varepsilon) \cdot (-\log N_0) + \log N_n / \log (1 + \varepsilon). \qquad (6.3)$$

Thus, the equation shows a linear relationship between n and $-\log N_0$. At the threshold cycle (Ct), the amount of amplicon produced is constant (as measured by the same fluorescent intensity) within the experiment. Hence, the efficiency of amplification (ε) of a target molecule can be calculated from the slope of the standard curve (a plot of Ct versus the negative log of the initial concentration of the template; Fig. 6.2b) where $\varepsilon = [10^{(-1/\text{slope})}] - 1$. Amplification of targets with high efficiencies has slopes approaching the value of 3.32 cycles for every 10-fold dilution of the target.

The accurate determination of the Ct values of an experiment is critical and relies on the identification of the exponential phases of the signal curves. Attempts have been made to model real time PCR data by sigmoidal functions and other mathematical models so as to reduce errors in the determination of Ct values. This has allowed the development of computational algorithms to automatically calculate the amount of a target, a necessary step in high-throughput real time PCR. The original method of relative quantification (fold change = $2^{-\Delta\Delta Ct}$) by Livak and Schmittgen, was based upon the assumption that the efficiencies of amplification of the target and the reference must be approximately equal.

In order to compare the specificity of any assay, it is critical to compare the differences in the Ct values of the target to the test templates with similar but not identical sequences (ΔCt), as well as the efficiencies of the amplifications. Hence, specificity (σ) can be defined by the equation: $\sigma = (1+\varepsilon)^{\Delta Ct}$, where $(1+\varepsilon)$ is $10^{1/slope}$. Here, the efficiencies of amplifying the test templates are assumed to be equal to or less than that of the target. This assumption is generally true as the primers and conditions are optimized for the amplification of target and not the test templates. Hence, the larger the σ values, the more specific the assays are in discriminating the target over test templates.

6.6. Normalization of Data

A number of variables can confound the accurate analyses of gene expressions and these include sample purity and integrity, reverse-transcription efficiencies and differences in transcriptional activities. Ideally, extracted high-quality RNA can be standardized to the number of cells. However, the accurate determination of cell number is often not feasible, as in the case with solid tissues. A frequently applied normalization approach is the use of the mass of RNA. A technical drawback to the use of this approach arises when the availability of nucleic acids is limiting. A strong argument against the use of total RNA mass for normalization is that total RNA is predominantly represented by the ribosomal RNA (rRNA), and that the assumption that the ratio of rRNA to mRNA is invariant is not always true, nor is the transcription of rRNA constant under different conditions. In addition, both 18S and 28S rRNAs are not present in purified mRNA samples. For a reverse transcription reaction, an alternative to the use of input total RNA as a normalizer is to use picogreen quantified cDNA, which is more accurate.

The use of endogenous (internal) and/or exogenous (external) genes for normalization of gene expression studies has been a subject of intensive studies. For any endogenous gene (often referred to as housekeeping genes) to be considered as a control gene, it must not vary in the tissues or cells under investigation, or in response to experimental treatment. Results from various studies suggest that the choice of suitable controls can be challenging and the ideal and universal normalization genes may not exist. Hence, the use of any control genes under any experimental conditions must be accurately reported and, if possible, standardized to allow reproducibility between laboratories. Practically, it is better to use more than one gene for normalization. If the availability of RNA is not limiting, a geometric averaging of multiple internal control genes can increase the accuracy of normalization.

6.7. Melt Curve Analyses

At the end of real time PCR, a melt-curve analysis of the products is a useful feature offered by most real time PCR instruments. A melt curve measures the changes of fluorescence intensity of a sample when heated (Fig. 6.2c). When the temperature is increased, the fluorescent signal decreases gradually as a result of a temperature-dependent quenching, and more abruptly at certain temperatures. A simple way to compare and visualize multiple melt curves is to use the negative first derivative ($-dF/dT$) of these curves, where distinct peaks can be identified. In a melt-curve analysis using fluorophores that are not incorporated into the amplicons, the measured T_m is not that of the native ds DNA. In this case, the T_m of a product is dependent on the concentration of the fluorophore used and the rate of temperature change. Complex melting curves have also been observed with short ds DNAs using SYBR Green I, and dye redistribution differences between different DNA binding dyes (SYBR Green I, LC Green, BEBO) has given rise to different melt profiles of the same samples. Nevertheless, when optimized, the melt-curve analysis is an informative homogenous method for the characterization of amplicons.

High-speed and high-throughput genotyping of mutations and SNPs using melt-curve analysis have also been reported. Certain formats of probes such as Taqman or Taqman-MGB probes cannot be used for melt curve analysis as the effective concentrations of the probes are reduced by degradation during PCR. Hence, by using just three colors for real time PCR and the ability to discriminate four T_m, it is possible to distinguish a total of 12 different products in a single experiment, thus increasing the versatility of the study.

6.8. Concluding Remarks

Real time PCR is a versatile and reliable technology for the quantification of nucleic acids. Some of the limitations of the technology are being addressed. The integration of automated sample processing and data handling will continue to be improved, hence enabling this technology to be more accessible to many diagnostic laboratories. Newer instruments of real time PCR will likely have smaller footprints, higher throughputs, lower cost and with more rapid cycling.

Miniaturization of the PCR process on microfabricated devices that enable rapid reaction time, greater throughput and lower sample consumption (nanoliters to subnanoliters) is an interesting innovation that has been adapted for real time detection as "lab-on-a-chip" concept. The current limitation of multiplexing in real time PCR due to the lack of suitable organic fluorophores may be addressed by the use of semiconductor nanocrystals (quantum dots), which have very narrow emission spectra. Online public primer and probe databases more than 600,000 commercially pre-designed Taqman Gene expression assays (Applied Biosystems) and the availability of a limited number of prevalidated genome wide real time sequence-dependent probes (ProbeLibrary™) will serve as invaluable resources for the rapid development of specific assays for a large number of genes. In view of these developments, real time PCR will continue to have major impacts on many clinical and biological disciplines, and in some circumstances, replace conventional PCR.

Further Readings

1. Alivisatos AP, Gu W and Larabell C. Quantum dots as cellular probes. *Annu. Rev. Biomed. Eng.* **7**: 55–76, 2005.
2. Didenko VV. DNA probes using fluorescence resonance energy transfer (FRET): Designs and applications. *Biotechniques* **31**: 1106–1116, 1118, 1120–1101, 2001.

3. Guttenberg Z, Muller H, Habermuller H, Geisbauer A, Piper J, Felbel J, Kielpinski M, Scriba J and Wixforth A. Planar chip devise for PCR and hybridization with surface acoustic wave pump. *Lab Chip* **5**: 308–317, 2005.

4. Heller M and Morrison L. Chemiluminescent and fluorescent probes for DNA hybridization systems. In: *Rapid Detection and Identification of Infectious Agents*, eds. DT Kingsbury and S Falkow. Academic Press, New York, 1985.

5. Higuchi R, Fockler C, Dollinger G and Watson R. Kinetic PCR analysis: Real-time monitoring of DNA amplification reactions. *Biotechnology (NY)* **11**: 1026–1030, 1993.

6. Holland PM, Abramson RD, Watson R and Gelfand DH. Detection of specific polymerase chain reaction product by utilizing the 5'-----3' exonuclease activity of *Thermus aquaticus* DNA polymerase. *Proc. Natl. Acad. Sci. USA* **88**: 7276–7280, 1991.

7. Lakowicz JR. *Principles of Fluorescence Spectroscopy*, 2nd ed. Plenum Publishing, 1999.

8. Livak KJ and Schmittgen TD. Analysis of relative gene expression data using real-time quantitative PCR and the 2(-Delta Delta C(T)) Method. *Methods* **25**: 402–408, 2001.

9. Matsubara Y, Kerman K, Kobayashi M, Yamamura S, Morita Y and Tamiya E. Microchamber array based DNA quantification and specific sequence detection from a single copy via PCR in nanoliter volumes. *Biosens. Bioelectron.* **20**: 1482–1490, 2005.

10. Mouritzen P, Norerholm M, Nielsen PS, Jacobsen N, Lomholt C, Pfundheller HM and Tolstrup N. ProbeLibrary: A new method for faster design and execution of quantitative real-time PCR. *Nat. Methods* **2**: 313–316, 2005.

World Wide Web Sites

- Instrumentations
  http://www.biocompare.com
- Primer and probe designs
  http://www.medgen.ugent.be/rtprimerdb/
  http://web.ncifcrf.gov/rtp/gel/primerdb/

Part II

Food Production Involving Microorganisms and Their Products

The fermented foods industry today is a big business and is the result of developments from traditional, small-scale production methods. Over the millennia, mankind has learnt to manipulate microorganisms and the environment of food to encourage the growth of desirable organisms that result in valuable foods which have good keeping quality and flavors.

Three key discoveries in the science of microbiology have advanced this area of food science. First is the discovery of microorganisms and the realization that they can be isolated and manipulated to control food fermentations, and in doing so, produce desirable fermented foods. Second is the discovery of enzymes produced by microorganisms and their role as biological catalysts that can be isolated and used to modify foods and develop non-traditional methods for modifying foods. Third is the recent development of genetic engineering, which enables a gene for a particular desirable property of an organism to be transferred to a microorganism in which the characteristic is absent.

In Chapters 7 and 8, the production methods of well-known traditional foods and alcoholic beverages are discussed including Japanese saké. In Chapter 9 the theoretical basis for solid state cultivations is expounded. Elucidation of flavors, fragrances and other food ingredients and their production happens in Chapter 10. In Chapter 11 enzyme modified food products are set forth: how enzymes are applied to produce new foods and modify old foods in new ways. The annex discusses protocols and procedures for assessing the safety of newly developed food products and for giving permission for their release to the market.

Annex

Regulation of Foods Involving Genetically Engineered Microorganisms

The European Union **Novel Food Regulation** (1997) provides a framework for the safety evaluation of foods developed and manufactured using novel technologies and also for new food sources. The regulation covers all foods and food ingredients that "have not hitherto been used for human consumption to a significant degree within the Community," even if they have a long history of safe use in other parts of the world. Thus, novel foods involving microorganisms can be categorized into three groups.

(1) Genetically modified viable microorganisms defined by the Directive 90/220/EEC. An example is genetically modified lactic acid bacteria used in fermented milk.
(2) Foods produced from, but not containing, genetically modified microorganisms (Art. 1.2. a). An example is extracellular products of genetically modified microorganisms.
(3) Foods isolated from microorganisms (Art. 1.2. d). These include new products based on single-cell protein, such as the fungus *Fusarium graminearum*, and omega-3 fatty acids derived from algae, such as *Crypthecodinium cohnii* and *Nitzschia alba*.

Novel foods can be put into the market without formal approval if they can be shown to be substantially equivalent to their existing counterparts as regards to their composition, nutritional value, metabolism, intended use and content of undesirable substances. Substantial equivalence can be demonstrated either on the basis of generally recognized scientific evidence or on the basis of an opinion delivered by one of the EU member states' competent food assessment bodies.

Non-equivalent novel foods would need the authorization of the member state in which the product is intended to be marketed, and a copy of the request must be submitted to the EC Commission (Art. 4). The request must contain all relevant information on studies performed to demonstrate that the food does not present a danger to, or mislead, the consumer, and also that it does not differ from food that it is intended to replace to such an extent that its normal consumption would be nutritionally disadvantageous. Thus, a case-by-case evaluation of their safety, to include toxicological, microbiological and nutritional considerations, is necessary.

The Novel Food Regulation cross links the Deliberate Release Directive for Genetically Modified Microorganisms (90/220/EEC Art. 9) and the Plant Variety Directives (70/457/EEC and 70/458/EEC Art. 3) to avoid duplication of procedures.

The Novel Food Regulation requires specific labeling of "any characteristic or food property such as composition, nutritional value or nutritional effects, or intended use which renders the food no longer equivalent" to its conventional counterpart (Art. 8). Thus, if a genetic modification leads to a novel protein, all foods and ingredients containing this protein should be labeled as such. However, labels are not required for hydrolyzed protein preparations, in which the modified protein is degraded, or for refined oils, which would no longer contain significant amounts of the protein.

The Directive 90/220/EEC also implies that a new *Lactobacillus* strain developed using traditional genetic tools (e.g. mutation and selection) falls completely outside the scope of the regulation and would require neither a formal safety evaluation nor a label. A strain with the same modification, but achieved using targeted genetic modification (e.g. recombinant DNA technology), would need both an authorization and a label.

The US Food and Drug Administration (FDA) requires that genetically modified foods meet the same safety standards for all other foods, based on existing food law. US law places responsibility on producers and sellers to offer only safe food products to consumers. Thus, pre-market approval for any new substance that is added to a food, irrespective of whether or not it is a product of genetic engineering is not required.

Further Readings

1. Ager B, revised Moses V. The regulation of biotechnology in Europe. In: *Biotechnology: The Science and the Business*, 2nd ed., Chapter 9, eds. V Moses and RE Cape. Harwood Academic Publishers, The Netherlands, 1999, pp. 123–132.

2. Huggett AC and Conzelmann C. EU regulation on novel foods, consequences for the food industry. *Trends Food Sci. Technol.* **8**: 133–139, 1997.
3. Nicholas RB, revised Springham DG. The regulation of biotechnology in the United States. In: *Biotechnology: The Science and the Business*, 2nd ed., Chapter 8, eds. V Moses and RE Cape. Harwood Academic Publishers, The Netherlands, 1999, pp. 113–122.
4. Robinson C. Genetically modified foods and consumer choice. *Trends Food Sci. Technol.* **8**: 84–88, 1997.
5. Robinson C. Understanding the commercial and regulatory issues for genetically modified and novel foods and food ingredients. *Trends in Food Science and Technology* **9**: 83–86, 1998.
6. Roberts SM, Turner NJ, Willetts AJ and Turner MK. *Introduction to Biocatalysis Using Enzymes and Microorganisms*, Cambridge University Press, New York, 1995.

World Wide Web Sites

- Agriculture and Agri-Food Canada
  http://aceis.agr.ca
- Australia CSIRO Library
  http://www.dfst.csiro.au/fdnet20a.htm
- Codex Alimentarius Home page
  http://foodnet.fic.ca/regulat/codex.html
- Colorants: An opinion on riboflavin as a coloring matter authorized for use in foodstuffs produced by fermentation using genetically modified *Bacillus subtilis*
  http://europa.eu.int/comm/food/fs/sc/scf/out18_en.html
- EU COST Action 99 (Food Consumption and Composition Data Research)
  http://food.ethz.ch:2000/cost99.html
- FAO 50th Anniversary Internet Forum on Food Security
  http://FAO50.FSAA.ULAVAL.CA/english/start.html
- Food and Agriculture Organization of the United Nations (FAO)
  http://www.fao.org

- FoodCom News
  http://www.foodcom.com/foodcom
- FoodNet, Canada
  http://foodnet.fic.ca
- INFOODS, the UN University Food and Nutrition Program
  http://www.crop.cri.nz/crop/infoods/infoods.html
- International Food Information Council
  http://ificinfo.health.org
- International Organization for Standardization (ISO)
  http://www.hike.te.chiba-u.ac.jp/ikeda/ISO/home.html
- Food composition data from around the world
  http://food.ethz.ch/
- Leatherhead Food Research Association, UK
  http://www.worldserver.pipex.com/lfra
- MATFORSK (Norwegian Food Research Institute)
  http://128.39.177.3/Matforsk/engelsk/summary.htm
- MEATNET(links to other websites of interest to the meat sector)
  http://www.meatnet.nl/links.htm
- New Scientist
  http://www.newscientist.com
- Paul Singh's On-line Food Engineering Teaching Course
  http://nachos.engr.ucdavis.edu/-rpsingh/index.html
- Royal Society of Chemistry's Nutrition page
  http://chemistry.rsc.org/rsc/nuts.htm
- UK Government Information Service (with links to all Government departments)
  http://www.open.gov.uk/index.htm
- UK Institute of Food Research
  http://www.ifrn.bbsrc.ac.uk/information/IFR.html
- UK Institute of Food Research (food safety research information)
  http://www.ifrn.bbsrc.ac.uk/Information/safety.html
- UK Institute of Food Science and Technology's Home Page
  http://www.easynet.co.uk/ifst
- US Department of Agriculture (USDA) Food and Nutrition Information center
  gopher://gopher.nalusda.gov:70/11/infocntr/fnic

- USDA Food and Nutrition Software List
  gopher://gopher.nalusda.gov:70/11/infocntr/fnic/software
- US Institute of Food Technologists (IFT)
  http://www.ift.org
- US Food and Drug Administration (FDA) Center for Food Safety and Nutrition
  http://vm.cfsan.fda.gov/list.html

Chapter 7

Fermented Foods

Desmond K. O'Toole
Department of Biology and Chemistry
City University of Hong Kong
83 Tat Chee Avenue, Kowloon, Hong Kong, China

Yuan Kun Lee
Department of Microbiology, National University of Singapore
5 Science Drive 2, Singapore 117597
Email: micleeyk@nus.edu.sg

Fermented foods are those that have been transformed by microbial action or enzymes to produce desirable biochemical and physical changes. Fermentation, from the perspective of the microorganism, is an oxygen free process (anaerobic) in which an organic compound endogenous to the organism acts as the final electron acceptor. However, in food fermentations some microorganisms use oxygen as the final electron acceptor. These processes are also referred to as fermentations (Fig. 7.1). In addition, another food fermentation commonly referred to is the tea fermentation that is not a microbial process at all but an endogenous enzymic oxidation process. The microorganisms responsible for food fermentation include bacteria, yeasts and molds.

For many microbial fermentations to occur the food must contain readily fermented sugar, usually either monosaccharides (hexoses but can be pentoses) or disaccharides. A good source of sugars is starch but the polymer must be broken down into its constituent parts before the

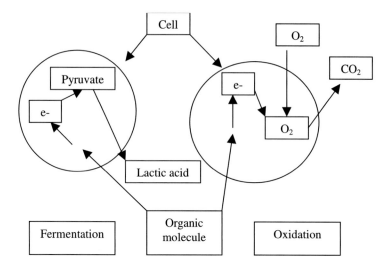

Fig. 7.1. The difference between microbial fermentation and oxidation. In fermentation, the energy is derived from organic molecules and the electron is finally passed to another organic molecule that is then expelled from the cell. In oxidation, the energy is derived from organic molecules and the electron is finally passed to an oxygen molecule and then expelled from the cell as carbon dioxide.

microorganisms can metabolize the sugars locked in the starch. Thus in some fermentations there is a preliminary step to break down starch to make it readily available for the microorganisms. This will be discussed later.

A further characteristic of many food fermentations is that they are two-stage processes that alternate between aerobic and anaerobic stages. The first stage is a true microbial anaerobic fermentation but aerobic organisms grow in the second stage. The aerobic organisms can be fungi or bacteria, although fungi are more common. In this second stage, the organisms produce proteolytic enzymes that breakdown the structure of the protein to soften the food and produce protein degradation products, and maybe, lipolytic enzymes that hydrolyze the fat to fatty acids resulting in esters that have strong flavors. In the other kind of two-stage process, the first stage involves an aerobic microbial growth stage, usually fungal growth, followed by an anaerobic stage.

7.1. Food Involving Acid Fermentation

The major food acids are lactic and acetic acids. However, propionic acid is also produced in small quantities as a flavoring compound. From the point of view of the microorganism, acid production is a very inefficient method for getting energy. So for a small amount of growth the microorganisms produce a large amount of acid.

7.1.1. Lactic acid bacteria

Lactic acid bacteria are those that produce lactic acid as the sole product or the major acid from the energy yielding fermentation of sugars. Traditionally it has been easy to define the group but in recent years, it has become more difficult so that opinions differ as to what should be called a lactic acid bacterium. They can be broadly defined as Gram positive; anaerobic, micro-aerophilic or aero-tolerant bacteria, either rod- or coccus-shaped, catalase negative, and fastidious in their growth requirements; many need B vitamins to grow. This definition can include an endospore forming bacterium (*Sporolactobacillus*), and a motile species but none of these are important for food production and preservation. Although considered fermentative organisms and unable to produce and use the respiratory chain found in the aerobic organisms, it is now known that *Lactococcus lactis,* when grown in the presence of hemin, produces a respiratory chain and can use O_2.

The major genera of lactic acid bacteria of importance in the food industry for food preservation (Table 7.1) and production (some cause food spoilage) are *Lactococcus* (in milk), *Lactobacillus* (in milk, meat, vegetables, and cereals), *Leuconostoc* (in vegetables and milk), *Pediococcus* (in vegetables and meat), *Oenococcus* (in grape wine), *Streptococcus* (in milk) and *Tetragenococcus* (in soy sauce). They are important for humans as well as for domestic animals in the production of many different foods that include milk and dairy products, meat products, vegetable products, grape wine, soy sauce and silage for animals.

7.1.2. Types of lactic acid fermentation

When these lactic acid bacteria ferment sugars, they produce varying amounts of lactic acid. This is due to two different metabolic pathways by

Table 7.1. Characteristics of the genera of lactic acid bacteria important in the food industry.

Genus	Shape	Tetrads formed	CO_2 from glucose	Lactic acid[a]	Growth at 10°C	at 45°C	in 6.5% NaCl	in 18% NaCl	at pH 4.4	at pH 9.6
Bifidobacterium	Pleomorphic rod	–	–	L	–	±	–	–	–	–
Enterococcus	Coccus	–	–	L	+	+	+	–	+	+
Lactobacillus	Rod	–	±	D, L, DL	±	±	±	–	±	–
Lactococcus	Coccus	–	–	L	+	–	–	–	±	–
Leuconostoc	Coccus	–	+	D	+	–	±	–	±	–
Oenococcus	Coccus	–	+	D	+	–	±	–	±	–
Pediococcus	Coccus	+	–	L, DL	±	±	±	–	+	–
Streptococcus	Coccus	–	–	L	–	±	–	–	–	–
Tetragenococcus	Coccus	+	–	L	+	–	+	+	–	+
Weissella	Short rod, coccus	–	+	DL,D	–[b]	–	+	–	+	–

[a]This refers to the optical isomer of the acid; [b]15°C.

which they may metabolize the sugars. The pathways result in different metabolic by-products. One pathway results in **homolactic fermentation**, while the other pathway leads to **heterolactic fermentation**. Consequently, the lactic acid bacteria are generally categorized on the basis of these fermentation patterns.

(1) Homolactic Fermentation
The hexose is fermented exclusively to lactic acid.

$$1 \text{ hexose} \rightarrow 2 \text{ lactic acid} + 2 \text{ ATP}$$

The bacteria use the glycolytic pathway and possess aldolase, but not ketolase. Consequently, they cannot metabolize pentoses or gluconate. Organisms that use this pathway include *Lactobacillus acidophilus*, *Lb. delbrueckii* subsp. *bulgaricus (Lb. bulgaricus)*, *Lb. helveticus*, *Enterococcus faecium* and *Streptococcus thermophilus*.

(2) Heterolactic Fermentation
In this metabolic pathway, hexoses are fermented to equimolar amounts of lactate, ethanol (acetate) and CO_2.

$$1 \text{ monosaccharide} \rightarrow 1 \text{ lactic acid} + 1 \text{ (ethanol} + \text{acetic acid)}$$
$$+ 1 \text{ } CO_2 + \text{formic acid (trace)} + 1 \text{ ATP} + 1 \text{ ATP/acetic acid}$$

Depending on the conditions, the bacteria switch between producing either ethanol or acetate.

The phosphogluconate pathway is used and hexoses and pentoses can be metabolized through it. Organisms that use this pathway include *Bifidobacterium breve*, *B. infantis*, *B. longum*, *B. thermophilum*, *Lb. casei*, *Lb. fermentum*, *Lb. plantarum*, *Lb. rhamnosus* and *Lb. salivarius*.

The bacterial cultures can be categorized further on the basis of these two types of metabolic activity because in some cultures the organism may switch from using one pathway to using the other pathway. The categories are:

- Group 1: Obligate homolactic — these organisms use the glycolytic pathway only.
- Group 2: Facultative heterolactic — these organisms use the glycolytic pathway and possess aldolase and ketolase that is repressed in the presence of glucose.

• Group 3: Obligate heterolactic — these organisms do not possess aldolase and use the phosphogluconate pathway.

7.1.3. Malolactic fermentation

A further characteristic of some lactic acid bacteria is their ability to use L-malic acid as an energy source (D-malate is not metabolized). Malic acid is present in many vegetable and plant products, including grapes and cucumbers. Bacteria that are able to use malate include some of the species from the genera *Lactobacillus*, *Pediococcus*, *Leuconostoc* and *Oenococcus*. Utilization, however, also depends on the presence of some fermentable carbohydrate.

The enzyme responsible is L-malate: NAD^+ carboxylase. The reaction is as follows:

This reaction usually occurs late in fermentation. It raises the pH and results in the production of some gas.

7.1.4. Antibacterial properties of lactic acid bacteria

The many situations in which lactic acid bacteria dominate can be explained by the production of lactic acid that lowers the pH and prevents the growth of many acid sensitive bacteria. However, in addition many lactic acid bacteria produce substances that kill other bacteria. The most important group is the bacteriocins. The first bacteriocin described was nisin, discovered in around 1947, and it is now produced commercially and incorporated in many foods. Many more have been discovered and they are now placed in four broad classes. Class 1 consists of the lantibiotics that are small peptides that contain unusual dehydroamino acids and thioether acids, lanthionin and 3-methyllanthionine. Examples are nisin A produced by *Lactococcus lactis* subsp. *lactis*, and carnosin produced by *Carnobacterium piscicola*. Class 2 bacteriocins are also peptides that are small heat stable non-lanthionine-containing membrane-active compounds. An example is lactococcin A produced by *Lc. lactis* subsp. *cremoris*. Class 3 consists of high molecular weight (> 30 kDa) peptides that are heat sensitive. An example is acidophilucin A produced by *Lb. acidophilus*. On the other hand, Class 4 contains complex compounds consisting of protein, lipid and/or carbohydrate. An example is lactocin 27 produced by *Lb. helveticus*.

7.2. Vinegar Production

This is basically a two-stage microbial process with the stages being significantly separated in time. In the first stage, alcohol is produced from glucose in fruit or from glucose/maltose derived from starch using plant or microbial enzymes (fungal growth). The prepared alcoholic product is then converted to vinegar by bacteria.

Oxidation of alcohol by *Acetobacter aceti* (optimum temperature 29–34°C), *Gluconobacter oxydans* (optimum temperature 20°C), and by *A. pasteurianus*, found in Japanese rice wine based vinegar production systems, results in the production of acetic acid. The biochemical event is depicted by the following equation:

$$2C_2H_5OH \xrightarrow{\ O_2,\ \text{vinegar bacteria}\ } 2CH_3CO_2H + 2H_2O$$
$$\text{ethanol} \qquad\qquad\qquad \text{acetic acid}$$

If we start with glucose then the theoretical yield of vinegar is as follows:

$$1 \text{ g glucose} \rightarrow 0.51 \text{ g ethanol} \rightarrow 0.67 \text{ g acetic acid}$$

Today, vinegar is produced from a variety of sources of ethanol, but historically in the West wine was the source of vinegar, probably because of the frequency with which it is produced. Unlike other alcoholic beverages, grape wine can only be produced at one time of the year, but other beverages such as beer can be produced at will at any time of the year. Any attempt to keep alcoholic beverages exposed to air for long periods, particularly wine, results in the slow oxidative processes that produce vinegar. Commercially, the source of substrate is often indicated in the name of the product, e.g. spirit vinegar, wine vinegar, cider vinegar, malt vinegar, whey vinegar and rice vinegar. The strength of vinegar sold to consumers is about 6% acetic acid but the strength of acetic acid that is produced by these microbial processes ranges from 10–15%.

The development of vinegar production methods through the years gives us a glimpse of the advances in fermentation technology.

7.2.1. Traditional "slow" Orleans process

This is the traditional method and is still used in Japan to produce rice wine vinegar. The bacteria are obligate aerobes and consequently they do not grow through out the substrate but at the air/liquid interface. If the wine is left undisturbed, a cell mat, or pellicle, forms on the surface of the batch of wine (mother of vinegar or "Mycoderma aceti"). As oxygen enters the wine through the mat, the bacteria use it and convert the ethanol to acetic acid. Therefore, vinegar is produced at the surface and diffuses through out the body of wine (Fig. 7.2). In a similar manner, the ethanol diffuses to the pellicle where it is used by the bacteria. The pellicle is the starter culture for vinegar production, and, in a **batch-fed system**, wine is carefully added weekly for four weeks while avoiding

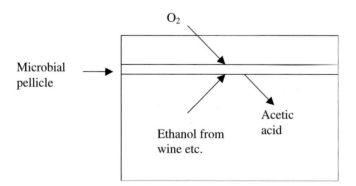

Fig. 7.2. Fundamentally the vinegar production system depends on oxygen supply. In the first production systems, a microbial pellicle formed on the surface of the wine and the oxygen and ethanol had to be transported by diffusion to the microorganisms that occupied a relatively small space. Modern systems increase the mass of microorganisms available in a production system and increase its interaction with ethanol and oxygen.

disturbance of the pellicle. A portion of vinegar is withdrawn in the fifth week, and the same amount of wine is added and the process is repeated resulting in a **semi-continuous variable volume fed-batch culture process** (see Chapter 2, Sec. 2.4.2).

The air-liquid interface area limits the production rate because the production relies on the diffusion of oxygen and ethanol to the interface where acetic acid is produced.

7.2.2. The "quick" process (German process)

The principle of this process is thought to date from a British patent from 1824 but in fact, it is based on a much earlier French process dating from at least 1670. In this process (Fig. 7.3) wine is trickled down packed pumice stone, ceramic chips or wood shavings, on which acetic acid bacterial cells adhere (**cell immobilization**) so it is a **fixed bed microbial film reactor**. In these generators, the volume of a typical commercial vinegar generator is 50–60 m^3. Air is allowed to move freely through

Fig. 7.3. The "quick" process for vinegar production. The mass of bacteria is increased in this system and ethanol and oxygen are transported to the cells.

perforations in the bottom of the packed container, or is forced through the trickling generator to increase the oxygen transfer rate. The alcohol is converted rapidly to acetic acid while the wine is trickling down the packed vinegar generator, and 98% conversion of ethanol (12% v/v wine) into acetic acid can be achieved in five days. During the process, the temperature in the generator increases from about 28°C at the top to about 35°C at the bottom due to the metabolic activity of the bacteria. Theoretically 1 l of ethanol requires 430 l of oxygen gas at 30°C for complete oxidation to vinegar. In practice the volume of air required is about ten times that volume because oxygen constitutes only about 20% of air. Eventually, the vinegar generator needs to be re-packed, due to excessive cell growth and clogging of the system.

Three factors increase the rate of vinegar production in this process when compared with the slow process. The amount of oxygen entering the wine per unit time is many times greater (higher oxygen transfer rate); the ethanol comes into contact with the bacteria more quickly due to the physical flow of the fluids; and there is a greater mass of bacteria to carry out the process. Diffusion, as in the traditional process, is a much slower process. The constraint remains that the bacteria are growing on a surface (the wood shavings, ceramic chips, etc.) and the oxygen and ethanol have to be brought into contact with the bacteria.

7.2.3. *Fringes Acetator® or submerged culture process*

Fringes Acetator was the first submerged culture process but others developed subsequently include Yeoman's cavitator used in the US and

---

Box 7.1

How to Immobilize Acetic Acid Bacteria on Solid Supports?

A viable acetic acid bacterial culture is trickled down the packed vinegar generator. The medium is absorbed into the porous supporting material when the cells adhere onto the surface of the material. The cells then grow to form a thin film on the surface and may eventually dislodge when the film of cells becomes too thick.

Japan, the Bourgeois process used in Spain and Italy, and the Fardon process used in Africa.

The serious development of the stirred tank fermentation process for aerobic bacterial processes, including vinegar production (Fig. 7.4), arose from the success of penicillin production during World War II (1939–1945 AD). The advantages of submerged cultivation over the trickling generator are as follows:

(a) It permits 30 times faster oxidation of alcohol, because of the high oxygen transfer rate.
(b) It requires a smaller reactor volume (16% of that needed for the trickling generator) to produce an equivalent amount of vinegar, because the dead space occupied by the solid support in the trickling generator is removed.
(c) It has greater efficiency with a yield 5–8% higher and > 90% of theoretical yield is obtained.
(d) It can be highly automated.
(e) It eliminates clogging of packed wood shavings and interruption of liquid flow.
(f) It allows an easy change in the type of vinegar being produced.

Fig. 7.4. The submerged culture process for vinegar production. Transport of oxygen is maximized in this system.

7.2.4. *Quality problems*

During vinegar production using the Fringes Acetator, bacteriophages can be a problem. As in the dairy industry, these bacterial viruses destroy the bacteria so they no longer produce acid. This is a greater problem if pure culture stains are used but is a lesser problem if mixed wild cultures are used. The solution is to use filtered, sterilized air and pasteurized substrate. In addition, the "quick" process vinegar producing units are a good source of bacteriophages for *Acetobacter* species so submerged fermentation production units should be separated from them and the exchange of material or personnel between the two kept to a minimum, or preferably banned completely.

In the "quick" and submerged processes, the vinegar is drawn off when the remaining ethanol reaches about 0.3%. The reason for this is twofold. Firstly, once the ethanol is depleted the *Acetobacter* strains undergo what seems to be an irreversible change in their metabolism and begin metabolizing the acetic acid to carbon dioxide and water. Consequently, the concentration of acetic acid decreases. In the presence of ethanol this metabolic pathway is repressed. If the culture undergoes this shift in metabolic activity, the only solution is to close down the generator, clean up and sterilize everything. Secondly, the presence of ethanol allows esterification to occur to improve the flavor of the final product.

Vinegar eels (*Anguillula aceti*, a small nematode) can also contaminate the vinegar and the packing in the "quick" process, but they are not so ubiquitous in modern vinegar producing plants.

7.3. Fermented Dairy Products

The traditional method for the production of fermented milk was by natural fermentation of surplus milk collected on the day. During the harvesting of the milk, microorganisms on the utensils, from the animals and from the environment find their way into the milk. The environmental conditions select bacteria that ferment the milk sugar, lactose. When the harvested milk leaves the animal its temperature is about 37°C and if it is not cooled quickly to about 4°C the contaminating bacteria grow in the milk. Traditionally, due primarily to the lack of proper temperature control, this resulted in sour milk that was tastier, possessed a thicker texture and was more stable than fresh milk. Eventually in traditional societies the primitive but simple technique was developed of keeping back a small portion of the milk, whey, etc. to add to the next day's supply of milk. It was a kind of culturing that is still used to this day for some fermentation applications.

With the development of the science of microbiology the bacteria responsible for the unique characteristics of each type of sour milk product were isolated and pure culture techniques applied to improve the quality of the sour milk products.

The chemical composition of milk used to produce fermented milk products and cheeses from a range of animals is shown in Table 7.2. Three chemical characteristics of milk aided in the development of fermented dairy products. The first is the lactose which is a disaccharide of galactose and glucose (β-D-galactopyranosyl-($1\rightarrow4$)-D-glucopyranose) and is a good source of energy for lactic acid bacteria. The second is fat. Bovine milk contains from 3.5 to 5.0% fat in the form of small fat globules. The third is a protein, called casein, or κ-casein. This protein has an isoelectric point at about pH 4.6 at which it clots. The clotted protein traps the fat globules in the matrix of the clot. The protein matrix then

Table 7.2. The chemical composition of milk from various animals used for production of fermented milk products and cheese.

Animal	Species	Composition, g kg^{-1}				
		Water	Fat	Casein	Whey protein	Lactose
Camel	*Camelus dromedarius*	865	40	27	9	50
Cattle	*Bos taurus*	873	39	26	6	46
Goat	*Capra hircus*	867	45	26	6	43
Horse	*Equus caballus*	888	19	13	12	62
Sheep	*Ovis aries*	820	72	39	7	48
Water buffalo	*Bubalus bubalis*	828	74	32	6	48
Yak	*Bos grunniens*	827	65	–	58[a]	46

[a]Casein plus whey protein.

shrinks and the water phase is expelled from the matrix by the process of synerisis. The shrunken protein matrix can then be separated from the water phase (whey) and stored as cheese. If the matrix crumbles instead of shrinking and the whey is not removed, the clotted mixture becomes very viscous giving us the sour milk products we have today. From the perspective of flavor development, a further ingredient of interest is the citrate in milk.

Box 7.2

Lactose in Milk and Its Uptake and Metabolism in Lactic Acid Bacteria

Lactose is a disaccharide comprised of a glucose and a galactose molecule. To metabolize the molecule it must be cleaved and several methods are used by these bacteria. One method is to transport the lactose into the cell via a phosphoenol-pyruvate phospho-transferase system (PEP-PTS) that results in an internal lactose-6-phosphate that is then split into glucose and galactose-6-phosphate. Glucose is metabolized via the EMP pathway and the galactose is shunted through the tagatose-6-phosphate pathway. Galactose-6-phosphate is isomerized to tagatose-6-phosphate, then phosphorylated to tagatose-1,6-di phosphate followed by cleaving to triose phosphate and metabolizing through the rest of the EMP pathway to pyruvic acid. The other method used by *Streptococcus thermophilus* is to transport the lactose via a permease that uses either a lactose/H^+ symport mechanism or a lactose/galactose exchange mechanism. The glucose is split from the galactose and metabolized via the EMP pathway and the galactose is excreted into the medium so that it accumulates in the product. Because of the rapid acid production by the yogurt bacteria, galactose concentrations up to 1% may accumulate in yogurt.

Galactose is a human nutrition problem because it is a poor sugar source for humans, but, more importantly, if too much is consumed, particularly while consuming alcohol, the galactose can be converted to galactitol and if it accumulates in tissue it can cause physiological disorders, including cataracts in the eye lens.

Box 7.2   *(Cont'd)*

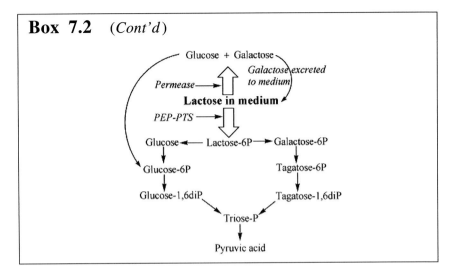

7.3.1. *Fermented milk products*

The general protocol for the production of fermented milk products is shown in Fig. 7.5. Most of the naturally fermented milks contain more than one type of bacterium and occasionally yeasts (Table 7.3) because the characteristic flavors are produced through the growth of microbial consortia (a mixture of different strains of bacteria and may be yeasts) present naturally in the environment. A typical example is yogurt, which contains at least two lactic acid bacteria, namely *Lactobacillus delbrueckii* subsp. *bulgaricus* and *Streptococcus thermophilus*. Amino acids released by *Lb. delbrueckii* subsp. *bulgaricus*, and carbon dioxide and formic acid released by *S. thermophilus*, mutually stimulate the growth of each bacterium (Fig. 7.6). Thus, co-metabolism, or associative growth, by the consortium produces more lactic acid than the individual bacterial strains. In addition, the two organisms also produce acetaldehyde, the dominant flavor compound in yogurt. Another different consortium is the starter culture for kefir. Kefir is produced through the metabolism of lactobacilli, lactococci, a leuconostoc species and two yeasts. However, unique amongst milk starter cultures, the organisms grow together and cause casein to clot around them to produce soft lumps of casein mixed with the microorganisms. These lumps are called kefir grains. To produce the kefir the grains are filtered from a previous batch and put into a new batch of

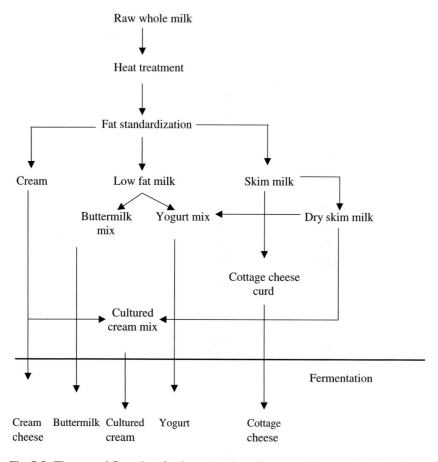

Fig. 7.5. The general flow chart for the production of a range of fermented milk products.

Fig. 7.6. The yogurt fermentation. The two bacteria co-metabolize to produce more lactic acid and acetaldehyde than either of them separately. The lactobacillus produces small amounts of amino acids that stimulate the streptococcus. Meanwhile the streptococcus produces carbon dioxide and formate that stimulate the lactobacillus.

Table 7.3. Some examples of fermented milk products and the cultures responsible for their characteristics.

Fermented product	Microbial cultures	Country of origin	Source of milk
Acidophilus milk	*Lb. acidophilus*	USA	Bovine
Buttermilk[a]	*Lc. cremoris* *Lc. lactis* *Lc. diacetylactis*	USA	Bovine
Cottage cheese	*Lc. cremoris* *Lc. lactis*	USA	Bovine
Cultured cream	*Lc. lactis* *Lc. cremoris* *Lc. diacetylactis* *Leuc. cremoris*	USA	Bovine
Cream cheese	*Lc. cremoris* *Lc. lactis* *Lc. diacetylactis* *Leuc. cremoris*	USA	Bovine
Koumiss	*Lb. bulgaricus* *Torulopsis holmii* (yeast)	Russia	Horse
Kefir	*Lb. brevis* *Lb. casei* *Lb. caucasicus* *Leuc. mesenteroides* *Lc. durans* *Torulaspora* *delbrueckii* (yeast) *Sacc. cerevisiae* (yeast) *Acetobacter aceti*	Russia	Bovine, ewe, goat
Labneh	*Lb. thermophilus* *Leuc. lactis* *Lb. acidophilus* *Kluyveromyces fragilis* (yeast) *Sacc. cerevisiae* (yeast)	Israel	Bovine
Laktofil	*Leuc. cremoris* *Lc. cremoris* *Lc. lactis* *Lc. diacetylactis*	Sweden	Bovine

Table 7.3 (*Continued*)

Fermented product	Microbial cultures	Country of origin	Source of milk
Leben	*Lb. bulgaricus* *Lc. thermophilus* *Lc. lactis*	Iraq	Ewe, goat
Leben raib	As for Leben	Egypt	Bovine, buffalo
Long milk	*Streptococcus* sp. *Leuconostoc* sp.	Scandinavia	Bovine
Quark	*Leuc. cremoris* *Lc. cremoris* *Lc. lactis* *Lc. diacetylactis*	Germany/ Europe	Bovine
Viili	*Streptococcus* sp. *Leuconostoc* sp.	Finland	Bovine
Yakult	*Lb. casei*	Japan	Bovine
Yogurt	*Lb. delbrueckii* ssp. *bulgaricus* *Lc. thermophilus*	Turkey, Caucasus	Bovine

[a]Traditionally buttermilk was the very rich fraction left over after churning "ripened" (fermented) cream into butter and contained a high level of fat globule membrane fraction. Modern consumer products are made with milk.

milk. The consortium of organisms produces lactic acid, carbon dioxide and ethanol.

The more recently developed fermented dairy products, such as Yakult® and acidophilus milk, are produced in controlled environment fermentation vessels using selected single strains of *Lactobacillus* species.

In addition to the cultures used, modern production methods are often modified from the traditional method to suit economic production methods. An example is yogurt production (Fig. 7.7). Traditionally the milk was boiled to reduce the volume and so concentrate the milk. This produced a thicker sour milk product with no macro structure as one finds in cheese. The boiling had the affect of destroying the ability of the milk to produce a smooth, firm curd. In modern production systems milk powder is added to fat standardized milk to increase the concentration of milk solids in the

Yogurt manufacturing

Fig. 7.7. Flow chart for the production of yogurt using modern production methods.

mixture. Then the milk mixture is heated to a high temperature, like in the traditional production system, to destroy the curd forming properties of the milk. As a smooth thick texture, particularly of the stirred yogurt types, desirable thickening agents may be added.

Table 7.4. Incubation conditions used for various fresh cheeses or freshly fermented milk products.

Product	Process conditions
Buttermilk	22°C, 18 h
Cottage cheese	22°C, 18 h or 35°C, 5 h
Cream cheese	22°C, 18 h
Cultured cream	22°C, 18 h
Kefir	18–22°C, 12 h or 10°C, 1–3 d
Koumiss	28°C, 2 h
Quark	31°C, 5 h
Yogurt	43–45°C, 3 h

Quality control of naturally fermented milk is generally through fermentation conditions and the use of substrate (type of milk) but in modern production methods suitable pure cultures are used. Table 7.4 shows the time and temperature conditions used for various fresh fermented milk products. Generally a low inoculum percentage is used in the manufacture of fresh cheeses made using a temperature below 30°C.

7.3.2. Cheese production

Production of cheese is a huge industry. A total of about 15 million tons in 500 varieties were produced in 1994, over a third of that production being from Western Europe.

(1) Production Process

The general protocol for producing cheeses is in Fig. 7.8. In general, milk casein can be clotted in four ways.

(a) The pH of the milk can be reduced through the growth of lactic acid bacteria that produce lactic acid throughout the milk.
(b) One can add to the milk δ-gluconolactone that then slowly hydrolyses to produce acid so producing the same effect as the growth of the lactic acid bacteria.
(c) The pH of the milk can be reduced rapidly to about pH 5.2 with exogenous acids, like vinegar and lemon juice, that destabilizes the casein so that when heated the casein is caused to clot.

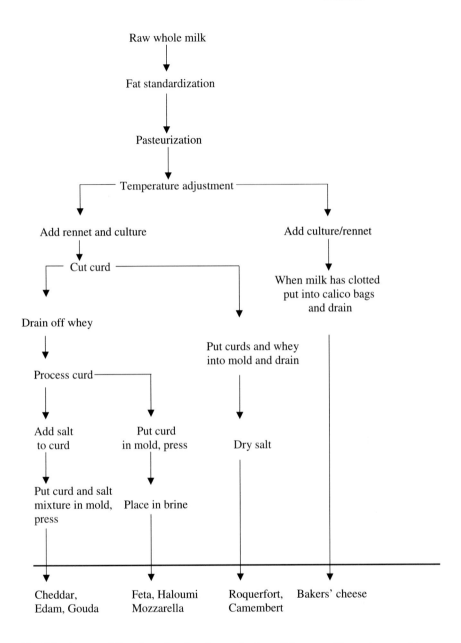

Fig. 7.8. A general flowchart showing the production of cheeses with varying consistency and characteristics.

(d) Finally, the most widely used method is to clot the casein with enzymes. The traditional enzyme source is rennet, a chymosin, obtained as an extract from calf stomach and stored in brine.

The κ-casein is responsible for the stability of the protein micelles in the milk, when treated with the enzyme part of it is hydrolyzed, and that action destabilizes the micelles and causes them to clot. The clotted curd produced by the enzyme has better handling properties than acid clotted milk and allows cheese makers to produce the various kinds of cheese that they make. Because of a shortage of animal chymosin, fungal chymosin is now available to the cheese maker. The fungal chymosin is used to make vegetarian cheeses, that is, those that contain nothing from a killed animal.

The casein fraction is important because its characteristics dictate the kind of fermented dairy product that is produced. Chymosin clots the milk from most animals but milk from the horse and the ass does not clot and so cannot be used for cheese production. Casein is also very sensitive to being denatured by heat. It is a common practice to pasteurize milk to kill spoilage microorganisms and, in particular, to kill pathogens, before processing it into dairy products. The heating profiles (time/temperature conditions) used to kill pathogens are very close to the profiles that reduce the clotting quality of casein. A high temperature causes the clotted casein to form a soft clot that crumbles easily. Consequently, if too high a temperature is used, the milk cannot be used to make good quality cheese but it can be used to make a clotted milk product.

Cheddar cheese is the most significant cheese economically and is a good example of the critical role that reliable lactic acid production at a predicable rate plays in production. The process begins with standardized and pasteurized milk at 30°C to which about 1–2% starter culture and rennet is added. The milk is allowed to stand still and it clots in about 45 minutes. At this point, it is cut with wire frames into small cubes that measure about 1.2 cm on each side. The cut curd and whey are then stirred continuously and the temperature of the curds and whey is slowly raised to about 37°C. During this "cooking" stage, the cheese maker determines the amount of lactic acid produced by titration and the fermentation is going well if the acidity rises about 0.2% within about 90 minutes. During this cooking stage, the gradually raised temperature and lowered pH, in addition to the physical contact between the curd particles through constant stirring, increases the expulsion of the whey from the curd. The whey is then drained from the curd and the curd is piled in heaps. The whey expulsion from the curd and acid production in the curd continue at this stage. The next three to four hours is the **cheddaring stage** in which acid production continues at a steady rate so that the curd fuses and forms the right texture for the cheese, a "chicken breast meat" characteristic. The acid content of the whey running from the curd has usually risen a further 0.4% at the end of the cheddaring stage. If acid production slows or fails at any point during this cheddaring stage, a poor quality cheese results, or the cheese may become a health hazard. The subsequent stages are milling of the curd (curd is cut into strips about 2.5 cm × 2.5 cm in cross section), addition of salt, and pressing of the curd. The pH of the final curd is about 5.2. The final pH alters the body of the final cheese. At about pH 5.2 the final cheese is more like typical cheddar cheese and the micelles therein are about 10 nm in diameter, but if the pH is below 5 the texture becomes "short" and crumbly and the casein micelles are about 3 nm in diameter.

(2) Second Stage Microbial Maturation Processes
During the maturation stage enzymes from the starter bacteria continue to breakdown proteins and other components and in some cases produce undesirable bitter tasting peptides (Fig. 7.9). The length of this maturation stage varies with the cheese. In the case of cheddar cheese, the time varies from about three months for a mild cheese to about two years for a well-matured cheese. The whole maturation depends on the starter enzymes in the cheese and fungi play no significant role in cheddar cheese maturation. In the case of other cheeses, however, molds and bacteria grow on or in the formed cheese and produce proteolytic and lipolytic enzymes. Examples of these are blue cheeses, white mold cheeses and bacterial slime cheeses. Fungal spores are added to the milk before formation of the curd in the manufacture of blue cheeses (Roquefort, Stilton, and Gorgonzola cheeses). Holes are then made through the formed curd with stiff wires to allow air to diffuse into the body of the cheese to promote the growth of the mold *Penicillium roqueforti* throughout the body of the cheese, and the production of proteolytic and lipolytic enzymes.

Fig. 7.9. A generalized scheme of the breakdown of milk components to produce cheese flavors. The lactic acid bacteria produce acid from the lactose and proteolytic and other enzymes from the bacteria, or from added bacteria and fungi, further reduce the peptides to flavor compounds. Added lipase, or lipase from fungi, breakdown the fats to fatty acids that can then react with alcohols to produce fruity esters.

In white mold cheese (Brie and Camembert) the mold (*P. camemberti*) grows on the surface of the formed curd and produces powerful proteolytic enzymes that diffuse into the body of the cheese. In bacterial slime cheeses (e.g. Tilsiter cheese) a red bacterium (*Brevibacterium linens*) grows on the surface and the enzymes penetrate the cheese and produce the typical flavors.

An exception to the above rule is the propionibacteria that give Swiss cheese its distinctive flavor and physical appearance. During maturation the propionibacteria grow anaerobically on the lactic acid so that the concentrations of lactic acid decreases, propionic and acetic acids increase, and gas holes, characteristic of the cheese, are produced.

Because of the long maturation times required for many hard to semi-hard cheeses attempts to accelerate ripening have been tried. The various approaches include the following:

(i)   Raise the temperature during maturation but the potential for microbial spoilage increases.

(ii) Add exogenous microbial enzymes. Added individual enzymes may produce a bitter flavor, but addition of a lactococcal extract has produced good flavored cheese after a short maturation time, probably due to the presence of a complex mixture of proteases and peptidases. However, the enzymes are added to the milk so enzymes are lost in the whey. An encapsulation step to trap the enzymes in the curd matrix for slow release would be a better solution.

(iii) Add extra attenuated starter cells from which enzymes are progressively released. The starter cells are either sub-lethally heat-treated or subjected to a freeze-thaw process to prevent growth and acid production. Encouraging results have been obtained with *Lactobacillus helveticus.*

(iv) Genetic modification to control expression of particular enzymes so that they are increased or decreased depending on whether there are multiple copies of genes and the presence or absence of strong promoters.

(v) Genetic modification to control release of lytic enzymes from the starter cells early in the maturation stage, or insert ripening genes into cells. For example, the gene for demethiolase from *Brevibacterium linens* can be inserted into lactococci.

(vi) The use of cheese slurries. Before the final cheese forming stage the cut curd (milled curd in cheddar cheese making) is mixed with about 5% NaCl and the mixture then stored at an elevated temperature for up to nine days before pressing. The cheese slurry can also be added to fresh milled curds to accelerate ripening. This technology has been applied to a range of different cheeses including blue cheese and Swiss cheese. However, the product is very vulnerable to microbial spoilage and the final stage of the cheese making occurs after the slurry stage, unlike normal cheese making.

7.3.3. *Quality considerations*

The underlying control mechanism is the growth temperature of the starter organisms as shown in Table 7.5. Hence, at high incubation temperatures (> 40°C), *Lb. delbrueckii* subsp. *bulgaricus* and *Streptococcus thermophilus* are the predominant bacteria, and yogurt is produced, or

Table 7.5. Growth characteristics, acid production and diacetyl production of lactic acid
bacteria used in fermented milk production.

Organism	Optimum temperature °C	Growth at		Acid production in milk, %	Diacetyl production
		10°C	45°C		
Lb. acidophilus	37	−	+	0.3–2.0	−
Lb. delbrueckii ssp. bulgaricus	42	−	+	1.5–4.0	−
Lb. casei	30–35	+	+	0.8–4.0	+
Leuc. cremoris	20–25	+	−	0.1–0.3	+
Lc. cremoris	22–30	+	−	0.8–1.0	−
Lc. diacetylactis	22–28	+	−	0.8–1.0	+
Lc. lactis	21–30	+	−	0.8–1.0	−
S. thermophilus	40–45	−	+	0.8–1.0	+/−

mozzarella (pizza) cheese and Swiss-type cheese are produced.
Leuconostoc and *Lactococcus* species are common bacteria found in
fermented products produced by cottage industry in temperate regions.
Lb. acidophilus and *Lb. casei* are widely distributed in fermented food
in the tropics.

In many freshly made dairy products, a desirable buttery taste is formed
due to diacetyl that is produced by some of the lactic acid bacteria. Its
production is dependent on the presence of citrate in the milk (Fig. 7.10). The

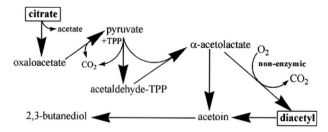

Fig. 7.10. Pathway for the production of diacetyl, responsible for the butter flavor, from
citrate.

organisms do not use citrate as an energy source but when citrate is present with a readily fermentable carbohydrate, such as lactose, then citrate is used and diacetyl is produced if conditions are right.

The growth of the lactic acid bacteria is essential to produce the high acidity of fermented dairy products that prevents the growth of pathogenic bacteria and other microorganisms, thus improving the shelf life and safety of fermented milk products. Raw fresh milk spoils at room temperature within 24 h, whereas fermented milk may be kept at room temperature for a week, until the product becomes too sour for consumption, due to the continuous production of lactic acid even when the bacterial cells have stopped growing (see Chapter 2, Sec. 2.2).

Besides improving the physical and sensory properties of milk, lactic acid bacteria enhance its nutritional quality by increasing the content of some vitamins (e.g. riboflavin, niacin, panthothenic acid, folacin, vitamin B12) in the fermented products as shown (Table 7.6).

In recent times the probiotic properties of lactic acid bacteria have become the subject of study. **Probiotics** are viable bacterial cell preparations or food containing viable bacterial cultures or components of bacterial cells that have beneficial effects on the health of the consumer (Table 7.7). Suitable probiotic bacteria have the properties of resistance to acid and bile, adherence to intestinal cells, transient intestinal colonization,

Table 7.6. Effect of fermentation on the vitamin content of some cultured milk products.

Vitamin	Uncultured products		Cultured products		
	Low fat milk	Cream (light)	Cultured cream	Yogurt (low fat)	Cottage cheese (low fat)
Ascorbic acid (mg)	1.0	0.8	0.9	0.8	Trace
Thiamine (μg)	36	32	35	44	21
Riboflavin (μg)	140	148	149	214	165
Niacin (μg)	88	57	67	114	128
Panthothenic acid (μg)	329	276	360	591	215
Vitamin B$_6$ (μg)	40	32	16	49	68
Folacin (μg)	5	2	11	11	12
Vitamin B$_{12}$ (μg)	0.38	0.22	0.30	0.56	0.63
Vitamin A (IU)	204	720	790	66	37

Table 7.7.   Documented health benefits from the following probiotic strains.

Probiotic effect	Genus	Species	Strain
Direct effect on intestinal complaints			
Rotavirus diarrhea shortened	*Lactobacillus*	*casei*	Shirota
		reuteri	SD2112
		rhamnosus	GG
	Bifodbacterium	*lactis*	Bb12
Relapses of inflammatory bowel disease fewer	*Escherichia*	*coli*	Nissle 1917
	Saccharomyces	*cerevisiae*	Boulardii
Helicobacter pylori colonization reduced	*Lactobacillus*	*johnsonii*	La1
Inflammatory bowel disease and irritable bowel syndrome relieved	*Lactobacillus*	*plantarum*	299v
		rhamnosus	GG
		salivarius	UCC118
	Bifidobacterium	*breve*	
Travelers' diarrhea incidence reduced	*Bifidobacterium*	*lactis*	Bb12
Antibiotic associated diarrhea reduced	*Lactobacillus*	*acidophilus*	La5
Effects on other aspects of health			
Oral vaccination improved	*Lactobacillus*	*johnsonii* La1	
	Bifidobacterium	*lactis*	Bb12
Modulation of immune system	*Lactobacillus*	*casei*	Shirota
		rhamnosus	GG
Treatment and prevention of allergies	*Bifidobacterium*	*lactis*	Bb12
	Lactobacillus	*rhamnosus*	GG
Recurrence of bladder cancer reduced	*Lactobacillus*	*casei*	Shirota
LDL-cholesterol reduced	*Lactobacillus*	*plantarum*	299v

Modified from Ouwehand *et al.* (2002).

and lack of any deleterious effects. Fermented milk has become a vehicle for the delivery of probiotic bacteria.

7.3.4. *Application of genetic engineering to dairy starter cultures*

7.3.4.1. *Phage control*

By far the most widely used pure bacterial cultures are the dairy starters. The application of pure bacterial cultures to the quality problems of the cheese (and butter) industries dates back to the very earliest application of the science of bacteriology. Dairy starters were being sold by Hansen's in the late 19th century and by 1929 the first pure strains of dairy starter cultures for cheese production were being used commercially in New South Wales and were subsequently adopted in New Zealand where the phage problem was discovered. With the advent of molecular biology much has been done to address dairy starter problems and their performance.

There are at least three major roles played by the lactic acid bacteria in dairy products; they are production of acid, of flavor and of texture. One of the characteristics of dairy starter cultures has been their inherent instability that has proved to be due to the technologically important traits being found on plasmids. These traits include lactose transport and hydrolysis, cheese maturation proteases, citrate permease and diacetyl formation, bacteriocin production and phage resistance.

Reliable acid production in the inoculated milk vat is crucial for predictable and good cheese quality and the greatest enemy of acid production is phage attack. The unpredictability of attacks, which restricts the range of strains that can be used in a given situation, is a problem so if cultures were available for which it was known that attack could not occur then cheese making would become more predictable and the range of available starter strains might be increased. Generally the ultimate source of phages is unknown but genetic studies have shown that the starter cultures themselves can be the source of the phages.

The first lactic acid bacterium for which the complete genome was elucidated is *Lactococcus lactis* subsp. *lactis* IL1403 that contains five complete prophages. In some bacterial strains the prophages may constitute 3–10% of the total genome.

The DNA sequences of over 100 phages of the lactic acid bacteria, in particular those that attack strains of *Lact. lactis* and *Strep. thermophilus*, are the most completely known. In addition, many mechanisms of resistance to phage have been characterized and as at 2002 they include: phage abortive infection systems (Abi) — 21 known; restriction modification (R/M) systems — 29 known; adsorption inhibition (Ads) mechanisms — 8 known; injection blocking systems — 2 known. These systems are predominantly linked to plasmid DNA, a characteristic that allows horizontal transfer via conjugation between starter strains so that phage resistance can be introduced to a phage sensitive strain that may have desirable flavor forming characteristics.

One plasmid encoded Abi system, namely pNP40, found in *L. lactis* subsp. *diacetylactis* DRC3, has been transferred to strains used industrially. The system seems to work by blocking phage adsorption to the cell wall by masking the receptor site. Changes in the cell wall lipoteichoic acid, or secretion of an external polymer, or altered cell wall proteins may be the reason for the blockage. A gene in a lactococcal prophage designated sie_{2009} produces a protein associated with the cell membrane. The phage attacking the bacterium adsorbs to the cell wall and transfects (in the sense of releasing its genetic material beyond the cell wall), however, plasmid transduction and phage replication are prevented by the membrane protein.

There are many phage abortive infection systems that can act in a variety of ways to prevent successful completion of a phage replicative cycle (Table 7.8) but they have only been found in *L. lactis* subsp. *cremoris*; and *L. lactis* subsp. *lactis* cultures so far. The lactococcal phages are divided into 12 species and the phage species controlled by these systems are the species 936, P335 and c2 that are members of the *Siphoviridae*. The Abi systems involve proteins that range in size from 628 to 127 amino acids long that are mostly coded on plasmids but some are on chromosomes. All affect phage species 936, four affect species P335 (Abi's A, K, T, U) and 13 affect C2 species.

The restriction/modification systems were the earliest systems known and they are classified into three types; Types I, II and III (Table 7.9). Type I and III systems cleave the DNA at a substantial distance from the

Table 7.8. Phage abortive systems in *Lactococcus lactis* subsp. *cremoris*; and *L. lactis* subsp. *lactis* and the action the systems have on the phage replication cycle.

Abi system	Action of the system on the phage replication cycle
A, F, K, R	Interferes with phage DNA replication
B, G	Interferes with RNA transcription
C	Reduces synthesis of the major capsid protein
D1	Interferes with a phage ORF essential for phage development
Q	Prevents maturation by accumulation of replicative form of phage DNA
S	Sequesters factors essential for phage development
U	Delays phage transcription

Table 7.9. Some restriction/modification systems, their types and their locations from *Lactococcus lactis* subsp. *cremoris* (LC), *L. lactis* subsp. *lactis* (LC) and *Streptococcus thermophilus* (ST).

Restriction/modification system		Bacterial host strain	Specificity[a]	Location: chromosome or named plasmid
Name	Type			
*Lla*1403I	I	LL IL1403	Undetermined	Chromosome
*Lla*90I	I	LL DPC721	Undetermined	pAH90
LLAKR2I	II	LL KR2	5' ↓ GATC 3'	pKR223
LLAFI	III	LL 42-1	Undetermined	pND801
LLA42OI*	I	LC IL420	Undetermined	pIL2614
*Lla*GI	I (var)	LC W10	Undetermined	pEW104
*Scr*FI	II	LC UC503	5' CC ↓ NGG 3'	Chromosome
LLAAI	II	LC W9	5' ↓ GATC 3'	pFW094
*Lla*BII	–	LC W56	Undetermined	pJW565
STHSFII	I	ST Sfi1	Undetermined	Chromosome
STH135I	I	ST 135	Undetermined	pER35
STH134I	II	ST 134	5' C ↓ CGG 3'	Chromosome
*Sth*8I	II	ST 8	Undetermined	pSt08

[a]The cleavage point where known is indicated by ↓

*Designated *Lla*2614I in REBASE.COM. Modified from Table 2 in Coffey and Ross, recognition site while Type II systems cleave DNA either within or very close to the recognition site (see Sec. 4.4.1).

7.3.4.2. Yogurt and neutraceuticals

The soft texture of yogurt and a lack of whey separation are important characteristics. Extra gelling or thickening agents may be added to the mix during manufacture to ensure this outcome but a more desirable situation is for the thickening agents to come from the bacteria. Some starter bacteria can produce exopolysaccharides that could fill that function. Two strains of *L. lactis*, namely NIZO B40 and SBT 0495, produce an identical polymer (Fig. 7.11) with a regular repeating unit, $\rightarrow 4)[\alpha$-L-Rha$p$-(1$\rightarrow$2)][α-D-Gal$p$-1-PO$_4$-3]-β-D-Gal$p$-(1$\rightarrow$4)-β-D-Glc$p$-(1$\rightarrow$4)-β-D-Gal$p$-(1$\rightarrow$, that is coded on a 42,180 bp plasmid, designated pNZ4000, that can be conjugally transferred between lactococcal strains. The gene cluster for the polymer is 12 kb and has 14 coordinately expressed genes, namely *epsRXABCDEFGHIJKL*. The polymer is synthesized on a membrane-linked lipid carrier.

Fig. 7.11. The repeating unit in the exopolysaccharide produced by *Lactococcus lactis* strain NIZO B40. If the gene for its production is transferred to suitable bacteria it may form the basis for a natural thickening agent in yogurt.

Because glycosyltransferase encoding genes are available, the possibility arises of oligo- and/or poly-saccharides with different repeating units being produced to suite particular needs. For example, it has been claimed that fructose-containing oligosaccharides stimulate the growth of bifidobacteria in the human intestine so if an organism could produce such an oligosaccharide it would be useful to promote the beneficial effects of probiotic organisms.

7.3.5. *Antibacterial factors in milk*

Normally the lactic acid bacteria grow well in milk but under some circumstances they might not grow, or they might not grow properly. There are three different kinds of antibacterial factors that can interfere with lactic acid production; natural factors secreted into the milk by the mammal, antibiotics in the milk in the udders of mammals treated for mastitis or other infections, and bacteriophage.

Natural factors include the lactoperoxide-thiocyanate-hydrogen peroxide system, lactotransferrin, and agglutinins. The first two factors can prevent the production of acid but the agglutinins cause the agglutination of the bacterial cells inoculated into unrenneted milk. The agglutinated cells form larger particles that fall more quickly than smaller particles to the bottom of the vat causing more acid to be formed on the bottom of the vat when compared with the top layer of the milk. Antibiotics prevent the growth of the lactic acid bacteria so no acid is produced. The phages can cause problems at three stages. At the first stage, they may infect the milk cultures being propagated daily for eventual use in the preparation of bulk starter. At the second stage, they may infect the bulk starter unit being used to prepare inoculum for the vats of milk. At the third stage, they may infect the vat of milk and the effect observed at this stage will depend on the cheese. In those cheeses requiring a slow setting time and low inoculum level gas-producing bacteria of the *Enterobacteriaceae* family may dominate. In cheese making procedures that require rapid acid production, acid production at the beginning of the process looks fine but as time passes the acid production slows down and stops due to the viral infections of the starter bacteria. Consequently the cheese produced does not reach the pH or degree of acidity required for it. This allows other contaminants to grow, such as *Staphylococcus aureus* that produces enterotoxins. The quality of the resulting cheese is poor and can be dangerous to health.

During a production day, the levels of bacteriophages built up in a factory can be very high in the whey within a relatively short period (Fig. 7.12). Bacteriophages in the cheese making industry are the major problem for the industry, and it is the classic example of an industrial problem caused by a virus.

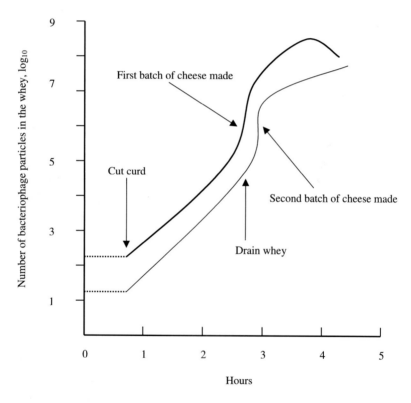

Fig. 7.12. A generalized diagram showing the growth and production of bacteriophage in whey produced during cheddar cheese manufacture. The amount of phage produced in whey from the first batch of cheese in the vat was more than that produced during production of the second batch. High concentrations of bacterial viruses build up quickly in the liquid phase and contaminate the factory environs.

Box 7.3

Traditional Techniques for Dealing with Bacteriophage Attacks in Commercial Bacterial Fermentations such as in Cheese Making

Traditionally mixed natural cultures were used but the quality of the product varied. To obtain consistent products, pure cultures were developed to make cheeses. However, when pure cultures are used, they sometimes fail due to bacteriophages that attack the bacteria. The reason for the problems is that during the growth of the fermenting bacteria, very large concentrations of bacteriophages build up in the environment of the factory. When a cheese maker tries to prepare mother cultures to make bulk starter cultures for the cheese making process, these cultures became infected with the bacteriophage and the bacteria are killed before they can grow. Where do bacteriophages come from? A primary source is the raw material (milk) that contains bacteria which are attacked by the phages that are resistant to the normal pasteurization temperatures used. Lysogenized cultures are also a source, the genomes of many starter bacteria code for prophages. Steps developed to control proliferation of the phage in a factory include: filtered air in culture preparation rooms, separation of personnel who handle cultures from those who work in the factory, daily rotations of known strains of phage unrelated cultures, the use of bulk starter media specially formulated to prevent bacteriophage attack and multiplication in the bulk starter medium, daily testing of whey for the presence of bacteriophage so as to change cultures, and purchasing of cultures from outside the factory. Frozen concentrated starter cultures are prepared elsewhere and shipped to factories for use directly in vats or for use in the preparation of bulk starter. Enclosed cheese vats and careful handling and disposal of whey also help in the control of phage.

7.4. Meat Fermentations

There are two broad applications of microbial fermentation to meat products. The first is in the production of hams and bacon; the second is in the production of fermented sausages. In the first case the portions of carcass are left in one piece while in the second case the muscle meat is comminuted.

7.4.1. *Production of ham, bacon and sausages*

Within the structure of muscle there are no bacteria. Bacteria collect on the surface of the carcass during the dressing of the carcass after slaughter. These bacteria can grow and cause spoilage, or disease, so meat is a perishable product that must be preserved for future use. The oldest technology used is salting and drying. Another method that was developed was to comminute the meat, add flavors and preservatives and stuff it into the cleaned intestines of animals before drying immediately, typically used for Chinese preserved sausages. In the comminuted form, bacteria are distributed all over the surfaces of the meat particles. Even though in modern Western production systems bacteria are used in meat fermentations, partial dehydration with fermentation still plays a role in the meat preservation.

The chemical qualities of meat influence microbial growth on the meat particles and surfaces. The pH of normal meat is about 5.5 (DFD — dark firm dry — meat is over pH 6.0). This arises due to the rigor mortis chain of metabolic reactions in the muscle. At death, the pH of the muscle is about 7.0 and glycogen present in the meat is broken down to lactic acid and the pH drops. Some glucose remains in the meat, depending on the animal species from which the meat came (Table 7.10) as well as other nutrients such as amino acids and peptides suitable for bacterial growth. In addition, meat contains hemoglobin, an iron-containing molecule that, under the right conditions, can reduce the redox potential of the meat so that anaerobic bacteria such as *Clostridium botulinum* will grow. *Cl. botulinum*, an obligate anaerobe, produces the most toxic substance known, botulinum toxin. Nowadays, the prime purpose of using additives in meat

Table 7.10. Typical glucose concentrations in various kinds of meat.

Meat	Glucose concentration $\mu g\ g^{-1}$
Beef	100
Sheep meat	400
Pork	800

preservation is to avoid the growth of this bacterium, while a secondary purpose is to prevent spoilage. Fortunately, heating the toxin to boiling point will destroy it, but unfortunately, many preserved meats are not cooked before consumption. In addition to this bacterium, under aerobic conditions *Pseudomonas* species grow on the meat, and under restricted oxygen conditions lactic acid bacteria can grow and suppress *Pseudomonas* spp. Further, because meat is a good source of food, bacteria grow rapidly at high temperature so low temperatures are used during preservation, usually about 20°C or less. Even though glucose is commonly found in meat, glucose is commonly added to meat fermentations to encourage the growth of lactic acid bacteria and other microorganisms. It is also common to add other flavoring agents such as pepper etc.

Nitrite is the key chemical in the preservation of meat, in addition to salt. Nitrite reacts with the heme (Fig. 7.13) in meat to produce nitrosyl myoglobin that then becomes nitrosyl myochromogen when the myoglobin is denatured, e.g. by cooking. This changes the red color of the heme to the characteristic color of cured meats. Usually it is added as either potassium nitrate or nitrite and when nitrate is used, the nitrite is produced by the action of nitrate reducing organisms in the fermentation.

In the manufacture of hams and bacon, the meat is placed in a saline solution containing (w/v) about 20% NaCl, and 0.5–0.7% potassium nitrate/nitrite and left for a few days to diffuse into the

Fig. 7.13. The heme molecule is the key in the formation of the typical cured meat color.

Box 7.4

How does Meat Become a Good Substrate for Microorganisms to Grow on?

Energy in muscle is stored as glycogen — a long chain polymer of glucose. In a live animal, the glycogen is hydrolyzed to produce glucose. The blood flow brings O_2 to the muscle and the glucose is usually metabolized via the EMP pathway and the citric acid cycle to produce mainly energy and to convert it to CO_2 and H_2O that are taken away by the blood flow in the animal. At slaughter the blood flow stops and no further O_2 flows to the muscle. At this stage the process leading to rigor mortis begins. In this process the metabolic pathways continue operating in the muscle while the glycogen and glucose are present. Thus the glucose enters the EMP pathway and is metabolized to pyruvate but because there is no O_2 it cannot enter the citric acid cycle as in the normal respiratory pathway. Consequently, the pyruvate, to gain further energy, is converted to lactic acid that builds up in the muscle and the pH of the muscle drops. Eventually, at about pH 5.5 in the muscle, no further glycogen is converted to glucose and the metabolic process halts. If there is little or no glycogen in the muscle at death the final pH will be above 6.0 and that influences the kind of microbial growth on the meat.

During these biochemical steps to rigor mortis, other energy compounds are released and eventually a significant amount of small molecular weight compounds are available for microbial growth. One key compound is glucose. The amount of glucose in meats from various domesticated animals varies as shown in Table 7.10. Because the compounds are in the muscle, the microorganisms cannot access them unless the meat is comminuted. Once the meat has been reduced to small particles the glucose and other small molecules can diffuse to the surface to replace those being used by bacteria on the surface. Under natural conditions the lactic acid bacteria come to dominate the microflora growing in the comminuted meat.

meat. In addition, brine solution may be injected into the muscle. The natural bacterial microflora in brine solution breakdown the nitrate to nitrite that then diffuses into the meat reacting with the hemoglobin. Following immersion in the brine, the hams and bacon are removed and left to dry out under controlled conditions. If the ham is dried for a prolonged period, say many months, the surface may be covered with a fat layer to slow down moisture loss and yeasts and molds grow on the surface. This fungal growth contributes to the development of flavor.

To make fermented sausages potassium nitrate or nitrite is added to the meat at low concentration. When nitrate is added several things begin to happen when the sausage mixture is made. Gram-positive facultative aerobes, such as *Staphylococcus* spp. and *Micrococcus* spp. convert the nitrate to nitrite. The nitrite then reacts with the hemoglobin and fixes the red color of the meat. It also reduces the redox of the meat mixture to a level above the level at which the obligate anaerobes can grow. The nitrate/nitrite mixture renders the meat safe and free from the risk of botulinum poisoning. The presence of the nitrite does not affect the growth of the lactic acid bacteria.

7.4.2. *Starter cultures in meat fermentation*

The production of fermented sausages is historically a relatively recent development as it is thought that it was only about 250 years ago that the first fermented sausages were produced. Originally these fermented sausages were made using various mixtures of ingredients and the manufacturer depended on the natural microflora for the microbial growth. However, in around 1954, lactic starter cultures were introduced for the manufacture of sausages. Today nearly all fermented sausages are made using added starter cultures (Table 7.11). In addition to bacterial cultures, fungal cultures are also used, for hams as well as sausages.

The first group of cultures now used in modern fermented meats production is the *Micrococcaceae*, such as strains of *Kocuria varians* (formerly *Micrococcus varians*), *Staphylococcus carnosus* and *Staph. xylosus*. These organisms help in the formation and stabilization of color as well as aroma formation. Their important nitrate reductase activities are influenced adversely by NaCl concentration and low temperature. Color stabilization is related to their ability to produce catalase, the enzyme that degrades H_2O_2. The nitrosyl myoglobin, arising from the heme-nitrite reaction, is susceptible to oxidation by H_2O_2 to other colored substances

Table 7.11. Cultures used in meat fermentation and their useful properties for fermented sausage production.

Culture	Some characteristics and useful properties
Kocuria varians	Nitrate reduction; catalase production; color formation.
Staphylococcus carnosus; *Staph. xylosus*	Nitrate reduction; catalase production; color formation; typical dry-salami aroma.
Streptomyces griseus	Better aroma, darker red color; uses O_2; does not grow in meat mixture.
Lactobacillus plantarum; *Lb. sakei*; *Lb. curvatus*	Homofermentative lactic acid production, but late heterofermentative metabolism occurs; hastens early moisture loss; some produce H_2O_2, destabilizes color, induces rancidity.
Pediococcus pentosaceus; *P. acidilactici*	Not natural to fermented sausages; produces lactic acid homofermentatively; hastens early moisture loss; dies off rapidly.
Debaromyces hansenii	Use O_2 rapidly, mainly active about meat/air interface of sausage; rapid exterior color formation; assimilates lactic acid giving slight rise in pH; produces Italian type flavor.
Penicillium nalgiovense, *P. chrysogenum*, *P. camemberti*	Reduces late moisture loss; consumes O_2 at surface, reduces O_2 penetration; anti-oxidative effect, reduces rancidity; mold penetrates whole of sausage, influences flavor and aroma; assimilates lactate, raises pH.

such as green verdoheme and yellow biliverdin. The H_2O_2 also initiates the oxidative processes that lead to the formation of rancid flavors. The H_2O_2 arises through the growth of certain lactic acid bacteria, in the presence of O_2, about the air/meat interface. These cultures of *Micrococcaceae* produce catalase to break down the H_2O_2 thus preventing the formation of off colors. We have to include here another unusual bacterial culture sometimes used, namely *Streptomyces griseus sensu* Hutter. Introduced in 1977, it produces a better aroma and darker red color. The organism is proteolytic, non-lipolytic, nitrate-reducing, catalase positive and non-pathogenic. It does not grow in sausage mixtures, nor does it survive well. However, when added to sausage mixtures, a good flavor develops and nitrate levels decline, indicating metabolic activity.

The second group of cultures is lactic acid bacteria. The species are from the genera *Lactobacillus* and *Pediococcus* and include *Lb. plantarum*, *Lb. sakei*, *Lb. curvatus*, *P. pentosaceus* and *P. acidilactici*. The desirable function of these bacteria is to produce lactic acid homofermentatively, that is, with no gas formation and little or no acetate/ethanol formation. The cultures of *Lb. sakei* and *Lb. curvatus*, although they grow at low temperatures, a desirable characteristic, they also tend to produce H_2O_2. On the other hand, pediococci are rarely found in naturally fermented sausages, but function well in the sausages because of their obligate homofermentative activity. The other lactic cultures used can switch metabolic pathways under certain conditions to produce formate, acetate and ethanol from lactate. The lactic acid produced lowers the pH of the meat to about 5.1–5.2 and decreases the water holding capacity of the meat so that free water is released and the sausage can dry more quickly.

The third group of cultures is the fungi — yeasts and molds. The species of yeasts commonly associated with natural fermentation are *Debaromyces*, *Candida* and *Torulopsis*. In commercial cultures, strains of *D. hansenii* are used, particularly for typical flavors of Italian type sausages, but they should only be used in conjunction with other cultures. The cultures when added at about 10^6–10^7 cells/g, use O_2 rapidly and are mainly observed at the periphery of the sausage where O_2 is available for growth. This results in rapid exterior color formation. The yeast is able to assimilate lactic acid so a slight rise in pH occurs.

The molds most commonly found on the surfaces of fermented meats are species of *Aspergillus* and *Penicillium*. However, most of the molds found on fermented meats are toxinogenic species so care has to be taken to select non-toxinogenic strains. The first white non-toxinogenic culture made available commercially in 1972 was a strain of *P. nalgiovense*. Other species used include *P. chrysogenum* and *P. camemberti*. The effect of the mold is to reduce moisture loss due to the mold covering, and it has anti-oxidative effects due to the catalase activity of the mold, and the consumption of O_2 that reduces O_2 penetration into the sausage.

Moreover, the mold penetrates the whole of the sausage and influences its flavor and aroma due to lipolytic and proteolytic activity as well as raising the pH, not because of ammonia production but because of lactate assimilation. The mold penetrates the sausage better if the casing for the sausage is natural and not made from artificial collagen.

Genetic engineering has been used to improve strains for meat fermentation. One approach has been to remove the need for the *Staph. carnosus* or *K. varians* strains by inserting the genes for catalase production into *Lb. curvatus*. The meat starter *Lb. curvatus* strain LTH683 was the base for the development. The *Lactobacillus* cloning vector pJK356 based on the cryptic plasmid pLC2 derivative of the meat starter strain LTH683 was isolated. The host was *Lb. curvatus* strain LTH1432, a plasmid cured derivative of strain LTH683. The catalase gene, *katA* from *Lb. sakei* strain LTH677, was then inserted and a recombinant strain of *Lb. curvatus*, strain LTH4002, isolated. Under aerobic conditions of growth this recombinant strain showed catalase activity four times greater than the strain of *Lb. sakei* used for the original source of the gene.

7.5. Vegetable Fermentation

A lot of vegetables are highly perishable and they are generally only available on a seasonal basis. In addition, some vegetable products are not at all highly perishable but have characteristics that make them unsuitable for human consumption. However, they can be made suitable by fermentation. In this section we will deal with these products.

7.5.1. *Fermented vegetables for preservation purposes*

These products can be broadly placed into two categories, those that are sliced before fermentation, and those that are fermented whole. Examples of the former are kimchi, the national dish of Korea, and sauerkraut, the fermented cabbage from Germany. Examples of the latter are fermented cucumbers, also from Germany and radishes from Japan.

Table 7.12. Fermentable sugars composition (%) in vegetables that are often fermented.

Compound	Cabbage leaves	Cucumbers	Olives
Sucrose	0.1–0.4	ND	0.1–0.2
Fructose	2.0–2.1	1.0–1.4	0.1–1.1
Glucose	2.2–2.6	1.0–1.3	1.0–3.0

The vegetables, as shown in Table 7.12, contain glucose, fructose and traces of sucrose. Also, the natural habitat of the lactic acid bacteria is the surfaces of vegetable matter. Under normal circumstances, the hexoses remain in the plant and the plant material repels attacks from microorganisms.

(1) Sauerkraut

Sauerkraut is fermented cabbage. The cabbage is shredded and about 2–2.5% (w/w) dry salt is incorporated into the sliced cabbage. The NaCl causes an osmotic shock to the plant cells and the carbohydrate-laden juices flow from the cells forming a brine. The shredded cabbage is then pressed down into the brined juice to exclude air. The lactic acid bacteria on the external surface of the plant materials, being halotolerant, then attack the glucose and produce lactic acid. The ideal temperature for bacterial growth is 18°C with a NaCl concentration in the brine of 1.8–2.25%. From the beginning of the fermentation, there is a distinct progression of bacterial species that carry it out. At first the hetero-fermentative organism *Lc. mesenteroides* dominates because of its short generation time and early initiation of growth. Due to the fermentative end products, undesirable Gram-negative bacteria die off at this stage. However, because *Lc. mesenteroides* is relatively sensitive to acid conditions, it declines in number and *Lb. brevis* and *Lb. plantarum* eventually dominate the fermentation. Initially *Lc. mesenteroides* produces lactate, acetate, ethanol and CO_2 from glucose, fructose and sucrose, but it can also use fructose as a final electron acceptor and produce mannitol. Finally *Lb. plantarum* dominates. At the end of the fermentation that is completed within 2 weeks to 2 months, the pH is ≤ 3.5 and total acidity about 1.6–2.3%. Fermentation temperatures above 30°C result in poor quality sauerkraut while low temperatures (≤ 10°C) favor *Lc. mesenteroides* exclusively and result in good quality product. Curiously this organism produces D (−) lactic acid that is not metabolized by humans so sauerkraut is not recommended for babies and small children.

Attempts to isolate and produce commercial starter cultures suitable for industrial production of sauerkraut have proved unsuccessful.

(2) Cucumber Fermentation
Cucumbers are fermented to produce pickles. The cucumbers are immersed wholly into open brine tanks containing 5–8% NaCl and incubated at ambient temperatures, usually in the range 15–32°C. One of the main defects of concern to pickle producers is the formation of "floaters" or "bloaters" that develop internal gas causing the cucumbers to float. The solution is to ensure the removal of CO_2 or the prevention of its formation. Acetic acid is often added to the brine to reduce the pH to ≤ 4.5. This facilitates the expulsion of CO_2 from the brine when it is purged with N_2 or air. Lactic acid bacteria and yeasts grow in the brine. The main lactic acid bacteria that complete the fermentation are *P. pentosaceus*, *Lb. brevis* and *Lb. plantarum*, the last being the most important species (Fig. 7.14).

Sources of CO_2 vary. A prime source is heterofermentative growth by lactic acid bacteria. In addition, there is sufficient malic acid (0.2–0.3%) in cucumbers for the *Lb. plantarum* and the other lactic acid bacteria to carry out the malolactic fermentation and produce CO_2.

The major fermentation control methods used include initial low pH, purging of CO_2, and NaCl level. At present no commercial cultures have been developed for this fermentation.

7.5.2. *Fermented olives*

Olives are processed either unripe (green) or ripe (black) and contain a bitter substance, the phenolic glucoside, oleuropein, that makes olive fruit inedible. It is either removed or hydrolyzed by soaking in water or brine or through alkaline hydrolysis (1–2% NaOH). Alkaline (lye) treatment of black olives does not involve fermentation, but brine treatment of black

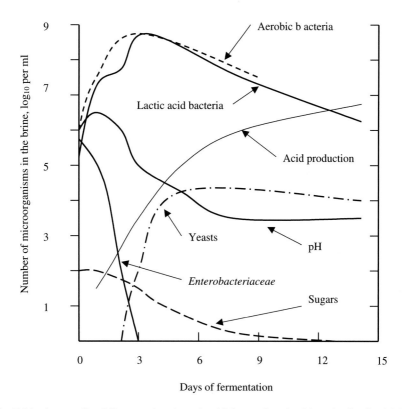

Fig. 7.14. A generalized diagram showing microbial growth and acid production in pickled cucumber. Lactic acid bacteria eventually dominate but the acid conditions also allow the yeasts to grow.

olives does. On the other hand lye treatment of green olives (for 5–12 hours) is followed by fermentation.

As a consequence of the lye treatment the microbial load is reduced and the pH is in the range of 7.5–8.5. The fermentation then goes through three stages. In the first stage, because of the pH, a wide variety of Gram-negative aerobes (e.g. species of *Pseudomonas*, *Flavobacterium* and *Aeromonas*) and facultative aerobes (e.g. species of *Enterobacter*, *Citrobacter* and *Klebsiella*) predominate for up to two days. The time is reducible by lowering pH to 6.0 by bubbling CO_2 through the liquid or by adding lactic or acetic acids. At this pH, the initial microflora die off and lactic acid bacteria start to grow in the second stage. The pH is lowered to about pH 4.5 over 10–15 days during which time *Lb. plantarum*, along with some yeasts and occasionally *Lb. delbrueckii*, dominate and reach a peak in numbers after about a week. Once pH 4.5 is reached, the third stage of the fermentation begins and lasts until the fermentable carbohydrates are depleted. *Lb. plantarum* dominates but also present are high populations of fermentative and oxidative yeasts that produce ethanol, ethyl acetate and acetaldehyde, that may contribute to the final flavor. Some of the yeast species involved include *Hansenula anomala*, *H. subpelliculosa*, *Candida krusei*, *C. parasilopsis* and *Saccharomyces chevalieri*. A fourth fermentation stage can occur but it usually results in spoilage.

Starter cultures for olive fermentation are generally obtained by the old method of re-inoculation from a previous fermentation. *Lb. plantarum* dominates in these fermented brines and the pH is usually about 4. Efforts to isolate suitable *Lb. plantarum* cultures have resulted in one culture, LPCO10, being isolated. When used in fermentation, its success depended on its ability to produce two bacteriocins: plantaricins S and T that prevented the growth of clostridia, a significant concern in olive fermentations, and propionibacteria as well as other lactic acid bacteria.

7.6. Soybean Fermentations

Soybeans are a good source of protein and are able to meet the essential amino acid needs of humans. However, like all dried legume seeds, the proteins are not readily digestible and contain anti-nutritional factors that can be destroyed by cooking and processing before consumption. When the cooked beans or processed beans are subjected to microbial fermentations a variety of diet-enhancing soybean products are produced. Fermented soybean products are most common in Asia.

7.6.1. *Sauces and pastes with meat-like flavor (soy sauce)*

Soy sauce is almost as old as recorded Chinese history. It was introduced to Japan in 552AD. During the 17th and 18th centuries, the Dutch exported large quantities of soy sauce from Japan to other parts of Asia and to

Europe. The Japanese established a soy sauce manufacturing plant in the United States of America in 1909.

The essential ingredient for soy sauce production is soybean. After mold growth on the grain, it is steeped in a brine solution. Soy sauce is the liquid extract from the mixture. However, the varieties of soy sauces on the market owe their differences to the addition of wheat, a source of carbohydrate. Soybean is high in oil and protein and contains about 30% carbohydrate that is comprised of a soluble and an insoluble fraction (Table 7.13). The soluble fraction, about 10% of the bean, is a mixture of sucrose (5%), raffinose (1%) and stachyose (4%), and the insoluble fraction consists of structural carbohydrates such as pectin, hemicellulose and cellulose, and less than 1% starch.

Starch added as wheat in the soybean mix promotes the growth of the fungus and production of enzymes. This is desirable as a greater amount

Table 7.13. Composition of soybeans, defatted soy flour, wheat grain and wheat flour, g kg^{-1}, used in the manufacturing of soy sauce.

Component	Soybean		Wheat	
	Whole bean	Defatted flour	Whole grain	Flour
Water	86	72	115–140	–
Protein	343	470	153	140–180
Oil	187	12	19	9–24
Ash	51	62	18.5	4–14
Carbohydrate	316	384	717	650–740

Table 7.14. Some of the major types of soy sauce, and the soybean: wheat ratios used in their production.

Soy sauce	Soybean:wheat ratio	Comment
Tamari	90:10	Most common type in South East Asia
Saishikomi	50:50	Koji is steeped in previously made raw soy sauce
Koikuchi	50:50	Most popular soy sauce in Japan, 83% of market
Shiro	25:75	High sugar content and very light color
Usukuchi	About 33:66	Color lighter than koikuchi
Kecap asin	100:0	Indonesian soy sauce

of enzymes enable better breakdown of the protein. The soybean: wheat ratios used vary from 100:0 to 25:75. In Japan, these ratios are formalized and the resulting soy sauces have various names (Table 7.14). Although the main Chinese type of soy sauce is similar to tamari, ratios of 66:33 and 50:50 are also common in China.

The carbohydrate in soybean and wheat is broken down into fermentable sugars with the aid of fungal enzymes from either *Aspergillus oryzae* or *A. sojae*. The soy sauce making process for koikuchi soy sauce is summarized in Fig. 7.15.

An important step is the cooking of the soybean. The yield of soluble nitrogen from the soybean increases with increasing cooking temperature and decreasing cooking time. Cooking temperatures of 118°C for 45 min and 170°C for 15 s result in soluble nitrogen yields of about 82% and 88%, respectively (Table 7.15). Defatted soybean is often used instead of whole soybean because the oil plays no significant part in the soy sauce. Wheat bran is often preferred over wheat kernels as it serves as a

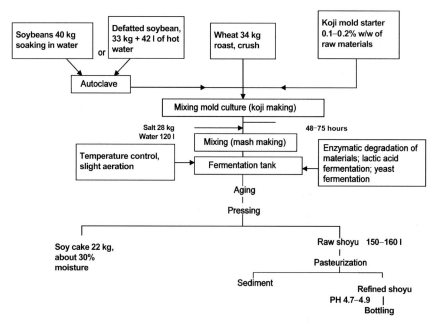

Fig. 7.15. The flow diagram setting out the process used to produce soy sauce. The process depicted here is for the production of koikuchi soy sauce. The other types depend on variations in the soy: wheat ratio.

Table 7.15. Effect of cooking pressure, time and temperature of soy beans on the yield in soy sauce of total and amino nitrogen and percent nitrogen yield, and protein digestibility of the beans.

Cooking			Effect of cooking on nitrogen fractions in soy sauce and on protein digestibility			
Pressure, kg cm^{-1}	Temperature, °C	Time, min	Total N, mg mL^{-1}	Amino N, mg mL^{-1}	Yield, %	Protein digestibility, %
0.9	118	45	16.53	8.17	82.05	86
2.0	132	5	16.97	8.37	84.24	92
6.7	151	2	17.52	8.39	86.87	–
7.0	170	0.25	17.78	8.78	88.26	95

source of ferulic acid that is converted to 4-ethylguaiacol. In addition, the pentose in wheat bran serves to enhance darker coloration and color stability, but results in reduced alcohol content in the soy sauce.

The vocabulary in soy sauce production is dominated by Japanese terminology. The molded soybean mixture is called "koji" which in Japanese means "bloom of mold". When the koji is mixed into the brine the resulting mixture is called "moromi".

As a preliminary step in koji making, steamed polished rice or wheat bran is often used to produce tane koji, the equivalent of a starter culture, to inoculate the prepared soybean and wheat for sauce production. The seed koji (0.1–1.0% w/w) is mixed with cooked soybean and wheat. The mixture of tane koji, soybean and wheat is spread out to a depth of about 3–5 cm and left to incubate for 2–3 days to allow mold growth and production of enzymes.

The temperature of the koji increases rapidly and reaches about 38°C. The optimal temperature for the growth of the mycelium of A. oryzae is between 30 and 35°C. The fungi produce amylase in the temperature range 35–40°C. In order to achieve faster mycelium growth and to prevent over production of amylases that cause sticky koji, the substrate needs to be cooled by mixing twice at around 18 hours and 27 hours of incubation (Fig. 7.16). Good ventilation is important, as it helps to prevent over-heating. However, over-cooling of koji can be detrimental, as it encourages the growth of undesirable yeasts (e.g. *Zygosaccharomyces rouxii* is

Fig. 7.16. The temperature and relative humidity in a traditional small tray method for preparing koji. The temperature increases and must be reduced by stirring the mixture. The optimum levels of enzymes are produced after about 48 hours. The harvested koji is then steeped in brine (after Fukushima).

undesirable at this stage but desirable in the moromi stage) and other fungi (e.g. *Rhizopus nigricans*). Common bacterial contaminants are *Bacillus subtilis*, and species of *Micrococcus*, *Leuconostoc* and *Lactobacillus*.

The growth of the mycelium eventually slows at around 36 hours, and the temperature in the koji gradually cools down to the room temperature. During the last stage of koji making, proteases begin to accumulate. The amount of proteases accumulated in koji is important as it determines the flavor of soy sauce. The process is stopped before fungal spore formation. Enzymes produced by koji molds are shown in Table 7.16.

The koji is subsequently steeped in brine water to give a salt concentration of 17–19% w/v. The high salt content inhibits the growth of *Micrococcus* and *Bacillus* species. Use of water containing low iron and copper to make the brine is important, as high mineral contents reduce color stability. The brine water is first adjusted to pH 6.5–7.0 and cooled to 0°C to maintain the temperature of the freshly prepared moromi mash

Table 7.16. Enzymes produced by koji mold and the levels of enzymic activities recorded.

Enzyme	Activity (units g^{-1} koji)
Total protease	1129
Acid protease (pH 3)	56
Neutral protease (pH 7)	89
Alkali protease (pH 7)	984
α-amylase	3920
Acid carboxypeptidase	0.46
Leucine aminopeptidase	0.36
CM-cellulase	21.6
Pectin transeliminase	12.3

at about 15°C for several days. The initial low temperature prevents a rapid drop in pH due to lactic acid fermentation. Alkaline proteases remain active for a longer period to further digest the 1–3% w/w of protein in the mash. Thereafter, the temperature is allowed to increase gradually to 28–30°C after 20–30 days. The temperature level is important because it controls the production of ethanol in the soy sauce. Japanese consumers prefer soy sauce containing about 3% ethanol. To obtain this ethanol, a starch source, e.g. wheat, must be included in the koji, and the temperature must rise to 30°C. If the temperature is kept below this level, ethanol is not produced. At the end of 2–3 months, most of the fermentable carbohydrate has been converted to lactic (1% w/v), ethanol (2–3% v/v) and simple sugars (e.g. glucose, xylose).

7.6.1.1. Enzymic processes

The protein in the soybean and wheat is released by macerating enzymes such as cellulase, hemicellulase, and pectinase (Fig. 7.17). Then the alkaline and neutral proteinases release polypeptides that are further digested by acid carboxypeptidase and leucine aminopeptidase to release amino acids. Glutaminase breaks down the glutamine to glutamic acid, preferably at temperatures below 25°C and in a low salt condition. Pectin transeliminase

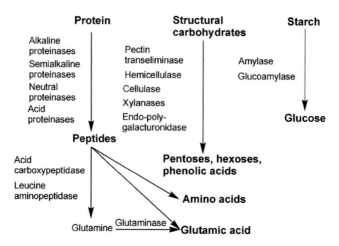

Fig. 7.17. The suite of enzymes produced by the fungi in the koji act on the three main components, the complex polysaccharides holding soy and wheat cells together, the proteins in the soybeans and wheat, and the starch in the wheat. The pentoses, hexoses and phenolic acids are further broken down to flavor compounds.

serves to reduce viscosity due to acidic polysaccharides derived from soybeans, so that the soy sauce is more easily pressed from the mash and clogging of the filter is avoided. The amylase enzymes release glucose from the starch.

7.6.1.2. *Microbial processes in the brine stage*

During the moromi stage, a complex microbial process occurs. The brine kills off aerobic bacteria and with the release of the glucose, lactic acid bacteria start to grow and lower the pH. When the pH drops to about 5.5, yeasts start to grow, in particular *Z. rouxii*, and its growth is followed by the growth of further important flavor producing yeasts (Fig. 7.18).

The soy sauce may be allowed to incubate a long time for aging, which involves darkening of color due to browning reactions.

Flavor compounds (Table 7.17) come via a variety of mechanisms and sources. *Z. rouxii* and *Candida* species such as *C. versatilis* produce 2-phenylethanol from *p*-hydroxy cinnamic acid and phenylalanine;

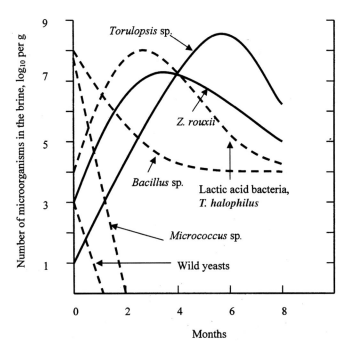

Fig. 7.18. The time course of the brine fermentation in Japanese style shoyu. The brine kills wild yeasts and species of *Micrococcus*, and any vegetative bacilli are killed off, leaving the endospores, and the lactic acid bacterium, *Tetragenococcus halophilus*, begins to grow lowering the pH. This then allows the yeast *Zygosaccharomyces rouxii* to begin to grow followed by the *Torulopsis* species, such as *Candida etchellsii*.

4-ethylguaiacol from ferulic acid; and isoamyl alcohol, iso- and n-butanol, 3-methylthio-1-propanol from amino acids; 2- and 3-methyl-1-butanol, 2- and 3-methylbutanoic acids from isoleucine and leucine; and methionol and methional from methionine. Roasting of the wheat allows the yeasts to produce furfuryl alcohol from furfural, and heating of the raw soy sauce promotes Strecker degradation reactions from which arise isovaleraldehyde, acetaldehyde and methional from dicarbonyl compounds and amino acids, as well as HMMF via the Maillard reaction, and HDMF from pentoses and amino acids (Fig. 7.19). The yeasts produce glycerol and *Tetragenococcus halophilus* produce acetic acid.

Table 7.17. Some major flavor components in Japanese soy sauce. The alcohols contribute to the volatile flavor of soy sauce. The HEMF is considered the **character impact substance** because of its intense sweet aroma. HDMF provides a caramel-like flavor. The characteristic soy sauce taste is also attributed to 4-ethylguaiacol produced by torulopsis yeasts. Methionol has a synergistic effect with 4-ethylguaiacol and it appears that 3.9 mg L^{-1} of methionol and 0.8 mg L^{-1} of 4-ethylguaiacol are the optimum concentrations.

Compound	Quantity, mg kg^{-1}	Compound	Quantity, mg kg^{-1}
Ethanol	31 501	Isobutanol	11
Lactic acid	14 346	Furfuryl alcohol	11
Glycerol	10 208	Isopentyl alcohol	10
Acetic acid	2107	Acetoin	10
HMMF	256	n-Butanol	9
2,3-Butanediol	238	HDMF	5
Isovaleraldehyde	233	Acetaldehyde	5
HEMF	239	2-Phenylethanol	4
Methanol	62	n-Propanol	4
Acetol	24	Acetone	4
Ethyl lactate	24	Methionol	4
2,6-Dimethoxyphenol	16	2-acetylpyrrole	3
Ethyl acetate	15	4-Ethylguaiacol	3
Isobutyraldehyde	14	Ethyl formate	3
Methyl acetate	13	γ-Butyrolactone	2
		ρ-Ethylphenol	Trace

HMMF = hydroxy-2(or 5)-methyl-5 (or 2) methylfuran-3(2H)-one
HEMF = hydroxy-2(or 5)-ethyl-5 (or 2) methylfuran-3(2H)-one
HDMF = 4-hydroxy-2,5-dimethylfuran-3(2H)-one

7.6.1.3. *Biotechnological approaches to soy sauce production*

One of the disadvantages of soy sauce production is the amount of time required to produce it, a problem common with all fermented foods with a long maturation time. Continuous processes and bioreactors to speed up the manufacturing time have been examined (Fig. 7.20).

One continuous method that ran for 80 days through a 280 l reactor involved immobilization of cells of *T. halophilus*, *Z. rouxii* and *C. versatilis* in calcium alginate gel to ferment the moromi. The soybean/wheat mixture was hydrolyzed enzymatically and passed through a column bioreactor containing immobilized *T. halophilus* cells for the lactic fermentation. Then the production stream was split into two, one stream went through a

R = H or CH$_3$

Pyrazine

4-hydroxy-2-methyl-5-ethyl-
3(2H)-furanone (HEMF)

4-Hydroxy-2,5-dimethyl-
3(2H)-furanone (HDMF)

Fig. 7.19. Important flavoring compounds in soy sauce (HEMF and HDMF) and natto (pyrazine).

Z. rouxii reactor and the other was put through a C. versatilis reactor. Each stream was then heated, filtered and combined to give a production time of two weeks, but the aroma pattern did not match traditional sauce. When the gel, unstable to heat and a high salt concentration, was replaced with a ceramic bead carrier for the whole cells, a satisfactory soy sauce was produced in eight days. A further advance was to use chitopearl to immobilize glutaminase, and alginate gel to immobilize the microorganisms. The koji mixture was then hydrolyzed in 8.5% NaCl at 45°C for three days, filtered and then put through the glutaminase reactor followed by the microbial reactors. Production took about two weeks and the system ran continuously for 100 d, but the product still did not match completely the natural product.

Fig. 7.20. A comparison of the traditional and a biotechnological approach for accelerating shoyu production. The difference between the traditional method and this biotechnological method is that the former requires up to eight months while the latter takes about two weeks.

7.6.1.4. *Variations in traditional soy sauce making*

The above process is basically the Japanese method of making shoyu, however, in other countries there is some variation in those methods used. In Indonesia only soybeans are used, the incubation period in the moromi (baceman) stage may be as little as one day but usually longer. Sugar may then be added to give a sweet soy sauce, kecap manis, as opposed to the normal sauce, kecap asin. In traditional soy sauce production in Hong Kong, the moromi stage is conducted in ceramic jars of about 40 L capacity and when the fermentation is completed the sauce is drawn off and fresh brine added, left for a few days and then the liquid drawn. Liquid is drawn about five times. This results in various grades of soy sauce. Further, the sauce is exposed to the sun for up to six weeks, a practice also in parts of China. In Korea, the koji is produced as a kind of

Box 7.5

Traditional Soy Sauce Production in Hong Kong

The photos show how soy sauce is prepared traditionally in Hong Kong. The soybeans are cooked, cooled and flour is added. The floured soybeans are then placed in a thin layer on trays to allow the mold to grow. Then the koji is added to the clay jars with brine and covered. They are left in the open for the moromi fermentation to take place. When the fermentation is judged to be complete the liquid is drawn off and more brine added. This is repeated several times so that up to five collections are made.

cake (meju or meiju). The soybeans are cooked and crushed to form a cake shape and inoculated with mold. The cakes are then incubated in the koji room at 15–20°C and 40–50% relative humidity up to 30 d. Better kanjang (Korean soy sauce) results if a temperature of 25–30°C and 80–90% relative humidity up to 30 d is used.

7.6.2. *Fermented soybean curd (furu, sufu)*

This is a cheese-like fermented bean curd, popular amongst Chinese people. It is also known as chao in Vietnam, tahuri in Philippines, taokaoan in Indonesia and tao-hu-yi in Thailand. Its production, outlined in Fig. 7.21, depends on the characteristics of the protein extracted as soymilk from soybeans. Beans are soaked overnight and then ground up into fine particles to form a slurry. The slurry is then filtered to produce soymilk and a solid waste product, okara. The soymilk contains protein and fat globules. Like casein, soymilk protein can be clotted to form a curd (tofu), but the clotting agent is usually Mg or Ca salts.

The dried surface of tofu encourages growth of fungi on the surface and hydrolysis of protein in the surface layer. The fungi used in commercial production of furu may include *Actinomucor elegans*, or *Mucor dispersus*, or *M. silvaticus*, or *Rhizopus chinensis*. However, in natural fermentations the furu fungi may, in addition, include *M. corticolus*, *M. praini*, *M. racemosus*, and *M. subtilissimus*, all of which are common microbial isolates found on the surface of rice or wheat straw.

During the ageing period, in brine, Na ions cause release of fungal proteases and lipases. The proteases penetrate the curd and hydrolyze the soybean protein component that results in a soft cheesy furu, and the lipase activity produces fatty acids. When ethanol is included in the brine, proteases activity is reduced and the fatty acids react with the ethanol to produce fruity flavored esters (Table 7.18).

7.6.3. *Whole bean fermentations*

For a substantial proportion of the human population, plants and seeds are the major sources of protein as animal products are not readily available.

Fig. 7.21. The production of furu. The raw material is soy extract from soybean (soy milk) that is precipitated and then fermented with microorganisms. The product has a very strong aroma.

However, plant proteins are lacking in taste and texture. Consequently, these populations often lack a plentiful supply of good quality protein. In addition, the cost of protein is high so to provide an adequate supply for the poor is a priority. One traditional source is fermented whole soybeans.

Table 7.18. Biochemical changes occurring during the maturation in brine or in brine with added ethanol, of furu made with *Actinomucor taiwanensis*.

Maturation time, days at 25°C	Brine, 12% w/v NaCl		Brine, 12% w/v NaCl, plus ethanol, 10% w/v	
	N ratio, amino N:total N	Free fatty acids, mg g^{-1} lipid	N ratio, amino N:total N	Free fatty acids, mg g^{-1} lipid
0	0.05	45	0.05	45
15	0.09	70	0.06	80
30	0.15	120	0.08	110
45	0.18	180	0.09	80
60	0.20	160	0.11	70
75	0.27	170	0.13	75

7.6.3.1. *Tempeh*

Tempeh kedelle is a popular fermented soybean product and meat substitute in Indonesia. After preparation it requires 3–4 min to deep fry or 10 min to cook.

The production of tempeh is outlined in Fig. 7.22. In the first step, the soybeans are soaked in water or acidified water at room temperature. During this stage two things happen, firstly a partial germination of the soybeans may occur depending on the amount of O_2 available to the seed, and secondly, acid produced by bacteria develops in the soak water and suppresses the growth of undesirable microorganisms. If a partial germination occurs, it can impact on the protein properties of the soybean and the subsequent fungal fermentation. Further, the bacteria that grow in the steeping water produce vitamin B_{12}, a significant nutrient in the final tempeh product. The most desirable bacterial species for this stage is *Klebsiella pneumoniae*, one of the *Enterobacteriaceae*, but by using pure cultures, other bacteria can perform the same function.

The soaked beans are then dehulled and cooked, a critical process in which a balance must be struck between over-cooking and under-cooking the beans. The cooked beans are then drained and cooled below 35°C followed by dusting with wheat flour to provide a good source of

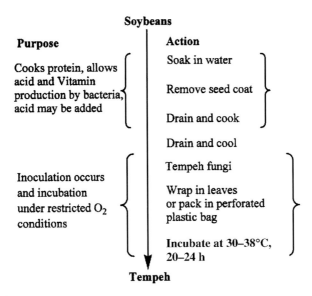

Fig. 7.22. An outline of the production of tempeh, an Indonesian soy based food produced using fungi. Careful control of oxygen supply during the fungal growth phase is critical for a good tempeh product.

carbohydrate for the fungi and inoculated. The desirable fungal species for successful tempeh production, whether from environmental inoculation or pure starter inoculation, are *Rhizopus oligosporus* (e.g. NRRL-2710), *R. stolonifer*, *R. arrhizus*, *R. oryzae*, *R. formosaensis*, and *R. achlamydosporus*. The spores of *R. oligosporus* are produced commercially in Indonesia for industrial scale tempeh production. During the fungal growth stage, the O_2 level must be controlled at a reduced level otherwise the fungus will grow too quickly and form black spore masses that are an indicator of poor quality. The traditional way to control O_2 is to wrap the inoculated beans in banana leaves but a modern innovation is the use of micro-perforated polyethylene plastic. The mycelia then grow and knit the beans into a firm cake, thus giving tempeh its meaty texture.

Proteases are most active at 50–55°C and pH 3.0–6.0. The enzymes from the fungi transform the soybeans making them more nutritious as shown in Table 7.19. Tempeh must be consumed quickly after purchase.

Defects include black patches due to sporulation, slime due to excessive bacterial growth because of too little O_2 or a temperature ≥ 42°C, or a

Table 7.19. Chemical changes in cooked soybean due to its conversion to tempeh.

Component	Cooked bean, %	Tempeh, %
Hemicellulose	2.0	1.1
Fibre	2.0	1.5–4.3
Total nitrogen	7.5	7.5
Soluble nitrogen	0.5	2.0
Free fatty acids	Trace	10
Vitamin B$_{12}$	Trace	$30\,\mu g\,g^{-1}$[a]

[a]Vitamin B$_{12}$ is produced by *Klebsiella pneumoniae* in the mixed inoculum.

yellow color due to growth of mycotoxic fungi. The yellow color indicates the tempeh is highly toxic and should not be eaten.

An Indonesian product similar to tempeh is ontjom in which groundnuts are used and the preferred fungus is *Neurospora sitophila* that has orange-red spores. The process differs in that sporulation to produce the color is encouraged.

7.6.3.2. Natto

This is produced in Japan and the Indian subcontinent. In Korea it is known as chung kook jang, and in Thailand as thua nao but the Thais sun-dry the product and a similar product is found in India. Traditionally the beans are soaked, boiled, cooled and wrapped in rice straw and left alone for one to two days at a warm temperature to allow bacteria to grow on the beans. The straw is said to impart a straw aroma to the fermented beans as well as being the source of the bacteria for the fermentation, and it absorbs some of the unpleasant aroma of ammonia released during the bacterial growth. During growth, the bacteria produce a viscous, sticky slime that pulls out into long strings when two soybeans from the final product are pulled apart. The viscous substance increases to about 2% of the natto after 48 hours and its crude protein content is about 80% of which about 20% is mainly the D isomer of γ-poly-glutamic acid. Free amino acids make up a large proportion of the remaining crude protein in the viscous material. The bacterium responsible for the characteristics of natto is an obligate aerobe *Bacillus subtilis* var. *natto*. The bacterial fermentation results in a high concentration of vitamin K_2 (menaquinone–7) that is important for carboxylation of osteocalcin, a bone protein. Some evidence suggests that natto consumption by Japanese women reduces the risk of hip fracture in postmenopausal women.

A further result of the fermentation is a reduction of the "beany flavor" and aroma of soybeans and an increase in alkyl pyrazines that are responsible for the characteristic odor of natto (Fig. 7.19). The amount of volatile compounds develops from around 35 μg/kg wet weight in cooked soybeans to 2–3.5 mg/kg wet weight in 72 h. Predominant volatiles in Japanese natto after 72 h of fermentation are 3-hydroxybutanone (acetoin), 2,5-dimethylpyrazine, trimethylpyrazine, and tetramethylpyrazine. Some compounds (acetoin) appear early in fermentation and then decrease, whereas others appear only towards the end of fermentation.

Further Readings

1. Beech GA, Melvin MA and Taggart J. Food, drink and biotechnology. In: *Biotechnology: Principles and Application*, eds. IJ Higgins, DJ Best and J Jones. Blackwell Scientific, Oxford, 1985, pp. 73–110.
2. Beuchat LR. Indigenous fermented foods. In: *Biotechnology Vol. 5*, Chapter 11, ed. G Reed. Verlag Chemie, Weinheim, 1983, pp. 477–528.
3. Bottazzi V. Other fermented dairy products. In: *Biotechnology: Food and Feed Production with Microorganisms Vol. 5*, ed. G Reed. Verlag Chemie, Weinheim, 1983, pp. 315–364.
4. Coffey A and Ross RP. Bacteriophage-resistance systems in dairy starter strains: Molecular analysis of applications. *Antonie van Leeuwenhoek* **82**: 303–321, 2002. (Note: Tables 1 and 2 were not included in the published paper).
5. Ebner H and Follmann H. Vinegar. In: *Biotechnology Vol. 5*, Chapter 9, ed. G Reed. Verlag Chemie, Weinheim, 1983, pp. 425–446.
6. Eck A and Gillis J-C, eds. *Cheesemaking: From Science To Quality Assurance*, 2nd ed. Lavoisier Publishing Inc. Paris; Intercept Ltd, London, 2000.
7. Fleming HP, McDonald LC, McFeeters RF, Thompson RL and Humphries EG. Fermentation of cucumbers without sodium chloride. *J. Food Sci.* **60**: 312–315, 1995.

8. Fox PF, ed. *Cheese: Chemistry, Physics and Microbiology Vol. 1, General Aspects, and Two Major Cheese Groups*. Chapman & Hall, London, 1993.

9. Frazer WC and Westhoff DC. *Food Microbiology*. McGraw-Hill, New York, 1988.

10. Friend BA and Shahani KM. Fermented dairy products. In: *Comprehensive Biotechnology: The Principles, Applications and Regulations of Biotechnology in Industry, Agriculture and Medicine Vol. 3*, eds. HW Blanch, S Drew and DIC Wang. Pergamon, Oxford, 1985, pp. 567–592.

11. Ghose TK and Bhadra A. Acetic acid. In: *Comprehensive Biotechnology Vol. 3*, Chapter 36, eds. HW Blanch, S Drew and DIC Wang. Pergamon, Oxford, 1985, pp. 701–730.

12. Han B-Z, Rombouts FM and Nout MJR. A Chinese fermented soybean food, *Int. J. Food Microbiol.* **65**: 1–10, 2001.

13. Jay JM. *Modern Food Microbiology*. Van Nostrand Reinhold Co., New York, 1992.

14. Lee YK, Salminen S, Nomoto S and Gorbach S. *Handbook of Probiotics*. John Wiley & Sons, New York, 1999.

15. Nout MJR and Aidoo KE. Asian fungal fermented food. In: *The Mycota: Industrial Applications Vol. X*, ed. HD Osiewacz. Springer-Verlag, Berlin, 2002, pp. 23–47.

16. O'Toole DK. Factors inhibiting and stimulating bacterial growth in milk: an historical perspective. *Adv. Appl. Microbiol.* **40**: 45–94, 1995.

17. O'Toole DK. Soy-based fermented foods. In: *Encyclopedia of Grain Science*, eds. C Wrigley, H Corke and C Walker, Elsevier, 2004, pp.174–185.

18. Ouwehand AC, Salminen S and Isolauri E. Probiotics: An overview of beneficial effects. *Antonie van Leeuwenhoek* **82**(1–2): 279–289, 2002.

19. Salminen S, Deighton M and Gorbach S. Lactic acid bacteria in health and disease. In: *Lactic Acid Bacteria*, eds. S Salminen and A von Wright. Marcel Dekker, New York, 1998, pp. 211–254.

20. Steinkraus K, ed. *Handbook of Indigenous Fermented Foods*. Marcel Dekker, New York, 1995.

21. Wood BJB, ed. *Microbiology of Fermented Foods Vols. 1 and 2*. Blackie Academic and Professional, 2–6 Boundary Row, London, 1998.
22. Yokotsuka T. Traditional fermented soybean foods. In: *Comprehensive Biotechnology Vol. 3*, Chapter 19, eds. HW Blanch, S Drew and DIC Wang. Pergamon, Oxford, 1985, pp. 396–427.

Questions for Thought...

1. Use vinegar production as an example to illustrate the technological development in microbial fermentations. Highlight the technologies developed at each stage to overcome the rate-limiting steps in the vinegar production processes.
2. Design a continuous flow fermentation process for the production of cultured milk. Incorporate a mechanism to use a portion of the final product as the inoculum.
3. Bacteriophages are viruses that attack bacteria. If you were setting up a new factory to produce a new product using a bacterial culture, what would you do to avoid the production problems that a bacteriophage attack may cause to the production system?
4. Discuss the measures taken in the various steps of soy sauce production to ensure the involvement of the right kind of microorganisms and enzymes.
5. "The production of fermented foods relies entirely on microbial anaerobic processes." Discuss this statement.
6. "Ultimately we will produce all of the fermented foods using enzymes alone". Discuss this statement.

Websites

http://www.soyfoods.com/index.html
http://en.wikipedia.org/wiki/Main_Page
http://www.japanweb.co.uk/listing/index.htm
http://www.foodreference.com/index.html
http://www.marukome.co.jp/mehome.html

http://www.saitoku.com/
http://www.ebs.hw.ac.uk/SDA/cheese1.html
http://biology.clc.uc.edu/Fankhauser/Cheese/Cheese.html
http://www.dairyfoods.com/CDA/ArticleInformation/features/BNP__
Features__Item/0,6775,144097,00.html

Chapter 8

Food Involving Yeast and Ethanol Fermentation

Desmond K. O'Toole
Department of Biology and Chemistry
City University of Hong Kong
83 Tat Chee Avenue, Kowloon, Hong Kong, China

Yuan Kun Lee
Department of Microbiology, National University of Singapore
5 Science Drive 2, Singapore 117597
Email: micleeyk@nus.edu.sg

Yeasts are a natural part of the microbial population on fruits and on ground cereal grains. When the fruit juice is released and the ground cereal grain is wetted, the yeasts (as well as other organisms) grow over a few days and produce ethanol. From these natural processes humans have learnt to harness the activity of the yeasts.

In ancient times alcoholic beverages were an important part of the diet. They provided a way of storing the food value in otherwise perishable agricultural products and provided important nutrients, including vitamins, in the diet. Today, these products are consumed for their flavors, their alcoholic content and its effects, and food value. The quality of the final beverage rests as much on the ethanol production as it does on the development of flavoring compounds in the final product. Production of ethanol may be the prime purpose of production methods but the final flavor does not depend on the ethanol producers alone. In addition to

293

yeasts other than *S. cerevisiae*, bacteria sometimes play a role in the development of the final desirable flavor of an alcoholic beverage.

Ethanol is also distilled from yeast fermented products and is used for a variety of purposes. These include:

- The production of alcoholic beverages to which flavors may be added or flavors may be allowed to develop by maturation processes;
- To preserve fruits and wines;
- To use as an industrial product;
- To use as a biofuel to replace up to 10% of the petrol (gasoline) used in internal combustion engines.

Ethanol is the end product of the anaerobic (without oxygen) glycolytic pathway (Embden-Meyerhof-Parnas cycle) for energy metabolism of yeast (*Saccharomyces cerevisiae*) and a bacterium (*Zymomonas mobilis*):

$$1 \text{ glucose} \xrightarrow{\text{Yeasts}} 2CO_2 + 2ATP + 2 \text{ ethanol}$$

Common alcoholic beverages include:

- Beer, containing about 5% alcohol v/v, the main substrate is barley;
- Wine, containing about 12% alcohol v/v, the main substrates are grape, rice and palm sap;
- Spirits (liquor), containing about 40% alcohol v/v, the main substrates are grape, barley, rice and/or sorghum;
- Fortified wine, distilled spirit is added to wine raising the ethanol content to about 18%, e.g. port and sherry are fortified wines.

The production of ethanol depends on the availability of glucose or other hexoses, and disaccharides for fermentation. Glucose is readily available in many fruits, for example grapes. However, by far the vast majority of glucose is in the form of polymers, namely the starches and cellulose. Starch is the energy storage compound of plant life but yeasts and *Z. mobilis* generally cannot use the glucose units unless they are removed from the starch molecules. This requires another step in the production system to make the glucose available to the yeasts and bacteria.

Two basic systems have been developed to release fermentable sugars from starch for the production of alcoholic beverages. The first is to begin sprouting the living cereal seed by steeping it in water and hydrating it under aerobic conditions. During sprouting the seed produces amylases to break down the starch so as to make the energy in the starch available to the growing seedling. A high level of amylase builds up. The sprouted grain is then dried to preserve the amylase activity before the seedling has grown significantly. The second system involves wetting or cooking the grain and then letting mould grow on the grain. The mold produces amylase (in addition to other enzymes) that is used to break down the starch in the grain.

Box 8.1

Starch is an important source of glucose for various fermentations

By far the greatest source of glucose for alcohol production is starch. However, starch is also the base for a range of other food products including MSG. Natural starch occurs in starch granules in plant material such as tapioca (cassava), maize, potato, barley, wheat, rice and sago palm, to name some examples of industrial sources of starch. The starch is not soluble in water and the granules are surrounded by a protein membrane. In most cases the granule must be destroyed and the starch gelatinized (hydrated) before the starch can be metabolized. The temperature of **gelatinization** varies from 65–80°C, depending on the plant source.

There are two forms of starch. Starch mainly consists of **amylose** that is comprised of long chains of unbranched glucose units with α-1,4 links. Waxy starch contains a high proportion of **amylopectin**, in addition to some amylose, that consists of chains of glucose units from which glucose units frequently branch through an α-1,6 link to form additional chains with α-1,4 links. Natural starch sources may also contain some β-glucans, such as β-1,3 and β-1,4-linked glucan polymers that are viscous and gum-like.

A range of enzymes from sprouting seeds and from a variety of fungi and some bacteria attack the links in the glucose chains. The enzymes from plant sources include α-amylases, β-amylases and α-1,6-glucosidases. The α-amylases breakdown the internal α $(1\rightarrow4)$ linkages in the chains. The β-amylases hydrolyze the penultimate α $(1\rightarrow4)$ linkages at the non-reducing ends of the chains thus producing maltose. Finally, the α-1,6-glucosidases breakdown the branch points resulting in a mixture of sugars including glucose, maltose, and maltotriose. On the other hand, fungi and bacteria produce amyloglucosidases that remove glucose units from non-reducing ends and can hydrolyze the α-1,6-glucoside links.

Box 8.1   *(Cont'd)*

These enzymes are more formally known as glucanases. The α (1→4) glucanases are further split into endo α (1→4) glucanase that hydrolyze the α (1→4) linkages anywhere in a chain and exo α (1→4) glucanase that hydrolyzes the non-reducing end. There are six enzymes that are classed as an exo α (1→4) glucanase.

Enzymes hydrolyzing the α (1→6) linkages are also known as the pullulanases and are α (1→6) glucanases of which there are two endo- α (1→6) glucanase and one exo- α (1→6) glucanase.

The expression of the amylase complex, as it is called in fungi, depends on catabolic induction from the substrate. Low basal levels of the amylase are produced that generate low concentrations of oligomers that feedback into the cell to induce amylase complex formation. Induction occurs due to the presence of maltose, isomaltose, and panose but glucose and glycerol repress complex induction.

Maltose (4-D-α-glucose-α-D-glucopyranoside)   **Glucose (α-D-glucopyranoside)**

8.1. Beer Production

Beer production originated in about 6000 BC in ancient Babylonia, but today, German terms dominate in the vocabulary of the beer brewing industry. Barley is the essential base material for the amylase and starch sources. After processing the barley, the sprouted and dried barley is called malt. The malting process used is expensive because of the heat energy needed to dry the sprouted barley. Brewers in the nineteenth century discovered that they did not need to use only malted barley to make beer because there are two kinds of barley, two-row and six-row, the latter produces more "diastase", has higher protein content and higher β-glucan level than the other. This allowed brewers to reduce the amount of malted barley they used and to increase the amount of other sources of cheap starch that are called **adjuncts**. There is sufficient amylase in the malted barley to break down starch in other grains, but the grain has to be cooked and the starch gelatinized.

Beer making is a complex process and is summarized in Fig. 8.1.

8.1.1. *Malting*

In the malting process, barley is soaked in water (steeped) to give 42–46% moisture content to release germination hormones, gibberellins, and to remove dirt, husk phenols, soluble organic matter, and soil micro-organisms. Oxygen is required by respiratory enzymes in the barley seed and is provided by forcing cool humid air through the grain bed. Germination occurs in the aerated germination chamber during 4–5 days at 13–16°C. Hot dry air, in a process called **kilning**, is then pulled through the grain bed to stop and kill the sprouting plant but retain enzyme activity. At the same time, malt looses its gherkin-like taste and changes to the typical malt aroma. Kilned malt can be kept for some time.

The quality of barley used for malting is dependent on a number of factors. Apart from the barley being of good quality (often defined in terms of enzyme production), there should be low contamination of the grain by *Fusarium* species, such as *F. graminearum*, *F. culmorum* and *F. poe*. Heavy *Fusarium* contamination of malting barley may lead to the presence of deoxynivalenol and other mycotoxins in barley, all water-soluble compounds that are removed during steeping. However, fusaria remain trapped within the mass of germinating grain, continue production of these heat resistant mycotoxins during germination of the barley, and can be found in the final beer. This leads to the fault in beer known as "gushing of beer", a quick uncontrolled spontaneous over-foaming immediately after opening the bottle or can.

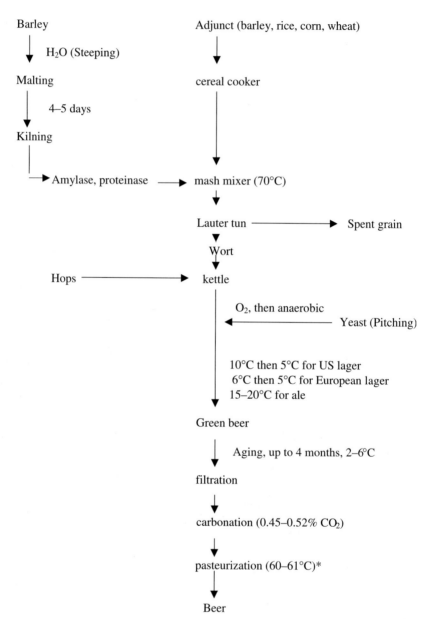

*In some more advanced production systems, beer is sterilized by filtration and packed aseptically.

Fig. 8.1. Flow diagram for the production of beer.

Fusarium contaminated barley is banned in some countries to avoid the problems associated with it. However, a new method for reducing the effects of fungal contamination and to improve malt quality is to add a **starter culture**, either lactic acid bacteria or *Geotrichum candidum*, a yeast, to the steeping barley to control microbial growth. Certain strains of *Lb. plantarum* and *Pediococcus pentosaceus* are especially efficient when added to the steeping water of barley at the level of about 10^7 cells/g. Addition of starter cultures ensures high quality malt, regardless of the natural variation of the microflora of barley, due to the microbiocidal compounds produced by the lactic acid bacteria and other enzyme activities. Addition at the steeping stage is important because fusaria grow rapidly during the first hours of steeping. The starter effect on this defect is proportional to the level of original fungal contamination in the barley, due to the negative effect on the fusaria. If there are no fusaria in the barley then there is no gushing defect. However, the added starters have other effects on other stages of the brewing process, and the lactic acid starter cultures also restrict the growth on the grain of harmful aerobic bacteria. In addition, the use of lactic acid bacteria in malting has led to significant improvements in the physical and chemical quality of malt including improved filtering (lautering) of wort (pronounced "wert") and beer, and reduced formation of beer haze during finished beer storage. The effect of lactic starter cultures on lautering performance can be explained partly by their own enzyme activity, namely the production of thermostable (60°C) β-glucanase. Further, the xylanase activities of malts produced with starter are higher. The net result is a higher extract yield.

The quality of malting barley is judged by the activities of various endogenous barley enzymes. These enzymes include α- and β-amylase, proteases and β-glucanase. While amylases are important in starch degradation, β-glucanase is also important because β-glucans in the grain are soluble and are extracted into the wort. The β-glucans interfere with the filtration of the wort and the beer and contribute to the formation of haze during the storage of finished beer.

8.1.2. *Adjuncts*

An adjunct, as the source of fermentable carbohydrate, can be corn free of germ and skin, rice, sorghum, barley, wheat grits, enzymically converted

Table 8.1. The main components in wort used for beer production.

Compound	Concentration, g L^{-1}
Maltose	40–55
Maltotriose	10–15
Glucose	5–10
Sucrose	1–3
Fructose	1–2
Non-fermentable carbohydrate	22–25
Free amino acids	1–2
Peptides and protein	1–3
Organic acids	0.175
Inorganic ions	1.5–2
Hops concentrate	0.05–0.1

corn syrup or sucrose. Adjunct is mixed with kilned malt to produce mash and is incubated at 70°C. This allows hydrolysis of carbohydrate and protein in adjunct by enzymes present in the malt. The **wort** (liquid part of the mash) is then separated from the **spent grain**. The wort contains the simple fermentable sugars as shown in Table 8.1 that are ready for beer fermentation.

8.1.3. *Hops*

Hops are then added into the wort. Hops, whose quality is important for the final beer flavor, were first introduced to beer brewing by St Hildegard (1098–1179 AD) in Germany who reported that hops promoted the stability of beer, and was good for health. Hops are unfertilized dried female cones of the hop plant *Humulus lupulus*. The small lupulin glands contain bittering resins and essential oils that contribute to the unique bitterness and flavor of beer. Polyphenolic compounds in bracteoles enhance the astringency and tea-like character of beer. The traditional baled hops have now been replaced by various concentrates, pellets etc. The preferred hops extract is that extracted using super-critical carbon dioxide or liquid carbon dioxide.

8.1.4. *The fermentation*

An important technical advance in brewing has been the use of high gravity or very high gravity brewing. The wort is produced and a sugar rich syrup is added to the wort kettle to give a wort with a substantially higher gravity, over 20° Plato (°Plato = weight of extract in a 100 g solution at 17.5°C), than needed for the final beer. The wort is then fermented after which the beer is adjusted to the desired concentration with oxygen-free water. This process increases the brewing capacity of a given brew-house.

The actual fermentation and maturation is the next stage in the process. Initial presence of oxygen is desirable for the yeasts because it is essential for membrane lipid synthesis and rapid cell growth before anaerobic growth begins. During ethanol formation and growth of the yeasts, the amino acids valine and leucine are produced and one of the intermediates in the metabolic pathway is α-acetolactate. This compound is excreted from the yeast cells into the brewing wort where it is then converted by a slow chemical process to diacetyl, a compound with a strong undesirable buttery flavor that has a very low taste threshold in beer of 0.05 mg l^{-1} or less. The diacetyl is subsequently degraded enzymically to acetoin and then to 2,3-butanediol. The long maturation process of lager beer is designed to allow the depletion of diacetyl in the beer.

The yeasts in beer do not breakdown α-acetolactate but some bacteria possess α-acetolactate decarboxylase so the exogenous enzyme can be added during fermentation. Alternatively, a gene encoding α-acetolactate decarboxylase can be inserted into the yeast genome. The genes of *Klebsiella terrigena*, *Enterobacter aerogenes* and *Acetobacter xylinum* have been inserted using either PGK or ADH promoters. Pilot scale studies have confirmed that such genetically modified yeasts can reduce the diacetyl concentration below thc taste threshold at the end of the main fermentation thus making it possible to reduce the maturation time which is usually about 14 days.

8.1.5. *Types of beer*

There are basically three categories of beer, the bottom fermented lager beer brewed by *Saccharomyces carlsbergenesis* and *S. uvarum*,

Box 8.2

Killer Yeasts

These are strains of yeasts that produce anti-yeast factors, also known as zymocins, killer factors or killer toxins, that kill other yeasts including strains of the same species. They are produced by yeast strains from many genera but industrially those produced by *Saccharomyces* strains are most important. The zymocin producing ability in a yeast strain confers resistance to zymocins produced by other yeast strains. In yeast cultures contaminated with a killer yeast strain, the latter becomes the dominant strain and will cause the production of off-flavors in addition to other defects. It is difficult to detect the presence of killer strains in a culture. In the case of the best known *Saccharomyces* zymocin, KT_1, to be effective, the zymocin requires a very narrow pH range (pH 4.3–4.7) as well as the right concentrations of some metal ions (K^+, Ca^{2+} and Mg^{2+}).

The KT_1 zymocin consists of a double stranded heterodimer protein in which the α chain (105 amino acids long) and the β chain (83 amino acids long) are linked by three disulfide bonds. The zymocin exerts its toxicity by absorbing to the cell wall of sensitive cells at a specific receptor-binding site containing a $(1\rightarrow6)$ β-D-glucan. This process requires no energy. Once bound to the cell wall, the next step requires energy to bring the toxin into contact with the plasma membrane. If the organism has no energy available then the cell membrane cannot be disrupted. Once in contact with the plasma membrane, ion-permeable channels that are large enough to allow leakage of ions open up. Consequently, protons are extruded which dissipate the membrane potential so that the proton-amino acid co-transport system is inhibited, potassium ions leak from the cell, and leakage of large molecules such as ATP occurs.

A yeast strain's ability to produce the toxin, designated K^+ phenotype, and the ability to resist it, designated R^+ phenotype, is due to a dsRNA called M (M = medium size). As there is more than one killer toxin, the others are designated with a subscript. To understand the yeast killer system, a very complex system, only the M_1 system will be briefly described. The M_1 dsRNA is a 1.8 kb unit that codes for the polypeptide and a resistance function. The strains carrying it are designated genotype KIL-k_1. For the killer system to work the cell needs to contain the cytoplasmic gene HOK (*h*elper *o*f *k*iller) that is located on certain forms of L-A dsRNA (L = large

Box 8.2   (*Cont'd*)

size). Furthermore, maintenance of the killer system in a cell requires possession of some chromosomal genes designated MAK (*m*aintenance of *K*IL-k$_1$), comprised of at least 29 genes. The L-A and M dsRNA strands are found in an intracellular virus-like particle (ScV particle) whose major coat protein (capsid) is coded for by the L-A strand. For the killer toxin to be expressed in a cell, the cell must also contain a further two chromosomal genes, designated *KEX*, needed to process the toxin precursor.

The ability of a killer strain to resist killer toxin offers the advantage of controlling wild infections of cultured strains. Consequently, attempts have been made to transfer the killer system to a desirable strain for wine or beer making. In one experiment, the killer system was inserted into the protoplasts of a non-killer strain of *S. cerevisiae* by electro-injection. The transformed culture was stable in culture for many generations. In another successful experiment, a DNA copy of one of the killer systems was cloned and put into an ADH1 expression cassette that was subsequently integrated into the yeast chromosome.

top fermented ale beer brewed by *S. cerevisiae*, and stout. At the end of beer fermentation, lager beer is siphoned off and the yeast cells settled on the bottom are discarded, whereas yeast cells floating on the surface of ale beer are skimmed off and re-used.

Ale beer tends to be sweet because *S. cerevisiae* ferments only the fructose moiety of the trisaccharide raffinose. The ale yeast forms a relatively high level of esters due to higher levels of coenzyme A activity. Esters constitute an important group amongst the beer volatiles due to their strong, penetrating fruity flavors.

Lager yeasts are able to utilize most of the sugars in the wort, thus lager beer is generally dry (not sweet). Moreover, rapid reduction of acetaldehyde to ethanol by alcohol dehydrogenase and reduced nicotinamide adenine dinucleotide coenzyme, results in a low level of acetaldehyde in the beer.

During aging, the flavor of beer matures. There is a reduction in the concentration of hydrogen sulfide, acetaldehyde, diacetyl and other undesirable flavors. Aging at high temperature in the presence of yeast cells results in autolysis of lager yeast and a change in the flavor profile of beer.

Recent developments in brewing science include brewing with unmalted cereals digested by microbial enzymes and production of low carbohydrate beer. As the malting process is difficult to control, use of microbial enzymes allows selection by the brewer of appropriate enzymes in respect to nature and quantity. Moreover, addition of these enzymes can be programmed, and it is not necessary for them to be present at the beginning of mashing. One possible consequence of the use of adjuncts is the presence of hemicelluloses that inhibit complete starch extraction or cause the formation of hazes in stored beer. Pentosanases that breakdown arabans, galactans, mannans and xylans, are added during either fermentation or maturation to eliminate them. Microbial enzymes used include bacterial or fungal α-amylases, in particular a thermostable one from *Bacillus licheniformis*, amylo-glucosidases, proteases, pullulanase and β-glucanases.

Light beer (also known as lite beer) has a caloric value that is a third, or more, lower than conventional beers. Besides having reduced alcohol content (20–50% of normal beer), the light beer has a reduced content of non-fermentable sugars, e.g. dextrins that emanate from the malt and adjunct. There are a number of ways to reduce the amount of non-fermentable sugars in beer.

(1) Dilution of the regular strength beer with water, but this would unavoidably dilute the flavor of the beer;
(2) Addition of microbial or cereal α- or β-amylase or glucoamylase, β-glucanase and pullulanase to the wort during mashing or fermentation;
(3) Use of a totally fermentable sugar such as flucose, fructose or sucrose as an adjunct;
(4) Use of a brewing yeast strain with amylolytic activity for production of low carbohydrate beers and for maximizing fermentation efficiency.

The last approach has been quite extensively studied. The yeast *S. diastaticus* produces the extracellular enzyme glucoamylase (regulated by genes *DEX1*, *DEX2* and *STA3*). Using classical hybridization techniques, a diploid strain containing the *DEX* and *STA* genes in the homozygous condition was constructed. The *DEX*-containing strain of *S. diastaticus* ferments wort to a greater extent than the brewing strain due to the partial hydrolysis of dextrins by glucoamylase, and is desirable in the production of light beer. However, the beer produced by this strain has a characteristic phenolic-off-flavor, due to the presence of

Box 8.3

The Role of Ferulic and *p*-Coumaric Acids in Food and Beverage Flavors

Ferulic and *p*-coumaric acids are widely distributed in plants, grains and fruits. Consequently they are commonly found in a variety of fermented products including wine, beer and soy sauce and they are the source of various flavoring constituents in these fermented food products.

The flavors or odors, often described as strong spicy, smoke-like, medicinal, clove-like, woody or phenolic, arise through the metabolic activity of yeasts and bacteria. The flavors are associated with the production of 4-ethylguaiacol, 4-ethylphenol, 4-vinylguaiacol and 4-vinylphenol as well as vanillin.

Ferulic and *p*-coumaric acids, originating from lignin, are transformed during the manufacturing process, to flavored compounds that may or may not be desirable in the product. Most wine yeast strains and some beer brewing strains of *Saccharomyces cerevisiae* are able to decarboxylate ferulic and *p*-coumaric acids to produce phenolic off-flavors (POF) during fermentation. In one study, 81 to 95% of 116 wine yeast strains produced these compounds while growing in spiked grape juice. However, in practice during wine making, *Brettanomyces* yeasts (sporulating form is *Dekkera*) are mainly responsible for the development of a phenolic character in red wines. The *Brettanomyces* species multiply by fermenting the very small quantities (275 mg of sugars l^{-1}) of residual sugars (glucose, fructose, galactose, and trehalose), much less than normal levels at the end of fermentation, but sufficient for the production of an excessive quantity of ethyl phenols. Other yeast genera that also decarboxylate ferulic and *p*-coumaric acids include *Rhodotorula*, *Candida*, *Cryptococcus*, *Pichia* and *Hansenula*. Some strains of *Lactobacillus brevis* and *Pediococcus pentosaceus* are able to decarboxylate *p*-coumaric acid to form 4-vinylphenol as actively as *S. cerevisiae*, but are unable to breakdown ferulic acid. In practice, the lactic acid bacteria have no impact on the development of phenolic off-flavors in wine. The critical concentration of these free forms of hydroxyphenolic acids in grape required for the production of the phenolic off-flavors is estimated to be more than $10 \, \text{mg} \, l^{-1}$. The lag phase of growth of *S. cerevisiae* is increased by as little as $50 \, \text{mg} \, l^{-1}$ ferulic acid, and higher concentrations can inhibit the organism.

In high salt conditions (15% NaCl), *Debaryomyces hansenii*, *Pichia subpelliculosa*, *P. anomala*, and *Candida famata* convert ferulic and

Box 8.3 (*Cont'd*)

p-coumaric acids to 4-vinylguaiacol and 4-vinylphenol in static culture, and *C. versatilis*, *C. halophila*, and *C. mannitofaciens* convert them to 4-ethylguaiacol and 4-ethylphenol. *C. etchellsii*, *C. nodaensis*, *C.halonitrato-phila*, *P. farinosa* and *Zygosaccharomyces rouxii* do not produce volatile phenols. These cinnamic acids are converted from ferulic acid and *p*-coumaric acid to vinylphenols by cinnamate decarboxylase, and successively from vinylphenols to ethylphenols by vinylphenol reductase. The soy sauce spoilage bacteria, such as *Bacillus subtilis*, *B. licheniformis*, and *B. megaterium*, as well as some *Staphylococcus* sp. (from koji), produce vinyl phenols at cell concentrations greater than 10^7 cells mL^{-1}.

Although some microorganisms can metabolize them, they are also inhibitory for some yeasts. Ferulic acid has an inhibitory effect at 100 mg l^{-1} on strains of *P. anomala*, *D. hansenii* and *S. cerevisiae*. At 500 mg l^{-1} it appreciably inhibits *P. anomala*, *D. hansenii* and *S. cerevisiae* and prevents detectable growth of odd strains of *P. anomala* and *D. hansenii*. Coumaric acid is less inhibitory than ferulic acid.

Box 8.3   *(Cont'd)*

The hydroxycinnamic acid decarboxylase enzyme, with an apparent molecular mass of 39.8 kDa by gel filtration, catalysing the decarboxylation of ferulic and *p*-coumaric acids was isolated from the yeast, *Brettanomyces anomalus* (NCYC 615). Its activity was optimal at 40°C and pH 6.0 and was enhanced by EDTA, Mg^{2+}, and Cr^{3+} while Fe^{3+}, Ag^+, and sodium lauryl sulfate completely inhibited the activity. The para-hydroxy group on the phenol ring is essential for enzyme activity. On the other hand, a strain of *Lb. plantarum* has shown substrate-inducible decarboxylase activity on *p*-coumaric and ferulic acids indicating that at least two phenolic acid decarboxylases are produced in this bacterium. On SDS-PAGE the enzyme band is 23.5 kDa but the native molecular mass is 93 kDa indicating that the enzyme is a homotetramer. This *p*-coumaric acid specific enzyme exhibited maximum activity at 30°C and at pH 5.5–6, and did not need cofactors or metal ions.

4-vinylguaiacol. A single dominant gene *POF* (phenolic-off-flavor) codes for a ferulic acid decarboxylation enzyme that is responsible for the conversion of ferulic acid, a wort constituent derived from the grain, to 4-vinylguaiacol. Thus, a haploid strain that was *DEX* and *POF* positive was mated with a *DEX* and *POF* negative haploid strain. When tetrad dissection of the resultant diploid strain was performed, the *DEX* and *POF* genes segregated independently of each other. Thus, a haploid strain that was *DEX* positive and *POF* negative was obtained, and subsequently a diploid strain with the genotype *DEX2/DEX2 pofo/pofo* was constructed.

8.1.6. The flocculating properties of yeast

When brewing lager beers the precipitation of the yeast cells is an important characteristic. Cells that flocculate are more easily removed from the green beer. However, it is not a simple relationship as ideally the yeast should remain in suspension while the sugar is fermented and at the end (the stationary growth phase) the yeasts should flocculate, that is, the yeast cells should adhere and sediment rapidly to the bottom of the brewing vessel. This helps reduce the cost of filtration or centrifugation to clarify the beer. The process of flocculation depends on an initial input of mechanical energy (stirring) to promote physical contact, and the presence of Ca^{2+}. However sensitivity to Ca^{2+} varies between bottom and top fermenting yeasts, possibly due to cell surface hydrophobicity. Bottom fermenting strains flocculate in the presence of high Ca^{2+} concentration and low sugar concentrations; mediated by a lectin-like mechanism. Top fermenting strains flocculate without added Ca^{2+}, but need a high ethanol concentration. However, the flocculating activity of some top fermenting strains is inhibited by mannose, but not by sucrose or galactose.

During yeast flocculation a lectin-like protein binds to a mannose chain receptor on neighboring cell walls. The lectins are synthesized continuously in the exponential phase of growth and are inserted into the cell wall, where they remain inactive for up to 14 hours, before being activated at flocculation onset by an unknown mechanism. The binding is inhibited by mannose but not by glucose. In addition, flocculation can be affected by the final pH as shown in experiments where flocculation was induced with a pH-upshift and de-flocculation occurred with a pH-downshift.

The flocculating properties are genetically controlled and are expressed as changes in the cell wall components. Experiments have shown a protein component is responsible and cell wall proteins have been identified. The gene *FLO1* that is located on chromosome I is the major gene as shown when a nonflocculent industrial polyploid yeast strain, *S. cerevisiae* 396-9-6V, was converted to a flocculent one by introducing the *FLO1* gene at the *URA3* locus. A homologue of *FLO1*, named *Lg-FLO1*, was isolated from a flocculent bottom fermenting yeast strain in which flocculation is inhibited by both mannose and glucose (mannose/glucose-specific flocculation).

The DNA sequence of *FLO1* showed that the open reading frame (*ORF*) of the intact gene, composed of 4611 bp, codes for a protein of 1537 amino acids. This Flo1 protein contains four families of repeated sequences and it has a large number of serines and threonines. The N- and C-terminal regions are hydrophobic. Another gene isolated from flocculating strains was *FLO5*. This gene has been shown by restriction mapping, Southern analysis, and chromosome mapping to be related to the *FLO1* gene.

8.2. Distilled Spirits

Once wines and beer are made they are subject to deterioration due to the activity of vinegar bacteria or chemical oxidation unless stored properly.

Distillation of the ethanol is one way of at least preserving the ethanol content that itself can also be applied to preserving wines and fruits. Historically distillation of alcoholic spirits from fermented products developed in both East and West. Western spirits will be dealt with here as sorghum liquor, an Eastern product, is dealt with elsewhere.

Western liquors can be broadly categorized as grape or fruit based (e.g. brandy, cognac, calvados), molasses based (e.g. rum) or grain based (e.g. whisky, bourbon). The grain based liquors are split into malt whisky and grain whisky, the former being made exclusively with malted barley kilned with smoke from peat fires, and the latter being based on other grains like maize and rye grain. The malted barley and grains are gelatinized and hydrolyzed in a manner similar to the beer making process except that no hops are added, the wort is not boiled, and the spent grains may not be separated from the wort before fermentation and distillation. For the actual fermentation a heavy inoculum ($5–20\times10^6$ cells mL^{-1}) of a cheap source of yeast is often used.

The secret of success in the production of whisky, rum and some brandies is thought to be in the actual distillation process followed by maturation in wooden casks, particularly for whisky. Pure ethanol boils at 78°C but the ethanol/water mixture results in a non-ideal mixture so that during distillation an azeotropic mixture comes off containing 96.5% ethanol. Generally higher alcohols in the mixture distill off more easily and methanol distills off less readily than ethanol. These characteristics are taken into account during the final distillation stage.

For whisky production the brew is distilled twice using two batch based **pot stills** in sequence. These pot stills are always made of copper, with a spherical base, a long conical neck and a long horizontal arm that takes and condenses the spirit mixture through a coiled tube cooled with water. The shape and size of these pot stills, and the high polish on the copper that reduces heat losses during distillation, are considered critical to maintain the flavor characteristics of a whisky from a single distillation unit. The first still is the "wash still" from which the output contains about 21% ethanol. (Spent grains from the first still are dried for cattle feed.) This output is then introduced to the "spirit still" and the output here is split into three portions, the first and last portions are fed back into the spirit still but the middle fraction, that contains about 60% ethanol, is collected as the product for maturation.

Whisky by law must be matured for at least three years and it is placed in wooden casks that may be new or have been used for aging other wines or liquors. During aging the water penetrates more rapidly than ethanol through the wood so that the ethanol becomes more concentrated. The flavor changes during maturation due to chemical reactions and to flavors seeping from the wood. Color, acids, esters, furfural and tannins all increase, while aliphatic sulfide compounds, e.g. dimethyl sulfide, decrease rapidly. Vanillin, syringaldehyde and β-methyl-γ-octolactone are extracted from the wood.

The final product for sale usually contains caramel to produce a desirable color and it is diluted to about 40% ethanol.

Distilled liquor is also produced for which no maturation period is used. Flavors, usually from plant sources, e.g. juniper berries used for gin, are added and blended to obtain a desired flavor.

8.3. Grape Wine

Fruit juices contain readily available glucose and fructose for the growth of yeasts. In addition, the pH of fruit juice is generally below pH 5. In the case of grape juice, the pH is in the range of pH 3.0–4.0, due mainly to the natural tartaric, the dominant acid, and malic acids in the juice. The natural habitat of sugar fermenting yeasts is the outer layer of the fruits. The most commonly fermented fruit is grape juice, which is used to produce grape wine, but other fruits include apples that are used to make cider. Records of wine production go back over four thousand years in ancient Egypt, where ancient wine making steps are depicted on the walls of the tombs of the Pharaohs.

The traditional way of making wine is as follows. The grapes are picked at the optimum ripeness and crushed. The juice is collected and placed in a suitable vessel. The natural yeasts on the grape skins begin to grow and ferment the glucose and fructose to ethanol and carbon dioxide. The fermentation period is usually 2–4 weeks. The fermented juice is then drawn away (yeasts cells are separated from the finished wine) to avoid yeast cell autolysis that can cause off flavors, and the finished wine stored in a wooden cask or stainless steel tank until bottling.

A critical factor in the process is the amount of sugar in the grape juice and that depends on the amount of sunlight that the grape vine received during the growing season. In northern climes, this becomes the critical factor restricting the places where wine can be produced. Other critical factors for wine quality are soil type and rainfall.

Wines are categorized according to a range of factors, one factor being color. Red wines are produced using red or black grapes. The color is in the skin, not in the juice so the color (anthocyanins) of the skin has to be extracted to color the red wine. The skins of the crushed grapes are not removed from the **must** (the grape juice for fermentation), and the fermentation begins while the skin is in the juice. When sufficient color is extracted, the skins are removed. White wines are produced from grape juice separated from the skins, but it can be produced on the skins if the grapes are white.

After the initial fermentation has been carried out, various other procedures are used to produce the wide variety of wines we see.

8.3.1. *The fermentation*

One key to the modern wine industry is the careful control of the fermentation and this begins at harvest. Two factors are important, chemical oxidation and the microbial activities of wild yeasts and bacteria. In the modern industry, grapes are harvested mechanically at night to ensure that the fruit temperature is kept low. In the field, the harvested grapes are swathed in sulfur dioxide to kill wild yeasts, a step only possible with mechanical harvesting. Following transport to the winery, the grapes are de-stemmed. After this, the steps depend on the wine being produced. Grapes for white wine are crushed and the must separated and pumped to the fermentation vessel where it is inoculated. Grapes for red wine are pumped into the fermentation vessel and inoculated. Volumes in the red wine fermentation vessel are kept to about 6,000 l so that the skins can be kept in contact, usually by pumping, with the fermenting must during the fermentation. During the crushing stage, sulfur dioxide may be added to kill wild yeasts and bacteria, and to act as an antioxidant. Modern wine making yeasts, usually *S. cerevisiae* var. *ellipsoideus* are resistant to this chemical and may even produce sulfur dioxide.

Another key factor is the temperature of fermentation. White wines are fermented at 7–16°C rising to 20–23°C to retain the fresh bouquet of the grapes. Red wines are fermented for a longer period at 21–27°C. In both cases the vats should be cooled to counteract the heat produced by the metabolising yeasts. At the end of the fermentation the red wine must is crushed to remove skins in the same way that grapes are crushed for white wine.

The fermentation and subsequent flavor of the wine is not due entirely to the fermentative activity of *S. cerevisiae*. When wild fermentations of grape juice are studied, there is a consistent pattern of microbial growth that occurs. Within the first two to three days the fermentation is dominated by *Candida* and *Kloeckeral/Hanseniaspora* species (Fig. 8.2).

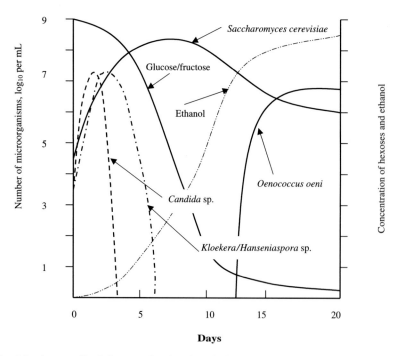

Fig. 8.2. A generalized diagram showing the relationship between the various microorganisms involved in wine fermentation, and the depletion of the fermentable hexoses, glucose and fructose, and production of ethanol. The *Candida* sp. and *Kloekera/Hanseniaspora* sp. are poor fermenters and grow in the early stage of the fermentation when oxygen is available. *Oenococcus oeni* begins growing if the batch of wine supports its growth, after the ethanol has been produced and it ferments the malate using the malolactic fermentation.

These organisms are obligate aerobes and poor fermenters and so are eventually overcome. They die off due to the lack of oxygen, and *S. cerevisiae*, which is a good fermenter, eventually dominates. In due course, the glucose and fructose are used up and the fermentation stops. At this stage a lactic acid bacterium, *Oenococcus oeni*, may or may not increase in number and begin fermenting the malic acid in the wine in the malolactic fermentation. This metabolic process raises the pH of the wine a little, produces carbon dioxide, and changes its flavor. If the fermentation occurs in a bottle of still wine ("still" or "table wine" means there is no gas in the wine) gas will be produced and give the impression that the wine has spoiled. One solution to this problem is to add *O. oeni* to the wine during the bulk storage stage to promote the malolactic fermentation so as to stabilize the wine before bottle storage. While the malic acid remains in a bottled wine, the possibility exists for a bacterium to ferment the malate and result in the formation of gas.

After the completion of the fermentation the wine is stored in suitable vats or wooden barrels. Flavoring compounds from the wooden barrels add further complexity to the flavor of the finished wine. Residual tartaric acid may remain in the wine. The tartaric acid precipitates out in the storage vessels and can be sold as food acid for use in baking. The wine maker then bottles the wine and the bottled wine can be stored for further maturation, or consumed immediately, depending on the wine.

Box 8.4

The Fermentation of Glucose by *Saccharomyces cerevisiae*

Yeasts are grown for three fundamental purposes, to produce gas, to produce ethanol and to produce yeast cell mass. When glucose is fermented for these purposes some unusual metabolic activity is seen. Glucose is metabolized by *S. cerevisiae* via the EMP pathway and tricarboxylic acid cycle so under aerobic conditions a large amount of cell mass is produced per gram of glucose because the metabolic pathway is very energy efficient. The major by-products from the metabolism are carbon dioxide and water. However, when grown under anaerobic conditions these yeast cultures do not produce a lot of cell mass because this ethanol producing pathway is not very energy

Box 8.4 (*Cont'd*)

efficient so mass production per gram of glucose is low. Although these yeasts can grow anaerobically they can only do so for a while, for a maximum of four to five generations, before they require a further brief period of aerobic growth. The reactions describing the major energy relationships are:

$$\text{Glucose} \Rightarrow 6CO_2 + 6H_2O \quad \text{energy } -686 \text{ kcal} \quad \text{large cell mass produced}$$
$$\text{Glucose} \Rightarrow 2C_2H_5OH + 2CO_2 \quad \text{energy } -54 \text{ kcal} \quad \text{small cell mass produced}$$

If sufficient oxygen is pumped into a broth medium culture so that it is saturated with oxygen, then the organism will only respire and produce no ethanol. This effect was observed by Pasteur and is known as the **Pasteur effect**. However, *S. cerevisiae*, along with other *Saccharomyces* species, is sensitive to a glucose concentration greater than 1.0 g l^{-1} and consequently, respiration is repressed for a while under those conditions and the organism reverts metabolically to anaerobic growth and ethanol production in the presence of oxygen. This is known as the **Crabtree effect**, or more generally as a case of glucose repression or catabolite repression. This effect does not occur with maltose, the main sugar in wort.

8.4. East Asian Alcoholic Beverages

Thousands of years ago, the Chinese developed a second method of processing the starch to make alcoholic beverages, namely a molded cereal based system. The starting point of the process is wet grain, cooked or uncooked, on which molds have been grown to produce carbohydrate degrading enzymes, the key constituent. In Japan, this molded product is called koji and in China it is called jiuqu. Apart from alcoholic beverages, koji is used to produce a variety of foods throughout East Asia. Early in Chinese history, the equivalent of beer brewing was carried out, however it seems to have been dropped in favor of the microbial based system. (Modern beer making methods were introduced by the Russians to Harbin in China before 1900.) This microbial-based starch degrading system has a long history, dating back with certainty to the Xia Dynasty from 4,000 years ago but archeological artifacts suggest an origin over 5,000 years ago. Specific mention of jiuqu in Chinese literature dates to the Zhou Dynasty (770–221 BC). As in the case of beer in the

West, in the early days in China, and even now, alcoholic beverages were perceived to have some beneficial effects for healthy living.

In China this mold based system is used to make grain based distilled liquors and wines, e.g. from rice, corn, millet, and it is also used in Japan to make saké, the Japanese rice wine. In the Philippines a wet form of koji, called binubudan, is mixed with sugar cane juice to ferment, as opposed to using rice. Study of the science of rice wine production is most advanced for the production of Japanese saké.

8.4.1. *Saké brewing*

Japanese rice wine production began around 300 BC and the system of production now depends on the use of pure fungal and yeast cultures as opposed to the traditional rice wine making systems mostly used in China. Saké production has three stages, the koji making stage, the moto making stage and the final brewing stage which is a **solid-liquid state fermentation**.

To make saké the rice must be polished resulting in reduction in the rice grain weight by 25–50%, and the degree of polishing affects the quality of the finished saké so saké is classified according to the degree of polishing (Table 8.2). Generally, ginjo saké is made with highly polished rice while honjozo saké is made with rice not polished below 60%. Less highly polished rice results in a higher level of inosinic acid that is responsible for the characteristic flavor of saké.

Table 8.2. The types of saké are based on the Seimaibuai classification that is related to the percentage of the original rice left after the rice has been polished.

Type of saké	Rice grain left after polishing, %	Comment
Junmai Daiginjo	50	Highest grade of sake, light dry type
Junmai Ginjo	60	Sweeter, heavier than above
Tokubetsu Junmai	60	"Special Junmai", a marketing product
Honjozo	70	Some distilled alcohol added, smooth, light more fragrant flavor
Junmai	70	"Pure saké", no added alcohol, has a heavier, fuller flavor

8.4.1.1. *Koji production for saké*

Koji is rich in fungal enzymes and is produced using a **solid-state cultivation** method. The enzymes breakdown starch external to the koji as well as in the koji so we can think of the added steamed rice as equivalent to adjuncts in beer production. Choice of rice variety is important as large rice grains with soft kernels are chosen so that water uptake is rapid and mold mycelia can penetrate the kernels more easily. Water quality influences the final saké quality, low iron and potassium levels and the presence of phosphorus and magnesium promote mold growth, as does a neutral to alkaline pH. The presence of calcium and high chlorination levels in the water improve fungal amylase extraction during the production.

Rice grains are steeped to take up water equal to 25–30% of the milled rice weight, then drained and steamed in shallow open vessels for half to one hour, in batch systems, or for a shorter period in continuous systems, so that the weight of the grains increase a further 8–12%. The cooled rice (34–36°C) is transferred to a koji making room with temperature and relative humidity control, and piled on an insulated bed and covered with cloth to equilibrate to the room conditions. The inoculum (10–100 g 100 kg^{-1} of polished rice), a seed culture of *Aspergillus oryzae* that has been grown for up to six days to maximize sporulation, is then spread evenly over the pile that is reheaped and covered with cloth. The temperature rises as the fungus grows so the inoculated rice must be stirred and reheaped to control temperature.

The way the mold mycelia grow through the resulting molded grain influences the quality of the saké produced. If mold growth is confined to the outer layers of the grains excessive breakdown of grain protein reduces saké quality. The fungus produces about 50 kinds of enzymes and of particular importance are α-amylase (amylo-1,4-dextrinase) and S-amylase (amyloglucosidase), the former breaks down the starch molecules to smaller fragments to decrease viscosity during the wine making stage, the latter removes glucose units (Fig. 8.3).

8.4.1.2. *Moto production*

This is the preliminary stage of the brewing process in which conditions for the growth of the yeasts are optimized. Although the process seems

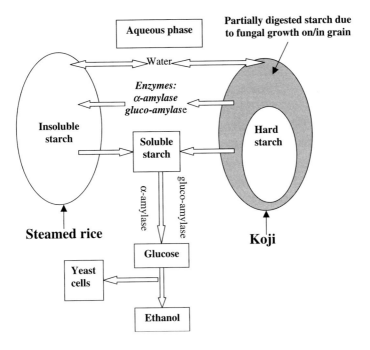

Fig. 8.3. During koji production the fungal enzymes in the koji penetrate the outer layers of the rice grain and begin hydrolyzing the starch. On mixing the koji and steamed rice in water, the water begins penetrating the rice and koji grains and absorbing into the starch and the mixture becomes very thick in consistency. The fungal enzymes and soluble starch are released from the koji and the enzymes begin hydrolyzing the steamed rice grains. As the soluble starch is released the water absorbed into the grains is also released and the mixture becomes more fluid. (Redrawn and represented from Yoshida, 2003.)

complex it is simpler than traditional methods. Water (200 L), rice-koji (60 kg) and steamed rice (120 kg) are mixed and held at 13–14°C for three to four days with occasional stirring, during which the temperature drops to about 7–8°C. The koji enzymes begin the breakdown of the starch in the rice, and some wild bacteria and yeasts begin growing in the mix. However, conditions favor growth of the lactic acid bacteria (*Leuconostoc mesenteroides* var. *saké* and *Lactobacillus saké*) which lower the pH. These acid conditions, along with the high levels of sugars, eventually kill off the wild bacteria and yeasts. The temperature rises to about 15°C and by about 15 days a pure culture of saké yeast, a strain of *Saccharomyces cerevisiae* with some unique characteristics (Table 8.3),

Table 8.3. Some cultural characteristics that distinguish saké yeast and *Saccharomyces cerevisiae*.

Characteristic	Response	
	Saké yeast	*S. cerevisiae*
Growth in vitamin free medium	+	−
Growth in biotin free medium	+	−
Tolerance to ethanol % concentration	20–23	16–19
Habitat	Saké brewing only	Widely distributed in fruits, grains etc.

is added to give a population of about 10^5 to $10^6 g^{-1}$ that rises to about $10^8 g^{-1}$ as incubation continues, and temperature rises to 20–23°C. Lactic acid may be added to the moto mix, instead of relying on microbial growth, to bring the pH down to 3.6–3.8, followed by the saké yeast inoculum. At the inoculation stage the water phase has a specific gravity of 1.124–1.128 due to the presence of 26–28% reducing sugar, 0.5–0.8% of amino acids as glycine, and a total acidity of 0.3–0.4% as lactic acid. The traditional production time for moto is 25–30 days.

8.4.1.3. *Saké production process*

The wine making process is basically a **batch-fed** system conducted as a parallel fermentation, preferably at low temperatures, beginning with the prepared moto (Fig. 8.4). In a **parallel fermentation** while the enzymes are releasing the glucose from the starch, the yeasts are fermenting it to produce ethanol. This is in contrast to beer where the soluble sugars are first released and then the fermentation is conducted.

Moto, at 12–13°C, is placed in a vessel six to eighteen kL in size along with steamed cool rice, rice-koji and water (Table 8.4) giving a mixture called **moromi**. The saccharolytic enzymes from the koji release the maltose/glucose from the rice starch thus making it available for the yeast to ferment. After the first addition of water, rice-koji and steamed rice to the moto there are two further additions to the vessel, and as each

Table 8.4. The raw materials added progressively to the prepared moto to ferment the rice for *saké*.

Preparation stage	Day	Cumulative totals			
		Total rice, kg	Steamed rice, kg	Rice-koji, kg	Water, L
Moto*	0	180	120	60	270
1st addition	1	510	360	150	500
2nd addition	3	1,020	830	210	1,230
3rd addition	4	1,710	1,350	360	2,329
Total		3,420	2,640	780	4,329

*Moto production takes two to four weeks to complete before the actual wine production begins.

addition is made the temperature is lowered by adjusting the temperature of the steamed rice and water so that by day 4 the temperature is 7–8°C. The lower temperature controls wild microbial growth. The moromi is agitated twice daily and by three to four days the specific gravity of the water phase is 1.051–1.058, the ethanol level is 3–4%, and total acidity, as succinic acid, is 0.06–0.07%. At each addition of steamed rice, koji and water, the ethanol and reducing sugar levels drop.

The yeast concentration reaches about $2.5 \times 10^8 \text{cells g}^{-1}$ and the fermentation becomes more vigorous and temperature rises to about 13–18°C in six to nine days where the temperature is held for a further five to seven days until the ethanol content is 17.5–19.5% and the temperature begins to decline.

The release of the sugars initially from the steamed rice occurs rapidly but the increased glucose concentration suppresses α-amylase activity so release of glucose declines until significant yeast activity reduces glucose and produces ethanol so that enzyme activity increases again (Fig. 8.5). The breakdown of the rice-koji in the moromi is more complex. The mold growth in the outer layers of the rice grain solubilizes the outer layer of starch rather quickly leaving variable amounts of a core of the rice grain that breaks down much more slowly under the action of the α-amylase. The breakdown of this inner core is dependent on the variety of rice used in the manufacturing.

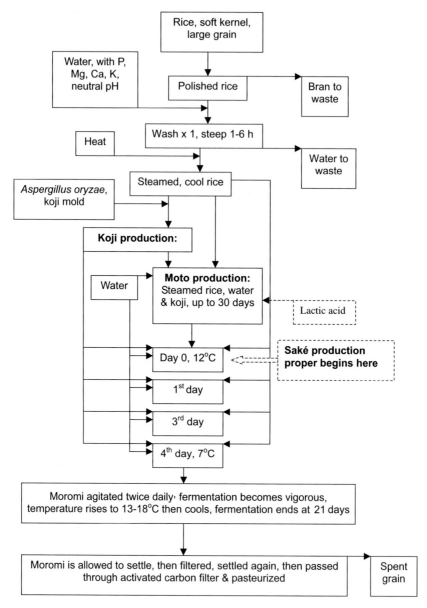

Fig. 8.4. An outline of the production stages and parameters used in the production of saké.

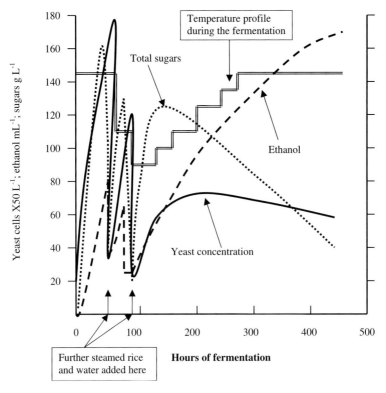

Fig. 8.5. After steamed rice and koji are mixed in the water phase a parallel fermentation takes place. The koji, steamed rice and water are mixed with the moto and the enzymic and microbial processes begin accelerating. After the initial mixing the concentration of total sugars increases rapidly, before the yeast concentration has significantly changed. The high sugar concentration temporarily reduces amylase activity until the yeast can significantly reduce the sugar level where upon amylase activity resumes. As further steamed rice, water and koji are added the concentration of yeast cells, sugars and ethanol are temporarily reduced accordingly. The temperature is controlled to minimize wild yeast and bacterial growth by adding cooled water and steamed rice. (Redrawn and represented from Yoshida, 2003.)

In the final stage pure distilled ethanol may be added to the moromi to bring ethanol to 20–22%, and 7–8% steamed rice may be added to encourage production of excess glucose to sweeten the wine by a natural process. However, pure glucose may also be added as well as lactic, succinic and glutamic acids. By 21 days the density of the water phase is 1.004–1.000, reducing sugar is 3–2%, ethanol is 18.5–19.5% and the acid content as succinic acid is 0.14–0.16%.

The moromi is allowed to stand for a few days to settle and it is then filtered and again allowed to settle for five to ten days. If the period is too long the quality begins to deteriorate. It is then passed through an activated carbon filter, pasteurized, and stored at 13–18°C to mature before bottling.

8.4.2. *Chinese rice wine manufacture*

Although Chinese rice wines are usually made with rice they are also made with maize and millet. When made with rice, about 230 kg of rice wine is produced from 100 kg of rice. Generally they are classified according to the residual sugar level, namely dry type, semi-dry type, semi-sweet type, sweet type and extra-sweet type (Table 8.5). They are also classified according to the production method. The four basic methods and their names (Table 8.6) are:

(1) Ling Fan Jiu, e.g. Yuan Hong wine that is also a dry type: the steamed rice is cooled by rinsing the rice with cold water then mixing it with a jiuqu starter where upon saccharification begins. It is commonly used as starter for inoculating more mash. It can be consumed after fermentation for several days.
(2) Tang Fan Jiu, e.g. Jia Fan wine, a semi-dry type: the steamed rice is spread onto bamboo rafts before mixing with jiuqu, and acidified rice-steeping liquid.
(3) Wei Fan Jiu, e.g. San Niang wine, a semi-sweet type: this rice wine is produced with a **fed-batch** fermentation method. The fermentation is begun as in Ling Fan Jiu but steamed rice is added intermittently during the fermentation period, usually in three batches.

Table 8.5. The Chinese classification of rice wines.

Wine type	Sugar (g 100 mL^{-1}, as glucose)	Ethanol (%, v/v)	Total acid (g 100 mL^{-1}, as succinic acid)
Dry type	< 0.5	> 14.5	0.45
Semi-dry type	0.50–3.0	> 16.0	0.45
Semi-sweet type	3.0–10.0	> 15.0	0.55
Sweet type	> 10.0	> 13.0	0.55

Table 8.6. The ingredients and chemical composition of four kinds of Chinese rice wine.

Component	Type of rice wine			
	Yuan Hong	Jia Fan	San Niang	Xiang Xue
Proportions of components used to make rice wine				
Glutinous rice, kg	144	144	144	100
Wheat koji, kg	22.5	25	25	10
Seed mash, kg	8–10	8–9	15	–
Acidified rice-steeping water, kg	84	50	50	–
Fresh water, kg	112	68.6	100*	–
Distilled water, kg	–	5	–	100[†]
Composition of rice wines				
Specific gravity	0.992	0.995	1.0349	1.073
Ethanol (%, v/v)	15.6	17.8	16.7	19.4
Extracts (g 100 mL^{-1})	3.325	4.45	15.65	24.44
Sugar (g 100 mL^{-1})	0.38	0.78	6.50	20.0
Total acids (g 100 mL^{-1})	0.48	0.46	0.46	0.28
Volatile acids (g 100 mL^{-1})	0.06	0.027	0.054	0.056
Non-volatile acids (g 100 mL^{-1})	0.42	0.43	0.406	0.22

*This is Yuan Hong wine.
[†]This is 50% distilled spirit.

(4) Fortified rice wine, e.g. Xiang Xue wine, a sweet type: aged rice wine or spirits distilled from rice wine are added before the main fermentation ceases. The final alcohol concentration is above 20%. Consequently, the microorganisms in the mash are inhibited, but the saccharification to sugar of the remaining dextrin in the main mash continues slowly, resulting in a high residual sugar content.

8.4.2.1. *Jiuqu*

Unlike koji, the jiuqu usually has two functions, the saccharifying function, due to the enzymes, and the fermenting function, due to the presence of yeasts in the final jiuqu. So during the many different procedures used to make jiuqu, the molds grow through the jiuqu producing saccharolytic enzymes that hydrolyze starch and produce soluble sugars that encourage yeasts to also grow in the jiuqu.

The actual number of types of jiuqu used for the production of alcoholic beverages in China is unknown. However, there are three broad types of jiuqu, based on the raw material used, and within each broad type there are subtypes based on the color, presence of additives and size of the molded masses or molded pancakes produced. The raw materials are rice, wheat and wheat bran. The manufacturing processes also vary, and this also influences the resulting jiuqu. These processes include the use of: pure or mixed cultures; a high temperature or a medium temperature; incubation of jiuqu in a thin or a thick layer; hanging in the air in a suitable small container; and wrapping in straw. Consequently, for a given mass of jiuqu the biological activity may vary widely.

The most active wheat based jiuqu, shenqu, is made by milling wheat then splitting it into three portions, one of which is roasted, another steamed, and the other left uncooked. The three portions are then mixed thoroughly with added water to a consistency that is less than a slurry. The mixture is transferred to an incubation room, and spread on the floor. Moisture content and temperature are critical and these are traditionally managed by opening windows to lower temperature and heaping the mixture to raise temperature. The first three days require high moisture content to promote growth of mycelia, and a lower water content is required for the subsequent incubation period up to 49 days. After this, the shenqu has a yellowish to yellowish-green color. The primary fungi for this preparation are *Aspergillus* species and pure cultures have been isolated and are used. In the preparation method with pure cultures a compart-ment similar to those used to germinate malting barley is used. The cooked wheat grits are inoculated with fungal spores and piled in a heap for three hours. The wheat is then spread in a 30 cm layer over the perforated plate in the compartment. The temperature and moisture content are controlled, and oxygen supplied by blowing suitably conditioned air (temperature and humidity) through the layer. Generally the temperature is kept to a maximum of about 30°C.

Daqu, meaning "large qu", is used for the manufacture of some Chinese spirits and its manufacture is different from shenqu. There are two basic kinds, depending on the temperature reached during manufacture. One kind is high temperature that is allowed to rise above 55°C, perhaps as high as 65°C, and the other is kept below 55°C. The raw materials are wheat flour, barley flour and pea flour mixed together in proportions, depending on the manufacturer. One manufacturer, Yanghe Distillery, uses the following proportions and procedures. The wheat:barley:pea proportion is 5:4:1; the moisture content is brought to about 40% and the resulting dough kneaded and eventually packed into a wooden mold (195×138×75 cm). The packing is done by human treading by ten people with a gradation in body weight to get the dough at just the right firmness in the mold (Fig. 8.6). The block is then removed from the wooden mold and placed in the incubation room, with other blocks, in two layers on bamboo rods giving a space of 2 cm between the layers covered with straw. Incubation is about four weeks and has three phases. In the first phase of about nine days, the temperature rises rapidly due to the growth of fungi and yeasts. The dominant fungi are *Aspergillus* species and *Rhizopus* species, namely *R. oryzae* and *R. chinensis*. The humidity is initially high and is reduced by opening room windows. The temperature of the whole batch is controlled within the range of 30–45°C by turning and re-arranging blocks about six times in the nine-day period. In the second phase, which lasts about eight days, the temperature of the batch

A picture showing a whole process
of DA-QU preparation

Fig. 8.6.   An old Chinese depiction of the process of making daqu. A line of ten workers treading the dough can be seen and the box like mold in which it is treaded is seen at the top.

is allowed to rise smoothly a little further but within the range of 30–45°C. In this phase, more moisture evaporates and its loss is controlled so that it is not lost too quickly. In the third phase, the room temperature is allowed to decline and moisture loss from the blocks continues. If at the end the moisture content has not been reduced to about 12%, the blocks are piled and re-covered with straw to raise the temperature and accelerate mois-ture loss.

Xiaoqu is another important jiuqu. The traditional method is similar to that used for daqu production except only steamed and powdered rice is used, the water used is flavored from steeped herbs, and the freshly made blocks are cut into smaller cubes and formed into balls. Then selected old jiuqu is used as a seed culture to inoculate the outer surfaces of the balls. As the dominant fungi involved have been isolated, a pure culture method is now used in its preparation as follows. The *Rhizopus* species are grown on rice meal or wheat bran. Yeast, *Saccharomyces* species, is cultured on an extract of a special jiuqu and harvested by centrifugation. Then the *Rhizopus* culture and yeast concentrate are mixed in a 50:1 ratio. This mixture is then used to make a seed mash for wine. This approach significantly reduces the time to make the rice wine.

8.4.2.2. *Rice wine making*

Ideally the water should have a hardness in the range of 2–6, pH around 7, and have a low iron content. The acidified rice-steeping liquid is traditionally prepared as follows. The polished rice is steeped in the water for a period of 10–20 days. During the steeping, lactic acid bacteria grow and lower the pH. This traditional liquid is separated out and reduced in volume by boiling; when added to the final brew, it helps to inhibit the germination of bacterial spores and the growth of vegetative cells, provides amino acids and vitamins and lowers the pH for yeast growth and helps improve the flavor of rice wines. However, alternatives to this include using diluted vinegar solution or a broth made with wheat flour, but lactic acid is now commonly added.

The actual production process begins with the preparation of **seed mash**. Traditionally the ingredients used to make this seed mash are 125 kg of glutinous rice that is soaked for several days and then steamed.

The steamed rice is then acidified to pH 4.0–4.5 with lactic acid or acidified steeping water, and after it cools to about 27–30°C it is spread in a ceramic vat and half of the xiaoqu jiuqu is mixed through the rice. The rice is then stacked around the sides of the vat to form a hollow in the middle and the rest of the xiaoqu powder is sprinkled onto the surface (0.187–0.25 kg total). This maximizes the surface exposure to encourage growth of *Rhizopus* species that produce lactic and fumaric acids to lower the pH, discourage the growth of undesirable organisms and encourage yeast growth. In addition, the amylase and glucoamylase in the jiuqu saccharify the rice. After about two days, a pool of liquid containing sugars etc., gathers in the hollow. Then 19.5 kg of jiuqu is added, mixed through the fermenting rice mix and fresh water is added to bring the total water, including that taken up by the rice during steeping and steaming, to 230 kg. A pure culture of yeast (1–2%) may be added to the seed mash instead to promote the development of yeast activity.

By **fed-batch** and **parallel fermentation** processes the fermentation proceeds for about 7 days. The timing of the first stirring is critical and it is followed fairly quickly by three more stirrings. After 7 days the fermented liquid is transferred to narrow-necked jars and sealed tightly to keep out O_2 where a **secondary fermentation** takes place over a 70–80 day period. The wine is then filtered, clarified, pasteurized and packaged.

8.4.3. *Sorghum liquor*

Maotai, Chu Yeh Ching (Bamboo Leaf Green), Ta Che, Fen Gue and many other Chinese liquors are sorghum spirit (53–55% alcohol) with characteristic aromatic fragrance due to the presence of 4-ethylguaiacol, vanillic acid and vanillin. Starch is hydrolyzed using exogenously produced fungal amylases in a **solid-state cultivation**. The production process as outlined in Fig. 8.7, varies. In Taiwan, three cycles of cultivation of the substrate and distillation are conducted. The fermentation stage is usually about ten days long.

A typical wine (yeast) cake (or jiuqu), produced with cracked wheat, contains five yeasts (including *S. cerevisiae*, *S. fibuligera*), four molds (including *Mucor* and *Rhizopus* species) and six bacteria, including a

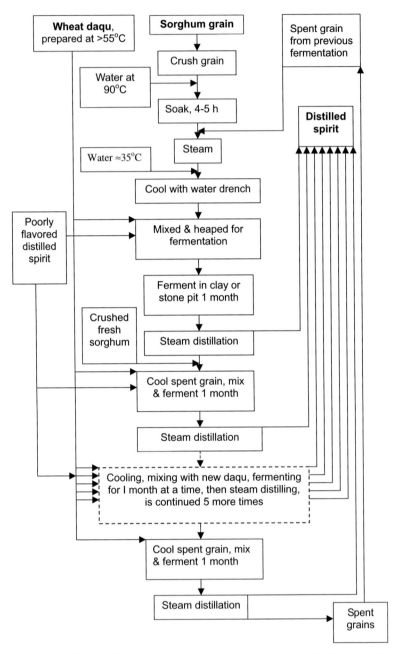

Fig. 8.7.   An outline of the production stages and parameters used in the production of sorghum liquor. Production depends on a repeated solid-state cultivation.

Clostridium species. The wine cake contains some hydrolytic enzymes, but activities are low.

To inoculate the steamed sorghum, it is spread out on the floor to a thickness of about 5 cm and powdered wine cake is mixed through it. The moisture content of the sorghum is about 55%. The fungi grow on the cool and dry surface, whereas bacteria and yeasts grow within the wet mass of sorghum. After 24 h the sorghum is piled into heaps that encourage growth of yeast in an anaerobic and warm condition (pre-fermentation). The size of the pile is restricted because the internal temperature of the pile should not be allowed to rise above 40°C, and the heat produced metabolically is not easily dispersed. In Taiwan, the inoculated sorghum is placed into stainless steel tanks or boxes and covered with polythene sheet to prevent moisture loss and reduce aeration. Microbial growth and metabolism is allowed to progress for up to 10 d after which the mixture is steamed to distil the ethanol. The pile of sorghum is re-inoculated to encourage further growth and when it is ready, the sorghum is distilled again. Heat treatment (steaming) causes partial hydrolysis of carbohydrate and protein. Steaming also serves to remove alcohol, which prevents the inhibitory effect of alcohol on cell growth in the thin film of water surrounding sorghum grains. Alcohol is collected by condensation of distillate. The yield of ethanol is about 22.53 kg per 100 kg of sorghum. In some procedures the processes of fermentation and steaming need to be repeated up to seven times to achieve complete conversion of carbohydrate to alcohol. However, in Taiwan, repeating a ten-day fermentation period and a distillation twice is sufficient to maximize ethanol production.

Microbial cells grow slowly on sorghum due to the low availability of fermentable substrates, and this leads to the production of secondary metabolites mentioned above, which contribute to the characteristic strong aromatic fragrance of sorghum liquor.

8.5. Dough Fermentations

The desirable qualities of baked flour dough products like bread depend on added yeast. Traditionally prepared dough was kept from day to day to provide active yeast but this encourages growth of lactic acid bacteria. Consequently two products of interest have arisen. Firstly, technology for producing baker's yeast to add to the flour was developed, and secondly, sour dough baked products are still being produced based on our understanding of the role of lactic acid bacteria.

8.5.1. Baker's yeast

Each year about two million tonnes are produced and sold as either cream yeast (solids 16–20%), compressed yeast (solids 30–35%), both of which may be frozen, or dried yeast. Consequently the fermentation to produce yeast for bakers (who prefer cream and dried yeast) must be controlled to produce cells that are very active and able to withstand the stresses involved. A broth medium of molasses and simple N sources as well as other nutrients are heat treated and added to fermenters with agitation and aeration. Optimal growth conditions rely on (i) nutrient feed rates (a batch fed system is used) that ensures carbohydrate levels are kept low, (ii) very high aeration levels that ensure respiration only occurs so maximizing yeast mass production, and (iii) pH, temperature and ethanol level. The respiratory quotient, $CO_2:O_2$ ratio, is also important.

At the end when nutrient addition ceases aeration is continued for up to an hour during which trehalose accumulates in the cells as a storage carbohydrate. Depressed carbohydrate level, high aeration rate and the removal of N sources favor the gluconeogenic pathway and synthesis of trehalose. The trehalose protects the yeast cells from stresses involved in freezing and drying.

8.5.2. Sough dough

Sour dough is produced when flour and water are mixed and left for up to 24 h during which yeasts and lactic acid bacterial grow. The soured dough is then used to make bread and other products. The production of rye bread relies on it. In Europe and many other parts of the world, sour dough products based on wheat flour, such as breads and soda cracker biscuits, are made. One example of a wheat dough product is a Cantonese deep fried product called Yau Chau Kwai. It is usually prepared in the mornings and people buy it for breakfast. In the USA, the San Francisco sour dough French bread process is used to make bread. The Cantonese product illustrates the basic process, used for thousands of years around the world. The maker of Yau Chau Kwai keeps the very wet dough aside and each day he takes out half of the mixture and adds an equal mixture of wheat flour and water back into the remaining wet dough. This stimulates the lactic acid bacteria and yeasts in the mixture to grow to produce sufficient microbial growth for the next day's production.

The process depends on the chemical and microbiological characteristics of the flour. The flour from wheat and rye grains contains monosaccharides and disaccharides, particularly maltose, amino acids, minerals, and fatty acids (Table 8.7). In addition, they contain lactic acid bacteria and yeasts from the growing environment of the grains. When the flour and water are mixed, the lactic acid bacteria begin to grow rapidly. They lower the pH of the mixture and this encourages the growth of yeasts that grow more slowly than the bacteria. In normal dough the yeasts constitute about 1% of the total number of microbial cells in the mixture. Factors that affect the process include temperature, water activity, redox potential (oxygen) and fermentation time.

The dominant organisms in sour dough are *Lactobacillus sanfranciscensis*, and *Lb. pontis*, and others found include *Lb. fructivorans*, *Lb. fermentum* and *Lb. brevis*. Most of them grow better on maltose. Homofermentative and heterofermentative rods and cocci are involved.

The yeasts associated with sour dough include *Candida milleri*, *C. holmii*, *Saccharomyces exiguous* and *S. cerevisiae*.

In the case of rye bread, the lowering of the pH of the dough is necessary for it to be made into bread. It inhibits amylase activity and improves the swelling of pentosans that take over the function of gluten in wheat bread. The pentosans in the flour at low pH (optimum pH 4.9) have improved water-binding and gas-retaining properties that produce

Table 8.7. Mono- and di-saccharides in wheat flour.

Sugar	Percent w/w
Glucose	0.02
Fructose	0.04
Sucrose	0.26
Maltose	0.12

better quality bread. In the case of rye and wheat based sour dough, the resulting breads show better keeping quality, and improved nutritional value. The sour dough prevents the production of ropey bread and the development of fungi on the stored bread.

The fermentation improves the flavor of the bread by non-enzymic browning during baking of the bread, fatty acid oxidation and products of microbial fermentation.

Further Readings

1. Ayres JC, Mundt JO and Sandine WE. *Microbiology of Foods*, Chapter 7: Alcoholic yeast fermentations. WH Freeman, San Francisco, 1980.
2. Beech GA, Melvin MA and Taggart J. Food, drink and biotechnology. In: *Biotechnology: Principles and Application*, eds. IJ Higgins, DJ Best and J Jones. Blackwell Scientific, Oxford, 1985, pp. 73–110.
3. Fugelsang KC. *Wine Microbiology*. Chapman and Hall, New York, 1997.
4. Hardwick WA. Beer. In: *Biotechnology Vol. 5*, Chapter 3, ed. G Reed. Verlag Chemie, Weinheim, 1983, pp. 165–230.
5. Jakobsen M, Larsen MD and Jespersen L. Production of bread, cheese and meat. In: *The Mycota: A comprehensive Treatise on Fungi as Experimental Systems for Basic and Applied Research*, eds. K Esser and PA Lemke. X, *Industrial Applications*, ed. HD Osiewacz. Springer-Verlag, Berlin, 2002, pp. 3–22.
6. O'Toole DK. Beverages: Asian alcoholic beverages. In: *Encyclopedia of Grain Science*, eds. C Wrigley, H Corke and C Walker. Elsevier, 2004, pp. 86–96.
7. Priest FG and Campbell I, eds. *Brewing Microbiology*, 2nd ed. Chapman and Hall, London, 1997.
8. Springham DG. The established industries. In: *Biotechnology: The Science and the Business*, 2nd ed., Chapter 16, eds. V Moses and RE Cape. Harwood Academic Publishers, The Netherlands, 1999, pp. 261–305.
9. Stewart GG and Russell I. Modern brewing biotechnology. In: *Comprehensive Biotechnology Vol. 3*, Chapter 17, eds. HW Blanch, S Drew and DIC Wang. Pergamon, Oxford 1985, pp. 335–382.

10. Tuite MF and Oliver SG, eds. *Saccharomyces: Biotechnology Handbooks Vol. 4.* Plenum Press, New York and London, 1991.
11. Vondrejs V, Janderova B and Valasek L. Yeast killer toxin K1 and its exploitation in genetic manipulations. *Folia Microbiol.* **41**: 379–394, 1996.
12. Yoshida T. Technology development of saké fermentation in Japan. In: *The First International Symposium on Insight into the World of Indigenous Fermented Foods for Technology Development and Food Safety*, organized by the Department of Microbiology Faculty of Science Kasetsart University, and the National Research Council of Thailand International Cooperation, Ministry of University Affairs, 13–14 August 2003 at the Microbiology-Genetics (MG) Building, Department of Microbiology, Kasetsart University, Bangkok, Thailand, I-4, 2003, pp. http://plantpro.doae.go.th/worldfermentedfood/.
13. Xu GR and Bao TF. *Grandiose Survey of Chinese Alcoholic Drinks and Beverages*, at http://www.wxuli.edu.cn/wine/umain.htm

Questions for Thought…

1. Production of sorghum liquor by solid state fermentation is a slow process. How can the process be speeded up without compromising the flavor and fragrant quality of the final product?
2. Compare and contrast the various ways that starch is used to produce alcoholic beverages.
3. Summarize the enzymatic processes occurring in rice grains during the manufacture of Japanese saké. How do these processes differ from traditional Chinese rice wine making processes?
4. What two environmental factors are used to control the growth of wild yeasts and bacteria during rice wine production? Why are they not used for that reason for beer and grape wine production?
5. Consider the advantages and disadvantages of producing rice wine using purely enzyme based methods or using microbes, either natural or genetically engineered, to carry out the processes.
6. Rice wine production in general is a batch fed system. What are the reasons for this?
7. Why is sulfur dioxide used in grape wine production but not in the production of other alcoholic beverages?

Websites

http://www.sytu.edu.cn/zhgjiu/umain.htm
http://www.brewingtechniques.com/
http://www.foodreference.com/
http://www.internationalrecipesonline.com/recipes/
http://www.beerinfo.com/vlib/commercial.html
http://dmoz.org/Business/Food_and_Related_Products/Beverages/Beer/
Tools_and_Equipment/
http://www.winepros.org/wine101/enology.htm
http://www.sake.nu/
http://www.esake.com/
http://www.faqs.org/faqs/beer-faq/part1/
http://www.esake.com/index.html

Chapter 9

Fungal Solid State Cultivation

Desmond K. O'Toole
Department of Biology and Chemistry
City University of Hong Kong
83 Tat Chee Ave, Kowloon, Hong Kong, China

A large number of industrial microbial fermentations are conducted in liquid media. Liquid media have the advantage that the growth medium is homogeneous and can be kept homogeneous, and is easily manipulated and handled in an industrial setting. However, other forms of microbial cultivation, solid state cultivations, are possible and have been used for centuries. The term "fermentation" is often applied to this form of microbial cultivation but the word has a specific meaning in microbial metabolism related to anaerobic growth, and as most of the processes dealt with here are aerobic in nature, the word "cultivation" is used here. However, even some of the liquid broth fermentations are aerobic in nature.

9.1. Fungal Solid State Cultivation

Fungal solid state cultivation can be defined as the growing of "microorganisms on moist, water-insoluble solid substrates in the absence or near absence of free liquid." The solid state is characterized by its "porous substrate with a gaseous continuous phase" that is usually aerobic in nature. It is used to produce microbial by-products, particularly enzymes (Table 9.1), and it is the basis of a number of Asian food production

Table 9.1. Enzymes and the fungi used to produce them on a commercial or pilot scale and the substrates used in their production by solid state cultivation.

Enzyme	Organisms	Solid Substrate
Pectinases	*Aspergillus carbonarius,* *A. sojae, A. saito, A. niger*	Wheat bran
Pectinases	*A. flavus, A. oryzae,* *A. fumigatus, Rhizopus* sp.	Wheat bran and other solids
α-amylase	*A. oryzae*	Wheat bran
Glucoamylase	*Rhizopus* spp.	Bran
Glucoamylase	*A. niger*	Wheat bran and corn flour
Rennet	*Mucor pusillus*	
Rennet	*M. meihei*	Wheat bran
Rennet	*R. oligosporus*	Wheat and rice bran
Cellulase	*Trichoderma reesi*	Wheat bran
Cellulase	*T. harzianum*	Wheat straw and bran
Xylanase	*A. terreus*	Wheat bran
β-xylosidase	*A. terreus*	Wheat bran

systems, e.g. soy sauce and rice wine. The system is ideal for aerobic, mycelium-producing organisms, like fungi and streptomycetes, and it is not commonly used for culturing single-celled organisms such as yeasts and bacteria, with the exceptions of a *Bacillus* species cultured on soybeans to produce natto, and silage fermentation with wild lactic acid bacteria that quickly becomes anaerobic and O_2 plays no part. Composting and commercial mushroom growing are further examples of solid state cultivations. However, the fundamental objectives of each of these processes differ (Table 9.2). During composting the objective is to **transform**, using natural microflora, a substrate that may be objectionable (e.g. animal manure) or of no other value so the end product can be used for another purpose. In mushroom growing the objective is to **grow** a particular fungus on the substrate so as to generate fungal fruiting bodies for human consumption. The substrate is of no direct value. (The mushroom industry is substantial because each year about 1,223,000 tons of *Agaricus bisporus* are produced and when the commercial production of all other fungal fruiting bodies is added in about 2,182,000 tons are produced.) On the other hand, in the solid substrate cultivations considered here the

Table 9.2. A comparison between applications of the principles of solid state cultivation.

Characteristic	Solid State Cultivations	Composting	Mushroom Growing
Substrate	Processed vegetative substrate, often of one kind, that favors fungal growth	Mixed material of vegetable origin; may include animal waste	Mixed or single type vegetative material that favors mold growth and fruiting body formation
Fungus/ microorganism	A pure strain, or wild mixture, usually not grown aseptically	Wild mixture depending on the environment	A pure strain, but other organisms may be needed to promote fruiting
Substrate for human consumption	Yes, or an extract of the fermented substrate	No	No
Fungus consumed	No	No	Fruiting body only
By-product isolated and used	Yes, in some cases	No	No

objective is either to transform the substrate so it can be **consumed completely** or **extracts** for human consumption or for human use can be produced, or to prepare and isolate **by-products** for further use. Much of what follows can be applied to all three broad applications because aeration, water activity and temperature level are common parameters that impact the outcome of the processes.

9.1.1. *Substrate characteristics*

Ideally the solid substrate should consist of solid particles (Table 9.3) that are of small dimensions, say 3 to 10 mm, either oblong or ovoid in shape, e.g. soybean, rice, or long in shape, that when packed together result in gaps between the particles to allow air to pass easily. Sliced material or flat material tends to pack down and close off gaps bringing about a reduction in the porous nature of the solid mass. However, geometry of the substrate particles affects the volume of the porous continuum and

Table 9.3. Substrates commonly used in fungal solid substrate cultivations.

Agricultural Waste	Processed Food Waste	Grains	Other Materials
Wheat straw	Wheat bran	Rice	Wood
Corn stover*	Sugar beet pulp	Soybean	Cassava
Rice stover	Banana meal	Corn	
Feedlot waste	Rye meal	Buckwheat seeds	
		Sorghum	

*Dry lignocellulose material that is not straw.

the movement of air in a substrate mass and eventually microbial growth as well. For example, in one comparative experiment the growth of *Aspergillus koppan* was improved on a rice medium as the mesh size of the particles was reduced from 8 mesh (2 mm) to 30 mesh (0.5 mm). However, when the mesh size was smaller and allowed closer packing of the particles and reduced air volume, growth was suboptimal.

The advantages of this system for producing microbial products, when compared with liquid fermentations, are claimed to include better product quality, and higher product yield and concentration that results in less expensive downstream processing, low capital and energy costs, and low wastewater output. Disadvantages include the heterogeneity of the substrate, restricted transport of the mass and heat energy which causes local overheating and dehydration, and, within the substrate, local depletion of oxygen and increase in carbon dioxide. In addition, when a solid state cultivation is leached to obtain the product, the extract is often viscous which cannot be easily vacuum concentrated resulting in a product that cannot be converted to a powder and is not easily blended or re-constituted.

9.1.2. The characteristics of fungal growth

The hyphal mode of growth by fungi gives the fungi a large advantage over the growth of single-celled organisms like yeasts and bacteria. Fungi grow by apical extension of the hyphal tips and by the generation of new hyphal tips by branching. The growth at these tips, however, is due to the transport of nutrients in the hyphae to the tip, and the growth is so quick that the hyphal tips extend at a greater rate than the rate at which nutrients in the substrate can diffuse to the hyphae. The rate of branching is related to the density of nutrient available in the substrate, but the rate of extension of mold hyphae, that is their colonization rate, is unaffected by nutrient availability. The highest hyphal extension rates are observed in **non-septate fungi** because the lack of septa allows rapid transport of nutrients from the rest of the mycelia to the growing apical tip. In **septate fungi** the nutrients must pass through septal pores that separate fungal cells and slow down nutrient transport. Streptomycetes, which have a similar growth habit to fungi and form mycelia, cannot grow as fast due to the formation of individual separate cells in the hyphal chain through which nutrient cannot be delivered to the growing tip consequently their expansion is limited.

9.1.3. *The fungal inoculant*

The usual form of the inoculum is a preparation of mold conidiospores or sporangiospores (the bacterial streptomycetes produce exospores) and not vegetative molds. The spores are produced aseptically by inoculating sterile rice or wheat bran in flasks with fungal spores and incubating them until growth terminates at maximum sporulation. The spores are of uniform size and are easily harvested and stored for a long time while being resistant to mishandling; vegetative mycelial fragments, on the other hand, are heterogeneous and lumpy and would require considerable force (that could damage the hyphae) to obtain a reasonable mixture that could be easily mixed with the substrate.

Following spore inoculation into the prepared substrate, the fungal growth is delayed because of the lag time required for the spores to germinate and start producing the hyphae which produce enzymes to act on the substrate from which nutrients for the growing fungus arise. Further, the optimum germination temperature may differ from the optimum growth temperature for the hyphae that must spread through the substrate to produce the mycelium.

The density of the inoculum should be optimized for each substrate and organism. Spore inoculation levels used range from 10^2 to 10^7 spores g^{-1} substrate. If inoculation levels are too high or too low, lag time, mold growth rates and product yield can be influenced. Early fungal sporulation during the production stage must also be avoided in some cases, e.g. for tempeh, but in other cases it does not matter.

9.1.4. Growth conditions in solid state cultivations

Three factors must be controlled in the midst of the substrate during the growth phase for a successful outcome, namely: temperature, water activity and air or oxygen. Another factor that might affect growth is pH, but filamentous fungi can grow over a wide pH range (2–9), and optimally from pH 3.8–6.0. As a consequence of the fungal tolerance to environmental pH, the pH of the substrate can be manipulated to a low level to restrict bacterial growth, but the pH level of the growth medium can influence enzyme production by the fungi, e.g. *Aspergillus niger* produces protease optimally in the pH range 3.55–4.3.

9.1.4.1. *Temperature*

In general, the fungi used grow in the temperature range 20–55°C but the optimum temperature for growth, and for product formation, can differ. The temperatures used for some production systems is shown in Table 9.4.

Temperature control is a key issue. Firstly, because of the aerobic metabolism high levels of heat are generated by the microbial growth. The maximum and average rates of heat generation by *Aspergillus oryzae* in a soybean/wheat mixture for soy sauce production was measured at $130 \, kJ \, kg^{-1} \, h^{-1}$ and $53 \, kJ \, kg^{-1} \, h^{-1}$, respectively, and the maximum and average rates of heat generation for *A. niger* grown on a starch/rice bran

Table 9.4. Examples of optimum temperatures used in some fungal solid state cultivations.

Microorganism	Substrate	Product	Temperature, °C
Rhizopus oligosporus	Wheat bran	Acid protease	25
Aspergillus niger	Wheat bran	Glucoamylase	30
Humicola sp.	Agribusiness waste	α-galactosidase	45
Penicillium candidum	Wheat bran	Lipase	29
Aspergillus carbonarius	Canola meal	Phytase	53

mixture for citric acid production was measured at $71\,kJ\,kg^{-1}h^{-1}$ and $33\,kJ\,kg^{-1}\,h^{-1}$, respectively. The fungi penetrate the mass of the substrate completely and grow over and into the particles making up the substrate. However, the substrate does not have good heat conducting properties, and convection currents to take away heat may be weak, nor can refrigeration be used as it can with broth-based cultivations. Consequently, temperature gradients may build up in the solid substrate and some parts of the substrate mass may reach temperatures sufficient to kill any fungus and stop product formation. The excessive heat may also dry out the substrate as evaporative moisture loss, through the latent heat of evaporation, is one of the key ways in which the temperature can be reduced.

Control in traditional systems is by using shallow trays so that the heat generated has only a short passage to travel to be dissipated to the environment. In modern systems, mechanically driven moisturized air is pushed through the solid state material but the air flow rate is another important parameter affecting temperature, water activity and oxygen levels. In reality, the whole process is not easy to completely control.

9.1.4.2. *Water*

If water content is too high the substrate particles may agglomerate and diffusion of air throughout the substrate may be restricted thus allowing undesirable bacteria to grow. However, water content alone is not a measure of the best moisture level for microbial growth. **Water activity** is the best measure because the level of water activity affects mold growth (Table 9.5) and in these systems the water activity is finely

Table 9.5. Some optimum water activity levels used in fungal solid state cultivations.

Microorganism	Substrate	Product	Water Activity
Trichoderma viride	Sugar beet pulp	β-xylanase	0.995
Candida rugosa	Coconut oil cake	Lipase	0.92
Penicillium roquefortii	Buckwheat seeds	Growth and sporulation	>0.96
Aspergillus oryzae	Various	Protease	0.982–0.986
Rhizopus oligosporus	Chickpea	Fermented food	0.92

balanced between the optimum and actual water activity level that may change during growth.

The water activity in a substrate is not directly related to the moisture content. It reflects the free water available for microbial activity and it ranges from 0 to 1 with 1 being the optimum for most organisms, and about 0.6 the minimum at which any microorganism will grow. It is equivalent to **relative humidity** but it is expressed as the **ratio** of the relative humidity over a substrate in an enclosed container, to the equivalent relative humidity over pure water, both at equilibrium with the air phase. The amount of water that evaporates into the air depends on the number of soluble particles that can bind water in the water phase, or the concentration of bound water molecules at the water surface; a low concentration of soluble particles that bind water molecules results in a high water activity, a high concentration of solutes results in a much lower water activity. In many of the solid substrates used in these cultivations the levels of available solute are relatively low so relatively low water content can result in a high water activity which favors the growth of most microorganisms. For example, when the water activity is just at 1 the moisture contents of grain sorghum, wheat straw and sugar beet pulp are about 40%, 63% and 77%, respectively.

A further factor is the flow of air through the substrate along with a rising temperature. The air flow and rising temperature increase the evaporation of water from the substrate causing the water content to go down which may further lower the water activity.

To control water activity levels in primitive systems the substrate is stirred and doors and windows of the incubation room may be opened and closed but in modern systems using mechanical delivery of air through the substrate, the relative humidity of the air can be controlled to balance all the factors to maintain optimum water activity levels.

9.1.4.3. *Oxygen*

Oxygen control methods are really about air control and depend on the design of the bioreactor system. Shallow substrate layers, stirring, tumbling and mechanically driven air are four possible ways to control the access of air to the substrate.

9.2. Bioreactor Systems for Fungal Solid State Cultivations

Five broad classes of bioreactors are used. The first is the traditional tray system, the second is the traditional small package system, the third is the packed column system, the fourth is the rotating drum system, and the fifth is the stirred substrate system.

9.2.1. *Trays*

In the **tray system** the bottom of the tray is perforated to allow natural air circulation through the inoculated substrate that is placed in a layer 2 to 4 cm thick in the tray that is incubated in a suitable room. The conditions, including temperature, aeration and relative humidity, in the cultivation room must be controlled during the incubation period. Temperature and humidity gradients develop in the substrate so, if needed, the substrate can be stirred to reduce temperature.

The main advantage of the tray system is its simplicity. During the cultivation period it is easily observed, and experimentation is easily conducted.

9.2.2. *Small packages*

The traditional **small package system** is similar to the tray system except the solid substrate is inoculated and then formed into small, flat packets, e.g. as used in tempeh production, or flat round cakes, e.g. as used for Korean soy sauce making and Chinese rice wine making, or small round balls (about 15–40 mm diameter) on and through which the fungi are allowed to grow.

9.2.3. *Packed columns*

The **packed column system** can be an enclosed system and it depends on forced air flow through the substrate in the column. Some advantages of the packed column system are:

(i) variations in temperature and gas concentration throughout the substrate can be eliminated;

(ii) water activity levels can be maintained with moisturized air;

(iii) aseptic conditions can be maintained, unlike other systems; and

(iv) the column can be converted to a **trickle bed extractor** for recovery of soluble product.

9.2.4. *Rotating drums*

In the **rotating drum system** the drum is laid horizontally. After the substrate is introduced the drum is rotated either continuously or intermittently so that the substrate is put in motion which promotes heat and mass transfer. Three disadvantages of the system include the development of particle agglomeration over time, mechanical damage to the vegetative hyphae of some fungi and the difficulty of controlling temperature. The addition of baffles in the drum facilitates aeration because they lift the substrate and drop it as opposed to allowing it to roll or slump.

9.2.5. *Stirred substrates*

The **stirred substrate systems** are more modern. They include a **fluidized bed system** which costs, it is claimed, only one-sixth of that for other types of solid state cultivation systems for culturing *Aspergillus oryzae*. **Continuous systems** with mechanical stirring have been developed for koji production for making Japanese shoyu products.

9.3. Modeling in Solid State Cultivation Systems

Various aspects of solid state substrate systems can be modeled; e.g. the effect of fungal mycelium on mixing in these systems, fungal growth and metabolism inside solid particles, and temperature and water relations. However, a further complication is the fact that fungal growth is also three-dimensional.

9.3.1. *Fundamentals of fungal growth*

The growth cycle of the filamentous fungi is far more complex than bacteria and yeasts. Generally they have a vegetative growth stage where the hyphae extend and branch into a complex three-dimensional mycelial structure. During this stage the fungi may produce enzymes and metabolic by-products that are linked directly to the growth, or the by-products of interest may be produced as a consequence of **secondary metabolism** that is not linked directly to vegetative growth. Most of the industrially important fungi then enter a sporulation stage where aerial spores are formed, often on specialized fruiting structures. The cytoplasm in some hyphae may migrate leaving portions of the mycelium empty — cell walls surrounding empty space — resulting in some metabolically inactive hyphal biomass along with the spores formed that also show no metabolic activity. These non-metabolizing components of fungal origin are known as **necromass** and can, in time, become a significant component of fungal-derived material in the substrate. Consequently, measuring the mass of fungal material and relating it to fungal growth is not as simple as for organisms that form single cells and grow in liquid culture. Furthermore, estimations of fungal mass may be confined to total biomass and necromass (includes all mycelia and sexual and asexual bodies), or growing biomass, or biologically active biomass, or reproductive biomass. When fungal growth occurs in a solid substrate the estimation of fungal growth becomes even more difficult.

During fungal growth in solid substrate cultivations other parameters are also changing. In an incubating mass of substrate the total dry mass declines as the total biomass increases. This is due to the consumption of O_2 and production of CO_2, the production of other volatile products, the evaporation of metabolically generated water and the consumption of water in the hydrolysis of polysaccharides, assumed to be 1 mol of water chemically bound per mol of glucose produced. Also, as the fungi grow and produce enzymes, the enzymes hydrolyze the proteins, starch, cellulose and hemicelluloses to water soluble solutes that then lower the water activity of the solution in the substrate.

9.3.2. *Measuring fungal growth*

Before any models can be developed to describe fungal growth in solid substrates, a measure of fungal mass is essential. Some of the methods devised include:

(a) Total biomass measurement; suitable for some simulation models based on substrates that can be solubilized.
(b) Total hyphal length; as km g^{-1} dry substrate, can be converted to biovolume with estimates of hyphal diameter.
(c) Propagule concentration; by a viable plate counting method, suitable for examining the inoculants used as starters.
(d) Cell components, chitin/glucosamine, this may also include necromass; glucosamine concentrations of about 114 mg g^{-1} fungal dry mass have been reported and this number is used in some studies, but glucosamine concentrations vary according to the age of the culture and growth medium; a conversion factor of 12 g dry mass g^{-1} glucosamine has been suggested for *Rhizopus oligosporus*.
(e) Cell components, ergosterol from cell membrane; ergosterol contents are usually of the order of 5 μg mg^{-1} fungal dry mass, but this value is affected by the lipid content of the substrate, O_2 level during growth, the analytical method used, growth conditions and growth phase.
(f) Cell component, protein (N × 6.25); N in the substrate complicates the estimate.
(g) Enzyme activity; laccase, reductase and esterase.
(h) Direct measure of cell activity; ATP, DNA.
(i) Compression/penetration pressures; the response of the matted mycelium to pressure is related to the mass of mycelium.
(j) Metabolic activity; O_2 consumption and CO_2 release can be used to follow the progress of the cultivation for process control purposes.

9.3.3. *Mathematical models*

The mathematical models suggested for fungal growth in solid substrate fungal cultivations are classified into four kinds.

(1) Branching models based on mycelial growth using a combination of elongation of hyphal tips and branch formation. This model has limited application.
(2) Black-box models in which the kinetics of experimentally observed growth are not related to any cause and effect relationships. These

models rely on a modified form of the logistic law used to describe skewed sigmoidal curves;

$$r_X = \mu_{max}[1 - (X/X_{max})^n]X$$

where r_X is the biomass production rate (kg h^{-1}), μ_{max} is the maximum specific growth rate (h^{-1}), i.e. the limit of the specific growth rate as the amount of biomass approaches zero, X is the amount of biomass present (kg), X_{max} is the maximum amount of biomass (kg) that can develop in the system, and n is an exponent. The results obtained depend on the measure of biomass chosen, and this approach does not take into account the gradual decline in the biomass production rate.

(3) Particle-level models, mechanistic in nature, in which intrinsic growth and maintenance kinetics, and space and transport limitations on or in the substrate particle are used to predict experimental results. The advantage of these models is that they improve understanding of solid substrate cultivation processes, and allow evaluation of substrate particle geometry and any added liquid nutrients.

(4) Reactor-level models, that predict the macroscopic behavior of the bioreactor based on reaction kinetics and conservation laws.

Satisfactory models for the mathematical description of the growth dynamics and product formation still need further work before they can rival the growth models developed for cultivation of single celled and other types of organisms growing in liquid media. Mitchell has summarized models published since 1980. Amongst the models, the following have proved useful for various aspects of microbial growth in solid state cultivations.

A. An empirical logistic equation was used to describe the growth of *A. oryzae* in koji.

$$dm/dt = \mu m[1 - m/N]$$

where m = mycelial dry weight g^{-1} dry matter, μ = specific growth rate during logarithmic growth, and N = maximum mycelial dry weight g^{-1} of dry matter.

Because of the difficulties of measuring dry weight, the model then assumed that CO_2 production, an easily measured parameter, was due to both growth and maintenance and wrote the following equation:

$$dA/dt = K_1 dm/dt - K_2 m$$

where $A = CO_2$ evolved mg^{-1} of dry matter, $K_1 = CO_2$ evolved mg^{-1} of dry mycelium formed, and $K_2 = CO_2$ evolved for endogenous respiration mg^{-1} of dry mycelium h^{-1}.

By combining both equations and using a respiratory quotient of 1.0 the O_2 consumption rate could be predicted.

B. Another model was based on a consideration of energy balance, to control water content during koji production with *A. oryzae* by taking into account the heat capacity of the air used to aerate the bioreactor. To maintain constant water content, temperature is taken into account:

$$T_{surr} = T_{out} + 1.03(T_{out} - T_{in}) - 131PCO_2$$

where T_{surr} = temperature of the surroundings of the bioreactor, T_{out} = temperature of outlet air, T_{in} = temperature of inlet air, and PCO_2 = difference in partial pressure of CO_2 between the inlet and outlet air streams, i.e. correlated with metabolic heat production.

By controlling the temperature of the surrounding air, according to the equation, the water content of the koji remains constant.

C. However, Mitchell claimed that only the logistic model and that developed by him and his co-workers actually describe a model based on growth kinetics. The model was based on the growth of *Rhizopus oligosporus* on starch-containing substrate and took into account five key components, namely biomass, glucoamylase, starch, glucose and distance within the substrate. From an understanding of what happens during cultivation, the events surrounding each of these components can be described mathematically, namely:

(i)   Release of glucoamylase by the mycelium at the surface of the sub-strate.
(ii)   Diffusion of the glucoamylase to the starch, using Fick's Law.
(iii) Hydrolysis of the starch to release glucose, using Michaelis-Menton kinetics.

(iv) Diffusion of glucose to the surface mycelium.

(v) Absorption of the glucose and its conversion into mycelial mass, using Monod kinetics.

When all combined 13 parameters must be measured or estimated, however, when applied to real growth experiments the model satisfactorily described glucose and starch concentrations, and glucoamylase activities during growth.

Further Readings

1. Doelle HW, Mitchell DA and Rolz CE, eds. *Solid State Cultivation.* Elsevier Applied Science, London and New York, 1992.
2. Gowthamam MK, Krishna C and Moo-Young M. Fungal solid state fermentation — An overview. In: *Applied Mycology and Biotechnology*, eds. GG Khachatourians and DK Arora. Elsevier, Amsterdam, New York, 2001, pp. 305–352.
3. Mitchel DA. Growth patterns, growth kinetics and the modeling of growth in solid-state cultivation. In: *Solid State Cultivation*, eds. HW Doelle, DA Mitchell and CE Rolz. Elsevier Applied Science, London and New York, 1992, pp. 87–114.
4. Nout MJR, Rinzema A and Smits JP. Biomass and productivity estimates in solid state fermentations. In: *The Mycota: A Comprehensive Treatise on Fungi as Experimental Systems for Basic and Applied Research*, eds. K Esser and PA Lemke; *Environmental And Microbial Relationships*, *Vol. 4*, eds. DT Wicklow and BE Söderström. Springer-Verlag, Berlin, 1997, pp. 323–346.
5. Raimbault M. General and microbiological aspects of solid substrate fermentation. *E. J. Biotechnol.* **1**(3): 174–188, 1998.
6. Terebiznik MR and Pilosof AMR. Biomass estimation in solid state fermentation by modeling dry matter weight loss. *Biotechnol. Tech.* **13**: 215–219, 1999.

Related World Wide Web Sites

http://www.wcfs.nl/webdb/OverviewProjectPages?readform&Page=C005&Prog=C

http://www.ejbiotechnology.info/content/vol1/issue3/full/9/bip/

Questions for Thought...

1. Compare and contrast the growth of a mycelial fungus, like a *Penicillium* species, when cultured in a broth medium and a solid state cultivation.

2. Describe how a fungus might grow in a solid substrate cultivation and think about what would happen if the fungus produced an enzyme that broke down the substrate.

3. Why do manufacturers of soy sauce not use the packed column system of solid state cultivation to make soy sauce?

Chapter 10

Food Ingredients

Desmond K. O'Toole
Department of Biology and Chemistry
City University of Hong Kong
83 Tat Chee Avenue, Kowloon, Hong Kong, China

Yuan Kun Lee
Department of Microbiology, National University of Singapore
5 Science Drive 2, Singapore 117597
Email: micleeyk@nus.edu.sg

With a better understanding of food flavor, texture, rheological properties, nutritional value and human food preferences, the food ingredient industry has developed to enhance the qualities of many foods. The ingredients industry is a huge industry, e.g. the food flavors market in 1994 was estimated to be worth US$4.5 billion, and food manufacturers rely on the ingredients to enhance food quality.

10.1. Flavors and Fragrances

Microorganisms play an important role in the flavor and aroma formation of many foodstuffs. For example, the flavor of mold-ripened cheeses is determined by the break-down products of fungi (e.g. *Penicillium roqueforti* in blue cheese), such as methyl ketones, short chain fatty acids, acetic and lactic acids, alcohols, aldehydes, lactones, ammonia, amines, sulfides and mercaptans. On the other hand, the earthy notes of pond cultured fresh water fish, e.g. tilapia, is due to geosmin (*trans*-1,10-dimethyl-*trans*-9-decalol) produced by actinomycetes (e.g. *Streptomyces odorifer*), myxomycetes and cyanobacteria. Lately, there is increasing interest in using microorganisms as the source of "natural" flavor and fragrance, and some examples are cited in Table 10.1.

Table 10.1. Fragrances and their chemical foundations produced by various microorganisms.

Flavor/fragrance	Chemical	Microorganisms
Banana, peach, pear	γ-, δ-decalactone	*Ceratocystis moniliformis*
Apricot, banana, pineapple	Butyric acid pentyl ester	*Clostridium acetylbutyricum*
Pear, banana, pineapple	Butyric acid isobutyl ester	*Clostridium acetylbutyricum*
Apple	Isovaleric acid ethyl ester	*Peptostreptococcus anaerobius* *Pseudomonas aeruginosa*
Apple, raspberry	Isovaleric acid isopentyl ester	*Peptostreptococcus anaerobius* *Pseudomonas aeruginosa*
Coconut	γ-Octalactone, γ-nonalactone 6-pentyl-2-pyrone	*Trichoderma viride*
Musk precursor	Ustilagic acids A and B	*Ustilago zeae*
Butter	Diacetyl	*Lactococcus lactis,* *Lc. diacetilactis,* *Lc. cremoris,* *Enterobacter cloacae*
Rose	β-phenyl ethanol, furan-2-carboxylate	*Ascoidea hylacoeti*
Rose	6-methyl-5-hepten-2-ol acetate Citronellol, linalool, geraniol Geranyl acetate	*Ceratocystis virescenes*
Pine, rose, mushroom	Thujopsene, 3-octanone, 1-octen-3-ol, nerolidol, β-phenyl ethanol	*Penicillium decumbens*

10.1.1 γ-Decalactone

An important flavor component in the flavors of strawberry, raspberry and peach/apricot is γ-decalactone. A combination of biological and chemical methods was developed in the early 1960s to produce this component. Baker's yeast (*S. cerevisiae*) was found able to reduce β-ketocapric acid to hydroxy decanoic acid, a precursor of γ-decalactone. Subsequently, species of *Candida* (e.g. *C. tropicalis*, *C. parakrusei*), and a strain of *Escherichia coli* were found that could produce 4-hydroxydecanoic acid from ricinoleic acid that is found in relatively high concentrations in castor oil. The process depends on releasing ricinoleic acid from the triacylglycerol through the action of a lipase. Then the yeasts, through an aerobic metabolic process of β-oxidation, reduce the C18 ricinoleic acid to the C10 acid, 4-hydroxydecanoic acid, that is then lactonized by heating at a low pH to γ-decalactone. Yields are low g L^{-1}, even after seven days fermentation.

10.1.2 *Monosodium Glutamate (MSG)*

The most common flavor compound produced by a microbial process is L-glutamic acid. The establishment of fermentative production of the flavor enhancer, L-glutamic acid, by bacteria belonging to the genera *Corynebacterium* and *Brevibacterium*, is a milestone in the development of industrial production of **microbial primary metabolites**. Primary metabolites are biomolecules directly involved in cell growth and reproduction, such as amino acids and tricarboxylic acid (TCA) cycle intermediates, whose production is tightly regulated in cells. Industrial production of microbial primary metabolites was considered impractical due to their low concentration in microbial cultures but today, almost all of the amino acids are industrially produced by fermentation.

Development of the production of L-glutamic acid by fermentation was based on the understanding that accumulation of L-glutamic acid in a culture is determined by its excretion from cells rather than by its biosynthesis. Thus, by limiting the supply of biotin, or including C_{16}–C_{18} saturated fatty acids and their esters, penicillin or Tween-60 in the broth medium results in an increase in the permeability of the cell membranes in the growing cells, and excretion of L-glutamic acid and other amino acids

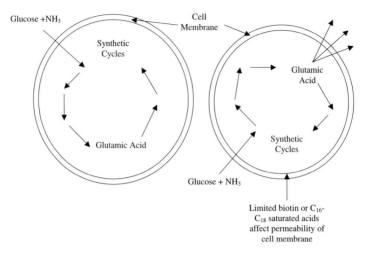

Fig. 10.1. The production of glutamic acid. By manipulating growth conditions, the cell membrane of the bacterium is weakened so that the permeability of the membrane increases. As the glutamic acid is produced, much of it is excreted from the cell. This helps to promote glutamic acid production in the cell.

(Fig. 10.1). Overproducing mutants were subsequently isolated which further increased the production of glutamic acid. Examples include monofluoroacetate resistant mutants derived from *Corynebacterium glutamicum* (formerly known as *Brevibacterium lactofermentum*) that have increased phosphoenolpyruvate carboxylase activity. Other mutants with decreased isocitrate lyase activity lead to increased carbon dioxide fixation and increased L-glutamic acid production.

To produce MSG a cheap source of starch is hydrolyzed enzymically to glucose. With ammonia as the N source and other nutrients the fermentation is conducted under aerobic conditions. Glutamine is produced that is then converted to glutamate under alkaline conditions. The MSG is then crystallized from the medium.

10.2. Antioxidants

One of the major causes of food deterioration during long term storage is the chemical oxidation of **unsaturated fatty acids** in the lipids present in all foods. These fats are present in fat deposits and in cell membranes.

End products of lipid oxidation, mainly aldehydes and ketones, contribute to the rancid off-flavors and odors in the foods. Other consequences of oxidation include discoloration; changes in the texture, flavor and taste; and the loss of nutritional value due to the destruction of essential fatty acids as well as vitamins.

The oxidation of a lipid involves the initiation of catalytic oxidation of lipid by the presence of oxygen and heavy metals, thermal or photodecomposition of peroxides, or by ultraviolet irradiation, resulting in the formation of a lipid radical ($L^\bullet$). Propagation of a chain reaction sees the formation of a lipid peroxyl radical ($LOO^\bullet$) and lipid hydroperoxide (LOOH). Lipid peroxyl radicals react with each other to form the complex LOOL. The addition of an antioxidant minimizes the extent of lipid oxidation by inhibiting the initiation of propagation steps, thereby prolonging the shelf life of the food products.

There is now a growing interest in the development of "natural antioxidants" for foods, because of consumer preference and some queries regarding the toxicity of chemical antioxidants such as butylated hydroxytoluene (BHT). The presence of highly unsaturated fatty acids in microbial cells and their relative stability have prompted many researchers to turn to microorganisms as a possible source of natural antioxidants.

The phospholipid fraction of algae showed antioxidant activity similar to that of BHT. Some of the phospholipids identified were phosphatidyl ethanolamine and phosphatidyl inositol. The bromophenols in the acetone-soluble fraction have been implicated.

Enzyme systems that could be used as antioxidants in food include superoxide dismutase (SOD) and glucose oxidase. SOD has been shown to prevent the oxidation of ascorbic acid and anchovy fat, and the enzymic darkening of sliced mushroom, apple, and potato. The enzyme scavenges oxygen free radicals and dismutates them into oxygen and hydrogen peroxide:

$$2O_2^- + 2H^+ \xrightarrow{\text{SOD}} H_2O_2 + O_2$$

The SODs isolated from the yeast *S. cerevisiae* var. *ellipsoideus* and *Lac. lactis* are relatively heat stable, and are particularly suitable for foods that are mildly acidic to alkaline pH.

Glucose oxidase catalyses the oxidation of D-glucose in the presence of oxygen, and, in the presence of catalase, the hydrogen peroxide, which may not be desirable in the food products, is broken down.

$$2C_6H_{12}O_6 + 2H_2O + 2O_2 \xrightarrow{\text{Glucose oxidase}} 2C_6H_{12}O_7 + 2H_2O_2$$

$$2H_2O_2 \xrightarrow{\text{Catalase}} 2H_2O + O_2$$

Overall: $\quad 2C_6H_{12}O_6 + O_2 \xrightarrow{\text{Glucose oxidase/catalase}} 2C_6H_{12}O_7$

Due to its mild action and a Generally Regarded as Safe (GRAS) status, glucose oxidase is now used in the food industry. It is used in the production of dried foods like instant coffee, cake mixes, milk powder and active dried yeast; alcoholic beverages, such as beer and wine; citrus drink; soft drink; mayonnaise and salad dressing.

10.3. Essential Amino Acids

Plant proteins are relatively low in essential amino acids, such as L-lysine and L-tryptophan. These amino acids are extensively used in the food and

Table 10.2. The organisms and their genetic characteristics, and substrates used for the production of various amino acids for various applications.

Amino acid	Microorganisms	Genetic characteristics	Major substrates	Applications
L-lysine	*Coryne-bacterium glutamicum*	Hom⁻Leu⁻AEC^r	Sugar cane molasses	Feed additive
	Brevibacterium flavum	Hom^{leaky}Thre⁻	Acetate, NH_4^+, biotin, thiamine	
L-tryptophan	*C. glutamicum*	Phe⁻Thyr⁻5MTrTrpHx^r 6FT^{r}4MT^{r}PFP^{r}PAP^r	Glucose	Antioxidant
L-aspartic acid	*Pseudomonas fluorescens*	Wild type	Fumaric acid, NH_4^+	Flavor, aspartame
	Escherichia coli	Aspartase mutant	Fumaric acid, NH_4^+	
L-phenylalanine	*C. glutamicum*	Tyr⁻PFP^{r}PAP^r	Glucose	Aspartame

Resistance: AEC: S-(β-Aminoethyl)-L-cysteine   6FT: 6-Fluorotryptophan
4MT: 4-Methyltryptophan   5MT: 5-Methyltryptophan   PAP: *p*-Amino-phenylalanine
PFP: *p*-Fluorophenylalanine   PheHx: Phenylalanine hydroxamate   TyrHx: Tyrosine hydroxamate

the animal feed industry as additives to achieve nutritionally balanced diets and animal feeds. Amino acids (e.g. L-arginine, L-isoleucine, L-leucine, L-phenylalanine, L-proline, L-tryptophan, L-valine) are used in medicine as ingredients of infusion solutions, and in the chemical industry as starting materials for the manufacturing of biopolymers. The latter includes polyalanine fibres and lysine isocyanate resins; poly-γ-methylglutamate is used as a surface layer for synthetic leather; N-acyl-serine is used as surface-active agent in cosmetics.

Demand for amino acids is estimated by Ajinomoto to be about 2 million tons per annum. Production by fermentation of major amino acids for food applications is summarized in Table 10.2.

Box 10.1

What is Aspartame?

Aspartame is an ester of L-aspartic acid and L-phenylalanine methyl ester. It is used as a low-calorie sweetener in drinks and beverages. On a per weight basis, aspartame is as sweet as cane sugar (sucrose) but has lower energy content. It is readily digested in the human digestive tract and is absorbed as amino acids. Biotechnologically, it is an interesting industrial compound because patents are lapsing and a number of methods are being explored for its commercial production. L-phenylalanine and L-aspartic acid can be produced fermentatively, the former using *trans* cinnamic acid, ammonia, and phenylalanine ammonia lyase. In addition, L-phenylalanine can also be produced from phenylpyruvate, formic acid and ammonia via ACA acylase. The gene for this enzyme has been cloned and sequenced and a promoter found for use in a mutant of *Corynebacterium*. L-aspartic acid is produced from fumaric acid with vermiculite immobilized cells of *Escherichia coli* with aspartase activity. In more recent research, genetically engineered organisms, based on *E. coli* and yeasts have been used. Complementary nucleotides have been designed to code for Asp-Phe-Asp-Phe. Following successful insertion into *E. coli*, one clone directed the synthesis of the fused poly(Asp-Phe) as inclusion bodies calculated to be 11.2% of the total cell protein. A chemically synthesized gene, which encodes a 64 or 128 times-repeated tripeptide aspartyl-phenylalanyl-lysine, has been cloned into a yeast that was able to synthesize the polypeptide so that it made up to about 30% of the total cell protein.

10.4. Essential Fatty Acids

Polyunsaturated fatty acids (PUFAs) are an important component of the human diet. In particular, omega-3 fatty acids (ω3FAs) are important dietary components required for the proper development in humans of brain and vision in infants. Some of the ω3FAs, e.g. γ-linolenic acid (GLA, 18:3), eicosapentaenoic acid (EPA, C20:5) and docosapentaenoic acid (DHA, C22:6) have been recognized as dietary components which prevent cardiovascular diseases. These PUFAs are also precursors for the industrial production of eicosanoids (e.g. prostaglandin, prostacyclins, thromboxane and leucotrienes), which are used in the treatment of high blood pressure, asthma, menstrual pain and induction of labor. The main traditional source of GLA is plant seeds (e.g. Perilla, linseed and red current), whereas EPA and DHA are extracted from aquatic animals (e.g. oils of Atlantic mackerel, Atlantic salmon, Pacific herring and cod).

Table 10.3. Fermentation times and production rates of omega-3 fatty acids by bacteria, microalgae and fungi.

Organism	ω-3 fatty acid content, % dry biomass	Fermentation time, days	ω-3 fatty acid productivity g/l/day
Bacteria			
Vibrio sp. (EPA only)	0.7	0.3	0.13
Microalgae			
Schizochytrium sp.	10.0	2.0	1.0
Schizochytrium mangrovei (DHA only)	20.9	2.2	1.3
Crypthecodinium cohnii (DHA only)	9.0	2.5–4.0	0.8–1.2
Nitzschia alba (EPA only)	2.3	2.7	0.25
Fungi			
Mortierella alpina	5.8	7.0	0.2
Saprolegnia sp.	0.4	3.0	0.1
Pythium ultimum	1.9	6.0	0.03

GLA in plants is synthesized from hexadecatrienoic acid ($16:3\omega3$) in the chloroplast membrane. Photosynthetic microorganisms, e.g. *Spirulina major*, a cyanobacterium, that accumulate approximately 1% GLA in dry biomass are being used as an alternative source, especially in the health food sector. EPA and DHA are synthesized by phytoplankton, which is consumed by fish, molluscs and crustaceans, and thereby are concentrated in the aquatic food chain. Thus, microorganisms are being considered sources of EPA and DHA. Some of the omega-3 fatty acid producing microorganisms are listed in Table 10.3. Another group of organisms, not listed in the table, are the *Thraustochytrium* sp., marine protists of Kingdom Chromista.

10.5. Vitamins

Vitamins are essential in both human and animal nutrition and their industrial production is a huge business estimated to be worth US$ 1–10 billion per year. The economic production method of choice is chemical synthesis but recently microbiological methods for the production of some of these compounds have become economically competitive (Table 10.4).

Table 10.4. Vitamins and vitamin-like compounds and the organisms used to produce them and their estimated yearly production levels.

Compound	Organism	Yearly world production, tons
Ergosterol*	*Saccharomyces cerevisiae*	?
γ-lenolenic acid*	*Mortierella isabellina*	1,000
Riboflavin	*Ashbya gossypii, Candida famata, Bacillus subtilis*	1,000–10,000
Pantothenate	*Escherichia coli*	1,000–10,000
Cobalamin*	*Pseudomonas denitrificans, Propionibacterium shermanii*	10,000
Orotic acid*	*Brevibacterium* sp.	100
Ascorbic acid	*Gluconobacter oxydans* ssp. *suboxydans*	100,000

*Only produced microbiologically, others in table are produced chemically and microbiologically. Table modified from Stahmann (2002).

Ergosterol, converted to vitamin D_2 under UV irradiation, is produced using *Saccharomyces cerevisiae*. The C source can be a carbohydrate or ethanol but adequate aeration in the fermentation is essential so that ergosterol may comprise up to 10% of the cell dry mass. To extract the ergosterol, cells are treated with hot alkali to saponify the fat followed by ether extraction.

The vitamin riboflavin is used in pharmaceutical products, in animal feeds and as a colorant (orange-yellow) in human foods. Three organisms are used for production. A biotechnology based production system for riboflavin failed in the late 1960s but in 1974 Merck began using the fungus *Ashbya gossypii*. Following improvements in cultures, production levels up to 15 g L^{-1} riboflavin have been reported. When *A. gossypii* is grown using glucose as a C source it produces ethanol until the glucose is exhausted. During this stage lipid droplets accumulate in the cells. On the exhaustion of the glucose riboflavin production begins and both the lipid and ethanol are metabolised. Lipid levels in the cells (up to 20% of the dry mass) can be enhanced, and riboflavin production increased, by adding vegetable oil to the growth medium; and the composition of the lipid can be changed by adding free fatty acid to the fermentation. High lipid levels can be added to a fermentation vessel without altering the water activity level of the medium. Production can be further enhanced by adding precursors, such as glycine to the fermentation mixture. A neutral pH and high O_2 levels are also beneficial to maximize production.

During the riboflavin production phase the lipid is catabolized via β-oxidation in the peroxisomes to acetyl units that are transported to the mitochondria. The actual synthesis begins with guanosine triphosphate (GTP) (precursors of GTP in the growth medium such as glycine and hypoxanthine enhance production) and ribulose-5-phosphate (R5P)

$$2 \times ribulose - 5 - phosphate + guanosine - triphosphate \rightarrow riboflavin$$

Following biosynthesis the riboflavin is either transported to vacuoles in the cells or it is exported to the cellular environment.

Gene manipulation has further enhanced production in three ways.

(a) Injection of a second copy of the *ICL1*-gene, that codes for isocitrate lyase, improves production when soybean oil is used in the fermentation.

(b) Delivery of glycine was improved. A copy of the *GLY1*-gene, that codes for threonine aldolase, under the control of the *TEF* promoter and terminator was inserted. Then when the medium was supplemented with threonine instead of glycine, riboflavin production increased ten-fold. This was due to the more rapid uptake of threonine by comparison with glycine uptake.

(c) Transport to internal cellular vacuoles instead of direct secretion into the medium slows production. To prevent vacuolar accumulation the vacuolar ATPase subunit gene, the VMA1-gene, was inactivated which ensured all the riboflavin was excreted into the growth medium.

For six years from 1990 BASF ran chemical and biological production methods side by side using *A. gossypii* and eventually dropped the former method. The other two cultures used for production are under patent protection. Strains of *Candida famata* have been genetically improved by Coors Biotech, Inc to produce >20 g L^{-1} riboflavin. Roche (Switzerland) are using a strain of *Bacillus subtilis* genetically engineered by Omnigene (USA).

Genetically engineered ascorbic acid (vitamin C) production has also been achieved. A gene for the enzyme 2,5-diketo-D-gluconate reductase from a *Corynebacterium* species was inserted into a culture of *Ewinia herbicola* resulting in a one step fermentation process. The culture could then ferment glucose to produce 2-keto-gulonate at a concentration of 20 g L^{-1}, after 72 h, to give a 50% yield based on the glucose input. The gulonate could then be chemically converted to ascorbate.

10.6. Polysaccharides

Polysaccharides are used as thickening and stabilizing agents in the food industry and plants and seaweed are the traditional sources for these polysaccharides (e.g. starch, agar and alginate). However, microbial poly-saccharides have become more widely used in recent years. Table 10.5 shows some of the important commercial microbial polysaccharides. Alginate is not included as it is currently extracted mainly from seaweed, but industrial production of alginate by the bacteria *Pseudomonas aeruginosa* and *Azotobacter vinelandii* is being investigated.

The extracellular polysaccharides are recovered by precipitation with alcohol, such as isopropanol or methanol. The precipitated polysaccharides are then dried and ground.

Table 10.5. Production of polysaccharides by microorganisms and the monomer structures of the polymers.

Polysaccharide	Sugar moiety/structure	C-substrate	Organism
Bacteria			
Curdlan	Glucose/β-1,3-glucan	8% glucose	*Alcaligenes* spp.*
Dextran	Glucose/α-1,6-, 1,2-, 1,3-, 1,4-glucans	Sucrose	*Leuconostoc mesenteroides, Leuc. dextranicum, Streptococcus mutans*
Xanthan	Glucose, mannose, glucuronic acid, acetate, pyruvate/pentasaccharide	4% glucose/ sucrose/ starch hydrolysate	*Xanthomonas campestris*
Fungi			
Scleroglucan	Glucose/β-1,3-, 6,1-glucans	Glucose	*Sclerotium glucanicum, S. delphinii, S. rolfsii*
Pullulan	Glucose/α-1,4-, 1,6-glucans	5% glucose	*Pullularia pullulans*

*Includes *Agrobacterium radiobacter*, formerly attributed to *Alcaligenes faecalis* var. *myxogenes*.

10.7. Microbial Meat Substitute — RHM Mycoprotein

A more recent modern source of a meat-like protein is fungi grown using a completely modern scientific approach to produce the commercial product Mycoprotein, also known as Quorn. In the 1960s, a lack of protein for human consumption seemed a real future possibility. Microorganisms can use simple nutrients, reproduce rapidly, and produce a far greater amount of protein per unit of energy input when compared with traditional land based protein sources (Table 10.6). Therefore, microorganisms were developed to supply microbial protein as a solution for future food supply for humans. However, unicellular microbial cells lack meaty texture. In addition, microbial cells contain a high level of RNA that can cause nutritional problems in those that eat large quantities. The organism chosen for the development of this product was the fungus *Fusarium*

Table 10.6. Feed conversion efficiency rates in protein formation.

		Biomass	
	Feed stock	Protein	Total
Cow	1 kg feed	14 g	68 g beef
Pig	1 kg feed	41 g	200 g pork
Chicken	1 kg feed	49 g	240 g meat
F. graminearum	1 kg $C_6H_{12}O_6$ + N	136 g	1080 g wet cell

graminearum ATCC 20334 (isolated from soil). The meat analogue developed from the fungal mycelium is now called Mycoprotein.

Mycoprotein is produced by a continuous fermentation process at a dilution rate of up to 0.2 h^{-1}, temperature of 30°C and pH 6, and controlled by the addition of gaseous ammonia into the inlet air stream, and using glucose, ammonia, ammonium salt and biotin. Cell concentrations of 15–20 g L^{-1} can be maintained.

After fermentation, the culture is heated (64°C for 20–30 min) to reduce the RNA content from 9% to < 2%. The mycelial mass gives a meat-like texture to the product, which is flavored and colored accordingly.

Microbial cells have lower fat and protein, but higher fiber and carbohydrate contents (Table 10.7). Mycoprotein is available in supermarkets as packaged vegetarian hamburger steaks.

Table 10.7. Composition of Mycoprotein compared to beef.

Component	Mycoprotein (% dry wt.)	Raw lean beefsteak (% dry wt.)
Protein	47	68
Fat	14	30
Dietary fiber	25	Trace
Carbohydrate	10	0
Ash	3	2
RNA	1	Trace

10.8. Organic Acids

It is possible to produce all organic acids by microbial fermentation processes. Examples of practical microbiological processes for organic acid production are listed in Table 10.8. Two acids of interest are citric acid and lactic acid and their production is illustrated.

10.8.1. *Citric acid*

Globally in 2000 AD about 900,000 tons of citric acid-1-hydrate, or anhydrous citric acid, was produced. It is useful because of its acidity and buffering capacity, taste and flavor, and ability to chelate metal ions. It is mainly used in the food and beverage industry (60%), for enhancing flavor of fruit juices, and for addition to marmalade, soft drinks, candy and ice cream, and as a food preservative. Citric acid is the only organic acid that is manufactured only by microbial fermentation and its fermentation is an example of a highly refined process based on scientific know-how. Fungi are used in the fermentation and while many fungi can produce citric acid only, *Aspergillus niger*, *A. wentii* and the yeast *Yarrowia lipolytica*, are used in commercial production due to their economic yields and minimal production of undesirable side products. *A. niger* can produce about 70% of the theoretical yield from carbohydrate sources.

Substrates used include sucrose, glucose and fructose. A high substrate concentration ($140–220$ g L^{-1}) is required to obtain satisfactory rates of production and yields of citric acid. Metabolically, citric acid production involves glycolysis and the TCA cycle. Hexoses are broken down to pyruvate via the hexose monophosphate pathway. The pyruvate is then decarboxylated to acetyl-CoA and CO_2 that is recycled to pyruvate via pyruvate carboxylase to produce oxaloacetate. Oxaloacetate and acetyl-CoA are then combined to produce citrate. To excrete citric acid, however, the feedback mechanism from excess intracellular citric acid, which acts on the phosphofructokinase enzyme, has to be inactivated. This occurs via an increase in the intracellular ammonium levels. Hence control of the nitrogen metabolism is also crucial for production. In submerged culture production systems, the nitrogen concentration is held between 0.4 to 0.6 g N L^{-1}. Other nutrients needed for growth include Mg^{2+} ($20–100$ mg L^{-1}), K^+, Zn^{2+} ($0.3–0.5$ mg L^{-1}) and phosphate ($0.4–1$ g L^{-1}).

Table 10.8. Practical production and use of organic acids, the organisms, and substrates used for production, and the process.

Organic acid	Organism	C-substrate	Process	Uses
Acetic acid	*Acetobacter aceti* *Gluconobacter oxydans*	Ethanol	35 hours, 30°C	Food
Citric acid	*Aspergillus niger* *Yarrowia lipolytica* *A. wentii*	Molasses, starch, n-alkane, glucose sucrose	8 days, 30°C	Food
Fumaric acid	*Rhizopus sp.* *Candida* sp.	Glucose n-Alkane	3 days, 33°C 7 days, 30°C	Resin
Gluconic acid	*Aspergillus niger,* *Gluconobacter* *suboxydans*	Glucose	24 h, high O_2 Pressure, 12–25% glucose	Food
Lactic acid	*Lactobacillus delbrueckii* *Lb. delbrueckii* ssp. *bulgaricus* *Lb. pentosus*	Glucose Whey Sulfite liquor	72 hours, 40°C	Food
Propionic acid	*Propionibacterium* sp.	Lactose, glucose, starch	8–12 days 30°C	Perfume, fungicide
Malic acid	*Leuconostoc brevis*	Fumaric acid, n-Alkane	24 hours	Food
α-keto-glutaric acid	*Candida hydrocarbofumarica* *Aerobacter spp.*	n-Alkane Glutamic acid	40 hours	
5-keto-gluconic acid	*Acetobacter suboxydans*	Glucose	5–6 days	L-tartaric acid
2-keto-gluconic acid	*Serratia marcescens*	Glucose	16 hours	Isoascorbic acid

citrate

Control of trace metals in the fermentation medium is also necessary for good citric acid production. A limiting concentration of manganese, $<1\,mg\,L^{-1}$, particularly in submerged culture conditions, is required. This requirement is linked to increased protein turnover, impaired DNA biosynthesis, and alterations to the composition of cell walls and plasma membranes in the cells. Iron, as a cofactor for aconitase, at a concentration of $1.3-1.5\ mg\ L^{-1}$, is necessary to achieve optimal growth for, and optimal production of citric acid from, *Aspergillus* culture. Thus, when invert sugar or starch hydrolysate is used as the carbon source, the high iron content of the raw material results in growth without citric acid production. In this case, a preliminary treatment with either precipitant, e.g. calcium hexaxcyanoferrate, sodium ferrocyanide, or cation exchangers, is carried out to remove iron and manganese. Likewise, molasses, a common cheap raw material, is treated with precipitants to remove metals. Otherwise, a large amount $(100\ mg\ L^{-1})$ of copper is added to the molasses mixture to reverse the inhibitory effect of manganese on citric acid formation during the production phase.

The optimal pH for growth of *Aspergillus* culture (trophophase) is about 5, but the pH is kept below 2 during the citric acid production phase (idiophase), in order to suppress oxalic and gluconic acid formation. If too much oxalic acid is produced, a yellowish pigment is formed, which later hinders the purification process of citric acid. Thus, ammonium is often used as the nitrogen substrate and the initial pH is set at 5–7 to allow germination of spores and optimal initial growth. In the first 30–48 h, during the trophophase, and as a result of the metabolism of ammonium ions, the pH of the culture falls to below 3. However, at that low pH heavy metals leach from normal steel fermenter walls and inhibit citric acid formation. Consequently fermenters must have plastic liners or be made of food grade stainless steel.

Citric acid is produced by surface or submerged culture of *Aspergillus*. The former process although low in volumetric productivity, still accounts for 20% of the world's production, because of the low capital investment (the fermenter), the low energy cost since there is no need for cooling, agitation and aeration, and low technology level. The surface culture method is usually used in developing countries where labour cost is low. In the process, spores are inoculated onto the surface of solid substrates, such as wheat bran or sweet potato pulp, or onto a liquid medium containing molasses, with depths of between 50–200 mm. Ventilation is essential to supply oxygen and to carry away CO_2 (> 10% v/v inhibits citric acid production) and fermentation heat (to maintain fermentation at 30°C). Yields from surface processes amount to 1.2 to 1.5 kg citric acid mono-hydrate per m^2 of fermentation surface per day.

A yield value of about 900 g citric acid m^{-3} culture medium can be obtained in the submerged process (usually in 120–220 m^3 fermenters). The form of mycelium appears to be an important factor in the production of citric acid by the submerged process. If the mycelium is loose, filamentous and has no chlamydospores, little citric acid is produced. Abnormally short, forked and bilbous hyphae are brought about by a deficiency of manganese in the culture medium. The deformed hyphae aggregate to form small (0.2–0.5 mm), hard, smooth surfaced pellets, and produce a high level of citric acid. In practice the culture is grown from spores in submerged culture in the laboratory to ensure that the pellet form of the culture is initiated. Once in the pellet form the culture is added to the production culture.

In advanced modern systems of production, agitation and aeration in deep fermenters is necessary to ensure a minimum dissolved oxygen concentration of 75% saturation of the growth medium with air (equal to 150 mbar partial pressure of O_2) below which citric acid production ceases. The culture is low in viscosity due to the formation of mycelial pellets, so an **airlift bioreactor** without stirring can also be used. In an airlift system, fine air bubbles are injected into the bottom of the culture. As air bubbles rise to the top, they are replaced by growth medium free of air bubbles. Larger air bubbles tend to rise faster than the water column falls, thus creating turbulence around them.

Apart from the classical genetic approach to strain improvement, some improvement attempts have been made using genetic engineering. Usually molasses is used as a cheap hexose source but reducing Mn^{2+} to a minimum to produce at a pH of <2 is essential. However, if the enzymes glucose oxidase and oxaloacetate hydrolase are inactivated citric acid accumulates at pH 5 with no need to control Mn^{2+}. The *goxC* gene in *A. niger* prevents the production of gluconic acid from glucose, and another gene, *prtF*, results in a lack of oxaloacetate hydrolase activity. When both genes are combined in one strain citric acid is produced at pH 5.

10.8.2. Lactic acid

The annual rate of world lactic acid production in 1982 was 24,000–28,000 tons. About 90% is produced fermentatively with yields of about 90 g lactic acid per 100 g glucose, and stereospecific lactate production, D(+) and L(−), is possible. It is used in food as an acidulant, a preservative (50%), as stearoyl-2-lactylate (20%) (an emulsifier for the baking industry), and in the pharmaceutical industry.

The desirable characteristics of the organisms used include an ability to completely ferment a cheap feedstock with minimal nitrogenous materials, and to give high yields of stereospecific lactic acid under conditions of low pH and high temperature with the production of low cell mass and negligible other by-products. The species used to produce from glucose include *Lb. delbrueckii*, and *Lb. leichmannii*.

Energy sources include glucose and starch. The starch is frequently degraded enzymically to glucose before fermentation. This breakdown step complicates the process so recent research efforts have been directed to developing bacterial cultures that can breakdown starch to glucose and then to lactic acid. Further, starch has to be gelatinized by cooking; so further research has been directed towards the use of raw starch sources as the feedstock for the fermentation.

An organism able to break down pure starch is *Lb. amylovorus* B-4542. After gelatinization and enzyme thinning of the starch, the organism metabolizes the starch directly without the need for enzymes in a saccharification step. This reduces the energy input, fermentation tank requirements and time needed. The rate of reaction and final level of lactic acid produced is the same as for *Lb. delbrueckii* B-445 using glucose as the substrate.

More recently, raw cassava and potato have been converted to lactic acid, either directly by the lactic acid bacterium used, or via an intermediate enzyme step using mould enzymes.

Another modern approach, using starch as the starting material, is to use a continuous combined enzyme and culture process involving a reversibly soluble-auto-precipitating amylase (D-AS) whose solubility is pH dependent. The D-AS is precipitated by the low pH of the lactic acid produced, separated and then fed back into the starch mixture. The culture used is *Lb. casei* entrapped in κ-carrageenan. The yield obtained is 3.1 g L^{-1} h^{-1} at a dilution rate of 0.1 h^{-1}, which is better than batch processes.

10.9. Colorant

Microorganisms are known to produce a variety of pigments, therefore they are a promising source of food colorants. Of all the microorganisms, the fungus *Monascus* and the algae are the most widely studied for their potential as a source of food colorants.

10.9.1. *Monascus pigments*

Monascus pigments have been used for centuries in the Orient, particularly China (known there as ang-kak), Japan (known as beni-koji, aga-koji), Indonesia and Philippines, to color food. Chinese red rice wines, red soybean cheese (Fu-ru), pickled vegetable, fish and salted meats are some examples of Oriental food colored red by monascus pigments. The main microbial producers used for commercial production of monascus pigments are the fungi, *Monascus purpureus* and *M. ruber*. Strain selection to screen out producers of citrinin, a nephrotoxic mycotoxin, is important, as some strains of *Monascus* have been reported to produce it. In one study, food grade ang-kak did not show death or toxic phenomena in mice at the level of 18 g kg^{-1} body weight. In the same study, the LD_{50} for peritoneal injection was 7 g kg^{-1}, and subacute toxicity tests revealed no abnormal phenomena. Some strains also produce antibiotics, an undesirable component for a food additive, but antibiotic formation can be reduced and pigment formation enhanced by using media containing acetate.

The monascus pigments are essentially a mixture of the red pigments monascorubin ($C_{23}H_{26}O_5$) and rubropunctatin ($C_{21}H_{26}O_5$), the purple pigments monascorubramine ($C_{23}H_{27}NO_4$) and rubropunctamine ($C_{21}H_{23}NO_4$), and the yellow pigments monascin ($C_{21}H_{26}O_5$) and ankaflavin ($C_{23}H_{30}O_5$) (Fig. 10.2). The red and yellow pigments are normal secondary metabolites of the growth of the fungus and purple pigments are thought to arise from chemical or enzymic modification of those pigments.

The monascus pigments are relatively stable to heat treatment, and stable in the pH range 3–10. The high affinity of the pigments for protein, particularly the red pigments that react easily with amino groups, makes them particularly attractive as colorants for protein rich food, such as meat, sausages, processed seafood (surimi) and dairy products.

The traditional method of pigment production involves the inoculation of *Monascus* seed culture (dry spores on rice grains) onto solid substrates, such as steamed rice. The resulting mass is then dried and ground to powder to be used as a colorant. In recent years, the pigments have been produced in a submerged culture of *Monascus*. Rice powder and tapioca starch are suitable carbon substrates for maximum pigment production, as

Yellow color compounds
monascin: R = C_5H_{11}
ankaflavin: R = C_7H_{15}

Red color compounds
rubropunctatin: R = C_5H_{11}
monascorubin: R = C_7H_{15}

Purple color compounds
rubropunctamine: R = C_5H_{11}
monascorubramine: R = C_7H_{15}

Fig. 10.2. Structure of the pigments produced by *Monascus*.

simple sugars are found to suppress the formation of pigments. Optimum conditions for pigment production are polished long rice at 50% moisture and 97–100% relative humidity, pH 5–6 and incubation at 30°C. Supplementation with Zn leads to increased red and yellow pigment by factors of 2.5 and 2.0, respectively. To overcome the initial high viscosity of the starchy medium and high oxygen demand for growth in the fungal culture as well as pigment production, starchy substrate can be partially digested by enzymes (controlled digestion is important to avoid formation of excessive simple sugars). Otherwise, suitable processes such as **solid-liquid state fermentation** can be used. In the solid-liquid state fermentation system, the solid starchy substrate is first coated onto the fermenter wall surface, by spraying onto it hot gelatinized non-water-soluble starch such as tapioca, and allowing it to solidify at a lower temperature. Then the liquid culture medium is added to the fermenter (Fig. 10.3). Amylolytic enzymes produced by the culture result in the

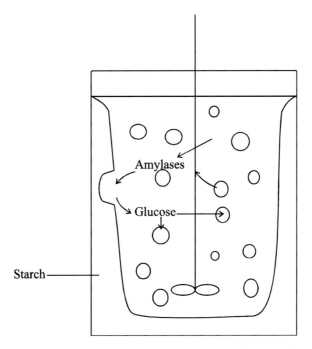

Fig. 10.3. A solid-liquid state fermentation system for the cultivation of *Monascus*. The suspended cells release amylase that releases glucose for metabolism from the surface of the gelatinized starch.

release of fermentable sugars from starch followed by metabolism in a **parallel fermentation**. The strain *M. purpureus* (ATCC 16365) is very suitable for these solid-state fermentations.

A recent novel approach to the production of these pigments involved a modified solid/liquid system in which a 400 g l^{-1} gelatinized tapioca starch cake in a liquid was used to increase yield. The organism grew in the liquid and slowly digested the starch block. A maximum cell concentration of about 37.5 g l^{-1} was achieved and a yield of about 145 OD units of both red and yellow pigments was obtained. This growth yield is significantly higher than on tapioca starch at 50 g l^{-1} on which the biomass usually reaches about 8 g l^{-1}.

10.9.2. *Algal pigments*

The need for natural, as opposed to synthetic, food colorants has led to the development of production systems based on microalgae. The pioneering work has been done with the halophilic green alga *Duniella salina* as a source of β-carotene. Other pigments include chlorophyll, of which *Chlorella* spp. are the major commercial source, as well as red and blue phycobiliproteins. These phycobiliprotein pigments are produced by red algae (Rhodophyta), cyanobacteria (Cyanophyta) and crytomonads (Cryptophyta). Phycobiliproteins are more stable in the pH range 5 to 9, and tend to precipitate at lower pH. However, stability of the pigments at low pH can be achieved by subjecting the pigments to hydrolytic reaction using proteolytic enzymes, such as pronase. While pigments extracted from most algae are relatively sensitive to heat, pigments obtained from thermophilic algae are quite stable to heat.

A blue colorant extracted from the cyanobacterium, *Spirulina platensis*, is commercially produced in Japan, Australia, and America. Uses of the colorants in chewing gum, soft drink, alcoholic drinks and fermented milk products, such as yogurt, have been patented in Japan. In the wild *S. platensis* grows at 27°C, in the pH range 7.2–9.0 and at a salt concentration in excess of 30 g l^{-1}. The red alga *Porphyridium cruentum* is grown in artificial seawater with added potassium nitrate at a temperature of about 21°C. Both organisms can be grown in either open ponds or closed polyethylene tubular reactors. Salinity and/or alkalinity restrict the growth of other microorganisms that might otherwise grow, particularly in open ponds.

Box 10.2

Open Pond vs. Enclosed Photobioreactor for Microalgae Cultivation

Open culture systems are usually race-way-like ponds with a divider at the centre, typically 100 m × 20 m × 10 cm. The cultures are circulated and agitated by a paddle wheel located at one end of the pond. The cell concentration is typically 0.5 g L⁻¹.

The construction of the systems is simple and low cost, but there is no control over the contamination of biological (microorganisms, insect fragments, bird matter and plant parts) and non-biological (sand, dust) contaminants. The Association of Official Analytical Chemists, USA, published a method to detect and enumerate these "light filth" in algal powder in 1989 (AOAC Official Method 970.66).

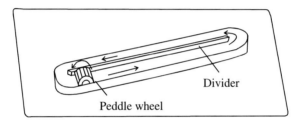

Divider

Peddle wheel

An enclosed photobioreactor can be a tubular or flat plate system made of glass or transparent plastic materials connected to a degasser. The culture is circulated by means of a pump or an airlift system. The system is closed, thus free from foreign materials. The area biomass output rate (g-cells m⁻² illuminated area) is comparable with that achievable in an open pond system, but the cell concentration is 10 to 40 times higher (i.e. 5–20 g L⁻¹), thus saving on down-stream processing costs.

Pump

Further Readings

1. Francis FJ. Less common natural colorants. In: *Natural Food Colorants*, eds. GAF Hendry and JD Houghton. Blackie, London, New York, 1996, pp. 10–342.
2. Jay JM. *Modern Food Microbiology.* Van Nostrand Reinhold Co., New York, 1992.
3. Solomons GL. Production of biomass by filamentous fungi. In: *Comprehensive Biotechnology Vol. 3*, Chapter 22, eds. HW Blanch, S Drew and DIC Wang. Pergamon, Oxford, 1985, pp. 483–506.
4. Stahmann K–P. Vitamins. In: *The Mycota. X, Industrial Applications.* Springer-Verlag, Berlin-Hiedelberg, 2002, pp. 231–246.

Questions for Thought...

1. What was the reason that industrial production of microbial primary metabolites had been considered impractical?
2. Today, almost all the amino acids, which are primary metabolites, are industrially produced by microbial fermentation. Outline the biotechnological developments that have made this possible.
3. Discuss the scientific know-how which lead to the industrial production of citric acid by microbial fermentation.
4. Discuss the metabolic processes that are the foundation for the production of riboflavin and consider the genetic engineering required to improve production.
5. Compare and contrast the biology of the microorganisms used in the production of citric acid and lactic acid.

Websites

http://ift.confex.com/ift/2001/techprogram/paper_7150.htm
http://aem.asm.org/cgi/content/abstract/60/7/2627
http://en.wikipedia.org/wiki/Lactic acid
http://www.fgsc.net/asilomar/citric.html
http://www.cyanotech.com/pdfs/axbul60.pdf
http://www.biomatnet.org/secure/Fair/S1015.htm

Chapter 11

Enzyme Modified Food Products

Desmond K. O'Toole
Department of Biology and Chemistry
City University of Hong Kong
83 Tat Chee Avenue, Kowloon, Hong Kong, China

Yuan Kun Lee
Department of Microbiology, National University of Singapore
5 Science Drive 2, Singapore 117597
Email: micleeyk@nus.edu.sg

The discovery of enzymes and their isolation opened up the possibility of using the enzymes to modify foods. Microbial enzymes can be produced cheaply in large quantities and because microorganisms grow under a wide range of environmental conditions, the enzymes can be used in relatively extreme conditions. Enzymes have been isolated to modify fats, carbohydrates and proteins as well as in other applications.

11.1. Oils and Fats

Oils and fats are triacylglycerols (three fatty acyl groups bound to a glycerol molecule by ester bonds). In the presence of water, microbial lipases (glycerol ester hydrolases EC 3.1.1.3) catalyze the hydrolysis of oils and fats to yield free fatty acid, partial glycerides and glycerol. The reaction is reversible under the condition of a limited amount of water, where lipases catalyze the formation of acylglycerides from glycerol and

Action of group I lipases

$$
\begin{bmatrix} A \\ B \\ A \end{bmatrix} + \begin{bmatrix} C \\ B \\ C \end{bmatrix} \rightarrow \begin{bmatrix} A \\ A \\ A \end{bmatrix} + \begin{bmatrix} A \\ A \\ B \end{bmatrix} + \begin{bmatrix} A \\ A \\ C \end{bmatrix} + \begin{bmatrix} A \\ B \\ A \end{bmatrix} + \begin{bmatrix} A \\ B \\ B \end{bmatrix} + \begin{bmatrix} A \\ B \\ C \end{bmatrix} + \begin{bmatrix} A \\ C \\ A \end{bmatrix} + \begin{bmatrix} A \\ C \\ B \end{bmatrix} + \begin{bmatrix} A \\ C \\ C \end{bmatrix} +
$$

$$
\begin{bmatrix} B \\ A \\ B \end{bmatrix} + \begin{bmatrix} B \\ A \\ C \end{bmatrix} + \begin{bmatrix} B \\ B \\ B \end{bmatrix} + \begin{bmatrix} B \\ B \\ C \end{bmatrix} + \begin{bmatrix} B \\ C \\ B \end{bmatrix} + \begin{bmatrix} B \\ C \\ C \end{bmatrix} +
$$

$$
\begin{bmatrix} C \\ A \\ B \end{bmatrix} + \begin{bmatrix} C \\ A \\ C \end{bmatrix} + \begin{bmatrix} C \\ B \\ C \end{bmatrix} + \begin{bmatrix} C \\ C \\ C \end{bmatrix}
$$

Action of group II lipases

$$
\begin{bmatrix} A \\ B \\ A \end{bmatrix} + \begin{bmatrix} C \\ B \\ C \end{bmatrix} \rightarrow \begin{bmatrix} A \\ B \\ A \end{bmatrix} + \begin{bmatrix} A \\ B \\ C \end{bmatrix} + \begin{bmatrix} C \\ B \\ C \end{bmatrix}
$$

Action of group III lipases

$$
\begin{bmatrix} A* \\ B \\ A* \end{bmatrix} + C* \rightarrow \begin{bmatrix} A* \\ B \\ A* \end{bmatrix} + \begin{bmatrix} A* \\ B \\ C* \end{bmatrix} + \begin{bmatrix} C* \\ B \\ C* \end{bmatrix} + \begin{bmatrix} C* \\ B \\ A* \end{bmatrix}
$$

A, B, C, represent different fatty acyl groups, A* and C* represent long-chain fatty

acids containing a *cis* double bond in the 9 position.

Fig. 11.1. The various reactions carried by the three groups of lipases that attack lipids resulting in inter-esterification of the fats. Through the action of these enzymes, fats can be tailored to meet specific characteristics.

free fatty acids. At a water content of about 10% of the weight of glycerol and free fatty acids, lipases catalyze the hydrolysis and re-synthesis of ester bonds but for some lipases the reactions can also occur under virtually anhydrous conditions. The process achieves an exchange of fatty acyl groups between glycerol molecules in a mixture of different fats or a mixture of fat and free fatty acids, and the reaction is called **inter-esterification**. Inter-esterification is used in the food industry to alter the composition and therefore the physical and nutritional properties of oils and fats.

The carbons in the glycerol in natural triacylglycerides are numbered due to the chiral nature of the molecules and the stereospecific reaction products. In natural triacylaglycerols at *sn*-1 (stereochemical number 1) saturated fatty acids predominate, at *sn*-2 unsaturated fatty acids are found and at *sn*-3 a variable range of fatty acids occur.

Based on their substrate specificity, microbial lipases can be classified into three groups as shown in Fig. 11.1.

(1) Group I
These are lipases that show no specificity both as regards to the position on the glycerol molecule and the nature of the fatty acid they attack. These lipases catalyze random exchange of fatty acyl groups between glycerol molecules in a mixture of two fats. Organisms known to produce them include *Candida cylindracae*, *Propionibacterium acnes* or *P. parvum* (formerly *Corynebacterium acnes*) and *Staphylococcus aureus*.

(2) Group II
These are lipases that catalyze reactions only at the *sn*-1 and *sn*-3 positions of the acylglycerols. That is they catalyze selective exchange of fatty acyl groups at the 1- and 3-positions. Organisms known to produce them include *Aspergillus niger*, *Mucor javanicus* and *Rhizopus arrhizus*.

(3) Group III
These are lipases that selectively attack esters of long-chain fatty acids containing a *cis* double bond in the 9 position, such as oleic, palmitoleic, linoleic and linolenic acids. An organism known to produce them is *Geotrichum candidum*.

Other organisms whose lipases have proved useful include *Candida antarctica*, "*Chromobacter viscosum*" (invalid name but published as strain NRRL B-3673; maybe "*Chromobacterium viscosum*" ATCC 6918), *Rhizopus delemar* and some *Pseudomonas* spp.

Lipase catalyzed inter-esterification reactions can be performed either batchwise or continuously in **stirred tank reactors** or continuously in **packed-bed reactors** and **membrane bioreactors**. The catalyst is often immobilized on solid supports, such as kieselguhr, and the fat is extracted by acetone precipitation or by solvent evaporation. The catalyst particles can be reused for up to ten successive batch reactions or up to 600 h in a continuous process, as long as the oils are saturated with water. Solvents such as petroleum ether are sometimes added to facilitate mixing in the oil phase.

11.1.1. *Cocoa butter equivalent*

An example in which lipase-catalyzed inter-esterification is used to modify edible fats and oils is in the production of the high value cocoa butter equivalent from cheap starting materials such as palm oil. Cocoa butter from cocoa bean has a high content of 1(3)-palmitoyl-3(1)stearoyl-2-monoolein (POS) and 1,3-distearoyl-2-monooleine (SOS) which gives the fat a sharp melting point of around 32–35°C. Consequently a bar of chocolate candy made with this fat will only melt in the mouth but not at room temperature or in one's hand. Supply of cocoa butter is limited so it is expensive. By mixing the mid-fraction of palm oil, which is rich in POP with the solid fraction of palm oil, which is rich in SSS (tristearine) in a suspension with Group II lipase, a product that is enriched in POS and SOS is produced as shown below. POS and SOS can be isolated by conventional fat fractionation techniques such as countercurrent liquid-liquid extraction and crystallization from solvent.

$$
\begin{bmatrix} P \\ O \\ P \end{bmatrix} + \begin{bmatrix} S \\ S \\ S \end{bmatrix} \rightarrow \begin{bmatrix} P \\ O \\ P \end{bmatrix} + \begin{bmatrix} P \\ O \\ S \end{bmatrix} + \begin{bmatrix} S \\ O \\ S \end{bmatrix} + \begin{bmatrix} P \\ S \\ P \end{bmatrix} + \begin{bmatrix} P \\ S \\ S \end{bmatrix} + \begin{bmatrix} S \\ S \\ S \end{bmatrix}
$$

A mixture of POP and stearic acid can also produce the same results (Fig. 11.2).

11.1.2. *Betapol*

This is the name of a human milk fat substitute developed by Loders Croklaan. The fat in cow's milk incorporated into infant milk formula is not as nutritious as human milk and results in reduced mineral and fat absorption as well as hard stools that can cause the development of constipation and blocked bowels in the bottle-fed babies. The fat in human milk dose not cause this because it contains a high quantity of palmitic acid at the *sn*-2 position and at the *sn*-1 and *sn*-3 positions the unsaturated oleic acid predominates, represented as OPO. Betapol is rich in palmitic acid at the *sn*-2 position and rich in oleic acid.

To produce OPO, tripalmitin rich vegetable oil and oleic acid are mixed and passed through a packed-bed reactor loaded with an immobilized *Rhizomucor miehei* lipase at 70°C under practically anhydrous conditions. At equilibrium, three triacylglycerols result, namely PPP, OPP and OPO,

Fig. 11.2. The manufacture of cocoa butter equivalent using a Group II Lipase immobilized on a substrate. The palm oil fraction and stearic acid dissolved in hexane with a minimum amount of water are passed through the bed at 50°C with a residence time of about 10 min. The hexane is recovered and free fatty acids from the palm oil separated out by vacuum distillation. Then using fractional crystallization the product is separated from other fats produced during the reaction.

the desired product. The time to equilibration depends on the substrate molar ratio (oleic to tripalmitin); the higher the ratio the longer the time and the higher the maximum incorporation. In the industrial setting two consecutive packed-bed reactors are used and after the first pass the free fatty acids released are removed by distillation and the product is passed through the second reactor with the addition of further oleic acid. Distillation, fractionation and refining follow to produce the final product.

11.1.3. *Low joule (calorie) fats*

Not all fatty acids are easily absorbed during digestion. A reduced calorie fat product has been produced by replacing some of the long chain fatty acid, behenic acid (C22:0) with short chain fatty acids. Procter and Gamble Company inter-esterify coconut, palm kernel and rapeseed oils to produce Caprenin. The resulting oil contains C8:0, C10:0 and C22:0 fatty acids. Because the behenic acid interferes with fat absorption, the actual calories delivered nutritionally are only 5 kcal per g versus 9 kcal per g for normal fat.

Diacylglycerols cannot be deposited in the adipose tissues. Designer cooking and salad oil based on diacylglycerols may prevent accumulation of fat in the body. Diacylglycerols can be produced by time-controlled hydrolysis of oil using Group I or Group II lipases.

11.2. Polysaccharides

11.2.1. *High fructose corn syrup*

Sweeteners derived from starch, particularly cornstarch, contribute to more than a third of the worldwide sweetener market. The process involves hydrolysis of starch by α-amylase and glucoamylase to dextrose monomer (glucose) and subsequent isomerization of glucose to fructose using immobilized glucose isomerase (Fig. 11.3). Fructose is twice as sweet as sucrose from cane and beet sugar.

In the first step, the heat gelatinized starch slurry (30–35% w/w) is liquefied and dextrinized by a bacterial thermostable α-amylase in a continuous two-stage reaction. Starch slurry (0.05–0.1% w/w starch) together with the enzyme (0.05–0.1% w/w starch) is incubated at

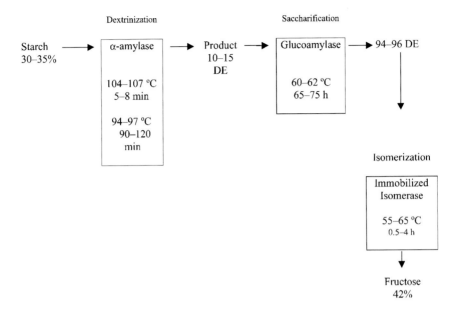

Fig. 11.3. Production of high fructose syrup from corn starch.

104–107°C for 5–8 min, resulting in the formation of 0.5–1.5 dextrose equivalent (DE) hydrolysate. Pure dextrose solution has a unit of 100. Lowering the temperature to 94–97°C and holding for 90–120 min, results in a hydrolysate of 10–15 DE.

During saccharification, the hydrolysate is further converted to dextrose by the action of a fungal glucoamylase (at 1 l/ton starch). Commonly, a temperature of 60–62°C and a residence time of 65–75 h are used to obtain hydrolysate containing 94–96% dextrose.

The 94–96% dextrose hydrolysate is then passed through a column of immobilized isomerase (55–65°C), with a residence time of 0.5–4 h. The activity of the enzyme decays in a nearly exponential manner over a period of several months (140–240 d). Thus, the residence time for a new enzyme column is short, and then the flow rate through the column has to reduce progressively to give a longer residence time compensating for the lower enzyme activity. As the conversion of glucose to fructose is a reversible reaction with an equilibrium constant of about 1.0 at 60°C, a practical conversion level of 42% fructose can be achieved.

11.2.2. *Non-digestible oligosaccharides (NDOs)*

Oligosaccharides are saccharides containing between 3 and 19 sugar moieties. Some of these large saccharides cannot be hydrolyzed or absorbed in the upper part of our gastrointestinal tracts. These NDOs eventually reach the colon and become substrates for a restricted group of bacteria, such as species of *Bifidobacterium* and *Lactobacillus*. Thus, NDOs are **prebiotics** that selectively enhance the growth of indigenous probiotic bacteria in the colon. They are classified, for food labeling, as dietary fiber, and are used by the food industry as bulking agents, sugar substitutes and fat replacers besides being functional food ingredients. Examples of commercially available NDOs and the enzymes used in their production are shown in Table 11.1.

Table 11.1. Production of oligosaccharides and the substrates and enzymes used to produce them.

Oligosaccharide	Substrate	Enzyme Used
Produced by hydrolysis of polysaccharides:		
Fructo-oligosaccharides, Fructofuranosyl (Fn) type and non-reducing glucosyl (GFn) type	Inulin	β-fructofuranosidase
Produced by transglycosylation reactions:		
Fructo-oligosaccharides (GFn) type	Sucrose	Fructosyltransferase
β-galacto-oligosaccharides	Lactose	β-glactosidase
Genito-oligosaccharides	Starch/maltodextrins	β-glucosidase
Glucosylsucrose	Maltose/sucrose	Cyclomaltodextrin-glucanotransferase
Isomalto-oligosaccharides	Starch/maltose	α-glucosidase
Lactosucrose	Lactose/sucrose	Levansucrase
Palatinose (isomaltulose)	Sucrose	Sucrose-6-glucosylmutase

11.2.3. *Plant pectins*

Pectin consists of linear polymers of α-1,4-linked galacturonic acid residues of which up to 70% are methoxylated and which have neutral sugar side branches. Pectins act as adhesives in plant cell walls and lamellae. Three groups of enzymes degrade native pectins; (i) polygalacturonases are hydrolytic enzymes that cleave linkages randomly and release galacturonate from non-reducing ends; (ii) pectin lyases cleave, by a trans-elimination mechanism, 1,4-α-glycosidic linkages; and (iii) pectin esterases cleave off the methoxy side groups.

Fruit juices and fruit by-products from the fruit processing industry contain pectins. Natural juice, e.g. from apple, cherry and raspberry, contains pectins that render the juices viscous and turbid so that the juices cannot be easily filtered. To change these characteristics fungal enzymes have been added to juices since the 1930s. The enzymes of choice, particularly from *Aspergillus* sp., are the pectin lyases because methanol (a toxin) is produced by pectin esterases and polygalacturonases tend to produce pectin precipitates.

Pectin extracted from apple pomace and citrus peel is used as a gelling agent and can be extracted by boiling in weak acid and eventually precipitating the pectin with salts. However, chemical extraction of pectin from mandarin orange peels is too harsh so the peels are inoculated with a fungus, *Trichosporin penicillatum*, that releases the pectin within about 24 h.

11.3. Proteins

Proteases that catalyze the hydrolysis of protein change the solubility and functional properties of protein and food. Three groups of microbial proteases are widely used in the food industry; they are the alkaline, acid and neutral proteases.

11.3.1. *Alkaline proteases*

Alkaline proteases are serine proteases that contain both serine and histidine at their active site. Alkaline proteases for food use are produced by bacilli, e.g. *B. licheniformis*, and by aspergilli. The enzymes have an optimum pH of about 8.0, are used to produce highly functional soy protein hydrolysate with good whipping characteristics, and a highly soluble hydrolysate, used in protein fortified soft drinks as a dietetic beverage. The enzymic reaction is stopped by reducing the pH to 4.0.

11.3.2. Acid proteases

The acid or carboxyl proteases generally have pH optima in the range of 3–4. An example is the microbial rennet produced by *Mucor miehei* and *M. pusillus*. Rennet causes coagulation of milk for cheese making, by hydrolysis of a specific bond in κ-casein. The traditional source of rennet is the inner lining of the fourth stomach of bovine calves.

Other important acid proteases used in food processing are pepsin and pepsin-like enzymes from *Aspergillus* and *Rhizopus* species. Pepsin in combination with alkaline proteases is being used to produce soy hydrolysates with good whipping characteristics.

11.3.3. Neutral proteases

Neutral proteases of *Bacillus subtilis* and *Aspergillus oryzae* are metalloproteases with a metal ion involved in their catalytic reaction. The enzymes have a pH optimum near 7.0. *B. subtilis* neutral protease enhanced flavor intensity of cheese without producing bitterness. Thus, this protease in combination with lipases is used to produce strongly flavored cheese by adding the enzymes to scalded curds and then curing at 10–25°C for 1–2 months. This flavor enhancing approach that uses proteases has led to the development of enzyme modified cheese (EMC). This is a cheese flavor enhancing process in which proteases and lipases are added to a slurry of cheese and incubated for several weeks under controlled conditions. EMC pastes of Blue, Cheddar, Cream, Edam, Mozzarella, Parmesan, Provolone and Romano cheeses are commercially available. The EMC pastes are incorporated into processed cheeses to improve flavor.

Proteases of *A. oryzae* are also used in bread making. The fungal proteases include both endo- and exo-peptidase. Endopeptidase modifies the viscoelastic properties of the dough by hydrolyzing interior gluten peptide bonds, thus improving the dough handling properties and increasing loaf volume. Exopeptidase releases amino acids from gluten, that react with sugars through the Maillard reaction during baking, thereby contributing to flavor and crust color.

11.4. Amino Acid Production

Apart from the straight fermentative production of amino acids, they can also be produced from non-food sources using microbial enzymes. Amino acids are racemic and contain a chiral center and the biologically important amino acids are the L- isomers. When chemical synthesis is used to produce amino acids, a racemic mixture inevitably results. Using microbial enzymes the racemic mixture can be converted to the L-isomer (Fig. 11.4).

11.4.1. *Methionine*

The first application of this approach was developed in the 1950s, to produce L-methionine. N-acetyl-DL-methionine is synthesized using conventional chemical methods. Cells of *A. oryzae* that produce an N-acylase specific for the L-isomer of methionine were used to hydrolyze the racemic mixture to produce L-methionine and N-acetyl-D-methionine. L-methionine was then precipitated from the supernatant by adjusting the pH to the isoelectric point of the amino acid. The remaining N-acetyl-D-methionine was then chemically racemized and the whole process repeated.

Fig. 11.4. The production of L-amino acids from a racemic mixture using an enzyme. Following chemical synthesis during which the DL mixture of the amino acid is produced linked to an acyl group, the product is then hydrolyzed by an N-acylase enzyme that releases the acyl group from the L-amino acid. The R-D-amino acid is unchanged but it can be isolated and chemically racemized resulting in the formation of another racemic mixture on which the N-acylase can act to release more L-amino acid. Spontaneous racemization is also possible thus eliminating a separate chemical racemization step.

The whole cells produced many more enzymes than the specific enzyme required so whole cells were abandoned and the requisite enzyme isolated and immobilized on a basic ion-exchange resin in a column. When the reaction mixture is passed through the column the pH of the mixture drops as the free amino acid is released, consequently the pH of the column must be managed to prevent precipitation of the free L-methionine in the column. A similar approach is used for transforming other racemic amino acid mixtures.

11.4.2. *Cysteine*

L-cysteine has traditionally been produced by acid hydrolysis of keratin, mostly from hair. This method is being replaced by a chemical synthesis method involving stereospecific hydrolysis followed by racemization. In the chemical synthesis stage methyl-2-chloroacrylate is converted to DL-amino-Δ_2–thiazoline-4-carboxylate. An enzyme from either *Pseudomonas thiazolinophilum* (invalid name), *Delftia acidovorans* (formerly known as *P. desmolytica*) or *Micrococcus luteus* (formerly known as *Sarcina lutea*), is then used to hydrolyze the compound and release L-cysteine. Racemization of the D-form occurs spontaneously in the reaction mixture. The gene for the enzyme involved in the hydrolysis has been isolated, characterized and cloned into an *Escherichia coli* strain.

Further Readings

1. Akoh CC and Xu X. Enzymatic production of betapol and other specialty fats. In: *Lipid Biotechnology*, eds. TM Kuo and HW Gardner. Marcel Dekker, New York, 2002, pp 461–478.
2. Crittenden RG and Playne MJ. Production, properties and applications of food-grade oligosaccharides. *Trends Food Sci. Technol.* **7**: 353–361, 1996.
3. Marangoni AG and Rousseau D. Engineering triacylglycerols: The role of interesterification. *Trends Food Sci. Technol.* **6**: 329–335, 1995.
4. Panyam D and Kilara A. Enhancing the functionality of food proteins by enzymatic modification. *Trends Food Sci. Technol.* **7**: 120–125, 1996.

5. Poutanen K. Enzyme: An important tool in the improvement of the quality of cereal foods. *Trends Food Sci. Technol.* **8**: 300–306, 1997.
6. Roberts SM, Turner NJ, Willetts AJ and Turner MK. *Introduction to Biocatalysis Using Enzymes and Microorganisms.* Cambridge University Press, Cambridge, New York, 1995.
7. Voragen AGJ. Technological aspects of functional food-related carbohydrates. *Trends Food Sci. Technol.* **9**: 328–335, 1998.

Questions for Thought...

1. Public health authorities are becoming more concerned about the presence of trans fatty acids in hydrogenated oils. Consider how you might overcome this problem using enzymes.
2. What role do enzymes that breakdown protein have in food production systems?
3. If you invented a new food based on the introduction of a foreign gene into a different microorganism that coded for enzymes that broke down proteins, what would you have to do to have it accepted in the USA and Europe?

Part III

Microbes in Agrobiotechnology

Molecular biological methods have becoming increasingly applicable to the diagnosis of infectious diseases and vaccines development in veterinary medicine. The commonly employed microbiology and virology assays and the conventional serological test to identify microorganisms are often not specific and sensitive enough to make the diagnosis in a short period of time. However, with advances in diagnostic methods based on immunology and molecular biology technology, these methods are rapidly invading into laboratories as the preferred choice for diagnosis of animal diseases. The reasons are that these methods are user friendly, safe, cost effective, reproducible and some are automated to facilitate the evaluation of large numbers of specimens.

Molecular biology procedures also have a major impact on the preparation of veterinary vaccines. These "modern" vaccines did not only overcome the limitations of the "classical" vaccines but are also much safer and efficacious. It also stimulates an immune response distinguished from that of natural infection, so that immunization and eradication may proceed simultaneously. Such a vaccine is also commonly known as marker vaccines since it is able to "mark" the vaccinated animals from the naturally infected animals.

Plant biotechnology has opened a new era of agricultural revolution by introducing genes from different organisms to generate new varieties of plants — genetically modified (GM) plants. Among them, the new food crops (such as grain crops, vegetables and fruits) will have higher yield, better nutritional qualities, resistance against pests (insects, diseases, nematodes, weeds and abiotic stresses), others will help us to produce new industrial raw materials (oils, starches and biodegradable plastics) and pharmaceutical products. Despite the great economical and humanitarian potential, plant biotechnology has caused great concerns about the safety of the genetically modified (GM) foods, the environmental impact of GM plants, and moral issues. In this chapter, both technical and regulatory aspects of plant biotechnology are discussed in detail. The future success of plant biotechnology depends on science-based facts, proper governmental regulations and public education efforts.

Chapter 12

Microbes and Livestock

Jimmy Kwang
Temasek Life Sciences Laboratory
National University of Singapore
1 Research Link, Singapore 117604

We have charted an amazing growth rate in the human population in the last two centuries. From 1804 to about 1900, the world population had increased from one billion to more than two billion people. However, by 1974, the human population had doubled to four billion people. To date, there are more than six billion people in the world, with the figure still growing daily (Fig. 12.1). This rapid increase signals some major problems, like the shortage of food to meet human demand.

Backyard farming used to be a common practice, but with the high cost of maintenance and spatial constraints on an individual, it is no longer practical for one to raise his own livestock for food. Due to these reasons, modernization has made way for large-scale industries to produce the necessary sustenance to cater for the growing population. Animals bred for consumption are kept in large-scale space-conserved housing, with automated animal farming to provide a more cost efficient method of rearing. However, this is where potential problems set in. When animals are kept densely, in close proximity to one another, the entry of an infectious pathogen into these facilities can result in rapid transmission of the pathogen from one animal to the next within a short time, even causing mass mortality. This can result in big losses for the farms involved and also a major setback to the livestock industry. In 1996, the infamous mad cow

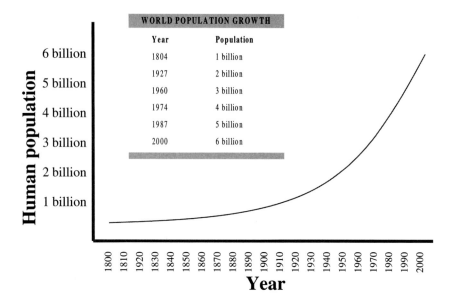

Fig. 12.1. World population growth from year 1804 to 2000.

disease or Bovine Spongiform Encephalopathy (BSE) led to the slaughter of six million cattle and inflicted worldwide paranoia over the consumption of beef from the UK. In 1997 alone, several epidemics had occurred, for instance Foot and Mouth Disease (FMD), which had resulted in the slaughter of four million pigs in Taiwan. Classical Swine Fever (CSF) caused millions of pigs in the Netherlands to be culled, and *Escherichia coli*: O157:H7 and *Salmonella* led to the loss of billions of dollars to the meat industry. In 1998, the emerging Nipah virus outbreak in Malaysia led to more than one million pigs being culled, and the banning of pork export to Singapore and other countries. The outbreak of FMD in the United Kingdom during early 2001 struck a devastating blow to the farm sector leading to the culling of millions of animals with a slump in exports and also the threat of spread to other European countries. This is in addition to the mad-cow disease that has already plagued some European countries.

These catastrophes could have been prevented with good management practices including effective vaccination procedures associated with the early detection of diseases before their transmission. The importance of vaccination and disease detection will be discussed.

12.1. Vaccination for Animals

Ancient Chinese observed that individuals affected by smallpox were resistant to repeated attacks of the disease. Therefore, attempts to protect against smallpox in those times had been practiced by inoculation using vesicle fluid from persons with mild forms of smallpox or by deliberate contact with diseased individuals. Subsequently, this technique called variolation spread westwards. "Variola" is the Latin word for "smallpox". During the second half of the 18th century, when smallpox claimed many victims during the epidemics in Europe, some of the milkmaids were found to be protected from smallpox. It was discovered that their immunity against smallpox was a result of their exposure to cowpox virus while milking the cows. Many years later, due to the efforts of the English physician Edward Jenner, who attempted the first immunization with cowpox in 1796 on a brave young man named James Phipps, smallpox was completely prevented. Jenner's experiments were followed up by Louis Pasteur who developed vaccines against fowl cholera and also a successful vaccine against rabies. The term "vaccination" derived from the Latin word "vacca" which means cow was introduced to replace the term variolation. Vaccination, which is fundamentally based on the concept that prevention is better than cure, was employed long before the mechanism of immune protection was known. To date, vaccination is the most successful veterinary measure, with more animal production safeguarded than through all other veterinary measures combined.

A range of microorganisms, including bacteria, ricketesia, viruses, fungi and protozoa, cause infectious diseases to animals. Within any one group of these pathogens, there is variation in the mechanism of disease production. Therefore, knowledge of the infectious mechanism of any pathogen is critical to understanding of the mechanisms of acquired resistance for formulating of an effective vaccine. As the pathogenesis and epidemiology of each disease varies, so does the role and efficacy of vaccination. While some vaccines may be highly efficacious, inducing an immunity that prevents clinical symptoms of the disease as well as infection and replication of the pathogen, others may only prevent clinical disease, but not infection and/or the development of the carrier state. In other cases, immunization may be completely ineffective or only able to reduce the severity of the disease. Therefore the application of vaccination as part of animal disease control strategy requires a thorough knowledge of the characteristics of the pathogen and its epidemiology, along with the characteristics and capabilities of various vaccines.

12.1.1. *The immune response*

The concept of immunity dates back many centuries, but the broad application of immunization is a more recent development and still a subject for continued experimentation. Vaccination is intended to induce an immunological defense response that will provide protection to the respective infectious agent. Therefore, as a general rule, the immune response should be of the same nature as that which results from a natural infection.

There are various cells in the blood, which include erythrocytes, granulocytes and lymphocytes. These cells not only have different morphology, but they also have totally different functions. Immune responses are carried out by lymphocytes, and the number of these cells only constitutes a small percentage (<0.2%) of the total number of blood cells. Immunological protection is provided primarily by humoral-mediated (B lymphocytes) and cell-mediated (T lymphocytes) arms of the immune system (Fig. 12.2). Two other major components of the immune system are complement and phagocytes.

All lymphocytes appear similar under the light microscope, ranging from 7 to 15 μm in diameter. The morphology, structure and staining characteristic of the cells do not provide clues to their complex roles within the body. Lymphocytes can only be differentiated and defined on the basis of their ontogeny, surface receptors, function and characteristic cell surface antigens. T cells are the predominant circulating lymphocytes, constituting 65–80% of the lymphocytes seen in peripheral blood. T cells are relatively long-lived cells, while B cells, in contrast, have a short life span.

B cells perform two important functions:

(1) They differentiate into plasma cells and produce antibodies. Antibody molecules are proteins called immunoglobulin (Ig) (Table 12.1). Ig may be classified on the basis of their electrophoretic mobility, their

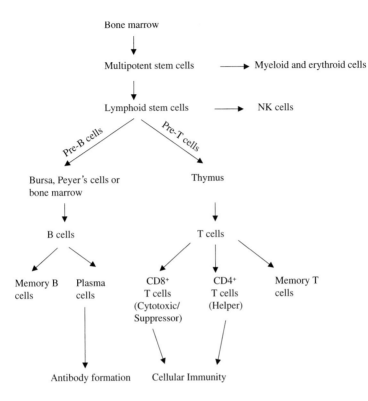

Fig. 12.2. Development of T and B lymphocytes in the body, which make up the arms of the immune system.

Table 12.1. Characteristics of the immunoglobulin isotypes in domestic mammals.

Property	Immunoglobulin Isotype				
	IgM	IgG	IgA	IgE	IgD
Sedimentation coefficient	19S	7S	11S	8S	7S
MW	900,000	160,000	360,000	200,000	180,000
Heavy chain	μ	γ	α	ε	δ
Mainly produced by	Spleen, lymph nodes	Spleen, lymph nodes	Intestinal, respiratory tracts	Intestinal, respiratory tracts	Spleen, lymph nodes

molecular weight, and their antigenic structure. Soluble antibodies (Fig. 12.3) are found in many body fluids but are present in highest concentrations and are most easily obtained from blood serum.

(2) They are antigen presenting cells (APC) (Table 12.2). B cells carry cell membrane-bound immunoglobulin molecules that function as antigen receptors to bind antigens (Fig. 12.4). In order for an antibody response to occur, the recognition of cell-membrane-bound antigens by T lymphocytes is required.

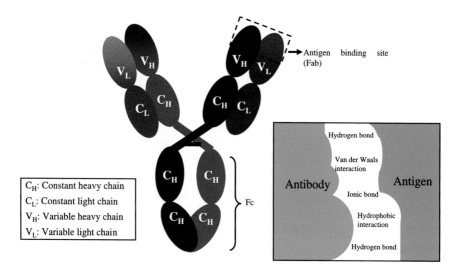

Fig. 12.3. Antibody structure and antibody-antigen binding reaction.

Table 12.2. Antigen presenting cells.

Cell Type	Distribution
B lymphocyte	Wide
Macrophage	Wide
Dentritic cells	Lymphoid tissues
Langerhans' cells	Skin
Kupffer's cells	Liver
Microglia cells	Central nervous system (CNS)

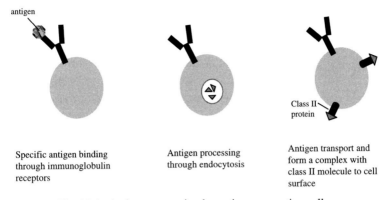

Specific antigen binding through immunoglobulin receptors

Antigen processing through endocytosis

Antigen transport and form a complex with class II molecule to cell surface

Class II protein

Fig. 12.4. Antigen processing by antigen presenting cells.

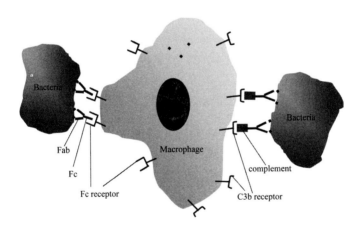

Bacteria

Fab

Fc

Fc receptor

Macrophage

Bacteria

complement

C3b receptor

Fig. 12.5. Two mechanisms by which antibodies enhance phagocytosis of bacteria: (1) Antibodies link the bacteria to macrophage by means of their Fc region. (2) Binding of the antibodies to bacteria initiate activation of the complement (C3b), which can then bind directly to the surface of the bacteria and enhance its phagocytosis.

There are two predominant types of phagocytic cells: neutrophils and macrophages. Macrophages exist both as free cells, e.g. monocytes and fixed in tissues, e.g. Kupffer cells of liver. Neutrophils and macrophages have been termed phagocytes because their membranes are equipped with two specialized receptors:

(1) Receptor for the Fc region of IgG molecules
(2) Receptor for the third component of complement (C3)

These receptors facilitate the process of phagocytosis more efficiently by assisting in the ingestion of microorganisms with IgG or activated C3 on their surface, a phenomenon known as opsonization (Fig. 12.5).

T cells perform several important functions, which can be divided into two main categories, namely regulatory and effector. The regulatory functions are mediated primarily by $CD4^+$ helper T cells, which have the ability to augment B cell response and to amplify the cell-mediated response through a group of mediators called interleukins (Table 12.3). The term interleukin (IL-1 to IL-20) is applied to the molecules that function as communicators between lymphocytes. Another regulatory function is mediated by suppressor $CD8^+$ T cells, which down-regulate the immune response. Failure of such regulation may cause autoimmune diseases.

Table 12.3. The interleukins.

Name	Mw (KD)	Mainly Produced by	Major Targets
IL-1	17	Macrophages	T cells, B cells
IL-2	15	T cells, NK cells	T cells, B cells
IL-3	28	T cells	Hematopoietic cells, stem cells
IL-4	20	T cells	T cells, B cells, mast cells
IL-5	18	T cells	B cells, T cells
IL-6	26	Fibroblasts, T cells	B cells
IL-7	25	Stromal cells	Immature lymphocytes
IL-8	8	Macrophages	T cells, neutrophils
IL-9	39	T cells	T cells, mast cells
IL-10	19	T_H2 cells, B cells	T_H1 cells, macrophages
IL-11	24	Stromal cells	B cells
IL-12	75	Macrophages, dendritic cells	T cells, NK cells
IL-13	12	T cells	B cells
IL-15	17	Wide variety of cells and tissues	NK cells, T cells
IL-16	14–17	$CD8^+$ T cells	$CD4^+$ T cells
IL-17	17.5	T cells	T cells
IL-18	22.3	Kupffer's cells	T_H cells
IL-19	?	Monocytes	T_H2 cells and/or B cells
IL-20	?	?	Keratinocytes

The effector functions are carried out primarily by cytotoxic ($CD8^+$) T cells, which kill virus-infected cells, tumor cells and allografts, as well as $CD4^+$ cells, which mediate delayed-type hypersensitivity (DTH) against intracellular organisms. T cells, B cells and macrophages must interact very closely to ensure that immune response occurs. In order for the cooperation between the three kinds of cells to occur, they must have common major histocompatibility complex (MHC) antigens. MHC antigens, encoded by three classes of genes, are highly polymorphic membrane glycoproteins expressed on body cells. Class I and II genes code for antigens associated with self or nonself discrimination and antigen recognition, and class III genes code for complement proteins. The complement is a collective term used to designate a group of chemically and immunologically distinct serum proteins that play a key role in the host defense process. Complement proteins are capable of interacting with each other, with antibodies and with cell membranes. These interactions lead to the generation of biological activity. The biological consequences of complement activation include:

(1) Lytic destruction of cells, bacteria and viruses, allografts and tumor cells.
(2) Regulation of the inflammatory and immune response.
(3) Opsonization.

Almost all nucleated cells express class I MHC molecules. On the other hand, class II molecules are only expressed in a few cell types, predominantly in antigen-presenting cells (APC) such as macrophages and B cells (Table 12.4). Although MHC class I and II molecules have subtle differences in their structural features, both molecules have highly polymorphic outer extracellular domains that form a long cleft in which peptide fragments are bound to. MHC class I molecules present peptides derived from endogenous proteins (e.g. fragments of viral proteins in infected cells) while class II molecules present peptides derived from exogenous proteins (e.g. pathogens that are internalized by phagocytic cells or pathogens living in vesicles of macrophages or B cells).

Class I and II MHC molecules are recognized by different subsets of functional T cells. Foreign peptides bound to MHC class I molecules on the cell surface are recognized by the $CD8^+$ T cells, which in turn

Table 12.4. Characteristics of T cells, B cells, and macrophages.

Features	T Cell		B Cell	Macrophage
	Helper	Cytotoxic		
Surface Markers				
CD4	+	−	−	−
CD8	−	+	−	−
Immunoglobulin (Ig)	−	−	+	+
MHC class I	+	+	+	+
MHC class II	−	−	+	+
Antibody synthesis	−	−	+	−
Cellular immunity	+	+	−	+
Receptor for IgG Fc	−	−	+	+
Receptor for C3b	−	−	−	+

differentiate into cytotoxic T cells (CTL) that kill the infected cells. Cytotoxic T cell response or cell mediated immune response is very important in host defense against pathogens, predominantly viruses that survive in the cytosol. Cytotoxic T cells kill infected targets with great precision to limit the spread of these pathogens.

Peptides presented on the MHC class II molecules are recognized by the CD4$^+$ T cells that can differentiate into T_H1 and T_H2 cells. Peptides derived from pathogens that are multiplying inside the intracellular vesicles tend to stimulate the differentiation of T_H1 cells, whereas peptides derived from extracellular pathogens tend to stimulate the differentiation of T_H2 cells (Figs. 12.6 and 12.7). T_H1 cells activate the microbicidal properties of macrophages by inducing the fusion of their lysosomes with the intracellular vesicles housing the pathogens and at the same time induce B cells to produce IgG antibodies that are effective at opsonizing extracellular pathogens for uptake by other phagocytic cells. T_H1 cells also release cytokines that attract macrophages to the site of infection and thus are also known as inflammatory CD4 T cells, or delayed-type hypersensitivity T cells. T_H2 cells play an essential role in humoral response by activating naive B cells to produce IgM antibodies and subsequently stimulating B cells to proliferate and differentiate into cells

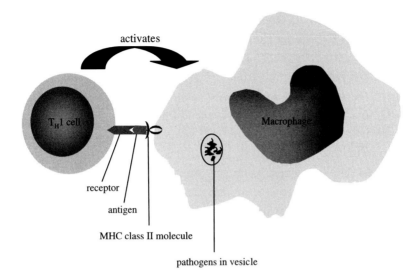

Fig. 12.6. T$_H$1 cell recognizes fragment presented by MHC class II molecules on the surface of a macrophage containing pathogens, and activates the macrophage.

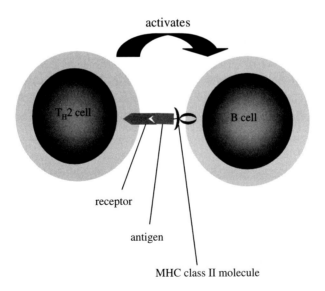

Fig. 12.7. T$_H$2 cell recognizes fragment presented by MHC class II molecules on the surface of B cells, and activates the B cell.

capable of secreting different antibody isotypes. Hence, T_H2 cells are also termed helper T cells for their ability to help B cells make antibodies in response to antigenic challenge.

Although helper T cells are required to induce B cell response to protein antigens, certain microbial antigens, such as bacterial polysaccharides, are able to activate B cells in the absence of helper T cells. These antigens are known as thymus-independent or TI antigens. TI antigens do not induce isotype switching or memory B cells but they certainly play an important role in host defense against pathogens whose surface antigens cannot elicit T cell responses.

The following are examples to illustrate these interactions among T cells, B cells, and macrophages to modulate immune responses. Certain organisms of veterinary importance, such as *Mycobacterium tuberculosis*, *Brucella abortus* and *Salmonella typhimurium*, are able to grow and thrive in the cytoplasm of macrophages after engulfment. These organisms are called facultative intracellular bacteria. When T_H1 cells encounter these bacterial antigens, they release a macrophage-activating factor that can enhance macrophage activity. Macrophages then destroy and digest these organisms into small pieces, and present the processed antigen in association with class I MHC molecules to T lympocyte. In addition, macrophages release biological mediators (such as interleukin), which further amplify the response of the T lymphocyte and thus greatly stimulate the immune response. Cell-mediated immunity associated with the activation of macrophages limits the process of infection. Since activated macrophages show a non-specific enhancement of their microbicidal activities, they confer enhanced resistance to bacteria in general.

Virus-host relationships are also very complex, which leads to pressure on the immune system to prevent clinical disease. Viral infections can sometimes be characterized as:

(1) Cytolytic, in which viral association with a host cell may lead to virus replication with death of the cell and release of mature virus particles into the extracellular environment.

(2) Steady-state, in which virus replicates and spreads from cell to cell without killing the host cell. In this case, virus may cause persistent infection, latency, or neoplastic transformation.

In general, humoral immunity mediates immunity against cytolytic viruses whereas cell-mediated immunity is of primary importance with steady-state viruses. However, a role for antibodies, especially in preventing an initial infection against steady-state viruses is recognized. For example, antibodies can effectively prevent influenza virus from infecting cells of the respiratory tract, but cell-mediated responses usually develop following infection and play a role in killing influenza virus infected cell. After cells are infected with the influenza virus, viral envelope glycoproteins, namely hemagglutinin and neuraminidase, appear on the surface of the infected cell in association with class I MHC protein. A cytotoxic T cell binds via its antigen-specific receptor to the viral antigen-class I protein complex and is stimulated to grow into a clone of cells by interleukin-2 produced by helper T cells. These cytotoxic T cells can then specifically kill influenza virus-infected cells by recognizing viral antigen-class I protein complexes on the cell surface (Fig. 12.8). Thus, in order to develop a vaccine against the flu virus, the viral envelope glycoproteins associated with the viral pathogenicity is a crucial ingredient.

The synthesis of antibodies directed against determinants on the pathogenic organism can protect the host in several ways. Sometimes the

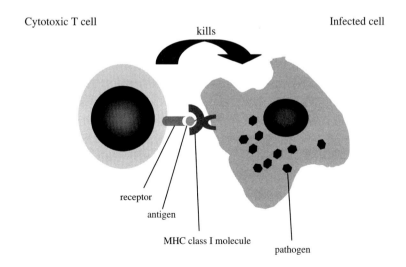

Fig. 12.8. CD8$^+$ cytotoxic T cell recognizes antigen presented by MHC class I molecules on the surface of an infected cell, and kills the infected cell.

simple interaction of antibodies with the antigen accomplishes the response such as toxin neutralization, virus neutralization and opsonization of bacteria. In other cases, the interaction of antibodies and antigens in turn activate complements. In this case, the function of the antibodies is to confer specificity on the immune response, and the actual effectors of the response are the complement proteins, which induce lethal injury to the membrane of the invading organism. A few examples of these mechanisms are discussed below.

Humoral immunity by antibodies alone is of primary importance in preventing diseases caused by toxin-producing bacteria. Toxins are major virulence factors in many bacterial pathogens and are either secreted (exotoxins) or derived by bacterial autolysis (endotoxins). Exotoxins are among the most toxic substances known. A disease-causing exotoxigenic bacteria, for example, *Clostridium perfringens (C. perfringens)*, is widely distributed in the soil and gastrointestinal (GI) tract of animals and is characterized by its ability to produce potent toxins that are responsible for enterotoxemia. The major toxins implicated in this infection have been identified and these are highly immunogenic. When treated with formaldehyde, these toxins lose their toxicity but retain their antigenicity. The inactivated toxins called toxoids stimulate protective antitoxin antibody production after injecting into animals. Vaccines based on toxoids are highly effective in controlling *Clostridal* infection in animals.

Another important virulence factor that determines the establishment of a disease is the ability of bacteria to adhere to surface of cells. The adherence mechanisms are essential for organisms such as enterotoxigenic *Escherichia coli (E. coli)* that attach to the specific receptors on mucosal epithelial cells. Humoral immunity expressed locally by antibodies can block adherence of *E. coli* to epithelial cell receptors preventing colibacillosis in nursing and weaning pigs. Bacterial adherence to epithelial cells is by means of pili or fimbriae which is the first step for successful establishment by *E. coli* on intestinal vili. Non-pathogenic strains do not possess pili and are washed by GI fluids and peristalsis and thus do not cause disease. There are multiple antigenic types of fimbriae of which K88, K99, 987P and F41 are most commonly recognized. Subunit vaccines containing the fimbrial proteins, responsible for attachment of the bacteria to the intestinal mucosal cells, when expressed and immunized to animals will promote antibody production to block bacterial adhesion. This is a novel approach to *E. coli* vaccination and indicates a shift away from the conventional whole bacterial vaccine to a more specific and defined product.

Mucosal immunity mediated by IgA antibodies is of particular importance as most animal pathogens initiate their infectious processes at mucosal surfaces of conjunctiva, nasopharynx, respiratory, gut, and genito-urinary tracts. The antibody response to the pathogens (e.g. viral particles) when directed against surface proteins can be an effective way to neutralize infectivity. This is brought about by binding of the neutralizing antibody to the surface protein responsible for adsorption of the virus to the target cells. A specific example to highlight the importance of mucosal immunity in the protection of piglets is the passive immunity derived from the sow against Transmissible Gastroenteritis Virus (TGEV). The virus destroys villous epithelial cells of the small intestine and causes vomiting, profuse diarrhea, with mortality reaching 100% especially in piglets less than two weeks of age.

Transmissible gastroenteritis is caused by TGEV, a member of the *Coronaviridae* family. The virus is enveloped and pleomorphic and has three major structural proteins: a nucleocapsid protein (N), a trans-membrane glycoprotein (M), and a peplomer glycoprotein (S). The S protein is of importance. It binds to cellular receptors, causes membrane fusion, and induces the production of complement-independent virus neutralization. Total immunity against TGEV is dependent on the infection of the small intestine, which stimulates antibody production. Active immunity against TGEV in the gut relates to stimulation of the secretory IgA (SIgA) immune system, with the production of intestinal SIgA antibodies by lymphoid cells within the lamina propia of the gut. In order to protect piglets, passive immunity from sow thus becomes critical against TGEV infection. However, conventional vaccines against TGEV are usually given parenterally (e.g. intramuscular injection, IM), which make it relatively difficult to achieving effective mucosal immunization. The humoral antibodies produced by IM injection mainly are of the IgG class, which protect the newborn piglet against systemic infection but not intestinal infection. The concentration of IgG antibodies is high in the colostrum but decreases rapidly during the first week of lactation, while SIgA concentration declines marginally and becomes the predominant

Ig in milk. Oral administration of attenuated TGEV vaccine thus offers better protection because infection of the gut elicits SIgA antibodies. These antibodies in the lumen of the intestine of suckling pigs will neutralize ingested TGEV and thus interfere with the virus adherence to the villous epithelial cells of the small intestine in an immunogenic mechanism called the lactogenic immunity.

Lactogenic immunity mediated by mucosal immunization provides greater protection against TGEV than parenteral inoculation because there is a higher level of IgA class in milk. IgA antibodies are more resistant to proteolysis and they selectively bind to gut enterocytes and confer protection. After exposure to the viral antigen in the gut, IgA lymphocytes from the lamina propia migrate to the mammary gland where they localize and secrete IgA antibodies into the colostrum and milk. This concept called the gut-mammary immunologic axis is important in vaccine design.

12.1.2. *Immunization procedures*

A non-immune animal can be artificially made immune to an infectious disease either by passive immunization or active immunization. Briefly, artificial passive immunization is produced by transfer of preformed antibodies from a resistant animal to a susceptible animal, resulting in a temporary protection against the pathogen (Fig. 12.9 and Table 12.5). For instance, antibodies against toxins of *C. perfringens* raised in horses are used to passively protect non-immune piglets, which have been exposed and are consequently at risk. These passively transferred antibodies give temporary immediate protection, but since they are gradually catabolized, the recipient will eventually become susceptible to re-infection.

Artificial active immunization, on the other hand, involves administering antigen in the form of vaccine to the animal so that it induces a protective immune response without causing the disease itself (Fig. 12.9 and Table 12.5). In any case, vaccination is based on the principle of artificial exposure of the body to an immunogenic agent that is harmless but shares antigenic determinants with the intact, fully virulent disease agent. Immunization with vaccines then provokes a primary immunological defense response of the body without causing the disease itself. Hence, exposure to this same antigen later in life will provoke a "ready to

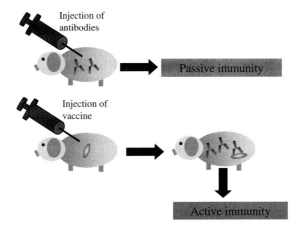

Fig. 12.9. Passive versus active immunity.

Table 12.5. Passive and natural immunity.

Type	Acquired	
	Natural	Artifical
Passive innunity	Maternal antibody	Serum antibody and cells
Active immunity	Infection	Vaccination

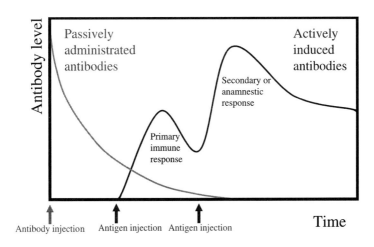

Fig. 12.10. Passsive immunity, primary and secondary immune response.

fight" secondary immune response. This response may be either antibody-mediated or cell-mediated, or both. Primary vaccination doses are usually followed by regular booster doses to maintain a continued immunity to the relevant pathogens (Fig. 12.10). Although protection conferred by active immunization is not immediate, the main advantages of active immunization over passive immunization are the prolonged period of protection achieved and the ability of recalling and boosting of this protective response by repeated injections of antigen or by exposure to infection.

12.1.3. *Vaccines*

An ideal vaccine for today's needs will not only be required to overcome the limitations posed by the classical vaccine, but should also be safer and more efficacious as compared to the classical vaccines. Besides, it should also stimulate an immune response distinguishable from that due to a natural infection, so that immunization and eradication may proceed simultaneously. Such a vaccine is also commonly known as marker vaccines, since it is able to "mark" the vaccinated animal from the naturally infected animal. Vaccination procedures usually require two injections; i.e. the first injection to sensitize the immune system and the second injection to boost the immune response. For the benefit of easy management, the vaccine should be capable of inducing a strong immunity after a single injection to save time, labor and cost for vaccination as well as induce less stress on the animal. Last but not least, it will be ideal if the vaccine can be administrated orally. This is especially useful for the vaccination of fishes as conventional vaccination procedures by injection can be quite a challenge.

12.1.3.1. *Conventional vaccines*

The market for veterinary biologicals in the US and Europe was approximately US$2.4 billion in 1998, an estimate that does not include large markets in Asia, such as China and India. Markets in the US and Europe are quite mature and future growth will most likely to be outside these regions. Animal vaccines are responsible for preventing diseases and reducing losses and thus have a major impact on animal industry; both attenuated and inactivated conventional vaccines have contributed enormously to animal health worldwide. Unfortunately, two prerequisites of an ideal vaccine, namely high antigenicity and absence of adverse side effects, tend to be mutually incompatible for two types of vaccines, namely attenuated and inactivated. Although attenuated vaccines stimulate a better immune response, it may present shedding hazards of residual and reverted virulence. On the other hand, inactivated vaccines may be safer but they are usually poorer immunogens. Table 12.6 gives a comparison of the primary characteristics of these two types of vaccines.

(a) Live or attenuated vaccines
Attenuated vaccines are usually formulated from alternative or mutant strains of the pathogenic organism that display reduced or no virulence (attenuated) in the host, while maintaining their ability of transient growth and antigenic integrity to induce an immune response in the vaccinated host.

Attenuated mutations can be introduced by adapting pathogens for prolonged periods under unusual conditions, which results in the pathogen losing its specificity for, and thus reducing its virulence in its target

Table 12.6. Limitation of classical vaccines.

Vaccine	Advantages	Disadvantages
Live Vaccines	Stimulate good immune responses	Risk of residual virulence
	Can be administrated in low doses	Shedding of live organisms
	Require no adjuvants	Risk of reversion to virulency
	Cheap to produce	Unstable and require constant refrigeration
Inactivated Vaccines	No risk of residual virulence	Low immunogenic
	Stable and easy to store	Need for adjuvants
		Requires administration of high multiple doses

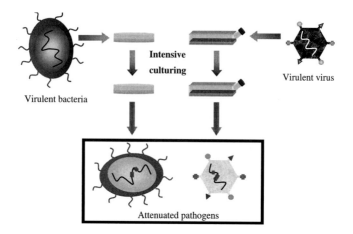

Fig. 12.11. Attenuation of virulent bacteria and virus.

animal species or tissue (Fig. 12.11). Bacteria can be attenuated by culture under abnormal conditions. For instance, the attenuated strain of anthrax currently used in vaccines was rendered avirulent by growing in 50% serum agar in a CO_2 rich atmosphere, resulting in its inability to form a capsule essential for infection.

Microbial pathogens are highly adaptable microorganisms that express complex and multifactorial virulence phenotypes. The factors required by a pathogen to cause disease are termed determinants of virulence. These determinants can be involved in toxicity, colonization, invasion and avoidance of host defenses. In order to design prevention strategies, it is important to identify virulence determinant genes and be able to differentiate them from normal house keeping genes. In order to identify the proteins that are involved in interaction with the host cells, it becomes necessary to mutate the gene coding for that protein or to change the conformation of that protein such that interaction with the host is affected. This is quite possible with the use of transposons. A number of bacterial pathogens like *Salmonella typhimurium*, *Vibrio cholerae* and *Edwardsiella tarda* have been effectively mutated using transposons.

In the case of *V. cholerae*, transposon mutagenesis was used to create a mutant with an altered viable but non-culturable response. Mutants defective in swarming motility as a result of transposon mutagenesis were also found to be defective in the synthesis of lipopolysaccharides that are crucial genetic components involved in virulence. Other mutants deficient in an outer membrane protein (OmpC) were found to be defective in adhesion to macrophages. Macrophages recognize, adhere to, and phagocytose *Salmonella typhimurium*. The major outer membrane protein OmpC is a candidate ligand for macrophage recognition. Transposon mutagenesis was used to develop an OmpC deficient mutant, which was compared along with the wild type for macrophage adherence and association assays. Radiolabeled wild type *S. typhimurium* bound to macrophages at levels five-fold higher than did the ompC mutant. The deduced OmpC amino acid sequence of *S. typhimurium* shares 77% and 98% identity with OmpC amino acid sequence of *E. coli* and *S. typhimurium*, respectively. Evidence from this study supports a role for the OmpC protein in initial recognition by macrophages and distinguishes regions of this protein that potentially participate in host-cell recognition of bacteria by phagocytic cells. With extensive experiments carried out to validate the efficacy of using transposon mutagenesis, the significance of the work lies in the extreme handiness of this tool. Transposons are highly mobile genetic elements that have unique inverted repeats at their ends. Apart from carrying an antibiotic resistance marker, transposons also bear a transposase gene that helps in the insertion of this transposon into the chromosome in the right frame. Mutations can be induced when transposons or insertion sequences are integrated into the bacterial chromosomes. These newly inserted pieces of DNA can cause profound changes in the gene into which they insert and perhaps even in the adjacent gene. These mutations may or may not have obvious defective phenotypes, which can be effectively screened using various assays.

Tn*phoA* is a transposon which, upon correct insertion into an open reading frame of a protein containing the N-terminal export leader sequence, will produce a fusion protein with bacterial alkaline phosphatase (PhoA). This can be detected by using a PhoA-specific chromogenic detection reagent (5-bromo-4-chloro-3-indolyl-phosphate) leading to the formation of a blue bacterial colony. The sequences flanking the Tn*phoA* can be amplified by PCR, cloned and sequenced to enable further genetic studies. Tn*phoA* is designed as a probe to detect genes whose products are exported across the cell membrane, as PhoA is only active when in the

Fig. 12.12. Tn*phoA* is a Tn5 derived transposon that can be used to identify mutants that lack major secreted virulence factors. Upon insertion into the host genome, PCR can be used to amplify the gene that has been interrupted due to transposon insertion.

periplasm and hence very useful for identifying exported gene products and protein regions that are periplasmic. A common factor of many virulence determinants is that they are exported across the bacterial cytoplasmic membrane to interact with the host. Thus, Tn*phoA* is an excellent tool for the identification of genes whose products are exported and therefore most likely to be putative virulence determinants (Fig. 12.12).

Edwardsiella tarda is a gram-negative bacterium that causes a systemic infection, edwardsiellosis, in fish. This pathogen also affects reptiles and is also responsible for stray incidences of gastroenteritis in man. Though some of the virulence factors of *E. tarda* have been elucidated like motility, adhesion, invasion and cytotoxicity, systematic studies regarding the pathogenicity and the genetic factors responsible for the virulence are lacking. Certain virulence factors of *E. tarda* were identified using Tn*phoA* mutagenesis. Mutants deficient in siderophore production, catalase secretion, motility and serum resistance were identified using transposon mutagenesis. Moreover, some of these mutants were also found to be effective as vaccine strains. There is sufficient evidence to show that transposon mutagenesis is a convenient tool to obtain mutants, which are deficient in major virulence factors, as well as to identify virulence genes and to study gene and protein functions.

Viruses, on the other hand, can be attenuated by growing in a non target cell culture line or species to which they are not naturally adapted (Fig. 12.11). For instance, canine distemper virus rendered avirulent by growing in birds was shown to be safe and immunogenic in gray foxes, bush dogs, maned wolves and fennec foxes. The flurry strain of rabies that underwent 178 passages in eggs lost its virulence for normal dogs and cats. In the case of the fowl influenza virus that normally infects chickens, attenuation was achieved by culturing in pigeon eggs. Nonetheless, the most commonly used method of virus attenuation is by prolonged culturing in cell types to which it is not adapted. For instance, canine distemper virus that normally infects lymphoid cells is attenuated by passage extensively in canine kidney cells.

Although attenuated vaccines are still commonly used for control of animal diseases today, they pose some risks of continued circulation at sub-clinical levels and as a result, their applications are not permitted in certain countries where eradication of the disease is being attempted.

Recombinant technology, which will be described later, permits irreversible attenuation of pathogen by selectively removing genes that are necessary for virulence.

(b) Killed or inactivated vaccines

Another common vaccine technology for control of animal diseases is to inactivate the pathogen so that it is no longer capable of replication in the host. It is important that the inactivated pathogen be antigenically similar to the living organisms to induce immune response similar to natural infection. For this reason, inactivation methods should maintain the structure of epitopes and surface antigen. Hence, crude methods of killing like heating are usually unsatisfactory because it causes extensive protein denaturation and some functional epitopes of may possibly be altered.

Knowledge of the molecular structure of antigens and the level of infectivity of pathogen is essential in determining the method for inactivation of organism. Chemical inactivation with various alkylating agents like ethylene oxide, ethylenimine, acetyl-ethylenimine and beta-propiolactone are among the most common choices in veterinary vaccines. These alkylating agents act by cross-linking nucleic acids chains while leaving the surface protein antigens intact, and hence do not interfere with antigenicity of the organism. Formaldehyde, on the other hand, acts on amino and amide groups of proteins and on non-hydrogen-bonded amino groups in bases of nucleic acids to form cross-links and confer structural rigidity. Although reduced antigenicity has always been a concern for inactivated vaccines, some of these vaccines confer immunity comparable to that induced by attenuated vaccines at their best. This is further enhanced by the use of special adjuvant formulations due to their synergistic activity. However, human carelessness in the production of chemically inactivated vaccines has raised safety concerns. The vaccine against Foot and Mouth Disease virus, though efficiently protects livestock from the disease, is linked to outbreak due to the release of virus from vaccine production plants, or the presence of residual live virus.

12.1.3.2. Recombinant vaccines

In simple terms, recombination is a process in which genetic material is broken and fragments of the genetic material are rejoined to give a new combination. In recent years, the application of recombination technologies resulted in the development of novel vaccines that overcame the limitations of classical vaccines. Besides addressing the safety issue and other limitations in relation to conventional vaccines, considerable improvements have already been achieved in certain fields, and some new approaches are widely commercialized. It has long been known that in controlling the spread and eradication of any disease, it is important to develop procedures that distinguish vaccinated from convalescent animals. This is because convalescent animals can become carriers of the pathogen, which consequently can become the source of new outbreaks, and possibly introduce antigenic variants into the population. A recombinant vaccine is able to stimulate an immune response distinguishable from that due to a natural infection, and hence is also known as a marker vaccine. This is something which a conventional vaccine is incapable of. Therefore, the use of a marker vaccine allows vaccination and eradication to proceed simultaneously. Indeed, the ability to differentiate between animals vaccinated with recombinant vaccines and those infected with natural strains of pathogen has already played a key role in eradication of certain diseases, like pseudorabies due to the development of companion diagnostic tests.

The following are types of recombinant vaccines that are either available for use or are under development against different animal diseases and have shown promising results. These are subunit, vector, gene-deleted and DNA vaccines. Each has different characteristics that allow it to be used for different diseases and situations (Fig. 12.13).

(a) Gene-deleted vaccines

Gene-deleted vaccines are engineered by deleting specific gene sequence, usually coding for its pathogenicity and virulence, from the genetic material of a pathogen, resulting in the gene-deleted organism to be irreversibly attenuated. Furthermore, given that such organism is permanently "marked" from the pathogenic strain, when it is used as a vaccine, antibodies can be distinguished from that induced by the pathogenic strain when tested serologically.

A classical example is the use of gene-deleted vaccine for the eradication of Aujeszky's disease. Aujeszky's disease or pseudorabies as

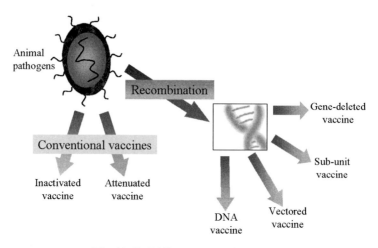

Fig. 12.13. Different types of vaccines.

it is otherwise called is found in most pig producing regions, is a herpes viral infection that affects the central nervous system leading to high mortality in piglets. There is an urgent need for its control in many countries. Vaccination with gene-deleted vaccine against pseudorabies is commonly practiced, and along with companion diagnostic tests, has successfully eradicated the disease in some countries. The pseudorabies vaccine is based on the deletion of the thymidine kinase gene from the virus, leading to its inability to replicate in the nerve cells to cause disease. The major glycoproteins gX, gI, gp50 and gp63 of pathogenic pseudorabies are not essential for either growth or virulence, and the virus appear to function normally when these genes are deleted (Fig. 12.14). Thus, this gene-deleted vaccine not only confers effective protection, but it also blocks cell invasion by virulent pseudorabies virus and prevents development of a persistent carrier state in the vaccinated animal. Besides, this vaccine induces an antibody response distinguishable from that caused by the pathogenic strains, enabling diagnostic differentiation for disease eradication.

(b) Subunit vaccines

Production of large quantities of purified antigen economically has often been difficult. Therefore genetic material encoding an immunogenic

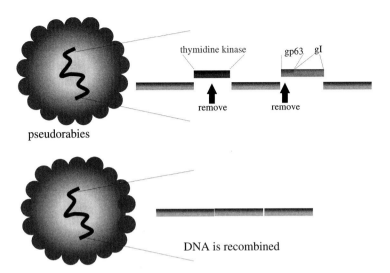

Fig. 12.14. Gene-deleted vaccine.

protein is recombinantly propagated in expression systems for effective extraction and purification of large and pure quantities of protein for use as subunit vaccines. To date, the most commonly used expression systems include the bacterial *E. coli* expression system, yeast expression system and the baculovirus expression vector system (BEVS). *E. coli* is a more preferred system because it is easier to work with and the expression level is high. Subunit vaccines are engineered with genes that code for part of the pathogenic organism, which are then introduced into the expression host system. The host cell manufactures the protein and if all goes well with the clinical trials, this recombinant protein will be available as a subunit vaccine. For instance, in order to prepare subunit vaccine for FMD, the first stage is to identify which portion of the virus can confer immunogenic protection. This genetic material can then be cloned into an expression system to produce the specific antigen for vaccination.

The first commercially available veterinary subunit vaccine was developed against Feline Leukemia Virus (FeLV). The major envelope protein of FeLV, gp70, is the main epitope that induces protective immunity in cats against feline leukemia virus. The gene encoding gp70 is expressed and the protein is produced in huge quantities from recombinant *E. coli*.

Once harvested and purified, the recombinant protein is mixed with a saponin adjuvant and used as vaccine (Fig. 12.15). Another example is the enterotoxin of enteropathogenic *E. coli*, which consists of two subunits, the alpha subunit that is toxic, and the beta subunit that binds the alpha subunit to enteric cells. The purified beta subunit is immunogenic and functions as an effective toxoid. The attachment pili protein of enteropathogenic *E. coli* is currently used as a vaccine to protect animals from infection by preventing bacterial attachment to the intestinal wall.

The recombinant expression system gives the capacity to produce large quantities of various bacterial or viral proteins that may be used as subunit vaccines. Vaccination with subunit vaccine produces antibodies only to the protein delivered. However, natural infection by a disease agent elicits responses to a variety of proteins. Therefore, the subunit vaccine itself should carry a marker, which distinguishes vaccinated animals from naturally infected animals (Fig. 12.16).

(c) Vectored vaccines

Genes encoding the protective antigens can be recombinantly inserted into an attenuated viral or bacterial "vector" that acts as a vaccine delivery system for the antigen into the animal body. The vectored vaccine is host-restricted so that it will only replicate itself within the vaccinated animal. Since this vectored vaccine can replicate in the host cells they overcome

Fig. 12.15. Subunit-FeLV vaccine.

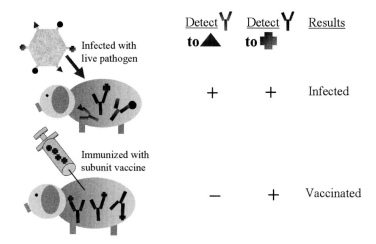

Fig. 12.16. Marker vaccines allow distinguishing vaccinated animals from naturally infected animals by serological methods.

the limited immunogenicity associated with sub-unit vaccines. Vectored vaccines possess many of the characteristics required for an "ideal" vaccine. These include being free of adverse side effects, high levels of stability, and its suitability for mass vaccination program. In addition, such vaccines exclude the use of adjuvants, and allow differentiation between vaccinated and naturally infected animals. Early data collected on these vaccines indicate strong immunity, no side effects, and no shedding into the environment when used as a "vector" to distribute antigen into the habitat of wildlife species.

Vaccinia virus, derived from the naturally occurring cowpox strain, is a large complex virus containing a genome of about 200 genes. It can carry several dozen foreign genes without impairing its capacity to infect host cells and replicate. However since vaccinia virus is not species-specific, concerns regarding the risk of subsequent expression in non-target animal were questioned. This problem can be overcome with the species-specific pox virus, in which vectored vaccine with a stable recombinant fowlpox virus had been made available recently, offering protection against Newcastle Disease (ND) and Infectious Bronchitis (IB). The canary pox vector protects dogs from canine distemper virus, and with canary pox vector containing a gene encoding rabies glycoprotein that offers protection for dogs and cats against rabies. Other vectors under investigation include the adenovirus vector systems, which have lower cloning capacity than pox viruses, but are generally stable, enabling ease of handling and induction of mucosal immunity of both respiratory and intestinal surfaces when administered orally. This makes them ideal candidates for the development of mucosal vectored vaccines.

To engineer a vectored vaccine against ND virus, the fusion gene and the hemagglutinin gene from ND virus are incorporated into a vaccine strain of fowlpox. The immune system of the host reacts against the recombinant pox virus, and the proteins encoded by the inserted genes of ND virus, inducing protection against both diseases (Fig. 12.17). In the recombinant vectored rabies vaccine, the gene encoding for the rabies glycoprotein was inserted into an attenuated vaccinia virus. Since this virus is not species-specific, cells of many different wildlife species can take up this vaccine non-selectively for protection against rabies. However, there are concerns about the used of vectored vaccines in immune deficiency animals, which even an attenuated vaccine can be potentially fatal.

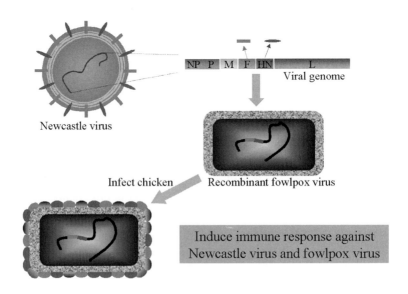

Fig. 12.17. Engineering a vectored vaccine against ND virus.

(d) DNA vaccines

Reports in 1990 that naked DNA encoding viral genes could induce protective immunity when delivered into the host came as a surprise to most scientists. In just a few years, the expanding number of examples where recombinant plasmid DNA induces immune response against the respective pathogens has developed from an interesting observation to a highly promising approach for the development of vaccine technology. DNA vaccination refer to the direct inoculation of eukaryotic expression vector encoding the antigenic protein(s) into the animal resulting in an *in situ* production of the encoded antigen within the tissue cells of the host and hence inducing immunity in the animal without disease. Commercially available mammalian expression vectors such as pcDNA 3.1 (Invitrogen, San Diego, CA) has been extensively used for DNA vaccine studies (Fig. 12.18). Over the last few years, DNA vaccines have been demonstrated to induce both humoral and cellular immunity against many different disease agents, and to be effective in a variety of bacterial, viral, and parasitic animal models. The use of pure plasmid DNA offers a number of potential advantages over the other commonly used methods of producing vaccines (such as genetic modification of pathogens, production of recombinant subunit proteins or attenuation of live organisms). These include simpler production, quality control, avoidance of contaminating pathogens and viability without cold storage during transport, as DNA material is stable over a wide temperature range. Other advantages included the ease of construction of recombinant DNA molecules expressing foreign genes, elimination of risk associated with use of live agents and problems associated with pre-existing immunity to the agent or vector, as well as low cost (Fig. 12.19). However, the antibody titer induced by this DNA vaccine is comparatively low to that induced by conventional vaccines and repeated injections at higher dosages are required to obtain the necessary levels of immunity.

An alternative strategy to boost the titer of antibodies induced by DNA vaccines is to deliver pathogen specific antibodies termed "intracellular antibodies" into the host instead of delivering pathogen-specific antibodies. The concept of intracellular antibodies is based on expression of an antibody fragment inside the cell that can bind to and inactivate a pathogen (Fig. 12.20). The intracellular antibodies delivered by DNA plasmids have shown significant potential against Human Immunodeficiency Virus (HIV) and a fish pathogenic rhabdovirus causing Viral Hemorrhagic Septicemia (VHS). To accomplish this task, a hybridoma cell line that produces a neutralizing monoclonal antibody was first raised against a neutralizing

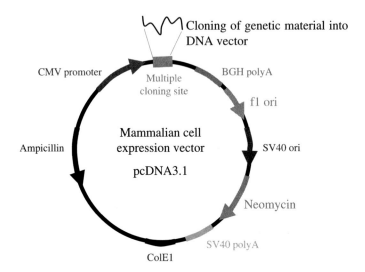

Fig. 12.18. A commonly used mammalian cell expression vector.

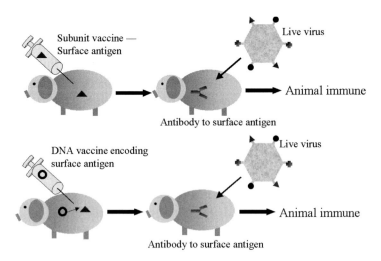

Fig. 12.19. Subunit vaccine and DNA vaccine.

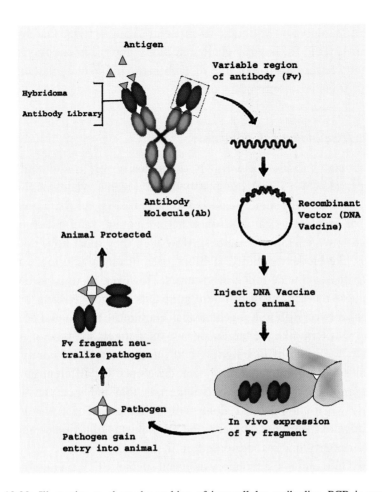

Fig. 12.20. Illustration to show the making of intracellular antibodies. PCR is used to amplify the variable region of the immunoglobulin chains (Fv fragment) from the hybridomas or antibody libraries. The sequences cloned into the mammalia expression vector allow *in vivo* expression of the Fv fragment. A pathogen gaining entry into the immunized animal will be neutralized by the expressed Fv fragment, thereby protecting the animal from infection.

epitope on a pathogenic organism. The variable domain genes of the monoclonal antibody were then cloned into a mammalian expression vector, which was then injected into the host. In an experiment with VHS virus, administration of DNA vaccine encoding the pathogen specific antibody demonstrated that intramuscularly injected fish produced high levels of recombinant antibodies with neutralizing activity that protected immunized fish from viral challenge. This approach circumvents the potentially hazardous use of live recombinant viral vectors that are often used for transfer of antibody genes.

12.1.4. *Immunological adjuvants*

Recombinant vaccines offer many advantages over traditionally used vaccines. However, most are poorly immunogenic when administered alone, especially for subunit and DNA vaccines. Therefore, a great need exists for immunological adjuvants that are potent, safe and compatible with the new-generation vaccines. Moreover, they must have additional features like biodegradability, stability, ease of manufacturing, cost, and applicability to a wide range of vaccines. However, there is no universal adjuvant as the actions of different adjuvants are not yet clear and they probably rely on different mechanisms. They must be adapted according to several criteria like the target species, the antigen, the type of immune response, the route of inoculation, and the duration of immunity. Thus, different types of adjuvants have been developed for different purposes.

Immunological adjuvants are materials that enhance the immune response when administered along with an antigen. These encompass a wide range of materials. Although the mechanisms of action of adjuvants are often poorly understood, based on their mode of action, they can be classified into immunostimulatory adjuvant such as BCG, muramyl peptide (MPP), saponin etc. and depot adjuvants such as aluminium salts, mineral oil etc. Although many adjuvants have been administered experimentally, few have been accepted for use in domestic animals. In following section, emphasis will be made on depot adjuvants (Table 12.7).

Depot adjuvants function as a slow-release deposit for antigen at the injection site into the body and thereby prolong the immune response. The aluminum based salts such as aluminium hydroxide, aluminium phosphate and potassium aluminium sulphate (alum) have been widely used in animal vaccination. In this technique, the antigen is precipitated with the salt before inoculation. Upon injection, the salt/antigen mixture forms a small deposit in the tissue that slowly releases the antigen into the body and provides a prolonged antigenic stimulus. Antigens that normally persist

Table 12.7. Adjuvants currently used in veterinary vaccines (from SEPPIC, Paris, France, with permission).

	Advantages	Disadvantages
Emulsions		

Water in oil	Long term immunity
	Stable
	Cost effective
		Local reactions with reactive
		    antigens (crude extracts)
		Safety is improved with
		    metabolizable oils

Oil in water	Short term immune
	    response (outbreak)
	Well tolerated
	    Cost effective
		Possible Pyrogenic effect

Water in oil in water	Short and long term immune responses Fluid, well tolerated	Less stable at 37°C
Mineral salts		
Al(OH)3	Well tolerated Cost effective	Poor efficacy with purified Ag No Th1 cell response No CTL response
Plant extracts		
Saponin extracts	CMI	Variability from batch to batch
Others		
Bacterial extracts	Induces cellular immunity	High cost Local reactions
Liposomes		Large scale production High cost

for only a few days may be retained in the body for several weeks by this technique. In addition, the adjuvants may influence their ability to cause inflammation, thereby intensifying the general reaction to antigen exposure. A combination of the above factors probably account for their activity. Aluminum salts emulsion has a good safety record, but is a weak adjuvant for antibody induction by protein subunits and cell-mediated immunity. This problem can be overcome by the use of oil-based adjuvants.

Fig. 12.21. Oil adjuvant (from SEPPIC, Paris, France, with permission).

Adjuvants based on highly refined pure mineral oil will render water in oil (w/o), oil in water (o/w) and water in oil in water (w/o/w) emulsions (Fig. 12.21). Varying percentages of emulsifying agents like Tween (polyoxyethylene sorbitan monoleate) or eumulgin and non-ionic emulsifying agents are added depending on the viscosity of final product required. Field trials conducted with FMD vaccines in swine using oil-adjuvants indicated their superiority when compared to vaccines using aluminium hydroxide adjuvants where the antigen was increased three to ten times. In the control of major poultry diseases since 1965, oil-adjuvant based vaccines have proved to be effective and hence are widely used. Adverse reactions to oil-adjuvants include pro-inflammatory reactions at the site of injection, sterile abscesses, cysts and granulomas but most of these reactions can be overcome by the use of highly refined standard ingredients.

In recent years, natural mediators of the immune systems such as cytokines were studied for use in new adjuvant technology. Most cytokines have the ability to modify and redirect the immune response. IL-12 performs a critical role in modulating the T_H1/T_H2 balance and in providing protective immunity in the host defense system against bacteria, viruses and intracellular protozoa. Co-administration of vaccine with exogenous IL-12 as an adjuvant has resulted in the induction of protective immunity against *Leishmania*, *Listeriosis* and *Schistosoma*. In the case of immunization of PRRSV killed vaccine, the vaccine alone was insufficient to induce a specific immune response against PRRSV. However, co-administration of the killed vaccine with porcine recombinant IL-12 induced a strong humoral and cell-mediated immune response. Other examples of adjuvants, such as recombinant carp interleukin-1β, when used as an immunoadjuvant for vaccination of fish, elicited a significantly higher antibody titer at three weeks of post-vaccination than the formalin-killed pathogen alone. These results reveal the potential application of cytokine in the role of immunoadjuvant in fish vaccination. Molecules that have been evaluated most extensively as adjuvants include IL-1, IL-2, IFN-gamma, IL-12 and granulocyte-macrophage colony stimulating factor (GM-CSF). It may be expected that these natural mediators will become successors of the empirically developed adjuvant.

12.1.5. *Vaccine delivery systems*

The simplest and most widely practiced method of delivering vaccines to livestock is primarily by intramuscular or subcutaneous injection. Although this conventional approach is excellent for relatively small number of animals and for diseases in which systemic immunity is important, it usually requires multiple doses, is time consuming and labor intensive and is stressful to the animal. Improved delivery systems require only a single injection, subsequently reducing the cost of vaccination and labor. Moreover, vaccination during early stages of life when the animal is young is easier to handle and thus reduces the stress level in the animal.

The mucosal system links the lungs, upper respiratory tract, gastro-intestinal tract, the urogenital tract, and the eyes. As more than 90% of animal pathogens gain access to their host cells through these surfaces, the ability to provide mucosal immunity offers enhanced disease protection. Furthermore, these responses may be circulated into the bloodstream to other tissues. For example, intranasal vaccines are effective in protecting cattle against systemic Infectious Bovine Rhinotracheitis (IBR) and poultry against Infectious Bronchitis (IB) and Newcastle Disease (ND). It has been shown that immunization of the gut or bronchial mucosa generates disseminated as well as local responses. For example, intranasal immunization of chicken with IBV vaccine induced a secretory antibody response in the GI tract and at other mucosal sites. Unfortunately, immunization techniques by the oral or intranasal route require handling of each individual animal. This induces the stress on the animal and also compounds labor and cost problems. Aerosolization of vaccines enables them to be inhaled by all the animals in a herd, group, or flock — an obvious advantage when the unit is large. Alternatively, the vaccine may be administered in feed or drinking water, which reduces some of the problems associated with the other methods. Edible or oral vaccines are particularly useful especially in the case of fish farms where herd vaccination can be very difficult and it would be ideal if the antigen may be added to the water in which they live.

Edible vaccines provide a viable strategy for the immunization of animals, as it facilitates compliance in the animals, and the process is less laborious and time consuming in comparison to the conventional method of injection immunization. This is especially evident for the case of mass vaccination of fishes where the conventional method is neither practical nor cost effective.

The use of transgenic plants as vaccine production systems has been researched extensively, and there have been some initial successful studies on edible vaccines delivered by transgenic plants. Through the delivery of antigens via genetically modified plant material, cheap and simple vaccination of livestock species may be possible in the future. There is also increased safety in terms of production, as plants do not generally serve as hosts for animal pathogens. Edible vaccines can be produced in plants in many ways. Antigens from infectious bacterial or viral diseases have been introduced into plants through plant virus-mediated infection or Agrobacterium tumefaciens-mediated stable transformation methods. Oral immunization with transgenic plant tissues that contain vaccine antigen proteins stimulates both systemic and mucosal immune responses in animals. Recent documentation in edible vaccine includes the expression of the *E. coli* heat labile enterotoxin (LT) in transgenic potato leaf and TGEV large spike peplomer glycoprotein in transgenic potato tuber.

12.1.6. *Vaccine microencapsulation systems — microsphere*

Development of single dose vaccines by microencapsulation of vaccines in biocompatible and biodegradable polymers has been investigated recently. It has been evident that recombinant vaccines encapsulated in biocompatible and biodegradable polymers, known as microsphere, are released in a continuous or pulsate manner with reliable and reproducible kinetics, concomitantly stimulating the immune system and obviating the need for booster immunizations in animal models. Results have shown that a single dose of antigen loaded microspheres induce antibody responses equivalent to those obtained with multiple doses of conventional vaccines. Microsphere vaccines have been shown to be effective in stimulating both humoral and cell-mediated immune response. To date, among the several polymers tested, co-polymers of glycolic and lactic acids, poly(DL-lactide-co-glycolide) (PLG), have yielded the most promising results with extensive records of safe use in human and veterinary medicine for drug-release. This polymer degrades by non-enzymic hydrolysis to yield the normal body metabolites lactic and glycolic acids that are easily resorbed or eliminated, and the rate of degradation can vary from a few days to a year. PLG can be processed as micropheres in a range of sizes that are optimal for uptake by macrophages or other antigen presenting cells (Fig. 12.22).

microspheres for
intramuscular use
(80 μm)

microspheres for
oral use
(5 μm)

Fig. 12.22. Microsphere delivery systems.

Orally ingested proteins are inefficiently taken up by the immune inductive tissues in the gastrointestinal tract, largely because they are degraded by both acid hydrolysis and proteolysis, and are rapidly excreted. Attempts to overcome these problems by increasing the amount of ingested protein, or the frequency of dosage, rapidly leads to the state of oral tolerance, i.e. unresponsiveness to vaccine. Particles size of less than 10 μm are known to be taken up by specialized epithelial cells in the GI tract, M cells, which transport the particles directly to the specialized immune inductive tissues, the Peyer's patches. Microspheres have also shown potential as carriers for oral vaccine delivery due to their protective effects on encapsulated antigens and their ability to be taken up by the M cells of Peyer's patches in the intestine. Following their uptake into Peyer's patches, a proportion of microparticles disseminate within macrophages to mesenteric lymph nodes and the spleen, where the vaccines are released in a controlled and sustained manner continues, stimulating the systemic immune response. Thus the oral delivery of microencapsulated vaccine can simultaneously induce mucosal and systemic immunity.

Although microsphere encapsulated vaccines have not been used in animal vaccination commercially, many studies have indicated promising

Fig. 12.23. Comparison of antibody response between conventional vaccination versus microsphere vaccination.

results in both human and animal medicine. Studies on PLG encapsulated formalin-inactivated *Salmonella enteritidis* demonstrated over six months of high serum anti IgG in chickens, in contrast to the short-lived response in those immunized with oil-emulsified *S. enteritidis* (Fig. 12.23). Both mucosal and systemic immune responses were detected very early after vaccination when this vaccine formula was immunized orally. Besides, both humoral and cell-mediated immune responses were demonstrated from microencapsulated vaccines such as HIV-1 gp120 to mucosal surfaces, and a single oral dose of PLG-encapsulated pertussis fimbriae protects against *Bordetella pertussis*-induced disease following intra nasal challenge. Thus, development of new microencapsulated delivery systems has made killed and subunit vaccines into more practical options for parenteral and mucosal immunization.

12.2. Diagnosis of Animal Pathogens

Diagnosis of animal pathogens has been one of the important methods recommended for the control of animal diseases. For example, as the demand for poultry products increases, the issue of public health and food safety assumes greater priority. *S. enteritidis* recorded outbreaks have occurred with frequency world-wide, including in the UK, Europe, and the United States. Whilst such occurrences and studies have been reported in Thailand, which has a large poultry export industry, information on outbreaks in Asia remain limited in terms of acquired information. *S. enteritidis* is accountable for a large percentage of food poisoning cases diagnosed, with some infections resulting in the death of young and old. Food poisoning cases such as these are attributed to the consumption of undercooked and infected poultry meat as well as contaminated eggs (Fig. 12.24).

Diagnosis and surveillance of infected flock have thus taken important roles in the management to reduce the likelihood that contaminated meat and eggs will reach consumers. Current techniques employed in the identification of *S. enteritidis* include bacteriological isolation and agglutination assays. In *S. enteritidis* infection, the pathogen colonizes the intestine of the chicken and causes brief mild diarrhea. The fecal material shed from the chicken is thus a means of detecting the pathogen. However,

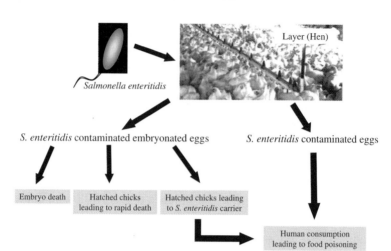

Fig. 12.24. Food poisoning caused by consumption of poultry meat and eggs contaminated with *S. enteritidis*.

some *S. enteritidis* strains are able to persist in the intestinal tract for a long period. This results in intermittent shedding of fecal material. This makes the conventional bacteriological isolation technique of *S. enteritidis* from environmental and clinical samples less reliable in identifying all infected flock. Moreover, the method requires lengthy procedures. On the other hand, the widely used *S. pullorum* slide-agglutination test has been adapted for use with *S. enteritidis*. However, the agglutination-based system has been proven less sensitive. It is advantageous to develop a more specific and purified antigen preparation for the detection of antibodies against *S. enteritidis* infection. Recombinant proteins of high purity are now accepted as reagents for serodiagnosis of *S. enteritidis*. Assays based on recombinant proteins provide a simple, highly sensitive, non-infectious and inexpensive tool for antibody detection. Compared to the conventional bacteriological examination and agglutination assay, this newly developed recombinant protein assay based on ELISA (which will be described in later sections) makes it a valuable tool for the reliable detection of *S. enteritidis* infection.

In another example, Ovine Progressive Pneumonia (OPP), also known as maedi virus or Ovine Lentivirus (OLV) pneumonia, is caused by a non-oncogenic, exogenous retrovirus of the *Lentiviridae* subfamily. The disease is characterized by increasing respiratory distress, chronic weight loss and wasting, and the inevitable death of infected sheep. The disease has been recognized in most of the sheep-producing countries and is endemic in the United States. OPP virus (OPPV) carriers are usually identified by detecting anti-OPPV antibodies in serum by the agar gel immunodiffusion (AGID) test. However this test employs a viral antigen that is produced by an inefficient and time-consuming tissue culture system. Therefore, the AGID test is expensive and insensitive for routine clinical laboratory use.

Second-generation of recombinant protein assays and synthetic oligopeptide assays have been developed to circumvent these problems. Properly selected and prepared recombinant and synthetic peptides used in solid-phase ELISA formats will increase the density of diagnostically relevant epitopes compared with viral lysates, resulting in increased test sensitivity. High compositional purity decreases the likelihood of nonspecific (false-positive) reactions. Furthermore, well-defined, uniform, and reproducible recombinant or synthetic peptides have distinct quality control advantages over virus culture-derived proteins when used as antigens in serodiagnostic assay. The requirement to maintain the virus itself is also eliminated by recombinant or synthetic antigen production.

While the initial development of recombinant proteins is a bio-technologically intensive and time-consuming process, it is a one-time investment of resources and effort. After that, it is possible to produce and purify sufficient recombinant antigen in a week to test several thousand serum specimens. Moreover, the growth and technical requirements for bacterial culture of *Escherichia coli* to produce recombinant proteins are simpler and less expensive than those for tissue culture of viruses to produce conventional antigens. Thus, given the inherently faster and simpler *in vitro* culture of bacteria than that of viruses, recombinant proteins offer the potential for reduced cost per animal tested. Similarly, identification of diagnostically relevant synthetic peptide epitopes is a one-time resource investment. Thus, the overall conclusion is that second-generation recombinant proteins offer the best combination of economy and test performance compared to conventional tests.

There are generally four types of detection for animal diseases, namely direct isolation of pathogens, detection of pathogens, detection of antibodies and antigens from pathogen-infected animals, and detection of DNA or RNA sequences from pathogens. Many immunological methods have been developed to detect for antibodies or antigens. Among these methods, the Enzyme-linked immunosorbent assay (ELISA), which detects for antibody or antigen is most commonly used for animal disease diagnostics. Foot and mouth disease of cattle is most effectively diagnosed by the presence of antibodies against the virus. Molecular biological methods have used to detect for pathogen nucleic acid. For example, the polymerase chain reaction (PCR), which detects for nucleic acid is gradually gaining popularity among clinical laboratories. The technique has been able to diagnose shrimp population for the presence of the White Spot Syndrome Virus (WSSV) by the amplification of a portion of a WSSV genome.

Progress in understanding the molecular biology of animal pathogens and the role of individual proteins in infection and immunity has been slow. In terms of viral pathogens, the low level of viral protein production in tissue culture systems and the difficulty in purification of the virus are probable reasons. As for bacterial pathogens, the probable reason lies in that batch production of antigens from crude extraction of bacteria requires specific growth conditions, and the purified antigens from the cultures are difficult to standardize. These factors make it difficult to obtain adequate quantities of pure antigens for experimentation.

Over the past several years, advances in diagnostic methods based on immunology and molecular biology technology are rapidly invading into laboratories to replace microbiology and virology assay methods, as the choice for diagnosis of animal diseases. Immunological and molecular biological assays greatly improved the diagnosis of many animal infections, especially diseases that cannot be practically detected in most clinical settings due to the difficulties in culturing and isolating of the pathogens. Besides, the main appeal of these methods over microbiology and virology assay systems is the speed that it can offer, in which identification of pathogens can be completed in a very short time, usually less than a few hours in certain cases. Aside from that, these methods are easy to perform, safe, reproducible and some are automated to facilitate the evaluation of large numbers of specimens.

12.2.1. *Direct isolation of animal pathogens*

Early diagnostic applications of animal diseases involved isolation of pathogens followed by various methods for identification and differentiation, in which clinical laboratories are generally able to isolate and determine the most routinely encountered pathogens. To date, direct isolation of pathogens remains the most accurate method of diagnosis for animal infection, with a few exceptions. The system used for pathogen isolation, which can either be *in vitro* or *in vivo*, depends on the nature of the organism being sought. Generally, *in vitro* methods are preferred for the obvious reason of convenience. Briefly, microbiological identification of bacterial diseases included isolation of the pathogens from clinical specimens by microbial culture, followed by identification of organisms, for instance, by growing on selective and differential media. On the other hand, diagnosis of viral diseases included isolation of pathogens from clinical specimens by tissue culture systems and examination by electron microscopy. When the identity of the pathogen is uncertain, multiple systems are used for identification. Several animal diseases are still preferably being detected by this method.

Direct isolation of pathogens, however is extremely laborious, time-consuming and not practical for large-scale diagnosis. Besides, pathogens like certain viruses, bacteria and protozoa present diagnostic problems since these organisms are difficult to culture *in vitro*, and efforts to isolate the pathogens are enormous. For instance, the *in vitro* isolation of shrimp pathogens like the White Spot Syndrome Virus (WSSV) is difficult due to the lack of cell lines compatible with the shrimp system. This has been a major impediment to the molecular biology of WSSV and the role of individual proteins in infection. Although many viruses have distinctive morphologies, and diagnostic virology can be done by electron microscope in major specialized laboratories with the appropriate equipment, it is not a confirmatory method and additional experiments need to be performed to confirm the identity of the virus. Also, for organisms like the mycobacterium species, *in vitro* culture systems exhibit extreme slow growth rate (positive cultures are obtained after 2 to 4 months). These concerns have been an impetus for the development of diagnostic methods that can accurately identify the pathogen without involving culturing work at all. Immunological diagnostic methods (such as ELISA) are developed for direct detection of pathogen related antigens in the animal and antibody responses induced against the pathogen by the animal. In addition, recent advances in the development of molecular biological technologies (such as PCR and specific nucleic acid probes) have been applied to the detection of pathogen genomes with sensitivity that is comparable to that achieved with detection by direct isolation of pathogen. Needless to say, immunological and molecular assays can function as a surrogate for detection of the infectious agent itself.

12.2.2. *Detection of pathogens*

For the detection of the pathogen, a classical case study would be that of the Bovine Spongiform Encephalopathy (BSE). The disease is believed to be caused by infectious agents called prions that have not been isolated as yet. A prion refers to a proteinaceous infectious particle, which is composed only of protein with no detectable nucleic acid. This is unlike all previously known pathogens like bacteria and viruses, which contain nucleic acids to enable them to reproduce.

Prion diseases, caused by a 33–35 kDa glycoprotein prion protein (PrP), are sometimes referred as transmissible spongiform encephalopathies (TSE), as the brains of infected animals were riddled with holes. Other forms of prion diseases include scrapie and Creutzfeldt-Jakob Disease (CJD). In mammals, PrP exists as a normal cellular protein in a stable shape. The cellular function of the PrP protein is not known, but it has been speculated that PrP involves in Cu^{2+} transport and Cu^{2+} homeostasis. The protein is easily degraded by cellular proteases and hence does not cause disease (cellular PrP). However, a variant form of PrP (scrapie PrP) exists, which is resistant to protease degradation. This form of PrP is believed to cause TSE. Further studies revealed that the difference between a cellular PrP and a scrapie PrP lies in the conformation. A scrapie protein contacts normal PrP proteins and somehow causes them to unfold and flip from their usual conformation to the abnormal shape of the scrapie protein.

This initiates a cascade, in which the newly converted molecules change the shape of other normal PrP molecules. In this way, the scrapie protein propagates itself. Thus prion diseases arise from aberrations of protein conformation.

It was also discovered that prions prefer to interact with PrP of homologous or similar compositions. Therefore, if the amino acid sequence of a variant PrP molecule resembles the PrP molecules in that of a host, it is likely that the host will acquire prion disease. Sheep PrP and bovine PrP share much homology in their amino acid sequence. This probably explains why cattle in UK contracted the prion disease when they were fed with sheep tissue-tainted feed.

The development of a diagnostic test for BSE test is particularly difficult because the specific infectious agent has not been isolated and there is no detectable immune response, which are the two prerequisites for developing diagnostic tests. Another problem which hinders the diagnosis of BSE is that it takes at least two and half years for abnormal PrP to accumulate, and eventually cause enough brain damage for the symptoms to be visible. Under such circumstances, there would be little point in testing the brain tissue of slaughtered cattle at abattoirs at between 18 and 24 months of age, because there would be no detectable infectivity even if they are infected. The test would give a false negative result and provide false reassurance. The current methods of diagnosis are histopathology and immunocytochemistry. The latter is adopted if histopathology does not give a definitive diagnosis. Hence, immunocytochemical tools using polyclonal or monoclonal antibodies against the prions are employed as an effective diagnosis for the presence of the pathogen.

12.2.3. Detection of antigens and antibodies from pathogen-infected animals

Detection of antibody response induced against pathogen by the animal is based on the fundament of antibody-antigen interactions. The interaction between antibody and antigen is highly specific, where each antibody molecule is able to detect for one molecule of the corresponding antigen out of more than 10^8 similar molecules. This profound specificity of antigen-antibody interactions has led to the development of a variety of assays for animal disease detection. These diagnostic methods differ in speed, sensitivity, and application, with some solely for detection of diseases, while others being able to detect as well as quantitatively determine the antigens or antibodies detected in the animals. Some of these diagnostic methods can only be used to detect for either antigens or antibodies, while others can detect antigens as well as antibodies at any one time, based on the presented format of the assay. The more commonly applied assays for animal disease diagnostics including immunodiffusion, immunofluorescence, and ELISA will be discussed.

12.2.3.1. *The nature of antigen-antibody interaction*

The interaction between an antibody and an antigen involves noncovalent interactions between the epitope of the antigen molecule, and the variable (V_H/V_L) domain of the antibody molecule, particularly the hypervariable complementarily-determining regions (CDRs). These noncovalent interactions that form the basis of antigen-antibody binding include hydrogen bonds, ionic bonds, hydrophobic interactions, and van der Waals interactions. Each of these noncovalent interactions generally occurs in close proximity, less than 1×10^{-7} mm in most cases, resulting in a very close fit between the antibody and antigen, which reflect the high degree of specificity characteristic of antibody-antigen interactions. Furthermore, in order to form a strong bond, large number of nonconvalent interactions is involved in any one antibody-antigen binding, since the strength of each noncovalent interaction is weak (Fig. 12.3).

Although antigen-antibody reactions are highly specific, an antibody elicited by one antigen can cross-react with an unrelated antigen in some cases. Such cross-reactions occur if two different antigens have shared epitope(s) or if antibodies specific for one epitope also bind to an unrelated epitope possessing similar chemical properties. For such reasons, detection for specific antigens is sometimes performed with monoclonal antibodies instead for polyclonal antibodies, while the detection of specific antibodies is done with an oligopeptide coding for the epitope of the pathogen, or a recombinant expressed polypeptide, which is derived from the pathogen. Generally, the level of specificity required for disease detection depends largely on the pathogen of interest.

Several viruses and bacteria possess antigenic determinants identical to or similar to the normal host-cell components. Indeed in some cases, these microbial antigens have been shown to elicit antibody that cross-reacts with host-cell components, resulting in a tissue-damaging autoimmune reaction. For instance, the bacterium *Streptococcus pyogenes*, expresses cell-wall proteins, known as M antigens. Antibodies produced against these antigens have been shown to cross-react with the glycoproteins of heart valves, antigens of joints and soluble components of glomerular basement membrane. This immunologic reaction resulting from the cross reaction has been implicated in rheumatic fever and acute glomerulonephritis following streptococcal infections.

12.2.3.2. *Assays based on immunodiffusion*

The purpose of an immunodiffusion technique is to detect the reaction of an antigen with an antibody by a precipitin reaction. Precipitin reaction is based on the formation of visible precipitate in solution from the antibody-antigen lattice, which involves the interaction of antibody-antigen complexes. Formation of an antibody-antigen lattice depends on the number of specific binding site on both antibody and antigen, in which the antibody must be bivalent, and the antigen must have multiple epitopes. When the ratio of antigens to antibodies is in optimal proportion (equivalent zone), extensive cross-linking formation occurs. Thus, formation of lines of precipitin in any immunodiffusion system is highly dependent on relative concentrations of the antigen and antibody. In the region of excess antibodies (prozone phenomenon), each antigen molecule is covered with antibodies, preventing cross-linkage and thus there is no precipitin reaction. In the region of antigen excess, each antibody molecule is bound to a pair of antigen molecules. Further cross-linking is impossible and thus no precipitin reaction can occur (Fig. 12.25).

Many diagnostic methods have been developed based on the precipitin reaction for qualitative and quantitative detection of either antibodies or antigens. The most basic format of precipitin reaction based assay is the agar gel immunodiffusion (AGID) and radial immunodiffusion. AGID is a qualitative assay, while radial immunodiffusion is a quantitative assay. The application of radial immunodiffusion is based on the principle that a quantitative relationship exists between the amount of antigen placed in a well that is cut in an agar antibody plate and the resulting ring of precipitin. In this method, antibodies are incorporated into an agar gel and precisely measured amounts of antigens are added into the wells. Following diffusion of the antigen into the agar containing antibodies, a ring of precipitin will form at the equivalent zone if the incorporated antibodies are specific for the antigens. The diameter of the ring formed represents the antigen concentration determined by comparing with antigen standards (Fig. 12.26). One major drawback of immunodiffusion assays is that at least 18–24 hours are required before precipitin lines appear. This limitation is overcome by using countercurrent electrophoresis, as seen in assays like electroimmunodiffusion, which exhibit greater speed and sensitivity than

Amount of antibody precipitation

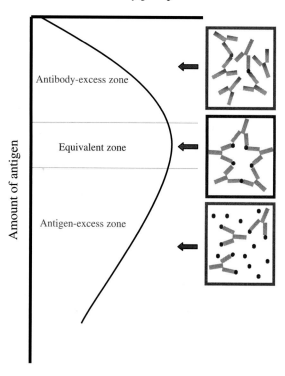

Fig. 12.25. A quantitative precipitation curve showing the effect of adding increased amounts of antigen to a constant amount of antibody.

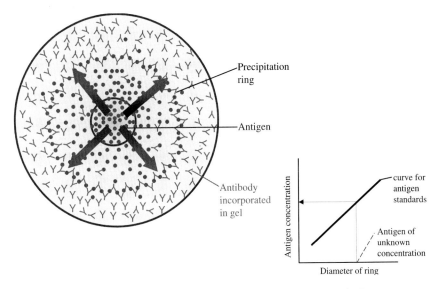

Fig. 12.26. Precipitin reaction — radial immunodiffusion, a quantitative assay.

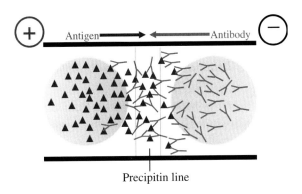

Precipitin line

Fig. 12.27. Precipitin reaction — electroimmunodiffusion. Migration of antigen and antibody towards each other in an electrical field results in the formation of a precipitin line.

in immunodiffusion assays. In the electroimmunodiffusion assay, positively charged antibody and negatively charged antigen are first added into separate wells in the gels. When directed by an electrical field, the antigen and antibody migrate towards one another, resulting in a sharp precipitin line within minutes (Fig. 12.27). Although these assays are seldom used for the detection of animal diseases today, a few pathogens like the causative agent of Rinderpest, Equine Infectious Anemia, Ovine Lentivirus, Bluetongue, Lumpy Skin Disease and Sheep Pox still rely on this assay as one of their diagnostic methods.

12.2.3.3. *Enzyme-linked immunosorbent assay*

Enzyme-linked immunosorbent assay, commonly known as ELISA, is the most popular diagnostic method among the existing immunological assays for animal disease detection, favored for its simplicity and sensitivity. In principle, ELISA employs an enzyme label for detection of antibody-antigen complexes formed on a solid phase. Alkaline phosphatase and horseradish peroxidase are the most commonly employed enzymes, which can be chemically coupled to antibodies under conditions that retain the biological properties of both components of the conjugate. Detection of antibody-antigen complexes is based on the enzyme catalytic activity of an appropriate colorless substrate to give a colored product, whose intensity is measured by the optical density. A variety of ELISA formats have been developed for detection and quantification of either antigen or antibody. The following are the most commonly used formats for the detection of animal diseases.

Indirect ELISA diagnoses for animal diseases through the detection and quantification of serum antibodies specific for pathogen related antigens. Serum or samples containing antibodies to be tested, referred as primary antibody, is added to an antigen-coated microtiter well to react with the bound antigens. If the antigen is being sought by the test sample, specific antibodies will be attached to the well, with unbound antibodies being washed away. Following this antibody-antigen interaction, secondary enzyme-conjugated antibody (e.g. horseradish peroxidase) specific for the primary antibody is added for reaction. After unbound reporter antibodies are washed away, the respective substrate for the enzyme is added, and the converted colored product is measured by specialized spectrophotometric plate readers, which can measure the absorbance of a plate of 96 samples in less than a minute (Fig. 12.28). The indirect ELISA has been the method

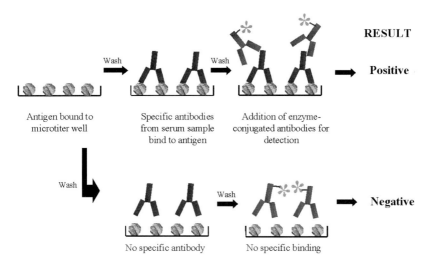

Fig. 12.28. Procedures for indirect ELISA.

of choice to detect for serum antibodies against many animal pathogens, including Porcine reproductive and respiratory syndrome virus (PRRSV), the causative agent of the porcine reproductive and respiratory syndrome which causes major losses in the pig industry. In this assay, either the whole virus or recombinant expressed capsid protein (ORF 7) of PRRSV is adsorbed as solid-phase antigen to microtiter wells. When an animal is infected with PRRSV, it may provoke the production of specific antibodies to epitopes on the whole virus, resulting a positive signal upon measurement.

A sandwich ELISA allows detection and quantification of pathogen related antigen. In this case, the antibody specific for the antigen is coated to the surface of microtiter wells to which the antigen containing test sample is added. After unbound antigens are washed away, the antibody-antigen complexes are detected by an enzyme-linked antibody, which is specific for a different epitope on the antigen. Next, unbound reporter antibodies are washed away, the substrate is added and the colored reaction product is measured (Fig. 12.29). The sandwich ELISA has been the method of choice for the detection of many animal pathogens such as Avian Leukosis Virus (ALV), Classical Swine Fever (CSF) and Feline Leukemia Virus (FeLV) (Fig. 12.30).

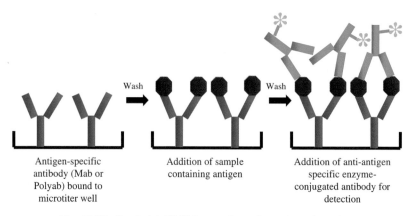

| Antigen-specific antibody (Mab or Polyab) bound to microtiter well | Addition of sample containing antigen | Addition of anti-antigen specific enzyme-conjugated antibody for detection |

Fig. 12.29. Sandwich ELISA procedures for antigen detection.

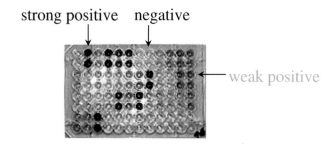

Fig. 12.30. Color development in an ELISA reaction correlates to the amount of antigen or antibody in the sample. A reaction that gives a colored product of strong intensity indicates a positive sample, while a reaction that gives no colored product indicates a negative sample. A colored product of intermediate intensity indicates a weak positive sample.

12.2.3.4. *Immunoblotting assay*

Immunoblotting assay, or Western blotting (WB), is a diagnostic method to identify either a pathogen-related antigen in a complex mixture of proteins, or a antibody specific for a pathogen related antigen of precise molecular weight. As its name implies, the principle of WB is similar to Southern blotting, which detects for DNA fragments, and Northern blotting which for detects mRNAs. In WB, proteins, following separation by electrophoresis under denaturing conditions, are transferred or blotted onto a solid support, usually a nitrocellulose or nylon membrane prior to being detected by either enzyme-labeled or radioisotope-labeled antibodies. The antigen-antibody complexes are then visualized either by a substrate conversion signal or autoradiography, depending on the reporter antibody used (Fig. 12.31). Similar to ELISA, indirect immunoblotting can be performed to detect antigen-antibody complexes by a secondary labeled anti-isotype antibody with improved sensitivity. Many different reporter systems are available for detecting antigen-antibody complexes in immunoblotting, with enzymatic-catalyzed methods being widely preferred over autoradiography. Enzymatic detection systems included chemiluminescence detection that provides signal that is captured by film darkening, and chromogen detection where enzyme-labeled reagents is used to catalyze a chromogenic substrate to give a color product, generating a visual signal. One drawback of the WB assay is that it maybe inconvenient to screen large numbers of samples for epidemiologic studies. However, WB is a sensitive and specific alternative to the ELISA. If an ELISA result disagrees with the known immunologic status of an animal, then WB or other serologic tests of equal or higher sensitivity should be used to validate the ELISA result. Pathogens like Human Immunodeficiency Virus (HIV), the presumptive diagnosis of the infection is made by the detection of antibodies in patients by ELISA.

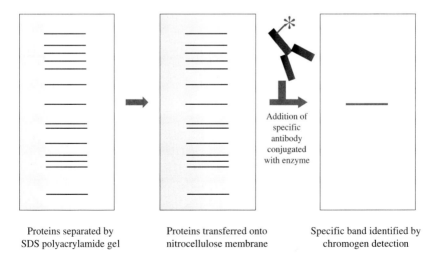

| Proteins separated by SDS polyacrylamide gel | Proteins transferred onto nitrocellulose membrane | Addition of specific antibody conjugated with enzyme | Specific band identified by chromogen detection |

Fig. 12.31. Western blotting procedure.

However, as there are some false-positives with this test, the definitive diagnosis is made by WB.

Mycoplasma hyopneumoniae is the etiological agent of enzootic pneumonia of swine, a chronic non-fatal disease resulting in poor food conversion, retarded growth and predisposition to bacterial pneumonia. The disease is spread worldwide and causes large economic losses in swine production. Transmission occurs through direct contact with respiratory secretions from carrier animals and via sow-to-piglet transmission.

Methods developed for the detection of *M. hyopneumoniae* have been based mainly on whole cell or membrane preparations and gene fragments. These include culture of *M. hyopneumoniae* from pig lungs, immunofluorescence assay, DNA probes, ELISA and PCR. Most of these methods lack specificity due to cross-reactions with the closely related swine mycoplasmas, *M. flocculare* and *M. hyorhini*. In addition, these methods have limited sensitivity and involve sophisticated laboratory procedures. Proteins that are predominantly antigenic in pigs infected with *M. hyopneumoniae* are important targets for the development of specific diagnostic tools. Research has shown that detection for *M. hyopneumoniae* specific antibodies by western blotting technique using either the L-lactate dehydrogenase or the ABC transporter ATP-binding protein analogue/ multidrug resistance protein homologue, PR2 of *M. hyopneumoniae*

Fig. 12.32. Western blotting detection of *Mycoplasma hyopneumoniae* infection in pigs using ABC transporter ATP binding protein. The brown band indicates that the animals have been infected. No band indicates that the animals are not infected. The blue bands on the left belong to the protein marker which are used to estimate the size of the protein.

were highly specific for *M. hyopneumoniae* (Fig. 12.32). These two proteins were chosen as potential diagnostic antigens based on their strong antigenicity and their roles in virulence of *M. hyopneumoniae*.

12.2.3.5. *Immunofluorescence assay*

Immunofluorescence assay is a technique where pathogen related antigens in tissue sections or cells are identified and localized by antibodies labeled with a fluorochrome (Fig. 12.33). Fluorescein isothiocyanate and rhodamine isothiocyanate are the most commonly employed flurochrome, which can be conjugated to an antibody molecule under conditions that retain the biological properties of both components of the conjugate. Each of these

Cells viewed under regular light.

Cells viewed under fluorescent light. The cells show fluorescent staining, indicating a positive sample.

Cells viewed under fluorescent light. The cells do not show any fluorescent staining, indicating a negative sample.

Fig. 12.33. Immunofluorescence assay.

dyes can be excited by a unique wavelength to emit a longer wavelength that can be viewed with the fluorescence microscope. Fluorescein absorbs blue light (490 nm) and emits an intense yellow-green fluorescence (517 nm), while rhodamine absorbs in the yellow-green range (515 nm) and emits a deep red fluorescence (546 nm).

Like ELISA, antigens in tissue section or cells reacted with primary unlabeled antibody can be detected with secondary fluorescein-labeled anti-isotype antibody by the indirect immunofluorescence method for increased sensitivity. Another reagent used for indirect immunofluorescence is the fluorochrome-labeled protein A of *Staphylococcus aureus*, which has high affinity for the Fc region of IgG, and can be used if the binding complex contains IgG antibody (Fig. 12.34).

Although not often used as the main assay, immunofluorescence has been applied as a confirmatory detection for several animal pathogens, including African Swine Fever virus, Pox Virus Disease of cattle and Equine Piroplasmosis. African Swine Fever (ASF) is an infectious disease of domestic pigs caused by a virus that produce a range of syndromes varying from acute to chronic. ASF cannot be differentiated from classical

Fig. 12.34. Immunofluorescence assay using fluorescent antibodies for antigen detection.

swine fever by either clinical or post-mortem examination and differential diagnosis between the two is essential. Direct immunofluorescence has been used to detect ASF antigen in tissues of suspected pigs in the field or those inoculated at the laboratory, and indirect immunofluorescence has been used to detect for ASF-specific antibodies.

12.2.3.6. *Quartz crystal microbalance*

Microgravimetric quartz crystal microbalance (QCM) is a measuring device that is based on the most frequently used acoustic wave sensor, the Piezoelectric quartz crystal. This combination of piezoelectric quartz crystal devices with biological active elements has brought about a new generation of immunoassay. This technique consists of a quartz crystal wafer sandwiched between two metal electrodes (Fig. 12.35). The electrodes provide means of connecting the device to an external frequency oscillator circuit that drives the quartz crystal at its resonant frequency. This frequency is dependent on the mass of the crystal and hence any changes in the mass on the surface of the electrode can be detected as a digital reading. For the detection of animal infection, antigen-antibody interaction results an alteration to the crystal mass, thereby causes a shift in frequency. Based on the frequency information generated from the immuno-interaction, QCM devices enable detection and quantification of antigen or antibody.

Most diagnostic methods for animal diseases are conducted in a laboratory environment and hence require the sample to be delivered prior

Fig. 12.35. Quartz crystal microbalance — the principle and instruments involved.

testing. Some animal diseases like *Salmonella enteritidis (S. enteritidis)* and Porcine Reproductive and Respiratory Syndrome Virus (PRRSV) require regular screening at the various stages of the growing cycles. Therefore, the QCM technique provides an on-site analysis that can be performed routinely with relatively low overheads. Like many immunoassay techniques, alternative assay designs and formats, for instance, direct assay, indirect assay and sandwiched assay, have been developed with the aim to increase detection sensitivity, specificity and the range of QCM applications. The veterinary applications of QCM developed included the Africa swine fever virus detection in pigs, adult worm antigen of *S. japonicum* detection in rabbits, and PRRSV infection in pigs.

   S. enteritidis is a major problem for the poultry industry and has been one of the dominant pathogens that causes food poisoning in humans. The QCM technique developed for the diagnosis of *S. enteritidis* infection in chicken were immobilized with recombinantly expressed *S. enteritidis* flagellin antigens that served to capture the pathogen-specific antibodies in infected sera. Binding of the specific antibodies to the immobilized antigens causes changes to the crystal mass, and thus resulting a shift in

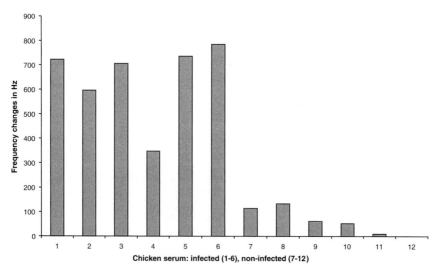

Fig. 12.36. Chart depicting the frequency changes in Hertz for infected and non-infected chicken serum. Lanes 1–6: serums from infected chickens; Lanes 7–12: serums obtained from non-infected chickens.

frequency, which is detectable by the oscillator. This procedure was used in a screening of chicken sera, noninfected and infected with *S. enteritidis*, and was found that the serological readings obtained from infected chicken and those from non-infected chickens could be distinguished (Fig. 12.36). Thus, this technology could provide another method of the screening of *S. enteritidis* infected flocks.

12.2.4. *Detection by pathogen nucleic acid*

The progress made in nucleic acid technology for diagnostics is enormous in relation to the period of time during which it has been applied to the development of assay systems as compared to that during which phenotypic *in vitro* analysis had developed. Indeed, nucleic acid technology has now assumed an essential role in various areas of *in vitro* diagnosis, including animal diseases. Basically, this approach employs genotypic materials (DNA and RNA) as diagnostic targets for identifying suspected pathogens. The rapid advancement of nucleic acid diagnostic is a consequent of two essential advantages: (1) nucleic acids can be rapidly and sensitively measured, and (2) the sequence of nucleotides in a given gene is highly specific and it can be used to distinguish closely related serotypes. Besides, nucleic acid detection offers many advantages over immunological assays. Nucleic acids are more stable than proteins to high temperatures, high pH, organic solvents, and other chemicals, hence clinical samples can be treated in a relatively harsh manner without destroying the nucleic acid for detection. Tools use for nucleic acid detection application like DNA primers and probes are more defined entities than antibodies and antigens, and the composition of these tools can be accurately checked by sequence analysis, and produced in DNA synthesizers whenever necessary. The more commonly applied nucleic acid assay technology in animal disease diagnostics include polymerase chain reaction and nucleic acid probes, and will be discussed in the following sections.

12.2.4.1. *Polymerase chain reaction*

The polymerase chain reaction (PCR) diagnostic method detects for specific nucleic acid sequence found in the genome of pathogens. It is based on DNA replication *in vitro* to produce large quantities of a target sequence of pathogen isolated from a complex mixture of heterogeneous DNA molecules (e.g. host genome, nucleic acids from other pathogens present). The reaction is set up using the following reagents: reaction buffer, DNA template, deoxynucleotides (dNTPs), primers, and thermostable DNA polymerase. A pair of primers is used for the reaction. The promers are single-stranded DNA molecules of about 20–30 nucleotides long (oligonucleotides), are designed from the sequences flanking the two ends of the target genome. These primers are specific to the pathogen to maximize the specificity of the assay. The amplification of DNA by the PCR is accomplished through cycling a succession of incubation steps of different temperatures optimized for DNA replication. Briefly, a sample containing mixture of DNA molecules is heat-denatured to separate complementary strands of DNA into single-stranded molecules for reaction. Next, the reaction is brought down to a lower temperature to allow pathogen-specific primers to search and anneal to the complementary sequence on opposite strands of the target DNA. This is followed by DNA extension at a higher temperature, with DNA polymerase adding on dNTPs using the annealed DNA sequence as a template to produce copies of the targeted sequence. The three steps make up one cycling reaction, with each cycle producing duplicates of the DNA targets. The cycle of steps is then repeated, in which the newly synthesized DNA strands can also serve as templates for primer extension. These steps are repeated 20–40 times, yielding exponential amplification of the target DNA sequences (Fig. 12.37). The amplified products can then be visualized on a DNA gel, or further assayed by nucleic acid probing for increased sensitivity and specificity. The technique offers a powerful approach to the diagnosis of infectious diseases that are defined at the nucleotide sequence level.

For RNA viruses, a complementary DNA (cDNA) copy of the RNA genome must be made with reverse transcriptase (RT) prior to the performing of PCR. One such example is the detection of nodavirus in marine fish. The virus is gaining more importance due to the high mortality it causes in larvae and juveniles of marine fish. In addition, it is wide spread in water bodies including the Pacific, Mediterranean and Atlantic oceans. In general, the earlier the commencement of the disease, the greater is the rate of mortality, which may even reach 100%. Therefore, it is

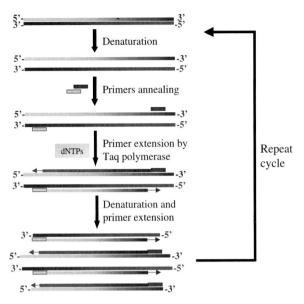

Fig. 12.37. Steps in each cycle of polymerase chain reaction.

essential to develop a specific and sensitive method for the detection of very early stage of nodavirus infection in diagnostic and epidemiological work. A reverse transcription polymerase chain reaction (RT-PCR) method to detect nodavirus based on specific primer sets has been developed and found to be more sensitive than histopathological and immunological methods.

PCR offers a major advantage of detecting pathogens in a single cell or viral particle. This saves the hassle of isolating the pathogens in culture. PCR is a highly sensitive procedure and may prove to be very useful in the diagnosis of chronic-persistent infections such as those caused by retroviruses. This disease presents serious problems in terms of diagnosis and prevention, as the infected animal often does not demonstrate clinical signs until it is at an advanced stage of disease. Furthermore, infected animals appear to be a constant potential source of transmission. However, due to the exquisite sensitivity of PCR, false-positives misidentified from contamination of samples can be possible. To overcome such a problem, a second set of primers that are complementary to the internal sequence of the target DNA can be used to amplify a subfragment from the PCR product of the first reaction. This increases the specificity and sensitivity of the assay, since spurious and nonspecific amplification products are eliminated. This procedure is commonly referred to as "nested PCR". Alternatively, diagnostic methods like nucleic acid probes, restriction fragment length polymorphism, or sequence analysis can also be used to confirm the identity of the PCR product and to further characterize the genome.

12.2.4.2. Nucleic acid probes

The fundament of the specific association of nucleic acid sequences by complementary base pairing through hydrogen bonding, which centralizes on the basis of adenine complementary to thymine and uridine, while guanine complementary to cytosine, has allowed the development of several nucleic acid diagnostic methods for the detection of animal diseases. Nucleic acid probes are short fragments of nucleic acid sequences that detect specific sequences associated with pathogen by nucleic acid hybridization through the use of highly conserved sequences. If pathogen contains sequences complementary to the probe, the two sequences can hybridize to form a double-stranded molecule. The probe can be chemically synthesized according to the target gene sequence, which can be derived if a short amino acid sequence of the protein encoded by the target gene is known. During synthesis of the probe, labeled nucleotides are incorporated, so as to enable detection of the probe after hybridization. The types of labeled probes include radioactive (which gives a radioactive signal), biotin (which generates a color) and chemiluminescent (which is enzyme-linked).

Nucleic acid hybridization is often performed in the form of Southern blotting which detects for DNA and Northern blotting which detects for RNA. As its name implies, the principle is similar to that of Western blotting, which detects for protein. In both Southern blotting and Northern blotting, denatured DNA and RNA samples are separated by electrophoresis and immobilized onto a solid support system, usually nitrocellulose or nylon membrane respectively, prior to detection with radioisotope-labeled or enzyme-labeled sequence-specific probe(s). Hybridization is allowed to occur at an optimized temperature in which considerable sequence homology between the target sequence and the probe is necessary to form a stable duplex. Following washing to remove unhybridized probe, the hybridization is detected according to how the probe was labeled.

Nucleic acid probes can be designed for identifying all serotypes or some important serotypes of a pathogen. However, when used alone probes may not be very sensitive. Although not practiced in the diagnosis of animal diseases, nucleic acid hybridization procedures can be combined with PCR to provide a powerful tool, which exhibit remarkable specificity and sensitivity. It is especially useful when the target pathogen cannot be cultivated, or is difficult to cultivate. On the other hand, nucleic acid probes in diagnostics is designed for use in the food industry, such as for the detection of *Salmonella* and *Staphylococcus* employ probe dipsticks, which remove hybridized DNA from liquid solution. For this method, a two-component probe is used, in which one serves to bind to the specific sequence of the pathogen, and the other serves to bind to the dipstick. Following hybridization of the two-component probe to target sequence, the dipstick is inserted into the hybridization solution to remove the hybridized DNA for measurement.

12.3. Conclusion

Animal diseases continue to affect global animal industries. In 2001, Foot and Mouth Disease (FMD) outbreaks in UK resulted in an estimated loss of 10 billion pounds (about 1.1% of the country's GDP). A better known example is the mad cow disease or Bovine Spongiform Encephalopathy (BSE), which caused great alarm among the consumers. The disease caught attention with worldwide paranoia over the consumption of British beef. Though the spread of the disease hit headlines across the globe, it was the grim tales of lost lives due to human variants of the disease that caught the attention of people. The UK witnessed about 100 cases of the medically relevant form of BSE, called the vCDJ or variant Creudzfeldt Jakob Disease that affects humans.

In Asia, the deadly H5N1 Fowlpox influenza virus killed six people in 1998 and wreaked havoc in Hong Kong's poultry industry. The Nipah virus cost the Malaysian government about one billion US dollars. All pig farms were cleared and a number of countries banned imports of pork from

Malaysia. All these examples illustrated that disease attacks on farm animals do not just simply affect the agricultural sector alone. In some cases, the massive outbreaks had rendered the farm products unfit for human consumption, leading to bans on the imports of these products in many other countries. Considering the vast market for cattle, swine and poultry products, the economic losses were definitely severe.

The reason for such losses in most instances was due to the absence of fast and early diagnosis of disease. If the infected animals were detected before clinical symptoms of disease were visible, action could be taken before the rapid spread of disease had occurred. Early diagnosis and detection is essential to maintain a disease-free and healthy environment for animals reared for human consumption. Reliable diagnostic tests that are rapid, sensitive and affordable should be developed for use by farmers and veterinarians. The veterinary diagnostic test kit market has experienced tremendous growth during the last 10 years. However, diagnostic testing is still not keeping pace with the magnitude of the infectious disease problem in either the efficiency required to economically test large numbers or in the flexibility to accurately test for different diseases in the veterinary market. Therefore, improved testing must include simplicity and economy, which integrates into a reduced reliance in laboratories and skilled technicians. In this respect, emphasis has to be placed on the development of user friendly approaches like those based on recombinant technology. By monitoring which pathogens persist in a herd and when transmission occurs, producers and their veterinarians can make informed decisions about health control of the herd.

Vaccines play a major role for disease control programs because highly intensive livestock farming has created problems associated with disease spreading, which contributes to the increase of new infections in a herd. In the United States, losses of cattle, swine and chicken cost the farmers billions of dollars annually. When talking about such a tremendous loss, one will then realize the importance and necessity of such vaccines that can salvage the lives of livestock. Vaccines that could prevent losses from infectious diseases would have a tremendous impact on animal industries. The new technology opens doors to biologicals that were never available before. Current advancement of recombinant technology has greatly improved vaccinology to meet today's need for an ideal vaccine. These recombinant vaccines include gene-deleted vaccine, DNA vaccine, subunit vaccine, vector vaccine, and so forth. In addition, adjuvant technology, vaccine delivery system and associated testing technology have since progressed drastically to help fulfill the criteria of an ideal vaccine. With these new technologies merging together, the search for the "new-age" vaccinology and companion diagnosis technology, which are essential for today's agricultural needs, should be near. There is a general acceptance that major advances in veterinary vaccine over the next few years will result from such approaches.

When a proper program of vaccination and early disease diagnosis is in place, its impact is tremendous. First, vaccination allows the animals to build up strong lasting immunity against prevalent diseases. With the present advances in vaccine technology, alternative methods of vaccination such as microencapsulated vaccines can be adopted to reduce stress in the animal. Secondly, a good farm management system should incorporate reliable diagnostic tests as a routine to screen the farm population for early signs of diseases. In actual fact, the preventive measure of vaccination, as well as early diagnosis of the disease through rapid testing are crucial in limiting the risk of diseases. Farmers will therefore benefit greatly as their profits are increased. For the consumers, they can enjoy a constant supply of safer animal products and need not fear contracting the disease directly from the products they eat.

Further Readings

1. Aiello SE, ed. *The Merck Veterinary Manual*, 8th ed. Merck and Co, New Jersey, 1998.
2. Calnek BW, Barnes JH, Beard CW, McDougald RL and Saif YM, eds. *Diseases of Poultry*. Iowa State University Press, Ames, 1997.
3. Leman AD, Straw BE, Mengeling WL, D'Allaire S and Taylor DJ, eds. *Diseases of Swine*, 7th ed. Iowa State University Press, Ames, 1992.
4. Lists of A and B: Diseases of Mammals, Birds, and Bees. *Manual of Standards for Diagnostic Tests and Vaccines*, 3rd ed. World Organization for Animal Health, Paris, 1996.

5. Schultz RD, ed. *Veterinary Vaccines and Diagnostics*, Advances in Veterinary Medicine, Vol. 41. Academic Press, San Diego, 1999.
6. Tizard I. *Veterinary Immunology: An Introduction*, 4th ed. W.B. Saunders Company, Philadelphia, 1992.
7. Wesley T. *Veterinary Vaccines: Products and Markets*, Animal Pharm Reports. PJB Publications, UK, 1999.

Related World Wide Web Sites

- Surveillance
  http://www.fao.org/WAICENT/FAOINFO/AGRICULT/AGA/AGAH/EMPRES/Info/other/surveill.him#Early%20Detection
- Animal Diseases
  BSE
  http://w3.aces.uiuc.edu/AnSci/BSE/
  http://sparc.airtime.co.uk/bse//welcome.htm
  FMD
  http://www.thepigsite.com/FeaturedArticle/Default.asp?AREA=FeaturedArticle&Display=305
  Poultry
  http://www.ianr.unl.edu/pubs/AnimalDisease/g1039.htm
- Directory to Veterinary Science
  http://www.mic.ki.se/Diseases/c22.html
  Immunology
  http://www-micro.msb.le.ac.uk/MBChB/ImmGloss.html
  http://golgi.harvard.edu/BioLinks/Immunology.html
  http://www.whfreeman.com/kuby/
  http://esg-www.mit.edu:8001/esgbio/imm/monoclonal.html
  Vaccine
  http://www.sciam.com/1999/0799issue/0799weiner.html
  http://www.sciam.com/2000/0900issue/0900langridge.html
  Recombinant DNA Technology
  http://esg-www.mit.edu:8001/esgbio/rdna/rdnadir.html
  Immunological Tests
  http://www.1rc.arizona.edu/courses/mmiweb/lake/immunology_modules/intro/immunology.html

Questions for Thought

1. How do the humoral-mediated and cell-mediated arms of immune response limit the spread of viral infection in the body?
2. When a chicken is infected by *Salmonella enteriditis*, how does the animal's defense mechanism function? Describe the interactions between B cells, T cells and macrophages that will bring about the immune response, as well as the other molecules are involved.
3. What are the important considerations in developing an effective vaccine for use in animal farms? Compare and contrast the differences between conventional and recombinant vaccines.
4. *Edwardsiella tarda* is a Gram-negative bacterium that causes a systemic infection, edwardsiellosis, in fish. In order to develop suitable vaccines, how can the virulence factors of this organism be identified and studied?
5. Describe how a recombinant vaccine was developed for the eradication of Aujeszky's disease or pseudorabies in pig farms.
6. Explain how large amounts of purified antigen can be produced economically for vaccination.
7. What are the principles behind the use of DNA vaccines?
8. Recombinant vaccines offer many advantages over conventional vaccines. How can they be administered into the animal to ensure their effectiveness?
9. With the use of examples, explain how cytokines may serve as immunoadjuvants.
10. What are the problems faced by fish farms in carrying out mass vaccination? How can they possibly be overcome?
11. What were the problems in diagnosing the White Spot Syndrome Virus (WSSV) in shrimps in the past? What are the alternative methods adopted nowadays?
12. Explain with examples why some methods of detecting pathogen in animals are time-consuming and may not be effective.
13. How can the principle of antibody and antigen interaction be applied to the immunological diagnosis of animal diseases?
14. How do the relative concentrations of antibody and antigen molecules determine the formation of precipitin lines in an immunodiffusion

reaction? How can this be applied to the quantitative detection of antigen or antibody?

15. How is the enzyme-linked immunosorbent assay applied in other formats for the detection and quantification of either antigen or antibody?

16. Describe how the principle of Immunoblotting can serve as a sensitive and specific diagnostic test for *Mycoplasma hyopneumoniae* infection in pigs.

17. Compare and contrast between enzyme-linked immunosorbent assay and Immunofluoresence assay.

18. How can the Quartz crystal microbalance method be used in screening of chicken sera for *Salmonella enteriditis* infection?

19. What advantages does polymerase chain reaction offer in the diagnosis of nodavirus in marine fish?

20. What is the significance of performing early diagnosis of disease and mass vaccination in managing an animal farm?

Chapter 13

Transgenic Plants

Jian-Wei Liu[*]
Institute of Molecular Agrobiology
National University of Singapore
1 Research Link, Singapore 117604
[*]Current Address: Syngenta Singapore Pte Ltd
6 Temasek Boulevard, #09-02 Suntec Tower 4
Singapore 038986

Plant biotechnology has opened a new era of agricultural revolution by introducing genes from different organisms to generate new varieties of plants — genetically modified (GM) plants. Among them, the new food crops (such as grain crops, vegetables and fruits) will have higher yield, better nutritional qualities, resistance against pests (insects, diseases, nematodes, weeds and abiotic stresses), others will help us to produce new industrial raw materials (oil, starches and biodegradable plastics) and pharmaceutical products. Despite the great economical and humanitarian potentials, plant biotechnology has caused great concerns about the safety of the genetically modified (GM) foods, the environmental impact of GM plants, and moral issues. In this chapter, both technical and regulatory aspects of plant biotechnology will be discussed in detail. The future success of plant biotechnology depends on science-based facts, proper governmental regulations and public education efforts.

13.1. Challenges in the Post "Green Revolution" Era

Crop yields increased dramatically in the 1960s and 1970s during the "Green Revolution". This success was mainly based on high yield hybrid rice and wheat varieties, application of agrochemicals (pesticides and artificial fertilizers) and improvement of irrigation technology. In the post-green revolution era, however, we face five upcoming crises facing world agriculture in the future.

Firstly, human population continues to grow at about 1.5% (86 million) a year, and is projected to reach 8 billion by 2020, and 11 billion by 2050. The growth is mainly (90%) in the poorer developing countries of Africa, Asia and Latin America. There are over 800 million people in the world today who are chronically under-nourished and 180 million children who are severely underweight. Secondly, the world's arable land, which comprises only about 3% of the earth's surface, is rapidly deteriorating and decreasing as a result of water- and wind-induced erosion, salinization, acidification, over cultivation, compaction, water logging, over-grazing, etc. at a net loss of 1.4% of total top soil each year. These factors, combined with the expected population increase, will actually decrease the global per capita arable land from the current 0.26 ha to 0.15 ha by the year 2050. In 1961, the figure was 0.44 hectares. Fresh water supplies are also becoming severely limited by increased human and agricultural use, and are being polluted by agricultural runoff and the widespread use of agrochemicals. Thirdly, the high yielding varieties (HYV) of wheat and rice resulting from the "Green Revolution" in the 1960s, which solved the problem of common food shortages of the early and middle years of the 20th century, have reached their production limitation. The average increase in grain yield (maintained at 2.1% until 1990) has since declined to only 0.5% a year and the 1940s agricultural technology is certainly outdated. Demand for better and more varied food products, particularly poultry and meat, require greater supplies of grain feed. Currently, 21% of the world's arable land and 35% of the world's grain is used as livestock feed. Fourthly, despite the heavy use of agrochemicals, 42% of crop productivity is lost to competition from weeds, pests and pathogens. Many pest insects and plant diseases have developed resistance against chemical pesticides. An additional 10–30% postharvest loss occurs due to a variety of factors including bad storage conditions. For fresh agricultural products, such as vegetables, fruits and flowers, the figure is 40–60%.

As a consequence of increasing demand and declining productivity, world food reserves currently stand at 48 days — the lowest level since the beginning of the Green Revolution — from a previous high of 77 days. India and China, the two most populous nations with chronic food shortages, became net food exporters during the Green Revolution. However, they will become net food importers in the future. China and India together are projected to import 260 million tons of food grain annually by 2030. Together with other populous parts of the world, demand for food will double by the year 2025, and nearly triple by 2050. Nobel laureate Norman Borlaug estimates that to meet projected food demands by 2025, average cereal yield must increase by 80% over the average value of 1990. Sufficient food supply is also a major factor for world peace and economic development. In the words of former US President Jimmy Carter, "There can be no peace until people have enough to eat … Hungry people are not peaceful people".

13.2. Importance of Agrobiotechnology

Plant biotechnology will not only help to solve the above mentioned crises and problems in agricultural production to achieve sustainability of the food supply system, but also provide powerful new ways to produce human and animal health care products and industry raw materials. Genetically engineered crops have been on the market for five years, and are rapidly gaining larger market shares every year. With the genomes of many organisms due to be completely sequenced over the next decade, and the concomitant explosion in biological knowledge, it is not surprising that this century has been predicted to be the "century of the life sciences". Agriculture is poised to be a major beneficiary of this new technological revolution.

Plant biotechnology, which has been developing over the past 15 years, together with animal biotechnology, is beginning to have a profound impact on global agriculture. The global sales of transgenic crop products have been estimated to be US$2.6 billion in 2000, with a projected growth of 10% per year for the coming decade. Currently the US, Canada, Brazil,

China and Australia are the major GM crop producing countries. To date, the predominant crops are maize, soybean, cotton and potatoes. In addition, there have been field trials of transgenics from at least 52 species including all the major field crops, vegetables, and several herbaceous and woody species. An increasing area of transgenic crops with higher yields, improved quality and enhanced resistance against herbicide, diseases and pests is being grown in the US and other parts of the world. In 1996, only 9.3 million ha was planted with genetically modifies seeds. In 2000, GM crops reached 44.2 million ha (107 million acres). In the US, around 24 varieties of genetically modified seeds, including maize, soybean, canola, potatoes, cotton, tomato and squash, have already been approved for planting. In 2000, 36% of soybean, 16% of cotton, 11% of canola and 7% of maize have been planted with genetically modified seeds worldwide. From 1986 to 1993, 1025 field trials of transgenic plants were released.

13.3. Plant Transformation

13.3.1. *Plant transformation*

There are several ways to bring in foreign genes into plant cells to generate transgenic plants — transformation (the process through which foreign DNA is introduced into plant cells), such as by using the *Agribacerium* Ti plasmid, by direct uptake of foreign DNA into protoplast, by high-velocity DNA-coated particles, by microinjection of DNA into plant cells, or by electroporation.

Agrobacterium-mediated transformation is the most widely used method for plant transformation. The tumor-inducing (Ti) plasmid originates from a soil phytopathogenic bacterium causing crown gall disease of dicotyledonous higher plants — the *Agrobacterium tumefaciens*. Ti plasmid is circular and double-stranded ring form DNA with the size of 200 kb and has an origin of replication, a 25 kb region named T-DNA, a virulence region for T-DNA transfer into the plant genome and a region for nopaline or octopine utilization. The transferred part, the T-DNA, codes for the production of tumors (uncontrolled plant cell division) and for the synthesis of opines such as nopaline by plant under the control of the plasmid. These opines are used by the bacterium, helped by the opine-utilizing genes on the plasmid.

For plant genetic engineering, parts of the plasmid are deleted, including the tumor-inducing region, and the real engineering is done with an intermediate vector that is inserted into the deleted Ti plasmid to form a cointegrated plasmid. A typical intermediate vector has the insert of interest in a cloning region with a variety of available sites for restriction enzymes, a selectable bacterial gene, such as a spectinomycin resistance gene (spr^R), or a bacterial kanamycin gene (kan^R) engineered for expression in plants, plus a region of the Ti plasmid enabling integration of the intermediate vector into Ti plasmid by homolohous recombination. Spectinimycin is used to select *Agrobacterium* cells, which have Ti plasmid containing the intermediate vector, which cannot replicate on its own in the bacterium. Cointegrate plasmids are being superseded by binary vector systems. These bacteria are then used to infect plant cells of explants (part of plant or tissue used for plant transformation), usually in cut leaf sticks or other tissues. The T-DNA of the Ti plasmid is transferred from the bacterium to exposed wound plant cells, integrating into plan chromosomes at random. Transfected plant cells are selected on medium with antibiotics such as spectinomycin or kanamycin, and callus clumps are produced in the subsequent tissue culture process. They can be induced to form shoots and roots for growing in soil as transgenic plants, i.e. plants with inserted external DNA. The DNA insert then behaves like a normal plant chromosomal gene in mitosis and meiosis, and is usable in conventional breeding.

The inserted genes may be useful in providing resistance to insects, diseases and herbicides, or they may be used to study the action and regions of expression of normal plant genes by enabling their actions to be "visibly" reported, i.e. by the use of a screenable marker — gene encoding a protein that results in a visible product, allowing identification of cells that express the gene, such as GUS gene, LUC and GFP. One system uses the bacterial β-glucuronidase gene (*GUS*), which when active, turns X-Gluc blue, or the bacterial β-galactosidase gene (*lacZ*), which turns X-Gal blue. Alternatively, there is the firefly luciferase enzyme, which produces visible green light from luciferin and ATP, and the green florescence protein (GFP), which produces visible green light under the radiation of a special blue light source. The main advantage of *Agrobacterium*-mediated transformation is single copy insertion of target

Fig. 13.1. *Agrobacterium*-mediated transformation of cotton. (1) Plant tissue explants incubated with *Agrobacterium* containing Ti-plasmid with target gene, kanamycin resistance gene (*NPTII*), and a visual reporter gene (*GFP*); (2) Callus initiation under tissue culture conditions. Only kanamycin-resistant calli can survive under the condition; (3) Callus on tissue culture plates; (4) GFP expression detected in callus indicating foreign gene integration into cotton genome; (5) Embryogenesis started after 6 month tissue culture; (6) Transgenic seedlings obtained from tissue culture; (7) Transgenic cotton seedlings were kanamycin-resistant while wild type untransformed cotton seedlings were kanamycin-sensitive; (8) GFP expression can only be detected in transgenic cotton seedlings, but not in wild type cotton seedlings.

gene. The method is more successful for transformation of dicotyledous higher plants. However, there are more successes in the transformation of monocotyledous crops such as rice, maize, etc. (Fig. 13.1).

Biolistics or particle bombardment is another way to get desired genes into plants. This is done by using a gene gun which fires DNA-coated gold or tungsten particles (about 0.6 to 1.0 μm) at high velocity into plant cells. The transforming DNA may be incorporated into a plant cell's DNA by recombination events, especially if there are regions of homology between the shot DNA and the plant cell. In the gene gun, helium gas is used to make an aerosol of gold particles in the DNA-laden water, and then the aerosol is accelerated rapidly by gas pressure (100–2,200 psi; 7–152 bar) in a Pitot's tube. The main advantage of biolistic transformation is that the method can be used for the transformation of both monocotyledous and dicotyledous higher plant species. The major disadvantages associated with the method is multiple insertion — the insertion of more than one copy of the target gene into a plant cell and gene silencing — the function of an inserted target gene is not shown because of interference in gene expression.

In planta Agrobacterim-mediated transformation of adult *Arabidopsis* plants is a direct transformation process without tissue culture process. In this method, seeds are treated with a suspension of *Agrobacterium* during the early stages of germination, entire plants are vacuum-infiltrated with the bacterial solution, or the flower parts of plants (inflorescences) are dipped in a bacterial suspension containing a surfactant that allowed the penetration of the bacteria into the intercellular space. The plants were grown under natural conditions to produce seeds by self-pollination. On a selective medium with antibiotics or herbicide depending on the selection marker genes used, some of the seeds gave rise to entirely transformed plantlets. This *in planta* method so far has been used on *Arabisopsis*, *Brassica rapa* L., *ssp Chinensis* and *Medicago truncatula*. Scientists in China have developed a technology of plant transformation through pollen tube-mediated approach. Although the detailed mechanism is still a mystery, foreign genes can be delivered into the target plants by this method. So far, cotton, soybean and barley have been successfully transformed by this method and the transformation efficiency ranges from 0.01–0.1%. The most obvious advantage of the technology is that it is free of tissue culture and germplasm-independent.

Chloroplast transformation has the advantage of preventing gene spread through pollen. Using the plastid genome as a site for transgenes has several advantages over the nuclear genome. These include the transfer of the genes only via the female line (for many but not all plant species), thus preventing the movement of the transgene to wild species via pollen; very high levels of expression (up to 20–25% of total cellular protein); and targeted homologous recombination into the plastid genome.

13.3.2. *Selectable marker genes*

Selectable marker genes are introduced alongside the insect-resistance gene to allow the separation of plant cells that have incorporated the new genes from untransformed cells. The majority of insect-resistant transgenic plants produced today — of which sufficient information has been published-contain antibiotic-resistance genes as the selectable marker genes. The most commonly used marker is the bacterial neomycin-phosphotransferase-II gene [nptII or APH(3')II]. A second antibiotic-resistance marker gene, used for rice and soybean transformation, is the hygromycin-phosphotransferase gene (hpt, hph or APHIV). Other selectable markers include herbicide-resistant genes (BAR, EPSPS etc.), which use different herbicides to select the true transgenic plants.

In addition to the antibiotic and herbicide selection markers that enable scientists to keep true transgenic callus and kill the untransformed plant ones on the culture medium, there is a visual selection marker based on chemical staining. After transferring the β-glucuronidase (*GUS*) marker gene into a transgenic plants together with the target gene (such as insect resistance Bt gene) and antibiotic resistance gene, part of the cultured tissue can be taken out for chemical staining in which only transgenic tissue will show the typical dark blue color compared to the yellowish or while color of the untransgenic tissue. The disadvantage of this method is that the issue will be killed after staining. There are two other visual selecting markers that emit green light under special light sources — the luciferase (LUC) from firefly and the green florescence protein (GFP) from marine jellyfish. Compared to GUS gene, their advantage is that scientists are able to observe and select transgenic tissue *in vivo*, that means the tissue does not have to be killed after selection and can be continuously used for tissue culture to transgenic plants.

Recent problems with the approval of transgenic maize in the European Community were because of risks connected with the antibiotic-resistance marker gene, rather than the insect-resistance gene itself. The current trend is therefore either not to use antibiotic-resistance marker genes or to deliver the marker gene in a different locus so that it can later be bred away. Although there have been a range of commercially successful transgenic crops, the technologies used for transformation have been relatively crude. There is a need to improve the efficiency of transformation, to limit the presence of unnecessary genes in the products, to direct insertion to specific sites, and to give more control over when, where, and how much expression of the transgene occurs.

One system uses the ipt (isopentenyltransferase) gene from the Ti plasmid of *Agrobacterium tumefaciens* to increase cytokinin levels, leading to the generation of shoots from transformed plant cells. However, these shoots retain the shooty phenotype and result in sterile plants. Using a construct in which the ipt gene is only active in the presence of a chemical inducer, the transformation and early cultivation of the cultures is carried out in the presence of the inducer. When shoots have formed, the inducer is removed and whole fertile plants can be recovered with high efficiency. The use of chemical induction cassettes in conjunction with the CRE-lox recombination system has enabled scientists to trigger the removal of transferred DNA from transgenic plants. This has considerable potential for crop biosafety through the ability to remove transgenes, which are no longer needed, before the product reaches the fields and the market. Herbicide-tolerance genes are an alternative to antibiotic-resistance markers, and genes coding for 5-enolpyruvylshikimate-3-phosphate synthase (EPSPS) and phosphinothricin-N-acetyl transferase (PAT) provide tolerance to glyphosate and glufosinate-ammonium herbicides, respectively, and have been used as selectable markers in maize. However, there remains a shortage of efficient selectable marker genes, and it is difficult to carry out repeated transformations using the same marker genes to introduce more than one desirable gene into the same plant line.

Recently, the Swiss-based multinational company Syngenta has developed another alternative for plant transformation — a novel marker gene system called Positech® as an efficient alternative to antibiotic resistance markers. Positech® provides a selection system based on a marker gene that enables plant cells to use a simple sugar (mannose) to grow and form new plants. Positech® is an alternative to antibiotic resistance genes as markers in future genetically enhanced crop varieties and will likely be welcomed by regulatory authorities responsible for assessing GM crops worldwide. Antibiotic marker systems do not create risks to farmers or consumers but such systems are often confused with antibiotic use in healthcare. Positech® has already been licensed to more than 100 advanced academic and industry research laboratories around the world that are developing more productive and healthier food crops. Positech® has been put at the license-free disposal, for example, of the agricultural research and development institute in Malaysia and the

International Rice Research Institute (IRRI) in the Philippines. In Malaysia, the system is being used in papaya projects concerned with resistance to virus diseases. At the same time, it is hoped that the stability of the fruit in storage can be improved and that the production of the papaya, which is so important to the local economy, will thereby be given a boost.

13.3.3. *Promoters and gene expression*

To express transgenes in plant cells, appropriate promoter sequences have to be introduced alongside the gene to ensure efficient transcription of mRNA. The first generation of transgenic plants almost all used constitutive promoters like 35S promoter, ubiquitin promoter and actin promoter. The same promoter, CaMV 35S (or derivatives of it), has been used in the majority of insect-resistant transgenic plants. This promoter originated from the cauliflower mosaic virus and, although not completely constitutive, it produces continuous gene expression in most tissues of the plant. However, levels of gene expression have been reported to vary between different species of plant and different parts of the plant. In transgenic cotton, for example, CaMV 35S resulted in gene expression in all tissues except mature petals and pollen. In maize, CaMV lines expressed high levels of toxin in pith and root, moderate levels in the kernel and no detectable toxin in the pollen or anthers. Continuous gene expression in all plant tissues is likely to increase the risk of the pests developing resistance, resulting in yield penalties as the plant directs more resources than necessary to its defense, causing gene silencing, and raising concerns about the safety of the foreign genes inside the edible part GMO crops, such as allergic reactions. Using constitutive promoters like CaMV35S to control the synthesis of antimicrobial or insecticidal products, plant growth and development are occasionally affected. In these cases, the use of chemically inducible promoters may prove helpful. Producers could then spray fields with an inducer that would activate synthesis of these products when pathogen/pest problems are detected.

Currently, the tetracycline-inducible promoter is the best-characterized chemically inducible system. However, various features, including the massive quantities of tetracycline required for gene induction, make it unattractive for agricultural use. Recently, a more promising inducible system based on glucocorticoid-mediated gene activation has been described. A chimeric transcription factor, consisting of the GAL4 DNA-binding domain, the VP16 transactivating domain and the receptor domain of the glucocorticoid receptor, activates gene expression in transgenic plants after treatment with glucocorticoids. A major advantage of this system is that gene expression is rapidly induced by concentrations of glucocorticoids that cause no adverse physiological effects on the plants. However, before such a system is approved for widespread use, the longevity and toxicity of the compounds involved will need to be evaluated on humans and other animals in the food chain, as well as on the environment.

For the second generation of transgenic plants, tissue-specific promoters will be widely used for leaves, roots, stems, fiber, grain and so on to improve efficiency for disease and pest control, to reduce the waste of resources in host plants, to enhance agronomic traits and/or to eliminate foreign gene products in eating parts. An example is the deployment of a phloem-specific promoter for genes providing resistance to phloem-sucking insect pests such as aphids. There is also potential for the use of wound-induced promoters, which lead to gene expression only when the plant is actually attacked. Other specific promoters that have been used with insect-resistance genes include seed-specific and pollen-specific promoters, and promoters especially suited for use in monocotyledonous plants. Fiber-specific promoters have also been isolated from cotton, which will be widely used to increase cotton yield, to improve fiber quality, and to produce naturally colored cotton fibers.

13.3.4. *Source of genes*

As genetic engineering has broken down the species boundary, genes from all living organisms can be used for genetic engineering. Microbial genes used for crop improvement through biotechnology include genes encoding avirulence factors of bacteria and fungi, viral coat proteins of plant viruses, insecticidal toxins from *Bacillus thurengensis*, chitinase from fungi, virulence factors of nematodes and mycoplasma, and herbicide resistance enzymes from bacteria. Map-based cloning, functional genomics and proteomics are powerful ways to identify numerous genes and promoters from human, animals (land animals, birds and water animals), plants and microbe that can all be used for genetic manipulation of plants. For example, the recent completion of complete sequencing of genomes of human, *Arabidopsis*, rice, yeast and several microbes will provide scientists vast information about genes based on DNA sequences. After that, functional genomics based on gene trap and enhancer trap technologies with transposons, microarray and proteomics will provide detailed information and the activities of genes and promoters in the developing process of a particular organism or responses to different environmental impacts. Once the functions of the promoters and genes are clear, scientists will have endless source of genes and promoters to create a lot of tailor made plants for higher productivity, reduced utilization of toxic chemicals, production of more nutritious and more tasty foods, production of raw materials for different industrial and pharmaceutical production as green factory, clean up organic and heavy metal pollution in the environment, and producing food-medicine nutraceuticals to reduce the medical bills and taking care the "grey population".

13.3.5. *Molecular checking of transgenic plants*

There are several ways to help scientists to make sure plants generated from the transformation process are really trangenic ones carrying all the genes designed in the process. The methods can be roughly classified in two classes: molecular checking and functional demonstration. Molecular checking is to use molecular biology techniques to demonstrate that the target genes are integrated into plant genome and properly expressed. These techniques include Southern blot analysis and PCR (polymerase chain reaction) to see if the target gene is inside the plant genome at the DNA; Northern blot analysis and RT-PCT (reverse transcriptase PCR) to see if the gene is expressed at the mRNA level, and Western blot analysis to see if the gene is expressed at the protein level. Functional demonstration is to use the transgenic plant to test if it shows the gained functions through genetic modification, such as insect resistance, disease resistance, herbicide resistance, stress tolerance (for heat, drought, cold and oxidative stress), increased vitamin A content in the rice seeds, or producing pharmaceutical proteins.

13.4. Applications of Transgenic Plants

13.4.1. *Plant disease resistance*

Microbes cause a lot of different diseases in plants. To ensure agricultural harvest, pesticides are heavily used in the field to protect crops from microbial pathogens in modern agriculture. Plant pathogens include viruses, fungi, bacteria, nematodes, mycoplasma and protozoa. Pathogenic microbes use different strategies to get into plants and propagate by using nutrients from plants.

Plants contain a vast arsenal of defenses whose strong and rapid activation after microbial infection is frequently governed directly or indirectly by a gene-for-gene interaction between the products of a plant resistance (R) gene and a pathogen avirulence (avr) gene. If either the plant R gene or the pathogen avr gene is lacking, these defenses are not induced in a timely and effective manner and the pathogen colonizes the plant. This gene-for-gene specificity appears to be mediated, at least in some of the better-characterized systems, such as the tomato — *Pseudomonas syringae* pv. *tomato* system, by a direct interaction between the products of the R gene Pto and the cognate avr gene. Over 20 R genes with recognition-specificity for defined avr genes have been isolated from seven plant species, including both monocots and dicots. These genes are effective against bacterial, viral, and fungal pathogens and in one interesting case, the Mi gene from tomato, against both nematodes and an aphid species. There is great diversity in the lifestyles and pathogenic mechanisms of disease-causing organisms. However, R genes were found to encode proteins with certain common motifs. Five classes of R genes are now recognized: intracellular protein kinases; receptor-like protein kinases with an extracellular leucine-rich repeat (LRR) domain; intracellular LRR proteins with a nucleotide binding site (NBS) and a leucine zipper (LZ) motif; intracellular NBS-LRR proteins with a region with similarity to the Toll and interleukin-1 receptor (TIR) proteins; and LRR proteins that encode membrane-bound extracellular proteins. Interestingly, a nematode resistance gene, HS1[Pro-1], has recently been cloned and shown to contain an extracellular leucine-rich repeat domain. Currently, the only resistance genes lacking any of these motifs are the maize Hm-1 and barley mlo genes, which are not associated with gene-for-gene-mediated resistance.

There are several different ways to various genes to confer crops for disease resistance (Table 13.1). Susceptible crops are transformed with cloned R genes, novel resistant lines can be generated rapidly. This approach is particularly useful in crops in which no effective resistance genes have been identified and allows interspecies transfer of R genes. At the same time, it avoids the genetic "drag" of negative characteristics, a frequent problem with classical breeding approaches. An inducible defense system would also confer higher-level resistance and utilize fewer plant resources than many of the constitutive defenses discussed previously. There are examples of successful transfers of R genes both within and between species. However, in other cases, interspecies transfer of R genes, particularly between more distantly related species, has failed. Thus, R genes isolated from one species may not necessarily be compatible with the resistance signaling components from another.

In the future, as more becomes known about the function of the different R gene motifs, it may be possible to engineer recombinant genes with novel resistance specificities. Of particular interest would be the creation of durable R genes that recognize conserved features of a broad spectrum of pathogens. The isolation of R gene — avr gene pairs, such as Cf9 and Avr9, will also allow the development of two-component systems that activate the HR and other defense responses after pathogen attack. Constitutive expression of the Avr9 protein has been shown to cause a developmentally lethal necrosis on Cf9-containing tomato. Therefore, expression of the avr gene must be regulated by an infection-inducible promoter that is active solely in the inoculated leaves. The major caveat to this system is the identification of an appropriate promoter. One candidate is the prp1-1 promoter from potato, which has been used successfully to induce the localized expression of barnase (and thus lesion formation) in transgenic potato plants infected with *Phytophthora infestans*.

Efforts to engineer broad-spectrum resistance are not limited to R genes but also include approaches around the plant defense responses elicited by these R-genes: the rapid and localized Hypersensitive response (HR) and the subsequent establishment of the systemic acquired resistance (SAR) response. "Hypersensitive response" was a term first applied by Stakman in 1915 to describe the rapid and localized plant cell death induced by rust fungi in rust-resistant cereals. HR, like the apoptosis is

Table 13.1. Disease resistance genes.

Gene	Conferred Resistance	Gene of Origin	Transgenic Plant Host
Chitinase	Leaf spot disease of peanut	Tobacco	Peanut
Recombinant antibodies (rAbs)	Bacteria, fungi, viruses	Animal	Crop plants
Defensin NP-1	*Fusarium oxysporum*	Rabbit	Tomato
Accelerated cell death 2 (acd2)	*Pseudomonas syringae*	*Arabidopsis*	*Arabidopsis*
Small GTPase Rac	Bacterial leaf blight	Rice	Rice
RPW8	Fungal mildew	*Arabidopsis*	*Arabidopsis*
Viral replicase	Potato leafroll virus	Potato leafroll virus	Potato
Antifungal peptide (alfAFP)	*Verticillium dahliae*	Alfalfa	Potato
Cationic peptide (MsrA1)	Bacterial or fungal phytopathogens	Synthetic	Potato
Cecropin B	Bacterial leaf blight	Bombyx mori	Rice
Ribozyme	Dwarf virus	Dwarf virus	Rice
Xa21	Bacterial leaf blight	Rice	Rice
Coat protein	Papaya ring spot virus	Papaya ring spot virus	Papaya
Mi	pathogens, nematodes	Tomato	Tomato
Defense-related Rir1b	rice blast fungus Magnaporthe grisea	Rice	Rice
Antizearalenone Fv (scFv)	Mycotoxin	Animal	*Arabidopsis*

Table 13.1 (*Continued*)

Gene	Conferred Resistance	Gene of Origin	Transgenic Plant Host
Beta-1,3-glucanase	Fungal pathogens	Tobacco	Tobacco
Chitinase	Fungal pathogens	Tobacco	Tobacco
Antifungal protein KP4	Stinking smut (Tilletia tritici)	Ustilago maydis-infecting virus	Wheat
RPP13	Downy mildew	*Arabidopsis*	*Arabidopsis*
Movement proteins	Tomato mottle virus	Geminivirus	Tomato
Cf-9	*Cladosporium fulvum*	Tomato	Tomato
Class II catalase (Cat2St)	*Phytophthora infestans*	Tobacco	Potato
Pto	Potato virus X (PVX)	Tomato	Potato
Sarcotoxin IA	*Pseudomonas syringae, Erwinia carotovora*	Flesh fly	Tobacco
Cryptogein	*Phytophthora cryptogea*	*Phytophthora cryptogea*	Tobacco
Calmodulin isoforms	Bacteria, fungi, viruses	Soybean	Tobacco
Pto	*Xanthomonas campestris, Cladosporium fulvum*	Tomato	Tomato
Antiviral protein II (PAPII)	Tobacco mosaic virus, PVX, *Rhizoctonia solani*	Pokeweed	Tobacco
RPS5	Downy mildew	*Arabidopsis*	*Arabidopsis*

Table 13.1 (*Continued*)

Gene	Conferred Resistance	Gene of Origin	Transgenic Plant Host
Coat protein-specific scFv	Beet necrotic yellow vein virus (BNYVV)	Animal	Tobacco
Viscotoxin	*Plasmodiophora brassicae*	Mistletoe species	*Arabidopsis*
Lipid transfer protein LTP2	*Pseudomonas syringae pv. tabaci 153*	Barley	Tobacco
NPR1	Bacteria, fungi, viruses	*Arabidopsis*	*Arabidopsis*
N	Tobacco mosaic virus	Tobacco	Tomato, Tobacco
Cf-2	Tomato laf mould	Tomato	Tomato
Viral RNA replicase	Cucumber mosaic virus (CMV)	CMV	Cucumber, tomato
H2O2-generating glucose Oxidase	*Erwinia carotovora subsp carotovora*	Fungus	Potato
RPM1	*Pseudomonas syringae*	*Arabidopsis*	*Arabidopsis*
RPS2	*Pseudomonas syringae*	*Arabidopsis*	*Arabidopsis*
Scorpion toxin	*Erwinia carotovora, Pseudomonas syringae*	Scorpion	Tobacco

mammals, is characterized by a rapid and localized cell death at the point of pathogen attack/recognition. One elegant strategy aims to alter regulation of the HR in such a way that this response is induced by both virulent and avirulent pathogens. The bottleneck of this approach is the identification of promoters that respond tightly, rapidly and in a cell-autonomous manner to infection. This study may greatly benefit from novel tools in genomics, which allow the identification of genes and promoters that only respond to very specific signals.

A variant of the above strategy is to generate a lesion mimic phenotype via either mutant selection or regulated expression of genes that trigger the HR. One of these key genes is the Npr1 gene, which encodes a putative transcriptional regulator. Overexpression of Npr1 enhances disease resistance levels against a broad variety of pathogens in *Arabidopsis*. Importantly, this resistance is not associated with any adverse plant phenotypes, such as stunting or undesired cell death. These findings make Npr1 an extremely interesting candidate for agricultural application. Experiments have been initiated to test the efficacy of either Npr1 or Npr1 homologues in crops. Preliminary results indicate that over expression of Npr1 in rice does lead to increased resistance against both the bacterial pathogen *Xanthomonas* and the fungal pathogen *Magnaportha*. Over expression of a Npr1 homolog in wheat, however, resulted in suppression of pathogenesis-related genes and enhanced disease susceptibility.

Another signaling gene that has been evaluated for its utility to enhance disease resistance levels is the Myb1 gene, which is induced by tobacco mosaic virus (TMV) in resistant tobacco plants, and encodes a transcription factor that binds to a promoter element of the pathogenesis-related gene PR1a. Modification of Myb1 expression levels in transgenic tobacco plants was shown to increase resistance against both a viral (TMV) and a fungal (*Rhizoctonia solani*) pathogen. Two additional highly interesting genes were identified through mutant screens in *Arabidopsis* but have not yet been isolated. These genes, named cpr6 and Ssi1, trigger not only genes associated with SAR but also genes that act in a second response pathway activated by jasmonic acid. Thus, both cpr6 and Ssi1 genes act as switches modulating cross-talk between different defense pathways. The robust resistance controlled by cpr6 and Ssi1 is unfortunately linked to severe stunting, and application of this technology for agriculture may require optimization. The very recently cloned *Arabidopsis* Pad4 gene may prove exceptionally interesting for the development of broad-spectrum disease resistance. The interesting aspect of Pad4 is that inactivation of this gene leads to extreme susceptibility against a wide variety of pathogens, including *Erysiphe orontii*, *Peronospora parasitica* and *Pseudomonas syringae*. Thus, over-expression of Pad4 in transgenic plants may enhance disease-resistance signaling. The Pad4 gene has been hypothesized to act by amplifying weak signals in disease-resistance responses through a positive feedback cycle with salicylic acid.

Some virulence factors of bacterial pathogens can also be sued to induce SAR in plants. The harpin of *Erwinia* spp., *Pseudomonas* spp., and *Xanthomonas* spp. will normally help the bacterial pathogens to infect host plants. However, it can also trigger the plant defense system SAR so that host plants will become more resistant to the pathogens after the initial infection. Through genetic engineering, harpin gene has been cloned and the harpin protein can be produced through fermentation. By applying the protein products directly onto crops, the plant SAR system is activated to fight off pathogenic pathogens. In April 2000, a biopesticide called Messenger based on harpin of *Erwinia amylovora* was approved by the US environmental Protection Agency (EPA, Washington DC). The product offers growers a novel, effective and environmentally sound toll to control plant disease, to enhance crop growth and to control pests like insects, mites and nematodes. The former Swiss company, Norvartis, now called Syngenta after the merger of Novartis Agribusiness and Zeneca Agrochemicals, also has a product that acts on SAR pathway. The synthetic product Actigard has been sold in Europe since 1996 as a plant immune system activator called Bion for plant disease control. Actigard is currently undergoing US approval.

Many of the genes induced by plant disease-resistance responses encode proteins with direct antifungal activity *in vitro*. One of the most important and straightforward strategies in enhancing disease resistance is based on the identification and expression of such antifungal proteins (AFPs). This is an extension of the paradigm that has worked extremely well for insect control genes based on insecticidal proteins from *Bacillus thuringienesis*. Reported AFP classes include defensins and other small cysteine-rich peptides, 2S albumins, chitin-binding proteins, lipid-transfer proteins, and hydrogen-peroxide-generating enzymes. Thionins are antimicrobial proteins that accumulate in a variety of plants resisting pathogen infection. Over-expression of the *Arabidopsis* Thi2.1 gene has been shown to enhance resistance to *Fusarium oxysporum*. The breadth of resistance conferred by overproduction of thionins is currently unclear. Plant defensins, best characterized from radish, are cysteine-rich proteins that efficiently inhibit the growth of a broad range of fungi. However, they are generally not effective against bacteria. Homologs of these proteins have been identified in several plants and shown to be induced upon pathogen infection, supporting their role in disease resistance. Moreover, expression of a radish seed defensin gene (Rs-AFP2) in tobacco confers enhanced resistance to the foliar pathogen *Alternaria longipes*. Similarly, expression of defensins in transgenic corn enhances resistance to Northern corn leaf blight. These proteins are promising tools for engineering fungal disease resistance.

13.4.2. *Insect resistance*

Today, insect-resistance transgenes, whether of plant, bacterial or other origin, can be introduced into plants to increase the level of insect resistance, a technology that has dramatically extended the scope of resistance genes available to plant breeders. The first reports of transgenic insect-resistant plants were only published in 1987, but technological development has been swift since then. Approximately 40 different genes conferring insect resistance have been incorporated into crops, and the first insect-resistant crops have been commercialized in several countries. This new technology is seen as an additional tool for the control of crop pests and could offer certain advantages over conventional insecticides, such as more-effective targeting of insects protected within plants, greater resilience to weather conditions, fast biodegradability, reduced operator exposure to toxins and financial savings. The use of transgenic plants may lead to a reduction in the use of broad-spectrum insecticides, thereby extending the useful life of these compounds and reducing the ecological damage they cause. However, there is concern that the constitutive expression of toxins will encourage the selection of resistance to such products in pest populations.

Bacillus thuringiensis (Bt) is a ubiquitous spore-forming soil bacterium that produces insecticidal protein crystals, also called Bt toxins, -endotoxins or crystal proteins, within its cells during the sporulation process. Spores and protein crystals of several strains of this bacterium have been used as microbial insecticides since the 1950s and, owing to their selectivity, now have an established role in some integrated pest-management systems. For the past ten years, endotoxins from different *Bacillus thuringiensis* (Bt) subspecies have been used to protect crops against insects. Currently, corn, potato and cotton plants expressing different synthetic Bt genes are commercially available; they show significant protection against European corn borer, Colorado potato beetle and bollworm infestations, respectively.

After ingestion by susceptible insects, the protein is solubilized and activated by proteinases in the insect midgut. The precise mode of action is not fully understood, but it appears that the activated toxins bind to receptors in the insect midgut epithelium and insert into the midgut membrane, leading to the disruption of the electrical, K+ and pH gradients by creating pores, resulting in irreversible damage to the midgut wall. Separate strains of Bt produce a variety of crystal toxins with distinct host ranges. At least ten genes encoding different Bt toxins have been engineered into plants: cry1Aa, cry1Ab, cry1Ac, cry1Ba, cry1Ca, cry1H, cry2Aa, cry3A, cry6A and cry9C. Most Cry proteins, even within the Cry1A subfamily, have a distinctive insecticidal spectrum. The cry1A and cry1C genes code for the Cry1A and Cry1C proteins, respectively, which are specific to larvae of lepidopteran pests such as the codling moth (*Cydia pomonella*), the European corn borer (*Ostrinia nubilalis*) or heliothine bollworms. In contrast, the Cry3A protein is toxic to coleopteran pests, including the Colorado potato beetle (*Leptinotarsa decemlineata*).

Bt toxins have been transferred and expressed in at least 26 different plant species. However, the level of resistance they confer will, in most cases, depend on whether native-bacterial or truncated, codon-optimized genes have been used. To date, codon-optimized genes have only been transferred into some of these crops, including cotton, maize, potato, broccoli, cabbage and alfalfa. Generally, these plants will express Bt toxins at levels sufficient to cause high mortality of target pests in the field. An alternative strategy has been to integrate the native cry1Ac gene into the chloroplast genome of tobacco plants, resulting in extremely high expression levels.

The insect-resistance genes transferred into plants to date mainly target the insect digestive system. Most have been derived from a single species of bacterium or a range of higher plants, although some insect-resistance genes from animals and other microorganisms have also recently been introduced into crop plants. However, the search for new genes is ongoing and aims to expand the range of insects affected, to combat the development of resistance in the target insects by identifying genes with different modes of action and to improve potency. Because Bt-endotoxins are not effective against all insects, efforts to identify other insecticidal proteins are also under way. Several such proteins have recently been identified, including Vip3A, cholesterol oxidases and the A1 amylase inhibitor. Various endogenous proteins, termed swrps (systemic wound response proteins), which are synthesized in response to insect attack, could also be used to engineer pest resistant plants. Of these proteins, serine proteinase inhibitors I (PI-I) and II (PI-II) reduce predation by inhibiting important digestive enzymes in the insect gut. Several components in the induction pathway leading to PI synthesis have been identified, including the 18-amino acid peptide systemin, various intermediates of the octadecanoid pathway and jasmonic acid (JA). Through manipulation of these signaling components, it might be possible to over express PI proteins and/or other swrps, thereby producing pest-resistant plants. Indeed, constitutive expression of a cowpea trypsin inhibitor in tobacco increases resistance to tobacco budworm.

Other microorganism-derived resistance genes include the isopentenyl-transferase gene (ipt) from *Agrobacterium tumefaciens*, which codes for a key enzyme in the cytokinin-biosynthetic pathway. Expression of ipt in tobacco and tomato by a wound-inducible promoter has resulted in a decrease in leaf consumption by the tobacco hornworm (*Manduca sexta*) and reduced survival of the peach aphid (*Myzus persicae*). However, the expression of ipt had negative effects on plant development, such as an underdeveloped root system and reduced total chlorophyll content. A cholesterol-oxidase gene from a streptomycete fungus has also been engineered into tobacco. Cholesterol is necessary for the integrity and function of almost all cellular membranes and cholesterol oxidase is highly toxic to larvae of the boll weevil (*Anthonomus grandis*) and retards the growth of the tobacco budworm (*Heliothis virescens*). Both species are major cotton pests, and observations have, so far, indicated that this protein damages the membranes of the insect-midgut epithelium. A different approach has been to transfer cDNA copies of the *Helicoverpa armigera* stunt virus (HaSV), an insect RNA virus belonging to the Tetraviridae, into tobacco. Expression of the RNA 1 and the coat protein of HaSV in tobacco caused severe stunting of neonate *H. armigera* larvae, another major cotton pest.

Two major groups of plant-derived genes used to confer insect resistance on crop are inhibitors of digestive enzymes (proteinase and amylase inhibitors) and lectins. Plant proteinase inhibitors are polypeptides or proteins that occur naturally in a wide range of plants and are a part of the plant's natural defense system against herbivory. Proteinases in insects include serine, cysteine, aspartic and metallo proteinases that catalyze the release of amino acids from dietary protein and so provide the nutrients crucial for normal growth and development. Different proteinases predominate in different insects; for example, serine-like proteinases are dominant in lepidopteran larvae, whereas coleopteran species have a wider range of dominant gut proteinases. Serine- and cysteine-proteinase inhibitors have been reported to inhibit the growth and development of a range of insects, mainly lepidopteran and coleopteran species, respectively. The anti-metabolic mode of action of these inhibitors is not fully understood: direct inhibition of digestive enzymes is not considered to be the main effect, and a more important factor might be the hyper secretion of digestive enzymes caused by the presence of the inhibitors, resulting in depletion of essential amino acids.

So far, at least 14 different plant proteinase-inhibitor genes have been introduced into crop plants. Most efforts have concentrated on serine-proteinase inhibitors from the plant families *Fabaceae*, *Solanaceae* and *Poaceae*, which are targeted mainly against lepidopteran species but also against some coleopteran and orthopteran pests. The most active inhibitor identified to date is the cowpea trypsin inhibitor (CpTI), which has been transferred into at least ten other plant species. Experiments with transgenic plants and artificial diets have shown that CpTI affects a wide range of lepidopteran and coleopteran species; CpTI-transformed tobacco, field-tested in California, caused significant larval mortality of cotton bollworm (*Helicoverpa zea*) larvae, but the protection provided by CpTI was less pronounced and consistent than that of tobacco containing a truncated Bt-toxin gene. The serine-proteinase inhibitors KTi3, C-II and PI-IV (from soybean) resulted in up to 100% mortality of first-instar cotton-leaf worms (*Spodoptera littoralis*) when expressed in tobacco. However, resistance levels achieved by the same constructs in potato were much lower, resulting mainly in retarded growth of *S. littoralis* rather than direct mortality. In addition to serine-proteinase inhibitors, one cysteine-proteinase inhibitor (OC-1, from rice) has been introduced into several other crops. To date, no crop expressing proteinase-inhibitor transgenes has been commercialized. Inhibitors of amylases are the second type of enzyme inhibitor used to modify crop plants. Genes for three amylase inhibitors have been expressed in tobacco, but the main emphasis has been on transferring the gene for the common-bean (*Phaseolus vulgaris*) amylase inhibitor (AI-Pv) to other legumes. AI-Pv is a thermostable glycoprotein that forms a 1:1 complex with insect and mammalian amylases but is not active against plant and bacterial amylases. AI-Pv inhibits-amylase in the midgut of coleopteran stored-product pests of the genera Callosobruchus and Bruchus and blocks larval development.

Lectins are carbohydrate-binding proteins, some of which are toxic to insects. Various lectins have shown some toxic activity against species of the insect orders *Homoptera*, *Coleoptera*, *Lepidoptera* and *Diptera*. The mode of action of lectins against insects remains unclear, but it has been shown that at least some bind to midgut-epithelial cells. However, some insecticidal lectins also show significant mammalian toxicity, including lectins from *P. vulgaris*, winged bean, soybean and wheat germ. Other lectins, for example those from pea and snowdrop (*Galanthus nivalis*), have demonstrated insecticidal activity but are less toxic to mammals. Recent interest has mainly concentrated on the lectin from snowdrop (GNA), because it has shown activity against aphids and the rice brown planthopper (*Nilaparvata lugens*). GNA has been expressed in nine different crops, including potato, oilseed rape and tomato. Laboratory tests with modified potatoes showed that GNA did not increase the mortality or development time of the glasshouse potato aphid (*Aulacorthum solani*) but reduced fecundity considerably, an effect that led to a much slower population buildup of this aphid in a glasshouse trial. Results of laboratory experiments with the peach potato aphid were similar but, in addition, the establishment of aphids on transgenic potatoes was reduced compared with untransformed plants. In a preliminary field trial, however, GNA potatoes did not show any clear effect on aphid numbers, and expression levels of GNA were highly variable. In the laboratory, GNA enhanced the resistance of potato to larvae of the tomato moth (*Lacanobia oleracea*). Leaf damage and larval weight were reduced by approximately 50% but larval mortality was only approximately 20%. It was concluded that the effect of GNA is antifeedant rather than insecticidal.

Chitinases have been introduced into plants, but have not shown a protective effect against larvae of the tomato moth and had only a very weak effect on the peach potato aphid. The tobacco anionic peroxidase, when over expressed in different species of tobacco, tomato and sweet gum, provided significant levels of resistance to several lepidopteran and coleopteran species as well as to the peach potato aphid. The mode of action of peroxidases is highly complex and somewhat controversial but their defensive effects are for the most part considered to be indirect and dependent on the effects of enzyme products. Peroxidase activities include crosslinking and polymerization, tying up of nutrients, inhibition of digestive enzymes and the generation of highly reactive, toxic species. Tryptophan decarboxylase (TDC) from periwinkle expressed in tobacco reduced the reproduction of the whitefly *Bemisia tabaci* by up to 97%. TDC converts tryptophan to the indole-alkaloid tryptamine, and the expression of tryptamine and tryptamine-derived alkaloids in plants may act as antioviposition and antifeedant agents or as inhibitors of larval and pupal development.

So far, genes derived from animals for insect resistance, involved primarily serine-proteinase-inhibitor genes from mammals and the tobacco hornworm (*Manduca sexta*). Based on *in vitro* screening of inhibition of proteolysis by midgut extracts of a range of lepidopteran larvae, bovine pancreatic trypsin inhibitor (BPTI), 1-antitrypsin (1AT) and spleen inhibitor (SI) have been identified as promising insect-resistance proteins and transferred into a range of plants. However, initial results with the

Table 13.2. Insect resistance genes.

Gene	Conferred Insect Resistance	Gene of Origin	Transgenic Plant Host
CrylAb	*Chrysomela tremulae*	*Bacillus thuringiensis*	Poplar
Agglutinin AHA	Aphid	*Amaranthus hypochondriacus*	Tobacco
Snowdrop agglutinin (GNA)	*Lacanobia oleracea*	*Galanthus nivalis*	Potato
Snowdrop agglutinin (GNA)	Rice brown planthopper	*Galanthus nivalis*	Rice
CrylAb	*Helicoverpa zea*	*Bacillus thuringiensis*	Cotton
Cry1A(c)	*Bollworm fall armyworm*	*Bacillus thuringiensis*	Cotton
CrylAb	Rice stem borers	*Bacillus thuringiensis Berliner*	Rice
CrylAb	*Ostrinia nubilalis, Helicoverpa zea*	*Bacillus thuringiensis*	Sweet corn
CrylAb	Indian meal moth, Angoumois grain moth	*Bacillus thuringiensis*	Corn
Cry2Aa2	Cotton bollworm, beet armyworm	*Bacillus thuringiensis*	Tobacco
Proteinase inhibitor cathepsins L & B	Alfalfa weevil	Tobacco	Potato
Trypsin inhibitor BTI-CMe	*Spodoptera exigua*	Barley	Tobacco
CrylA(b)/CrylA(c)	Leaffolder, yellow stem borer	*Bacillus thuringiensis*	Rice
Cry3Ca1	Colorado potato beetle (CPB)	*Bacillus thuringiensis*	Potato
Cry1Ab, Cry1Ac, Cry9C	European corn borer	*Bacillus thuringiensis*	Corn
Scorpion neurotoxin (AaIT)	*Lymantria dispar*	Scorpion	Poplar
Cry1C	Diamondback moth	*Bacillus thuringiensis*	Broccoli
Cry1Ab, Cry1Ac, Cry9C	*Plodia interpunctella*	*Bacillus thuringiensis*	Maize
Trypsin and CHYMOTRYPSIN proteinase inhibitors	Fall armyworm	Soybean	Tobacco
Cry1Ac	Corn earworm, soybean looper	*Bacillus thuringiensis*	Soybean
Proteinase inhibitor	*Lepidotera, Coleoptera* and *Diptrea*	Arrowhead	Wheat

potato tuber moth (*Phthorimaea operculella*) and potatoes expressing either SI or 1AT did not indicate an increased level of resistance in transgenic potatoes — larval weight gain on transgenic plants was even greater than on untransformed controls. In contrast, *M. sexta*-derived proteinase inhibitors expressed in cotton and in tobacco were found to reduce reproduction in *B. tabaci* 7677 (Table 13.2).

13.4.3. *Nematode tesistance*

Nematodes predate on many crops worldwide, and the economic losses caused are estimated to be in the order of US$77 billion each year. Plant-parasitic nematodes are either ectoparasites (living outside the root) or endoparasites (living inside the root). In addition, nematodes would plant root system that create opportunities for fungal infection, such as the severe fungal diseases in cotton production caused by *Verticilum* sp. and *Fusarium* sp. The tiny parasites obtain their nutrients from the cytoplasm of living plant cells and comprise many species including actoparasites and endoparasites. The sedentary endoparasites of the family *Heteroderidae* cause the most economic damage, and this group is further divided into two groups: the root-knot nematodes (genus *Meloidogyne*) and the cyst nematodes (including the genera *Heterodera* and *Globodera*). Toot-knot nematodes forming of the characteristic root galls or "root-knots" on host plant infect a broad range of plant species and cause severe losses in yield in many crops world wide. Cyst nematodes, although have more narrow host ranges, do cause severe losses in certain crops. For example, the soybean cyst nematode (*H. glycins*) is one of the most economically important pathogens of soybean in the US. Potato cyst nematodes (*G. pallida* and *G. rostochiensis*) cause major losses in potato growing areas in many parts of the world.

Nematode resistant plant species have the capabilities to reduce nematode reproduction levels. Nematode resistant genes exist in several crop species and are an important component of many breeding programs for tomato, potato, soybeans and cereals. So far, three nematode resistance genes have been cloned, and additional genes have been either mapped to chromosomal locations/linkage groups or are likely to be cloned in the near future. The Hs1^{Pro-1} was cloned from a wild relative of sugar beet that confers resistance against the beet nematode *Heterodera schachtii*. Mi gene of tomato can confer effective resistance against several root-knot nematode species. The *Gpa2* was cloned by a positional cloning strategy from wild potato species *Solanum tuberosum* subsp. *Andigena* CPC1673 that confers resistance against some isolates of the potato cyst nematode *Globodera pallida*. A common feature of the three nematode resistance genes is that they share several typical functional domains with disease resistance *R* genes, i.e. the leucine-rich repeat (LRR), leucine zipper (LZ), nucleotide binding site (NBS), and a possible transmembrane domain (TM). An alternative strategy for conferring resistance to nematodes is to introduce genes with a nematicidal impact, and this approach is also having some success.

For use in breeding programs, PCR assays diagnostic for RFLP marker alleles closely linked to alleles of Gro1 and H1, a second resistance locus, have been developed. The function of H1 is according to the gene-for-gene model of resistance, leading to necrosis around the feeding site. Another candidate for positional cloning is the Hero gene from *Lycopersicon pimpinellifolium*, a wild relative of tomato. It confers resistance to *G. rostochiensis* and was mapped in tomato with tightly linked RFLP markers. Meanwhile, yeast artificial chromosomes (YACs) have been isolated from the gene-bearing region. This gene is also of interest for potato breeding because its resistance mechanism is different from any of the previously described genes of potato. In addition, the cloning of the Gpa2 gene, conferring resistance to *G. pallida*, should soon be completed. The gene has been fine mapped on chromosome 12, bacterial artificial chromosome (BAC) clones covering the gene region have been isolated, and genetic complementation experiments are ongoing.

The soybean cyst nematode (*H. glycines*) is economically the most important pathogen of soybean in the US. Different types of resistance to *H. glycines* have been mapped with molecular markers. A major partial resistance locus (rhg1) is located on chromosome G of soybean, explaining up to 50% of the genetic variation. A 620 kb BAC contig (contiguous segment of DNA) was constructed around this locus with the help of tightly linked markers, making rhg1 a candidate as the first nematode resistance gene to be cloned from this species. Closely linked markers are also available for the second resistance gene, Rhg4, on linkage group A of soybean, and these were used to identify clones from a BAC library. The Southern root-knot nematode (*M. incognita*) also belongs to the major group of pathogens of soybean, but resistance is inherited in a quantitative manner. Two major quantitative trait loci (QTLs) for resistance to this nematode have been mapped with RFLP markers.

Two genes for nematode resistance have been cloned on the basis of their chromosomal position and identified by genetic complementation. The first was the $Hs1^{Pro-1}$ gene from sugar beet, which confers resistance to the beet cyst nematode (*H. schachtii*). Juveniles of this nematode are able to invade resistant roots as well as susceptible ones. After sex differentiation in a compatible system, male nematodes leave the root to fertilize the females, which become cysts filled with numerous eggs. Only in rare cases can females be observed in resistant plants expressing the $Hs1^{Pro-1}$ gene. These females are not able to complete their life cycle and no eggs can be produced. Their feeding cells are comparably small and markedly different from normal syncytia. There is no resistance in cultivated Beta species. The wild relative *B. procumbens* carries three genes (Hs1, Hs2 and Hs3) that each confers complete resistance. The gene from chromosome 1 ($Hs1^{Pro-1}$) has been transferred to cultivated sugar beet by interspecific crosses and cloned using a positional cloning approach. The gene was identified by genetic complementation in roots of susceptible beet, with expression under the control of the CaMV35S promoter. The same incompatible reaction as in resistant plants was observed, demonstrating for the first time the potential of natural disease resistance genes for breeding resistant crops. It can thus be expected that resistant beet varieties will be available in the future with the help of transgenic technology. Recently, conventionally bred varieties that have been selected with the help of molecular markers tightly linked to $Hs1^{Pro-1}$ have been brought to the market. The second gene that has been cloned is Mi-1, which had been introduced into cultivated tomato by conventional breeding from the wild relative *Lycopersicon peruvianum*. It is responsible for the hypersensitive reaction of tomato roots after infection with *Meloidogyne* spp. (*M. incognita*, *M. javanica* and *M. arenaria*) resulting in the death of juveniles of the J2 stage. The gene has been tightly linked to RFLP, RAPD and AFLP marker loci and cloned from its position on tomato chromosome 6. Genetic complementation using whole cosmids resulted in resistant plants. The gene is located within a cluster of at least eight genes and has features typical of previously cloned disease resistance genes, such as a nucleotide-binding site and a leucine-rich region. However, there is little similarity between the gene products of Mi-1 and Hs1[Pro-1].

An alternative molecular strategy for establishing nematode resistance in plant species has been pursued with the development of artificial resistance. The aim of this approach is to generate a durable and efficient resistance that controls a broad range of plant-parasitic nematodes. Each different step of the attack of the nematode on the root tissue that is dependent on the individual life cycle of that species could serve as a putative target for the construction of an artificial defense system. This can be achieved by introducing effector genes into the host plant that have a nematicidal impact. Such transgenes can encode enzymatic inhibitors that block physiological processes within the nematode, toxic compounds that are then ingested, compounds that bind to signal molecules, enzymes that interact with the nematode, and substances that cause the breakdown of specific feeding structures.

The effective use of anti-nematode genes to generate artificial resistance relies on the specificity for the nematode, proper and sufficient expression at the target site and non-phytotoxicity to the host. Thus, the employment of compounds produced by the plant itself will be the most promising. In this respect, proteinase inhibitors (PIs), often found in plants in response to wounding or herbivory, have been investigated for their nematicidal activity. Cysteine proteinases are involved in protein metabolism and digestion of dietary protein, and could be shown to be present in nematodes. They are known to be inhibited by cystatins, which are small PIs. Therefore, the oryzacystatin-I gene from rice was modified by mutagenesis to improve the efficacy of the inhibitor, and was subsequently introduced and expressed, under the control of the CaMV-35S promoter, into tomato hairy roots and *Arabidopsis* plants.

The expression of cystatin in the transgenic plants prevented female nematodes from developing properly, causing reduced size and fecundity. However, it still has to be proven whether the reduced fecundity is sufficient to confer complete resistance against plant-parasitic nematodes in the field. The advantage of this type of approach is that, because the

PIs are non-toxic, they can be expressed in every cell — leading to the control of a broad range of nematode species. However, the evolution of PI-tolerant nematodes has to be taken into consideration. Lectins are another class of putative anti-nematode proteins. These carbohydrate-binding proteins could be targeted to interact with the nematode at different sites: within the intestine, on the surface coat, or with amphidial secretions. The gene encoding the snowdrop (*Galanthus nivalis*) lectin GNA has been introduced into potato plants and expressed at a maximum of approximately 0.5% of total root protein, which led to a reduction of *G. pallida* females of up to 80%. By contrast, a higher expression of GNA resulted in an increased number of females, indicating that only a critical level of GNA expression might have an effect on the nematode.

The degradation of structural components of the nematode catalyzed by enzymes like chitinases or collagenases is an additional objective for engineering nematode resistance. Chitinases are known to be part of the plant defense system and are antifungal. Thus, a possible target is the nematode eggshell, which largely consists of chitin. However, to date, no attempt has been made to construct transgenic plants over expressing chitinases for this purpose. Human collagenase cDNAs have been expressed in tobacco plants (*P. havstad*) for breaking down the collagen-containing cuticle of the nematode. For successful application of this approach, two main obstacles have to be overcome. First, there is evidence that the structure of the collagen from nematodes is different from that of vertebrates, questioning the efficiency of this enzyme; and second, the collagens are not present on the outermost surface coat of the nematode, and are therefore not exposed to the environment.

The nemato-toxic properties of Bt-toxins from *Bacillus thuringiensis* strains were thoroughly studied with the free-living nematode *Caenorhabditis elegans*, revealing a different mode of action compared with insecticidal strains. A preliminary study with transgenic tomato plants expressing the Bt-endotoxin CryIab after inoculation with a *Meloidogyne* sp. resulted in a reduction in egg mass per gram of root of about 50%. Further studies and subsequent field trials have to be performed to corroborate the impact of Bt-toxin on plant-pathogenic nematodes.

Expression within the plant of monoclonal antibodies that are directed against nematode-specific proteins should lead to a specific inhibition compared with PIs, chitinases, collagenases and Bt-toxins. This concept has been tested by expressing a monoclonal antibody specific to stylet secretions of *M. incognita* in tobacco plants and by transient expression of single-chain antibodies in tobacco protoplasts. However, although the antibody was expressed in the appropriate plant tissue, it appeared to have no influence on the parasitism of the nematode, indicating that antibodies might not be a sufficient means of control.

Another strategy for engineering an artificial resistance system is to promote the breakdown of specific feeding structures. The most realistic approach is the introduction of genes encoding phytotoxic compounds that disrupt these specialized cells. This can only be realized by selecting promoters exclusively induced in the cells of the feeding structure, thus limiting the expression of "suicide" genes to the target tissue. Therefore, the promoter-trapping technique has been applied to detect up- and down-regulation of genes within, or in the vicinity of, feeding cells. However, such regulating elements are difficult to find, because except for natural resistance genes, a plant might not have evolved genes specific for a pathogen-induced tissue. To circumvent this problem, a two-component system consisting of a plant-toxic gene under the control of a nematode-responsive promoter and a second detoxifying gene under the control of a constitutive promoter has been developed. Thus, "leaky" expression of the cytotoxic gene can be neutralized by the antagonistic gene. This system makes use of the barnase and barstar genes first applied to generate and subsequently to restore male sterility in plants. Meanwhile, different strategies have been applied in the search for nematode-responsive promoters that are up-regulated in the feeding tissue. Differential cDNA screening led to the identification of Lemmi, a tomato gene that is up-regulated in giant cells after *M. incognita* infection. The tobacco gene TobRB7 is up-regulated, and deletion analysis of the promoter revealed that a 300 bp fragment conferred high GUS expression solely in giant cells, but not in syncytia. GUS fusion analysis of hmg2, which encodes a key enzyme of isoprenoid metabolism, indicated that the promoter of the gene might be up-regulated in tobacco giant cells. However, employment of this system with the specific promoters described here still remains a challenge.

Isolation and characterization of key pathogenicity factors will be very important to combat these complex pathogens. The isolation of several beta-1,4-endoglucanase genes from two species of plant- parasitic cyst nematodes, *Globodera rostochiensis* and *Heterodera glycines*, was a major progress in the field. These proteins are secreted by the subventral glands, which suggests that migration through the cell walls of plant roots involves enzymatic softening. Inhibition of these enzymes would probably hinder nematodes from reaching their destination. The same could be true for certain enzymes secreted by fungi.

13.4.4. *Plantibody for pest management*

Plantibody would provide a novel approach to control plant diseases, and at the same time, to reduce pollution from synthetic pesticides and to make agriculture more sustainable. In transgenic tobacco producing antibodies against a viral coat protein, they were resistant to artichoke mottled crinkle virus or tobacco mosaic virus (TMV). Antibody-mediated virus resistance is an attractive alternative to the various forms of pathogen-derived resistance. In the later case, unintended side effects, such as heteroencapsidation and recombination of viral genomes, cannot always be excluded. Most plant viruses are RNA viruses that replicate in the cytocol. Therefore, the highest resistance is expected by targeting the antibody to this cell compartment. When a TMV virion-binding scFv fragment is expressed either in the cytosol or the apoplast of the transgenic tobacco, the most dramatic reduction of necrotic lesion numbers upon virus infection was observed in plants accumulating scFv fragments in the cytosol (infectivity was reduced by more than 90%), although much higher accumulation levels of scFv were obtained in the apoplast. Polycolonal antibodies against pectate lyase of the fungus *Colletotrichum gloeosporioides* could inhibit symptom development in the host plant.

Plantibody is a new strategy for developing powerful ways to control pest insects in agriculture. For insects, immunization of vertebrate hosts with antigens from the parasite was showed to protect against hematophagous arthropods. When such vertebrate insects ingest antobodies from their host, a large fraction remains functional in the gut of the insect and a small fraction passes into the hemoglymph, both enabling the antibodies to bind their antigen and to interfere with the metabolism of the insect. These results suggest strongly that the plantibody approach would also work against insects.

Transgenic plants can efficiently produce complete antibodies or antibody fragments, which are often referred as "plantibodies", that can be used not only for medical sciences, but also for *ex planta* applications in the field of molecular farming. A strategy called immunomodulation based on this technology has been developed to produce antibodies and antibody fragments to modulate the function of corresponding antigen. As such, immunomodulation can be used to study the function of the antigen or even of the epitope in plants, to change agronomic traits, or to "immunize" the plant against pathogen infection is a molecular technique that allows the interference with cellular metabolism or pathogen infectivity by the ectopic expression of genes encoding antibodies or antibody fragments. There are two ways to isolate antibody genes: monoclonal hybridoma technology or using phage technology. Hybridoma technology is expensive and time-consuming. In addition, plant laboratories do not have the facilities to work with animal cell culture. Phage display technology, on the other hand, allows the isolation of antigen-binding scFv-encoding sequences directly from established libraries or from the spleen of immunized mice. Phage display makes use of generally applied recombinant DNA techniques. Five scFv fragments against plant enzyme dihydroflavonol-4-reductase were successfully isolated and expressed in transgenic *Petunia hybrida*. Antibodies or antibody fragments in trangenic plants can be targeted not only to the cytosol, the endoplasmic reticulum (ER) and the apoplast, but also to the nucleus, mitochondria, or chloroplast. There are several ways to modulate antigen-antibody interaction *in vivo*: (1) Binding of the antibody on the substrate or ligand-binding site of an enzyme or receptor blocks interactions between enzyme and substrate or between receptor and ligand directly by competitive inhibition; (2) binding of the antibody on other sites of the enzyme or receptor can cause allosteric inhibition by inducing structural changes in the active site or ligand-binding site; (3) antibody binding can interfere with the protein folding, either directly or by hindering the interaction with chaperones; (4) the formation of quaternary protein structures, such as multienzyme complexes, can be prohibited; (5) the antigen can be mistargeted as a result of antibody binding, such as retention in the ER of ER-translocated secretory proteins or metabolites.

13.4.5. *Herbicide resistance*

Weeds compete with forest trees, especially at the early stage of plantation, for nutrients, water and sunlight. Therefore, extensive weed control throughout the whole growth season is vital for the success of re-forest after tree cutting. For herbicide resistance, direct herbicide detoxification and reduction of target enzyme sensitivity to the applied herbicide are the two major approaches for herbicide resistance in transgenic trees and crops. Three different strategies for protecting plants from the action of specific herbicides through genetic engineering include: (1) introducing an additional insensitive form of the gene for the target site of the herbicide; (2) over-producing a sensitive target site enzyme to swamp out or titrate out the effects of the herbicide; and (3) introducing a novel detoxification system into the plant to break down or metabolize the herbicide before it has acted at its targeted sites. Herbicides are designed specifically as inhibitors to target biochemical pathways (such as photosynthesis and biosynthesis of amino acids) of plants in order to minimize their toxicity to human and animals.

Glyphosate inhibits aromatic amino acid biosynthesis, and is the main active component of many branded herbicides that are used for pre-emergence weed control. It is a specific competitive inhibitor of the enzyme 5-enol-pyruvylshikimate phosphate synthase (EPSPS). The sixth enzyme in the shikimic acid pathway leading to the aromatic amino acids tryptophan, tyrosine and phenylalanine. When EPSPS was targeted in chloroplast in transgenic crops, such as cotton, soybean, canola and corn, these crops gained resistance against glyphosate (Roundup Ready^{TM}) and thus became a major commercial success for transgenic crops. Another class of very potent and broad-spectrum herbicides is the sulfonylurea herbicides inhibiting the enzyme acetolactase synthase (ALS), the first enzyme in the pathway for biosynthesis of the essential amino acids isoleucine, leucine and valine in plants. Sulphonylurea tolerant tomato, tobacco and cotton were developed by using a mutated resistant form of ALS from tobacco by changing two amino acids in the encoded protein.

The bialaphos resistance gene (*bar* and *pat*) from the *Streptomyces* species, on the other hand, can detoxify any activated glufosinate herbicide. When corn expressing the *bar* or *pat* gene under the 35S promoter, transgenic tobacco, potato, canola exhibited strong resistance to glufosinate ammonium (Liberty® herbicide). A specific nitrilase enzyme gene isolated from the soil bacterium *Klebsiella ozaenae* has also been used to develop a resistant cotton variety against the broadleaf herbicide bromoxynil (BXN™).

13.4.6. *Stress tolerance and environmental clean up*

Under the natural environment, plants have to endure different environmental conditions, such as heat, drought, cold or freeze and salination. Human activities have made the situations even worse. For example, fume gas produced by cars and other industrial activities by burning fossil oil and coal has significantly contributed to the "greenhouse" effect on the climate, i.e. the average temperature in the atmosphere and on earth has increased dramatically. Its adverse effect on earth is that some parts of the earth become extremely hot while other parts become colder, one region experience drought while another region suffers from flood. Modern agriculture has also contributed to the worsened situation of drought and salination by pumping out ground water with higher mineral content for irrigation that caused the decrease of ground water level and increased mineral content in soil after evaporation. In addition, using artificial fertilizers in agriculture and some other industry activities cause the increase of heavy metal ions (such as Pb^{2+}, Cd^{2+}, Cu^{2+}, Hg^{2+}, Al^{2+}, etc.) content in the soil of agricultural land. Genes isolated from plants, microbes and animals have been used to confer resistance against cold, freeze, drought, heat, oxidative stress and heavy metal (Table 13.3). In addition, transgenic plants can also been used to suck out harmful heavy metals from soil so that after several season's plantation, the soil is basically cleaned up for normal crop production. Transgenic plants expressing certain enzymes can help to speed the process of break down of organic pollution caused by spaying pesticides and organic chemical manufacturing.

Table 13.3. Stress resistance genes.

Gene	Conferred Resistance	Gene Origin Organism	Transgenic Plant Host
Aldose/Aldehyde Reductase	Lipid peroxidation	Alfalfa	Tobacco
Aminophospholipid translocases (ALA1)	Cold	*Arabidopsis*	*Arabidopsis*
Anti-Proline Dehydrogenase (AtProDH)	Salt, Cold	*Arabidopsis*	*Arabidopsis*
AtHAL3	Salt, Cold	*Arabidopsis*	*Arabidopsis*
C_2H_2-type zinc finger protein (SCOF-1)	Cold	Soybean	Arabidopsis, tobacco
Ca2+-dependent protein kinase (CDPK)	Salt, Drought, Cold	Rice	Rice
CAP160	Cold	Spinach	Tobacco
CAX2	Mn^{2+}, Pb^{2+}	*Arabidopsis*	Tobacco
CBF3 transcriptional activator	Cold	*Arabidopsis*	*Arabidopsis*
Calmodulin-binding transporter	Ni^{2+}, Pb^{2+}	Tobacco	Tobacco
Chloroplast glutamine synthetase	Salt	Rice	Rice
Choline dehydrogenase (betA) and betaine aldehyde dehydrogenase (betB)	Salt, Cold	*E. coli*	Tobacco
Choline Oxidase (codA)	Salt, Cold, Oxidative stress	*Arthrobacter globiformis*	Rice, *Arabidopsis*
Citrate synthesis enhancing enzymes	Al^{2+}	Tobacco	Tobacco, Papaya
Cold-regulated genes (COR15a)	Cold	*Arabidopsis*	*Arabidopsis*
Dehydration response element (DRE)	Drought, salt, cold	*Arabidopsis*	*Arabidopsis*
EREBP/AP2 DNA binding motif (Tsi1)	Pathogen, Osmosis Stress	Tobacco	Tobacco
Fe-superoxide dismutase (FeSOD)	Oxidative stress	*Arabidopsis*	Tobacco

Table 13.3 (*Continued*)

Gene	Conferred Resistance	Gene Origin Organism	Transgenic Plant Host
Ferritin	Oxidative damage, pathogens	Alfalfa	Tobacco
Glutathione S-transferase (GST) and Glutathione peroxidase (GPS)	Stress	Tobacco	Tobacco
Glutathione synthetase	Cd^{2+}	E. coli	Indian mustard
Glyoxalase I	Methylglyoxal, salt	Brassica juncea	Tobacco
HAL1	Salt	Yeast	Tomato
Heat shock factor (HSF)	Heat	Arabidopsis	Arabidopsis
Heat shock protein 101	Heat	Arabidopsis	Arabidopsis
Iron-superoxide dismutase	Photosynthetic oxidative stress	Arabidopsis	Alfalfa
Metallothionein-I	Cd^{2+}	Tobacco	Tobacco
Metallothionein-like gene	Cd^{2+}	Tobacco	Tobacco
Mitogen-activated kinase	Oxidative stress	Arabidopsis	Tobacco
Mn-superoxide dismutase (Mn-SOD)	Cold	Arabidopsis	Alfalfa
NtCBP4	Ni^{2+}, Pb^{2+}	Tobacco	Tobacco Seed-specific
Transcriptional activator (ABI3)	Cold	Arabidopsis	Arabidopsis
Superoxide Dismutase (FeSOD)	Oxidative stress	Arabidopsis	Maize
Trehalose-6-phosphate synthase (TPS1)	Drought	Saccharomyces cerevisiae	Potato
Vacuolar Na+/H+ antiport	Salt	Arabidopsis	Arabidopsis

Cold stress on plants induces changes in the transcription of cold response genes. A cDNA clone encoding C2H2-type zinc finger protein, SCOF-1, was isolated from soybean. The transcription of SCOF-1 is specifically induced by low temperature and abscisic acid (ABA) but not by dehydration or high salinity. Constitutive over-expression of SCOF-1 induced cold-regulated (COR) gene expression and enhanced cold tolerance of non-acclimated transgenic *Arabidopsis* and tobacco plants. Other genes can help plants to fight against freeze and cold include the anti-freeze protein from Atlantic salmon fish and tomato.

Oryza sativa L. cv. *Kinuhikari* was transformed with a chloroplastic glutamine synthetase (GS2) gene from rice. When exposed to 150 mM NaCl for 2 weeks, the control rice plants completely lost photosystem II activity, but G39-2 plants retained more than 90% activity after the 2-week treatment, whereas G241-12 plants lost these activities within one week. In the presence of isonicotinic acid hydrazide, an inhibitor of photorespiration, G39-2 showed the same salt tolerance as the control plants. The intracellular contents of NH_4+ and $Na+$ in the stressed plants correlated well with the levels of GS2. Thus, the enhancement of photorespiration conferred resistance to salt in rice plants. A salt-induced transcript Tsi1 (for Tobacco stress-induced gene) was isolated from a tobacco cDNA library. *Agrobacterium*-mediated transient expression and transgenic plants expressing Tsi1 demonstrated that over expression of the Tsi1 gene induced expression of several pathogenesis-related genes under normal conditions, resulting in improved tolerance to salt and pathogens. A rice gene encoding a calcium-dependent protein kinase (CDPK), OsCDPK7, was induced by cold and salt stresses. Over-expression of OsCDPK7 in transgenic rice enhanced induction of some stress-responsive genes in response to salinity/drought, but not to cold. Over-expression of the HAL1 gene in yeast has a positive effect on salt tolerance by maintaining a high internal $K(+)$ concentration and decreasing intracellular $Na(+)$ during salt stress. *Arabidopsis thaliana* AtHAL3: a flavoprotein related to salt and osmotic tolerance and plant growth. Trans-genic *Arabidopsis* plants, with gain of AtHAL3a function, show altered growth rates and improved tolerance to salt and osmotic stress.

HSF3, a new heat shock factor from *Arabidopsis thaliana*, depresses the heat shock response and confers thermotolerance when over expressed in transgenic plants. Organisms synthesize heat shock proteins (HSPs) in response to sublethal heat stress and concomitantly acquire increased tolerance against a subsequent, otherwise lethal, heat shock. Heat shock factor (HSF) is essential for the transcription of many HSP genes. Over-expression of HSF3 causes HSP synthesis at the non-heat-shock temperature of 25°C in transgenic *Arabidopsis*.

Heat shock protein 101 plays a crucial role in thermotolerance in transgenic *Arabidopsis*. Genetic engineering of drought resistant potato plants by introduction of the trehalose-6-phosphate synthase (TPS1) gene from *Saccharomyces cerevisiae*. In yeast, trehalose-6-phosphate synthase is a key enzyme for trehalose biosynthesis, encoded by the structural gene TPS1. Trehalose affects sugar metabolism as well as osmoprotection against several environmental stresses, such as heat and desiccation. The TPS1 transgenic potato plants showed significantly increased drought resistance. These results suggest that the production of trehalose not only affects plant development but also improves drought tolerance.

Over-expression of a tobacco glutathione S-transferase with glutathione peroxidase activity (GST/GPX) in transgenic tobacco (*Nicotiana tabacum* L.) enhanced seedling growth under a variety of stressful conditions. Over-expression of GST/GPX in transgenic tobacco seedlings provides increased glutathione-dependent peroxide scavenging and alterations in glutathione and ascorbate metabolism that lead to reduced oxidative damage. Rapid accumulation of toxic products from reactions of reactive oxygen species (ROS) with lipids and proteins significantly contributes to the damage of crop plants under biotic and abiotic stresses. A stress-activated alfalfa gene encoding a novel plant NADPH-dependent aldose/aldehyde reductase that also exhibited characteristics of the homologous human enzyme. Ectopic synthesis of this enzyme in transgenic tobacco plants provided considerable tolerance against oxidative damage caused by paraquat and heavy metal treatment. These transformants could also resist a long period of water deficiency and exhibited improved recovery after rehydration. Over-production of *Arabidopsis thaliana* FeSOD confers oxidative stress tolerance to transgenic maize.

Metal-binding proteins and peptides in bioremediation and phyto-remediation of heavy metals — the expression of metal-binding proteins or peptides in microorganisms and plants in order to enhance heavy metal accumulation and/or tolerance has great potential. Several different peptides and proteins have been explored. This review focuses on cadmium (Cd) because of the significant importance of this metal and because of its global presence in many food materials. Expression of *Arabidopsis* CAX2 in tobacco altered metal accumulation and increased manganese tolerance. Metal transport from the cytosol to the vacuole is thought to be an important component of ion tolerance and of a plant's potential for use in phytoremediation. Expression of CAX2 in tobacco increased Cd(2+) and Mn(2+) transport in isolated root tonoplast vesicles.

Breeding mercury-breathing plants for environmental clean up — when a bacterial mercury detoxification pathway was introduced into plants, they showed enhanced mercury tolerance and can convert highly toxic forms of organic mercury to less toxic elemental mercury, which is volatile. Expression of a truncated tobacco NtCBP4 channel in transgenic plants and disruption of the homologous *Arabidopsis* CNGC1 gene confer Pb2+ tolerance. When over-expressed in transgenic plants, NtCBP4 confers Pb2+ hypersensitivity associated with enhanced accumulation of this toxic metal. Cadmium tolerance and accumulation in Indian mustard is enhanced by over expressing gamma-glutamylcysteine synthetase. Plants ectopically expressing the iron-binding protein, ferritin, are tolerant to oxidative damage and pathogens.

Aluminum tolerance in transgenic plants by alteration of citrate synthesis — aluminum when in soluble form, as found in acidic soils that comprise about 40% of the world's arable land, is toxic to many crops. Organic acid excretion has been correlated with aluminum tolerance in higher plants. Over production of citrate was shown to result in aluminum tolerance in transgenic tobacco (*Nicotiana tabacum*) and papaya (*Carica papaya*) plants.

13.4.7. *Plant shape and agricultural traits*

Plant morphological characteristics, such as the height, the shape of the plant and leaves, the thickness of the stem and the strength of the root system, can be modified to improve productivity and reduce production related costs. For example, if the form and thickness of the rice plant is changed, it photosynthesis efficiency can be increased greatly. The hair on leaves and stem called trichome can be increased or decreased by genetic engineering either to strengthen the insect resistance in some crops or to reduce trichome contamination of cotton fiber.

In *Arabidopsis*, the technology gene trap has been used to block normal functions of genes so that we can learn which genes can be used to modify plant shapes. For instance, when the function of a particular P450 gene in *Arabidopsis* is blocked, the *Arabidopsis* plant cannot maintain its original shape and the branches will keep divide and a very "bush" plant with thousands of branches. This "supershoot" gene can then be used to produce pharmaceutical important proteins since the proteins will be easily extracted from the green parts of the plants. Another gene controlling the opening of the seedpod was also blocked by the technology. Instead of opening after maturation, the seedpod remains closed. Traditionally farmers have to harvest the seedpod of soybean oils seed rape when the plants are still green and therefore the seeds are not fully matured. The seed yield and quality are affected. Transgenic pants with this "anti-shattering" gene can be harvested by machine when the seedpods are fully matured without the fear of losing seeds in the field. The third gene NAC1 found in *Arabidopsis* can facilitate the root system development. NAC1-transgenic plants have much lager and stronger root systems and become more resistant to heat, drought and can survive better in poor soil conditions.

In plants, bioactive gibberellins (GAs) affect a number of plant developmental processes, including seed germination, leaf expansion, stem elongation, flower initiation and flower and fruit development. One of the main contributions for the so called "Green Reviolution" during the 1960s and 1970s, was that scientists and breeders developed new varieties of wheat and rice that were shorter, higher yield due to reduced straw biomass and more resistant to damages caused by wind and rain. The wheat *Reduced height-1* (*Rht-1* and *Rht-D1*) contributed to reduced response of wheat to gibbberellin and orthologues of the *Arabidopsis Geberellin Insentitive (GAI)* gene. The three genes encode proteins resembling nuclear transcription factors and participate in the gibberellin signal transduction. Transgenic rice with the mutant *GAI* gene showed reduced responses and was dwarfed. The *GAI*-similar genes can be sued to engineer field crops like wheat, rice, corn and sorghum to further increase their grain yield and strengthen their resistance against wind and rain.

13.4.8. *Food with better taste and more nutrients and nutraceuticals*

13.4.8.1. *Vitamins and minerals*

Milled rice is deficient in important nutrients, causing serious public health problems among people in many countries of Africa, Asia, and Latin America, who rely exclusively on rice as their staple food. As many as 5 million children in the developing countries suffer from vitamin A deficiency and 3 million of them eventually suffer from irreversible blindness. This is because human and vertebrate animals do not synthesize carotinoids, and they depend on dietary carotinoids for making their retinoids (such as the main human visual pigment retinal, the vitamin A retinol and the substance controlling morphogenesis retinoic acid). The main precusor of retinoids is β-carotene, a β-carotenoid containing two sustituted β-ionone rings at the two ends of molecule. β-carotene is cleaved into retinal by a dioxygenase. The Recommended Dietary Allowance (RDA) for vitamin A is 1000 retinol equivalents, equal to 6 mg β-carotene per day. Rice endosperm does not contain any provitamin A (β-carotene). The introduction of a combination of transgenes (daffodil phytoene synthase, lycopene β-cyclase and *crtI*) into rice under the control of the endosperm-specific glutelin and CaMV 35S promoters, the transgenic "golden rice" contains up to 200 μg β-carotene per 100 g rice (a tenth of the RDA in a daily intake of 300 g rice). Biochemical analysis confirmed that the color was due to provitamin A. The genetically engineered rice lines provide sufficient vitamin A in a typical Asian rice diet. Using similar technical approach, β-carotene has also been introduced into canola seeds and tomato fruits.

Iron deficiency has also been approached by a multigene strategy in which genes for phytase, ferritin, and a Cys-rich metallothionin-like protein were transferred and expressed in rice endosperm. Current lines have around twice as much iron as the wild type. This modified rice now has to go through the normal biosafety and agronomic tests prior to release into Asian fields. Transgenic rice and wheat can also significantly reduce the allergic reactions caused by the seed protein gluten and therefore enable more people consume the grain products without health-related concerns. Over-expression of the iron storage protein ferritin in transgenic plants as an approach for treating iron deficiency in humans, can achieve a threefold increase in iron content in seeds. As oral administration of plant ferritin can alleviate iron deficiency anemia in rats, these findings have promulgated the idea that iron-fortified transgenic plants over expressing ferritin could be used to mitigate iron deficiency in the human diet.

13.4.8.2. *Proteins and amino acids*

The seed storage globulins of legumes are low in sulfur-containing amino acids, cysteine, and methionine. Thus, in improving seed nutritional quality, the emphasis has been on introducing sulfur-rich proteins. Spectacular success was achieved with the methionine-rich 2S storage albumin of Brazil nut. Expression of the corresponding gene has resulted in substantial increases in seed methionine in transgenic canola and Vicia (beans). However, the Brazil nut protein proved to be a major allergen preventing commercial development. More recently, expression of a related 2S albumin in transgenic lupins has resulted in increased amounts of cysteine and methionine in protein, largely at the expense of free sulfate.

In cereal grains, the major deficiency is lysine, which contributes to the poor nutritional value of foods to human beings and feeds to domestic animals. Although lysine-rich proteins do occur in seeds and have demonstrated increased amounts in one barley mutant, these have not been used to engineer increased seed lysine. When a gene was designed to code for a protein containing 31% lysine, its expression in tobacco seeds resulted in a significant increase in lysine (by up to 30%). An alternative strategy for increasing the lysine content of grain for animal feed rather than ramping up the content of high-lysine proteins is to change the regulatory properties of the enzymes that synthesize the amino acid. For instance, the expression of bacterial enzymes of the lysine biosynthetic pathway (aspartate kinase and dihydrodipicolinate synthase), lack the feedback sensitivity of the endogenous plant enzymes. If plants are transformed with bacterial genes for these enzymes, it would be possible to increase the total lysine in canola and soybean two and fivefold, respectively. Similar approaches might be used for other amino acids or a wide range of other products whose synthesis is regulated.

Proteins also confer functional properties, which allow seeds to be processed in food systems. These are particularly important in soybean (for texturing and gel (tofu) formation) and in wheat. Such functional properties are biophysical rather than biochemical and may involve complex interactions with other components as well as changes with temperature. They are more difficult to define in molecular terms than other targets for manipulation.

Considerable progress has been made, however, in altering the quality of wheat for making bread and other processed foods. The processing properties of wheat are largely determined by the gluten storage proteins. These form a continuous network in dough, conferring the viscoelastic properties necessary to entrap carbon dioxide released during the proving of leavened bread. The protein network also provides the cohesiveness required for other foods. The quality of wheat is determined by genetic and environmental factors, with poor quality generally resulting from low gluten elasticity. Although gluten comprises over 50 proteins, one group of proteins — the high molecular weight subunits of glutenin — are particularly important. There is convincing evidence from biochemical and genetic studies that variation in gluten elasticity is associated with differences in the number, amount (about 6%–12% of the total protein), and properties of these high molecular weight subunits. Attention has therefore focused on manipulating the amount and properties of this group by inserting additional genes encoding mutant and wild-type proteins. Increasing the number of expressed subunits from two to three and four results in stepwise increases in dough elasticity, mirroring the effects of manipulating gene dosage by classical breeding. However, it has been possible to go beyond the gene dosage obtainable by conventional means. This results in gluten that contains over 20% of high molecular weight subunits and a massive increase in dough elasticity. The resulting flour is actually too "strong" to be used for bread making but may be valuable for blending to fortify poor quality wheat.

Improvement of nutritive value of crop plants, in particular the amino acid composition, has been a major long-term goal of plant breeding programs. The seed albumin gene AmA1 from *Amaranthus hypochondriacus* is nonallergenic in nature and is rich in all essential amino acids, and the composition corresponds well with the World Health Organization standards for optimal human nutrition. In an attempt to improve the nutritional value of potato, the AmA1 coding sequence was successfully introduced and expressed in tuber-specific and constitutive manner. There was a striking increase in the growth and production of tubers in transgenic populations and also of the total protein content with an increase in most essential amino acids. The expressed protein with balance amino acids was localized in the cytoplasm as well as in the vacuole of transgenic tubers.

Transgenic plants used as feed can also help to improve quality and quantity of products from animal, such as sheep wool. The postruminal supply of the sulfur-containing amino acids, methionine and cysteine, has been reported to be a major limitation to wool growth in sheep. Wool growth rates were significantly increased when feed was supplemented with approximately 1 to 2 g of sulfur amino acids daily. A modified albumin rich in sulfur amino acids was engineered into alfalfa by plant transformation. The highest level of sunflower seed albumin found in transgenic alfalfa leaves was estimated to constitute 1% of soluble leaf protein. This level of accumulation of the foreign protein would be predicted to supply an extra 40 mg of sulfur amino acids daily to sheep fed the modified forage. In addition, another approach to increasing plant nutritive value of pasture is to reduce the content of indigestible lignin using similar approaches in engineering forest trees.

Expression of the human milk protein β-casein in transgenic potato plants opened the way for reconstitution of human milk in edible plants for replacement of bovine milk in baby foods for general improvement of infant nutrition, and for prevention of gastric and intestinal diseases in children. The storage proteins of maize are a group of alcohol-soluble polypeptides called zeins. These proteins are synthesized in the developing endosperm, where they form protein bodies within the rough endoplasmic reticulum. Because they account for more than half of the total seed protein, zeins are the primary determinants of the amino acid composition of the seed. All of the zeins are devoid of lysine an essential amino acid for monogastric animals. The genes encoding zeins were modified so that they encode proteins that contain lysine and tryptophan. Analysis of the synthesis and processing of these modified zein proteins in transgenic maize indicates that the content of lysine and tryptophan was increased without interfering with their association into protein bodies.

13.4.8.3. *Fats and oils*

It is now possible to engineer "designer oilseeds" tailored to specific end-uses. Plants produce a wide diversity of fatty acids, the majority of which accumulate in the seed as triacylglycerols. Plant lipids have different end-uses, from industrial applications as lubricants and detergents to food and nutrition. Plant lipids also have important roles as nutraceuticals and pharmaceuticals. The possibility of "designer oilseeds" stems from ground breaking work that allowed the definition and isolation of key genes involved in the biosynthesis and modification of plant fatty acids. The global market for plant-derived oils is immense. Consequently, much of the research has been driven by the major plant biotechnology companies.

There are, in essence, two sectors in the market for plant derived oils: "high volume, low value" products on one hand and "low volume, high value" on the other. An example of the "high volume, low value" category is the production of lauric acid (a 12-carbon saturated fatty acid) in transgenic canola. Laurate, which is used in confectionery, is normally obtained from either coconut or palm oil. Although both plants yield relatively high levels of the fatty acid, they are limited in their agricultural utility. Researchers at Calgene have demonstrated the feasibility of engineering canola to produce lauric acid by introducing the gene encoding an acyl-ACP thioesterase from the California bay plant. This enzyme prevents the production of long-chain fatty acids by cleaving the fatty acids from the enzyme complex that synthesizes it after it reaches the 12-carbon stage. This results in the accumulation of lauric acid. Calgene has produced a number of canola lines that accumulate over 40 mol% laurate. The company has shown that laurate is incorporated into the triacylglycerols of the canola oil.

The novel fatty acids are recovered from the transgenic canola by standard processing methods. These oils are now marketed for use in confectionery in North America under the Laurical trademark. This work has also demonstrated the feasibility of producing large amounts of transgene-modified plant oils to supplement or replace fluctuating "natural" sources. This has caused controversy because of possible adverse effects on the economies of countries that grow palm or coconut crops.

The "low volume, high value" oils sector has recently attracted great interest. There is now evidence that polyunsaturated fatty acids such as gamma linolenic acid are essential for healthy human metabolism. Gamma linolenic acid is also a registered pharmaceutical for the treatment of eczema and mastalgia, and may also have wider applications for antiviral and cancer therapies. The compound is unusual in that it only accumulates in a limited number of species, including evening primrose and borage. Oils from these plants are widely marketed as health food supplements. However, borage and evening primrose have poor agronomic properties, with low yields and relatively low levels of gamma linolenic acid. There is, therefore, a good deal of current research devoted to trying to produce such compounds in plants that are both easier to grow and more widely consumed.

The primary substrate for the synthesis of eicosanoids is linoleic acid, an essential fatty acid for human nutrition. This is initially converted into gamma linolenic acid, followed by subsequent elongation and further desaturation. A group at IARC Long Ashton has isolated a cDNA from borage encoding the enzyme (a 6 fatty acid desaturase) responsible for the synthesis of gamma linolenic acid, and introduced this into tobacco (used as a model crop). It resulted in the accumulation of gamma linolenic acid at levels equivalent to or greater than those in borage.

However, this is only the first step in the eicosanoid pathway. Additional genes must be isolated to reconstruct the whole metabolic pathway. Further work at IARC recently led to the isolation of the gene encoding the 5 fatty acid desaturase responsible for synthesis of arachidonic acid, the last fatty acid in the eicosanoid pathway. The final goal, the production of novel high value pharmaceutical oils and eicosanoid precursors, may be difficult to achieve but will probably require transgenic plants producing both the 5 and 6 desaturases, and other as yet undefined enzymes.

13.4.8.4. Starch

Cereal starches are widely used in the food industry to produce glucose and fructose syrups for beverages, conferring functional properties for food processing. Starch is primarily used as a thickener but also as a binder, adhesive, gelling agent, film former, and texturizer in many snacks. At present, the demand for variation in functional properties is satisfied by using starches from different plant sources (potato and cassava tubers, wheat and maize grain); by chemical modification; and, in the case of maize, by exploiting naturally occurring genetic variants.

Although knowledge of the biophysical and molecular basis for the variation in functional properties is still incomplete, it is generally accepted that several factors are important. These include the proportions of amylose and amylopectin, the fine structures of these polymers (notably the branching pattern of amylopectin), their organization within the granule, and granule size. The basic pathways of amylose and amylopectin synthesis are well understood and the genes readily cloned. Thus, there is the potential to produce cereal starches with a wider range of structures and properties. The targets are varied but include "mutant" wheat starches to mimic those from maize, phosphorylated starch (currently produced from potato), resistant starches for healthy diets, thermoplastic and biodegradable starches for packaging, eliminating the need for chemical modification. In addition, those in the food processing industry may want to fine-tune the amylose and amylopectin structures to produce new functional properties for food. This may be aided by the use of enzymes from other sources.

13.8.4.5. *Nutraceuticals or functional foods*

Nutraceuticals or functional foods are food or parts of food (genetically modified or otherwise), that provide medical or health benefits, including prevention and treatment of diseases. Nutraceuticals are poised for a healthy slice of the healthcare market. The nutraceutical market is estimated to be about US$250 billion in the US and a similar US$250 billion in Europe. In Japan, 47% of the population consumes nutraceuticals, which are responsible for a quarter of the US$6 billion total annual sales of food. Science could eventually find ways to stave off heart diseases and cancers, which account for US$232 billion of America's US$1 trillion annual health care bill. Both phamaceutical and biotechnology companies have been investing over the past few years in nutraceuticals. Drug companies that traditionally shied away from nutraceuticals because of the lower profit margins are now becoming increasingly interested because managed care has been eroding margins for phamaceuticals (which were typically 60–100%). In addition, nutraceuticals generally do not need to go through clinical trials, and can be brought to market for a fraction of the cost and in much less time. For example, instead of 10 years and US$250 million to bring a drug to market, a nutraceutical can take only a few years and a few million dollars.

Food generated from transgenic plants will have much better nutritional value and will help to prevent heart diseases and cancer. Transgenic canola which are otherwise exactly like regular canola plants produce seed with oils (a) that are modified in average fatty acid carbon chain length; (b) that are modified in content of saturated fatty acids (both lower and higher); or (c) that contain structured lipids. In principle, although the gene target may not be obvious, the relative amount of lipids compared to other nutrients can be decreased or increased in foods like peanut or soybean. The oil content of some food products might be modified to enhance levels of medium chain triglycerides or to contain "fish oil" fatty acids — without the cost or olfactory disadvantages. On a broader scope, the amino acid composition of proteins in basic grains is being pursued by several groups. Specific vitamin contents such as vitamin A or E might be enhanced in basic foodstuffs; type and content of fiber may eventually be manipulated. Specific components such as caffeine or phytic acid conceivably can be eliminated in the source plant, negating the need for processing steps that add cost and that lessen flavor and nutrition. Clearly, the biochemical composition of foods is complex and varied. Moreover, tailoring foods to better meet our needs is an expensive and lengthy process. The best possible understanding of how nutrition promotes good health is necessary to direct and prioritize our efforts to improve plants as food sources.

It is now becoming technically possible to use genetic manipulation to improve the quality of plant products consumed by humans and animals. This has already yielded more nutritious grains, process-improved flour, and "designer oilseeds" tailored to particular end-uses. There is also near-term potential in producing health-promoting "nutraceuticals", healthy starches, and starches adapted for packaging and processing.

Fruits, seeds, and other plant organs provide the majority of plant products that humans consume. They are the major targets for modification of quality traits. Quality is usually defined as fitness for use but the scientific basis of this fitness in plant products is not always well defined. In such cases, gene technology provides very powerful and novel approaches to define quality.

Quality may be improved both by addition and subtraction of properties and components. The development of the FlavrSavr tomato by Calgene (Davis, CA) — a transgenic fruit that ripened on the vine without going soft — was achieved by removing the action of an enzyme. Constructing changes and additions in end products has led to both entirely new products and to the transfer of properties from one species, usually not agriculturally suited to a particular region, to others well adapted. For example, canola (oil seed rape) has been transformed using a range of genes to produce different profiles of fatty acids.

13.4.9. *Forest biotechnology*

Forestry has two major contributions: commercial uses (such as timber for construction and furniture, raw material for paper and pulp production and as a energy source) and maintaining and preserving the Earth ecosystem. In order to improve forest productivity, there are several obstacles associated with the current classical breeding programs, such as the long generation time, lack of suitable parents, difficulties of crossing, maintenance problem because of high heterozygosity, and filed trial problem due to large size of forest trees. Biotechnology can help to overcome these problems by (a) clonal propagation of superior genotypes through tissue-culture techniques; (b) somatic hybridization using protoplasts, haploidization and use of somaclonal variations; (c) artificially induced mutations; (d) marker-assistant breeding and finally; (e) introduction of new traits into trees by genetic engineering.

The manipulation of internal phytohormones can influence the growth rate and biomass production of trees. Gebberellins (GA) are a group of more than 100 tetracyclic diterpences, some of them are essential endogenous regulators that influence growth and development processes throughout plant life cycle, such as shoot elongation, expansion and shaping of leaves, flowering and seed germination. In hardwood species and conifers, GA has impact on the differentiation of xylem fibers, the length of secondary xylem fibers, and on both longitudinal and radial growth. When a GA 2-oxidase gene was over-expressed in a hybrid aspen, the transgenic tree had faster growth in height and diameter, larger leaves, more numerous and longer xylem fibers, and increased biomass.

Lignin contributes up to 25% wood biomass and is therefore the second most abundant organic compound on earth after cellulose. For the pulp and paper industry, lignin must be removed in an energy-consuming process that uses environmental polluting chemicals. Through an antisense approach targeted at O-methyltransferase (OMT) in the shikimate and phenylpropanoid pathways, lignin content in trees can be reduced significantly, without any obvious side effects. Cinnamyl alcohol dehydrogenase (CAD) is another key enzyme for lignin biosynthesis. Transgenic poplar threes expressing sense or antisense constructs showed reduced CAD activity and red xylem. Although the lignin content was not reduced, more lignin could be extracted from transgenic poplar threes. Again with antisense approach, the expression level of the gene *Pt4CL1* encoding 4-coumarate:coenzyme A ligase (4CL) in the lignin biosynthetic pathway was down regulated in transgenic aspen with 45% reduction of lignin content and 15% increase in cellulose.

Weeds compete with forest trees, especially at the early stage of plantation, for nutrients, water and sunlight. Therefore, extensive weed control throughout the whole growth season is vital for the success of re-forest after tree cutting. For herbicide resistance, direct herbicide detoxification and reduction of target enzyme sensitivity to the applied herbicide are the two major approaches for herbicide resistance in transgenic trees and crops. Transgenic *Populus* hybrid plants with aroA showed a lower sensitivity to the commercial herbicide glyphosate, which is low toxicity and blocks 5-enolpyruvylshikimate-3-phosphate *aroA* (EPSP) synthase in the aromatic amino acid biosynthesis pathway. Over-expressing the *bar* gene, which encodes the PAT enzyme that directly inactivates the herbicide phosphinothricin) in transgenic popular trees conferred resistance to the herbicide Basta.

Insects cause severe damage and losses in forest-tree species and that sometimes contribute to the limiting factors for three growth and survival.

Bark beetles damage more mature trees, spruce budworms harm more buds and gypsy moths eat mainly *Populus* leaves. Intensive forest plantation practices also decrease forest tree's capability against root-system-feeding insects. Bt toxins from *Bacillus thuringensis* and proteinase inhibitors against insect digestive systems have been successfully engineered into European-larch trees, poplar and conifers with enhanced insect resistance.

13.4.10. *Green production of vaccines and biopharmaceuticals*

Biopharming is a term used to describe the use of transgenic plants to produce pharmaceuticals. The first success led to the synthesis in tobacco leaves of antibodies effective against dental caries. Subsequently, it has been possible to synthesize high-affinity monoclonal secretory antibodies that can prevent microbial infection in humans. The technology can thus be considered for other immunotherapeutic uses, especially in mucosal tissue. Plants could be particularly valuable as commercial production systems for antibodies that are needed in large amounts. Current pharmaceutical production is expensive and limited. The total Western hemisphere fermentation production capacity is 500 kg. If antibodies are to be used for prophylaxis, production would need to be in excess of 5,000 kg/year. Plant production systems could provide this, but the current limitation is devising suitable purification procedures.

Transgenic plants will become an alternative to produce vaccines and biopharmaceuticals with technological scope and economic scale. The industrialized countries can afford a number of effective vaccines against serious infectious diseases. However, in developing countries, whose health care budget is only a fraction of that of the industrialized ones, to control infectious diseases has become an overwhelming burden on national resources. In addition, the current high cost medical care system supported by sufficient young and middle-aged workforce cannot cope with the aging population, which started from Japan and industrialized countries and now become a common feature worldwide. There is an urgent need not only for new vaccines, but also for cheaper version of existing vaccines. The later can be achieved by lowering production costs and reducing the cost of immunization, such as eliminating refrigeration or oral vaccine, these are new strategies plant biotechnology can offer for health care for both human beings and domestic animals.

The list of medically important recombinant molecules that can be expressed in GM plants is growing rapidly, and it is likely that the market release of the first recombinant pharmaceutical from transgenic plants will occur within the next five years. Biopharmaceuticals have traditionally been produced using a variety of transgenic systems, including cultured mammalian cells, bacteria and fungi. Producing therapeutic proteins in plants has many economic and qualitative advantages, such as reduced health risks from pathogen contamination, relatively high yields and production in seeds or other storage organs. The cultivation, harvesting, storage, and processing transgenic plants will also use existing infrastructure and require little capital investment. The cost of producing recombinant proteins in plants could be 1–50 times lower than producing the same protein through *E. coli* fermentation. Plant-derived products, whether purified or not, are less likely to be contaminated with human pathogenic microorganisms than those derived from animal cells, because plants are not natural hosts for human infectious pathogens.

Plants used for biopharmaceutical production include tomato, banana, turnip, black-eyed beans, oil seed rape, Ethiopian mustard, potato, rice, wheat, maize, tobacco and *Arabidopsis*. Because extraction of proteins from fresh tissues containing a high proportion of water could results in unavoidable proteolysis, the efficiency is low and the costs are high. In contrast, proteins in seeds could remain intact for long periods, making seeds a convenient method of storing, distributing and administering biopharmaceuticals. Hepatitis B surface antigen (HbsAg) was expressed in transgenic tobacco and potato, and the later could elicit humoral immune response when orally administered to mice. Ultimately, oral vaccines will be produced in banana in a fraction of the price of traditional vaccines. In the future, banana vaccines can be produced against a range of human diseases, including measles, polio, diphtheria, yellow fever and some viral diarrhea. Many of the antibodies produced in transgenic plants have applications for human and animal health care, and they may also prove useful for other applications such as bioremediation. Recombinant antibodies produced in transgenic plants include fully assembled whole immunoglobulins, antigen-binding fragments of immunoglobulins, and synthetic single-chain variable fragment gene fusions (scFv). Plantibodies could be of particular interest for topical immunotherapy. Passive immunization of the mucosal sites could be effective against bacteria, fungi and viruses. Plantibodies against cell surface antigens of *Streptococcus mutans* have been shown to reduce tooth decay in animal models and humans. Repeated large doses of antibody are required for topical passive immunotherapy, and transgenic plants could be used to produce such large quantities. Biopharmaceuticals, such as enkephalins, α-interferon, human serum album, and two of the most expensive drugs (glucocerebrosidase and granulocyte-macrophage colony-stimulating factor), have also been successfully produced in transgenic plants (tobacco, potato, Ethipion mustard, oil seed, rice, turnip and maize).

13.4.11. *Production of industry raw materials*

Chemical companies have been searching for new sources of growth in the beginning of 1990s. Although the world market today for biotech-based basic and intermediate chemicals, fine chemicals, polymers and specialty chemicals accounts for only 2%, or about $25 billion in revenue, biotechnology does play a dominant role in production of enzymes, certain organic acids and some vitamins. For example, over 95% of all riboflavin was chemically synthesized in 1994. Today, 95% of it is produced by genetically enhanced fermentation with 50% cost reduction and 40% less capital expenditure. Biotechnology has been already used to make fine chemicals, dyes, food additives and textile chemicals. Laboratories and pilot plants have been built to produce pigments, polymers and surfactants. Other areas using transgenic plants to produce raw material include production of modified starch and biodegradable plastics in grain crops such as corn and root tube crops like potato and sweet potato, production of bio-diesel by transgenic oil crops (such as canola, oil palm and sunflower), based on the extensive knowledge and understanding of plant metabolism mechanisms. Poly-hydroxyalkanoates (PHA) are a class of polymers produced by numerous bacterial species as carbon and energy reserves with thermoplastic properties. They have received much attention as biodegradable alternatives to petrochemical plastic.

So far, PHAs have been successfully produced in transgenic *Arabidopsis*, canola and cotton fiber.

In 2010, the price and quality of biotech-based products will make them competitive in 30% of the chemical market as a whole. Biotech-derived fine chemicals will win a market share of some 60% thanks to their purity, ease of production and enhanced efficacy. The next area to be affected will be some 30% of the market for special chemicals. By 2010 to 2015, the raw-material costs of biopolymers produced from transgenic plants should be no higher than that of petropolimers. Currently, it costs about US$1.10/kg to produce polyesters from oil. Production costs of biopolymers made from corn would be on average US$0.75/kg. Dow Chemicals and DuPont plan to start world-scale production of the biopolymers PLA (polylactic acid) and PTT (polytrimethylene terephthalate) in 2002 and 2003, at an estimated cost of just US$1.00/kg.

13.4.12. *Field trials and commercial releases*

Many biotechnology companies and universities are evaluating the performance of AFP-based transgenes in the field. In fact, hundreds of release permit applications for field trials to test the efficacy of transgenes for disease control are submitted per year. Field trials do not only include efficacy tests of the transgenes but also rigorous tests on agronomic characteristics and yield of the transgenic plants. Prior to commercialization, transgenic plants are also assessed for nutritional composition, and transgenes are evaluated for any food safety issues. This rigorous process leads to deprioritization of many transgenes. For example, expression of the *Aspergillus* glucose oxidase (AGO) gene in transgenic potato plants increased disease resistance in growth chamber experiments but failed to provide commercial levels of disease control in the field. In addition, AGO gene expression was shown to be correlated with a slightly altered tuber phenotype. A second hydrogen-peroxide-generating enzyme, the barley oxalate oxidase, is under advanced field evaluation for control of *Sclerotinia* in soybean, canola and sunflower. Some groups are evaluating the simultaneous expression of two different AFPs in plants. Over-expression of an intracellular chitinase and an extracellular 1,3-glucanase resulted in a synergistic effect providing disease control against *Fusarium oxysporum* in tomato. An abundance of studies on the efficacy of AFPs in transgenic plants can be expected to be published in the near future. A major issue faced by the AFP approach is one of identifying AFP proteins that have fungicidal activity against multiple races of the pathogen as well as the durability of resistance. In combination with R-genes, however, these genes may prove extremely useful opening up the possibility of a multi-genic approach, which could provide efficacious and durable resistance against these pathogens.

Bt cotton has been released commercially in the US, China, Australia and South Africa. Bt potato has been approved for commercialization in the US, Canada and Japan, and Bt maize has been cleared for commercial release in the US, Canada, Argentina, Japan and the European Union. However, the first Bt maize to be approved has given rise to considerable controversy in the European Union, and several member countries have challenged the decision. In addition, permits have been granted for experimental field trials of a wide range of crops, expressing Bt toxin, protease inhibitors and other unspecified genes, in many countries, especially in the US. However, owing partly to patents held by some companies, it is not clear which of these plants will reach the market. In addition, there are several lawsuits under way to clarify ownership on patenting rights to certain technologies involved.

13.5. GMO Concerns and Regulations

13.5.1. *Public concerns*

In the past tree years, there has been worldwide concern about safety issues of GM food and products, starting in Europe and later spreading to other regions. The major concerns include: toxicity and allergic reactions caused by GM food, the environmental concerns of GM crops such as killing untargeted insects and transfer of targeted genes into wild plant species (gene pollution), ethical and religious issues of consuming GM foods, and so on. This anti-GM sentiment has been fuelled by problems ranging from political and trade issues between the USA and EU, some widely publicized food safety issues in Europe unrelated to GMs, such as mad cow disease in the UK and Coca Cola episode in Belgium, which have led to general distrust of governmental regulations, to anti-MNC sentiment in some of the developing countries such as India.

There are several concerns about GM foods, such as toxicity, allergenicity, gene pollution, ethical issues and bio-terrorism. Western Europe has become resistant to food produced from genetically modified (GM) crops. Almost all of the European supermarket chains have removed GM crops from their own brands. There have been extremely active protests from a number of non-governmental organizations, including Green Peace, Friends of the Earth, and Christian Aid. The EU and its member states are backing down from the approval of GM crops. The protests are spreading around the globe to the US and other actual or potential producer countries. The major ones being food safety, transfer of genes from crops to wild species through cross-hybridization, and the effect of crop husbandry on impoverishing the countryside. The general agreement from critics and proponents alike was that although there is a need to be careful, there are no major concerns about food safety of GM products, and that the UK, the US, and other countries have put into place effective regulatory procedures that are much more rigorous for GM than for non-GM foods. For some crops in some regions (but not for a whole range of others such as wheat, maize, and rice, over most of their global distribution), cross-hybridization of crops with wild species does occur. It is not easy to demonstrate that accession of transgenes by wild relatives would provide a selective advantage for the recipient in the wild, such as gene flow from transgenic sugar beet to wild beet populations.

In contrast, public perception is in favor of GM plants for pharmaceuticals because of the potential direct benefits are also clear. In indeed, many campaigners against GM foods are also generally supportive of medical applications, because whether they are administered orally or not, pharmaceuticals derived from transgenic plants are developed and tested as pharmaceuticals, but not as foods and herbal remedies. It is hoped that the eventual market release and safety evidence from GM plant vaccines and pharmaceuticals might allay many safety concerns surrounding GM foods.

There is a view that, since the West is chiefly suffering from too much rather than too little food, it can afford to put biotechnology aside and GM crops could only be used in the developing world. There are a number of dangers in this. First, there is the argument of resource efficiency: using more land, materials, and energy than we need to produce food is a misuse of resources. Second, the technology and know-how are being developed with commercial and public resources in the West, and if the technology is blocked, the flow of funds for research will inevitably diminish severely as shareholders and the public ask why money should be spent on developing this apparently unwanted technology. Third, the food supply chain is international and the decisions as to what is grown are taken by a few people in charge of the major supermarkets and commodity traders. They are more likely to support a GM-free chain for the richer customers rather than a special GM chain for the poorer customers. Finally, the globalization of pressure groups is leading to universal opposition to GM crops. Thus, there is a danger that the research and development of solutions that GM crops could provide will slow or stop, despite the needs of millions of people for sufficient locally produced food. A food shortage caused by an empty R&D pipeline in the long-term is lethal, preventable, and immoral.

13.5.2. Scientific evidence

The reasons why non-government organizations (NGOs) oppose the introduction of GM crops include issues related to politics and to our scientific knowledge about the crops. Genetic engineering creates unprecedented new genetic life forms, that genetically modified organisms (GMOs) cannot be recalled, that DNA exchange is not understood, and that the long-term evolutionary effects of introducing GMOs were unknown. Although NGOs did not wish to say that they would never accept GMOs, the organization has such a long list of questions that he saw little chance of these being satisfied in a decade or two. How many of these questions can ever be answered without large-scale experimental field releases, which NGOs also oppose, is completely unclear. NGOs feel that the big picture is political. Its concerns are the industrialization of agriculture, the movement of people from rural to urban areas, unlimited growth rather than sustainable growth, public good versus private business, and genetic engineering allowing agrochemical companies to "mutate" into life science companies. However, it was not clear that GMOs have had much to do with many of these issues, most of which predate the technology. In contrast, many others think that GMOs have the promise to address some of the real concerns raised by NGOs and others.

No technology is perfect, and the benefits and drawbacks of every technology are relative. GM crops need to be judged against the conventional technology they replace. Growers of GM cotton containing a gene for an insecticidal toxin derived from *Bacillus thuringiensis* (Bt) use one-half as much insecticide as growers of conventional cotton. It has been estimated that total insecticide use in cotton in 1998 was around 1,000 tons less than that used before Bt cotton was introduced. The grain levels of mycotoxins (aflatoxin and fumonisin) produced by fungi that infect the grain following damage to the ear by corn borers is far less in certain Bt maize lines.

The concerns about food-safety however, have little scientific evidence to back them. In the past 5 years, with the increase of growing acreage GM crops of more than 100 million acres worldwide annually and more than 60% of processed foods in the US containing ingredients derived from recombinant organisms, there has not been a single mishap resulting in injury to a single person. There is little evidence to suggest that GM crops and food will cause serious environmental or and animal health problems. According to a study released recently by the US National Council (NRC), the research arm of the National Academy of Sciences, there was no scientific evidence that crops that have been genetically modified to resist posts pose special health or environmental risks. There are a few cases that stirred the greatest misunderstanding and concerns about GMO issue that is worthwhile to have a detailed discussion to see if these concerns are science-based or media hot-fried.

13.5.2.1. *GM potato and safety*

Dr Arpad Pusztai at the Rowett Research Institute (RRI) in Aberdeen, Scotland, did an experiment to testing whether the GM potato containing the lectin "concanavilin A" (Con A) was harmful to rats. Instead of sending the results to a scientific journal to get an expert review before publishing, he released his finding in the TV interview program "World in Action" in August 1998. He claimed that the 5 rats fed with the GM potato grews lower and their immune systems were damaged.

This statement triggered widespread concern over the safety issue of GM foods. Just days later, the RRI suspended Putztai and banned him from speaking to the media, saying his claim lacked a scientific basis — a verdict later repeated in an internal review. After examining the data provided by Putztai, an international group of 23 scientists (none of them had any transgenic working experience) supported Pusztai's claims. The row continued when the Royal Society reviewed Putztai's data and concluded that the study was based on flawed design, uncertainty about the chemical composition of the potatoes, dietary differences, small sample numbers used, and inappropriate statistical techniques.

The Royal Society produced a review of what was known scientifically about the use of GM plants for food in September 1998. In April 1999, it established a working group to examine the issues related to the possible toxicity of genetically modified potatoes, reviewing all available data relating to work at the Rowett Research Institute in this area. The report's main conclusions were as follows. First, On the basis of information available, it appeared that the reported work from the Rowett Research Institute was flawed in many aspects of design, execution and analysis. Therefore no conclusions should be drawn from it. Second, the Society found no convincing evidence of adverse effects from GM potatoes. Where the data seemed to show slight differences between rats fed predominantly on GM and non-GM potatoes, the differences were uninterpretable because of the technical limitations of the experiments and the incorrect use of statistical tests. Third, the work concerned one particular animal, when fed with one particular product modified with the insertion of one particular gene by one particular method. However skillfully the experiments were done, it would be unjustifiable to draw from them general conclusions about whether genetically modified foods are harmful to human beings or not. Each GM food must be assessed individually. Fourth, the whole episode underlines how important it is that research scientists should expose new research results to others able to offer informed criticism before releasing them into the public arena.

The data of Pusztai was later published by the medical journal *The Lancet*, a leading British medical journal. *The Lancet* justified its move as the best way to combat public suspicions of a cover-up. Dr Richard Horton, its editor, rejected accusations of behaving irresponsibly by reproducing work deemed deeply flawed by rival scientists. He said *The Lancet* was not endorsing the controversial findings of Dr Arpad Pusztai and Professor Stanley Ewen. "Ewen and Pusztai's data are preliminary and non-generalizable, but at least they are now out in the open for debate. Only by welcoming that debate will the standard of public conversation about science being raised" (*Financial Times*, October 15, 1999).

The response from the Royal Society was again to insist on its previous statement that the experiments were seriously flawed, because (1) too few animals were used to give statistically significant results for the complex phenomena being examined; (2) the diets used were incompletely controlled; (3) no control group of rates was fed a reduced protein diet — this omission is important because the GM potatoes contained less protein than normal potatoes; (4) there was a lack of rigor in the experiment design and statistical interpretation.

13.5.2.2. *GM corn and Monarch butterfly*

On May 20, 1999, three scientists from the Cornell University published a paper in *Nature* titled as "Transgenic pollen harms monarch larvae". This laboratory-based study evaluated the impact of Bt corn pollen i.e. corn containing genes from *Bacillus thuringensis*, on the survival of Monarch butterfly, whose larvae feed solely on the milkweed common around corn fields. Bt corn was designed to eliminate harmful insect pests like the European corn borer, which annually costs US growers more than US$1 billion. They showed that milkweed leaves dusted with Bt-corn pollen were harmful when fed to the larvae of Monarch butterfly, as the larvae died.

These experiment results sent out an alarm, as the Monarch population already faces a loss of habitat, so any additional threat could have important implications for the survival of this endangered insect. However, there were several problems associated with the publication. First, the method the scientists used was not quantitative, i.e. they did not measure how much pollen the Monarch butterfly larvae had eaten. Second, the dosage of Bt pollen was very high, in fact so high that the milkweed leaves appeared yellowish as they were almost fully covered by the pollen. It is not possible in the natural environment to find so much pollen fallen onto the milkweed leaves inside the cornfield, not to mention the milkweed outside the cornfield. Third, there were no choice experiments and appropriate controls: The Monarch butterfly larvae should have been given both milkweed leaves dusted with Bt pollen and leaves without pollen. This is so that the larvae would have a choice to eat or not to eat the pollen dusted leaves. Finally, the situation in the laboratory was irrelevant to the field situation, as there would be no over-dusted leaves and no forced feeding in the real field environment.

In a previous more relevant and realistic field study, scientists found that the pollen levels were the highest within the field (where monarch larvae are scarce), but even there, the mortality of monarch larvae was only 16%. These results mean that the survival of Monarch butterfly larvae, living predominantly on milkweeds outside the cornfield, should be insignificantly affected by the Bt pollen. Twenty new studies that were presented in November 1999 showed that earlier characterizations of peril to the butterfly were likely to be overstated. These findings were presented by the Agricultural Biotechnology Stewardship Working Group — a consortium of agribusiness firms — in response to the Cornell University study. 60% of the studies were funded by the industry group, the remainder being funded by individual institutions or the USDA. There were several pieces of evidence suggesting GM corn presented little threat to Monarch larvae. For example, data showed that pollen concentration dropped off rapidly several meters from the cornfield, harmful pollen was often washed off or blown off milkweed leaves, and 95% of corn pollination was complete before the first Monarch eggs hatched. In an earlier *Nature* paper, it was also shown that adult Monarchs avoided laying eggs on milkweed leaves near corn plants and those covered with pollen.

13.5.2.3. *Transgenic soybean with Brazil nut protein gene*

In an attempt to increase the sulfur-containing amino acids (methionine and cysteine) in soybean, which is known for its lack of the essential amino acid, the gene encoding the 2S albumin of Brazil nut was introduced into soybean by genetic engineering. As Brazil nut is a well known source for food allergy to a small human population, the transgenic soybean was under test for allergic reaction on patients sensitive to the nut. It was shown that the transgenic soybean indeed triggered the allergic reaction, and therefore the transgenic soybean was never approved for commercialization by the US Food and Drug Administration. These studies indicate that, when the gene source is from a known allergen or if the recipient contains allergens, it is possible to determine whether the allergen content of the transgenic line is altered relative to the nontransgenic varieties.

In another case, Kraft Foods announced the recall of 2.5 million boxes of "Taco Bell" taco shells from US supermarket shelves in October 2000 after the environmental group Genetic Engineering Food Alert (GEFA; Washington, DC) reported finding traces of this GM corn in food products "Taco Bell". The corn shells contained a protein from the bacterium *Bacillus thuringiensis*, Cry9C, which does not occur in the human food chain. Corn containing the protein, which is marketed by Aventis under the brand name Star Link, is approved only for animal consumption. When Star Link was registered, agency officials specified that it and other types of corn grown within 660 feet be used "only in animal feed, industrial non-food uses such as ethanol production, and for seed increase".

In March 2001, officials at the Environmental Protection Agency (Washington, DC) announced that GM crops would no longer be licensed for animal feed unless also approved for human consumption. In addition, agency officials invited public comments on a draft document describing how food processing can affect levels of the insecticidal protein in finished products: wet milling removes essentially all such protein from finished foods (such as in oils), whereas dry-milling procedures (used to make cornmeal) do not. Also, in March, the United States Department of Agriculture (USDA; Washington, DC) agreed to purchase seeds that inadvertently contain genes encoding the Cry9C protein. Although industry estimates say less than 1% of such corn seed is currently in the system, the USDA expects this seed purchase program will cost between US$15 million and US$20 million.

Aventis petitioned EPA to grant a "time-limited exemption" to allow Star Link in food products for the next four years — a change in status that would relieve some of the economic and legal pressures facing the company. In response, EPA officials quickly assembled a scientific advisory panel (SAP), asking its members to assess the public health risks if Star Link was consumed. There was a series of 35 reports involving 44 individuals who claimed to have experienced allergic reactions following consumption of corn-containing foods, at least seven of those cases might represent genuine allergic food reactions. Because the biotechnology industry has waved the banner of "sound science", granting an exemption to Aventis "would erode public trust". In addition, to allow animal feed into the food supply sends the wrong message to consumers. Speakers from the industry, however, described the current situation facing Aventis as "intolerable" because it effectively imposes a "zero tolerance" on the amount of Star Link that may enter the food supply, essentially disregarding how low the public's exposure to this material actually might be. He also points out that new results from tests commissioned by BIO indicate that Cry9C is substantially degraded during food processing, further reducing any likelihood that it could trigger food allergies. In June 2001, the scientific panel announced that Star Link is safe even for human consumption and there is no evidence that the Cry9C will cause human allergic reaction.

13.5.3. *Substantial equivalence*

Food safety is defined as providing assurance that food will not cause harm to the consumer when it is prepared and/or eaten according to its intended use. The report of the 1990 joint FAO/WHO consultation established that the comparison of the final product with one having an acceptable standard of safety provides an important element of safety assessment. The Organization for Economic Cooperation and Development (OECD) has elaborated this concept and advocated that the concept of substantial equivalence is the most practical approach to address the safety evaluation of foods or food components derived by modern biotechnology.

Substantial equivalence embodies the concept that if a new food or food component is found to be substantially equivalent to an existing food or food component, it can be treated in the same manner with respect to safety (i.e. the food or food component can be concluded to be as safe as the conventional food or food component). Account should be taken of any processing that the food or food component may undergo as well as the intended use and the intake by the population.

An assessment of substantial equivalence can be carried out at the level of the food or food component, which will be used as human food. Where possible, a determination of substantial equivalence should consider comparisons as close to the species level as possible in order to allow the flexible use of many types of food products from the species in question (e.g. modified soybean). This will entail consideration of molecular characterization, phenotypic characteristics, key nutrients, toxicants and allergens. For certain food products, comparison at the food product level would permit a conclusion of substantial equivalence even though a comparison at the species level would consider the product to be substantially equivalent except for a defined difference (e.g. oil from herbicide tolerant canola). The approach is to compare the food or food component from the genetically modified organism to the range of values obtained for the traditional counterpart, taking into account the natural variation in range for the host organism and for foods and food components obtained therefrom. The data required to demonstrate substantial equivalence may come from a variety of sources including existing databases, the scientific literature or data derived from the parental and/or other traditional strains/varieties. Regardless of the techniques used to produce new food organisms, attention must be paid to the impact of growth conditions on levels of nutrients and toxicants; for example, in the case of new plant cultivars, attention must be paid to the impact of different soils and climatic conditions.

When a food product is determined to be substantially equivalent to an existing counterpart except for defined differences, it was concluded by the consultation that further safety assessment should focus only on those defined differences. Typically, the defined differences will result from the intended effect of the introduction of genetic material that encodes for one or more proteins that may or may not modify endogenous components or produce new components in the host organism. This category could also include products from genetic modification and that have been shown to produce an unintended substance(s) if that unintended substance(s) is clearly defined. The safety of introduced DNA and messenger RNA (mRNA) *per se* is not an issue. However, the stability of introduced genetic material and the potential for gene transfer are relevant issues in the assessment. The potential for gene transfer is addressed in Sec. 13.3.2. Stability of introduced genetic material should be addressed during both the molecular characterization and the performance evaluation of the genetically modified organism in the development process. These processes minimize the likelihood of unintended effects in subsequent generations.

Up to the present time, and probably for the near future, there have been few, if any, examples of foods or food components produced using genetic modification, which could be considered to be not substantially equivalent to existing foods or food components. Nevertheless, it is conceivable that with future developments in biotechnology, products with no conventional counterpart and for which substantial equivalence could not be applied could be developed. For example, there could be products derived from organisms in which there has been transfer of genomic regions, which have perhaps been only partly characterized.

If a food or food component is considered to be not substantially equivalent to an existing food/component, it does not necessarily mean it is unsafe and not all such products will necessarily require extensive testing. Such a food or food component should be evaluated on the basis of its composition and properties. However, a sequential approach to assess the safety and wholesomeness of these foods should be considered. First, it will be necessary to characterize the product based on information as described in Sec. 13.5.1. This includes details of the host organism, the genetic modification and inserted DNA, as well as properties of the modified organism/product with respect to phenotype and chemical and nutritional composition. Where a modification has involved the insertion of genomic regions, which have been poorly characterized, it will also be important to consider the donor organism.

The results of this initial characterization and the role that the product is to play in the diet will determine whether further safety testing is needed. Although many protocols are in place for testing food ingredients, these methods were not designed to test the safety of complex whole foods. In particular, the use of animal feeding studies has many limitations because of the insensitivity of the test system in the detection of low level effects, problems of diet balancing and the problem of assigning adverse effects to specific major foods or food components. Despite these limitations, there are no alternatives at the present time, and if animal studies are deemed appropriate, their objectives should be clear and care should be taken in the experimental design.

Because of these difficulties, a tailored testing program should be devised for these products on a case-by-case basis depending on the information generated during the initial characterization. This could involve testing for products which possess new inserted traits such as new proteins, fats or carbohydrates. A combination of *in vitro*, and specific *in vivo* animal models may need to be employed to further assess the safety of the product. Particular attention needs to be paid to the bioavailability of new food components as well as wholesomeness. With respect to nutritional aspects, human studies may need to be carried out, especially if the new product is intended to replace a significant part of the diet. These should only be carried out when animal studies have shown that the product is not toxic. Attention should be paid to sensitive segments of the population. Furthermore, it will be important to take into account regional variations for foods, which will have international distribution.

13.5.4. *Precautionary principle*

"Precautionary Principle" has been used to cripple research activities in both public and academic sectors and the biotechnology industry, and it had been used in history to destroy several industries. In the late 1980s, environmental activists were trying to convince water authorities around the world of the possibility that carcinogenic by-products of chlorinating made drinking water a potential cancer risk. Peruvian officials caught in a budget crisis used this supposed threat to public health as a justifying reason to stop much of their country's drinking water. That decision contributed to the acceleration and spread of the 1991–1996 cholera epidemic in Latin America, which afflicted more than 1.3 million people and at least 11,000 were killed. More recently, some people used the precaution principle to attack phthalates — a liquid organic compound added to certain plastics to make them softer. These soft plastics are used for important medical devices (such as fluid containers, blood bags, tubing and gloves), children's toys (e.g. teething rings and rattlers) and household and industrial items (e.g. wire coating and flooring). The activists claimed that phthalates could have numerous adverse health effects — even in the significant scientific evidence to the contrary. Governments have taken these unsupported claims seriously, and several formal and informal bans have been implemented around the world. The industry has been heavily affected, consumers have been deprived of product choices, and doctors and their patients have been denied lifesaving tools.

Several European countries have used the precautionary principle to justify paralyzing restrictions on agricultural and food biotechnology, and the European Commission (EC) has used it to justify a moratorium on the approval of new recombinant DNA-modified products. Many scientific organizations, such as the British Royal Society, the American National Academy of Sciences, the World Health Organization, and the American Medical Association have analyzed all the aspects of GMO and come to the remarkable conclusion that genetic engineering is an extension and refinement of classical and less precise breeding techniques. Adding genes to plants and microorganisms does not make them less safe either to the environment or to be consumed by both human beings and animals. The risks associated with recombinant DNA-modified organisms are the same in kind as those associated conventionally modified organisms, and regulation should be based on the risk-related characteristics of individual products, regardless of the techniques used in their development. Recombinant DNA-modified plants are now grown worldwide on more than 100 million acres annually, and more than 60% of processed foods in the Unites States contain ingredients derived from recombinant organisms. There has not been a single mishap resulting in injury to a single person.

It is precisely the anti-technology nature of the precautionary principle that makes it the darling of many non-governmental organizations. One of the principle advocates of the precautionary principle, Green Peace, wrote in its 1999 Internal Revenue Service filings that the organization's goal is not the prudent, safe use of recombinant DNA-derived foods or even their labeling. It demanded nothing less than GMO products to be "completely eliminated [from] the food supply and the environment". Many of these groups had also carried out vandalism in the GMO crop field trials.

13.5.5. *GMO testing*

GM crops are adapted by the agricultural industry very quickly due to the advantages of more flexible crop management, higher productivity and safer environment through decreased use of pesticides. With the fast adaptation of GM plants by farmers, it may soon come to a time when most foods made with soybean, corn or cotton may contain some genetic modified ingredients. Considering soy protein can be found in about 60% of all processed food and corn starch provides the high-fructose sweeteners found in many sodas, fruit drinks and other products, a lot of different foods will be affected. Consumers, however, are not fully ready for such products. The lack of direct benefits and the remote possibility of risk are all deterring factors for consumers to accept GM foods. There is the worldwide trend of consumers requesting to be informed to make a choice. In a recent survey conducted by the UK Consumers' Association, 92% of the respondents wanted GM food to be labeled. Likewise, 87% of the Singaporeans surveyed recently wanted labeling of foods with GM ingredients in order to make a choice. In EC, food labeling is dealt with by the EC regulation 1139/98 requiring mandatory labeling of GMO products if not considered to be substantially equivalent, or if a new protein and/or the genetic modification of the DNA can be detected. Worldwide, more and more countries are in the process of enacting legislation to require labeling of GM foods. The Singapore government has set up Genetic Modification Advisory Committee to draft regulations on research, import of GM plants and food labeling issues. Eventually, a decision will be made as to what to label and thresholds set up for the purpose of labeling. In any genetic modification of plants, one or few genes conferring a target plant certain desirable phenotypes were introduced. Marker genes were also used for the purpose of screening for transgenic plants. To detect GM plants and products, three techniques are currently used for detecting either utility gene(s)/marker gene(s) or their protein products and biological functions (Table 13.4).

The first is the biological technique to detect biological activities conferred by transgenes (either antibiotic resistance or herbicide resistance). Generally, it requires the presence of full-length protein and biological activity retained in the materials, thus it is not suitable for processed foods.

Table 13.4. Main GMO testing methods.

	Biological Testing	PCR I: Qualitative	PCR II: Quantitative	ELISA
Principle	Detect biological activity conferred by foreign proteins	Detection for the presence of foreign genes	Detection and quantification of foreign genes with the use of proper reference materials	Detection of the interaction of foreign proteins with antibodies
Advantages	Function testing	a. Sensitive b. Fast c. Cost effective d. Easy to setup and conduct e. Able to detect the presence of gene fragments, thus suitable for processed products	a. Sensitive b. Quantitative, with the use of proper references, able to tell the mixing ratio c. Able to detect the presence of gene fragments, thus suitable for processed products	a. Fastest method available b. Reliable method c. Easy and cheap to conduct
Disadvantages	a. Limited to those proteins whose biological activities are easy to detect b. Unsuitable for processed products c. Laborious and time consuming	a. Reliability is the concern b. Provides only yes/no answer	a. High initial setting up cost b. High running cost c. Technically more complicated d. Internal reference very important	a. Provides only yes/no answer, only semi-quantitative in the best case b. Less sensitive c. Has problems in detecting processed products dueto protein denaturation
Turn-around time	7–9 days	2 hours to 2 days	1–3 days	5 minutes to half a day
Cost per test (US$)	25	1–20	100–350	15–20
Major commercial players	ICIA	Genetic-ID	IdentiGEN, Genetic-ID, Eurofins Scientific, Qualicon, RHM Technology	Strategic Disgnostics, Alchemy Laboratories

This technique is limited to few GM products and is mainly used for the detection of fresh and unprocessed materials. The second method is the polymerase chain reaction (PCR): the use of specific primers to detect presence of foreign genes after polymerase amplification. The technique is the most sensitive method available. It can be used to detect either the presence of the whole foreign gene or gene fragments, thus suitable for detection in processed products. Generally, the technique is much faster than biological activity detection methods and typically takes a few hours to 3 days to obtain results. Regular PCR technique is qualitative. Several modified PCR methods are now available for the purpose of quantitation of transgenes with proper controls and references. The third detection method is the ELISA method: making use of a specific antibody to detect the presence of foreign proteins in plants or products. With a specific antibody and the right formulation of a detection method, it can be very reliable and fast: results can be obtained within a few minutes. However, it is generally qualitative or at best semi-quantitative if known standards are used. Moreover, it is not as sensitive as PCR technology and fails to detect very minute amounts of foreign proteins. It is also not suitable for processed foods due to protein denaturation and degradation.

The two quantitative PCR methods: quantitative competitive PCR (QC-PCR) and real time PCR are regarded suitable for the enforcement of threshold values. In QC-PCR, a competitive amplicon is included in the reaction and quantitation is based on the relative amplification of target versus defined amount of competitive amplicon. In real time PCR, quantitation is made during exponential phase of amplification in reference to standard curves of control samples. In 1998, twelve Swiss laboratories did ring test with QC-PCR of eight food samples. By using external RRS (Roundup Ready Soybean) reference materials as control, GMO content could be classified as negative, below 0.5%, between 0.5% and 2%, and above 2% and there were neither false positive nor false positive results and inter-laboratory differences were found to be small. Quantitative methods represent the future of GM food testing and becoming a hot spot for technology development. However, these methods demand higher investment of equipment and more skillful technicians to perform such testing.

13.6. Conclusion

From the research and development in the past 10 years, we can clearly see the potential and benefits for both agricultural producers and consumers. The GMO concerns are mainly based on "fear of the unknown", political reasons, trade-related issues and ideological beliefs, rather than scientific evidence. So far, we have seen no clear evidence that GMO will cause any adverse effects to both human beings and the environment. Instead, we have seen clearly that unsuitable GMO products for consumers such as the transgenic soybean with Brazil nut protein and SlarLink corn were properly regulated to be kept out of the dining table. With proper scientific education to gain public confidence, genetic engineering will in future be a powerful tool for our societies to further increase agricultural production, improve nutritional values, reduce environmental pollution or even offer alternatives for environmental clean-up, and as the basis for setting up green factories to produce pharmaceutical proteins and molecules and industrial raw materials.

Further Readings

1. Bennetzen JL. Transposable element contributions to plant and genome evolution. *Plant Mol. Biol.* **42**: 251–269, 2000.
2. Dempsey DA, Silva H and Klessig DF. Engineering disease and pest resistance in plants. *Trends Microbiol.* **6**: 54–61, 1998.
3. Dunwell JM. Transgenic approaches to crop improvement. *J. Exp. Bot.* **51**: 487–496, 2000.
4. Dutt MJ and Lee KH. Proteomics analysis. *Curr. Opin. Biotechnol.* **11**: 176–179, 2000.
5. Ellis JA, Dodds P and Pryor T. The generation of plant disease resistance gene specificities. *Trends Plant Sci.* **5**: 373–379, 2000.
6. Giddings G, Allison G, Brooks D and Carter A. Transgenic plants as factories for biopharmaceuticals. *Nat. Biotechnol.* **18**: 1151–1159, 2000.
7. Giuliano G, Aquilani R and Dharmapuri S. Metabolic engineering of plant carotnoids. *Trends Plant Sci.* **5**: 406–409, 2000.

8. Hansen G and Wright M. Recent advances in the transformation of plants. *Trends Plant Sci.* **4**: 226–231, 1999.

9. Hood EE and Jilka JM. Plant-based production of xenogenic proteins. *Curr. Opin. Biotechnol.* **10**: 382–386, 1999.

10. Jaeger De G, Wilde De C, Eeckhout D, Fiers E and Depicker A. The plantibody approach: Expression of antibody genes in plants to modulate plant metabolism or to obtain pathogen resistance. *Plant Mol. Biol.* **43**: 419–428, 2000.

11. Jung CA, Cai DG and Kleine M. Engineering nematode resistance in crop species. *Trends Plant Sci.* **3**: 266–271, 1998.

12. Le Gall, Bove F and Garnier M. Engineering of a single chain variable fragment (scFv) antibody specific for the stolbur phytoplasma (mollicute) and its expression in *Escherichia coli* and tobacco plants. *Appl. Environ. Microbiol.* **64**: 4566–4572, 1998.

13. Ma JKC. Genes, greens and vaccines. *Nat. Biotechnol.* **18**: 1141–1142, 2000.

14. Merkle SA and Dean JFD. Forest tree biotechnology. *Curr. Opin. Biotechnol.* **11**: 298–302, 2000.

15. Miflin B. Crop improvement in the 21st century. *J. Exp. Bot.* **51**: 1–8, 2000.

16. Miller JA and Nagarajan V. The impact of biotechnology on the chemical industry in the 21st century. *Trends Biotechnol.* **18**: 190–191, 2000.

17. Mor TS, Gomez-Lim MA and Palmer KE. Perspective: Edible vaccines — a comcept coming of age. *Trends Microbiol.* **6**: 449–453, 1998.

18. Parinov S and Sundaresan V. Functional genomics in *Arabidopsis*: Large-scale insertional mutagenesis complements the genome sequencing project. *Curr. Opin. Biotechnol.* **11**: 157–161, 2000.

19. Riemenschneider DE. Genetic engineering of horticultural and forestry crops for herbicide tolerance. In: *Biotechnology of Ornamental Plants*, eds. RL Geneve, JE Preece and SA Merkle. CAB International, 1997.

20. Rommens CM and Kishore GM. Exploiting the full potential of disease-resistance genes for agricultural use. *Curr. Opin. Biotechnol.* **11**: 120–125, 2000.

21. Schulera TH, Poppya GM, Kerrya BR and Denholmb I. Insect-resistant transgenic plants. *Trends Biotechnol.* **16**: 168–175, 1998.

22. Springer PS. Gene traps: Tools for plant development and genomics. *Plant Cell* **12**: 1007–1020, 2000.

23. Walmsley AM and Arntzen C. Plants for delivery of edible vaccines. *Curr. Opin. Biotechnol.* **11**: 126–129, 2000.

24. Whitelam GC and Cockburn W. Antibody expression in transgenic plants. *Trends Plant Sci.* **1**: 268–272, 1996.

25. Zhu JK. Plant salt tolerance. *Trends Plant Sci.* **6**: 66–71, 2001.

26. Zivy M and de Vienne D. *Proteomics: A Link Between Genomics, Genetics and Physiology*, 2000.

Part IV

Microbes in Medical Biotechnology

The staggering breakthroughs in medical and microbial biotechnology, notably in molecular biology, have led to great strides in the understanding and treatment of human diseases. Monoclonal antibodies, gene probes and polymerase chain reaction offer improved diagnosis of infectious diseases and other ailments. Microbial and animal host cells, transgenic animals and plants generate genetically-engineered, high value pharmaceuticals and vaccines in large quantity. Sophisticated structural studies and computer modeling of complex molecules permit the design of novel drugs. Previously intractable infections are now preventable by novel immunization strategies using recombinant DNA technology, and amenable to new anti-microbial agents discovered from nature. Complemented by the enormous Human Genome Project, gene therapy is at the threshold of a new dimension in medical science. Recognizing the impact of these advances on human health and economic development, scientists are harnessing these enabling technologies to meet the new challenges in medicine, including the disciplines of medical microbiology and infectious diseases.

Chapter 14

Diagnostic Clinical Microbiology

Vincent T. K. Chow[*]**, Tim J. J. Inglis**[†]** and Keang Peng Song**[*]
*Department of Microbiology, National University of Singapore
5 Science Drive 2, Singapore 117597*
†*Division of Microbiology and Infectious Diseases*
Western Australian Centre for Pathology and Medical Research
Locked Bag 2009, Nedlands WA 6009, Australia

The 21st Century is hailed as a potentially significant milestone in human history that will witness the next new wave of technology, namely the Biotechnological Revolution whose seeds were sown with the advent of many critical advances in molecular biology. Biotechnology is the integrated and practical application of bioscientific and engineering disciplines to the industrial processing of materials by biological agents to provide useful products and serve other desirable purposes. Recent advances in biotechnology have tremendous impact on numerous fields of human endeavor, including the food and agricultural industry, environmental management, and notably medicine. Fundamental scientific research in medicine, genetics, immunology and communicable diseases accelerated the discovery of new biotechnological innovations, especially in molecular biology over the last 30 years culminating in this "golden age of biology". This leads to a better understanding of the principles of human health, disease etiopathogenesis, prevention and management, with important implications in clinical medicine and health care. For example, significant advances in unraveling the molecular mechanisms of cancer, a major global health problem, have been accomplished through the study of oncogenic viruses.

14.1. The Challenges of Emerging and Re-Emerging Infectious Diseases

About two fifths of all deaths worldwide are caused by diseases of the circulatory system and cancers which are more prevalent in developed countries with increasingly aging populations, and are linked to genetic and lifestyle changes such as dietary habits, obesity and stress. These factors also contribute to the rising incidence of other noncommunicable ailments such as Alzheimer's disease, allergic and autoimmune disorders in developed societies.

The myth that infectious diseases have all been eradicated still persists in this era of high technology medicine. In fact, these diseases remain the leading cause of mortality, accounting for one third of deaths worldwide, especially in underdeveloped countries.

Multiple predisposing factors, often in combination, provoke new epidemics of communicable diseases, including:

(a) Evolution and adaptation of human pathogens (e.g. mutability to virulent strains), evasion of host defences, and decreasing herd immunity to certain microbes.
(b) Socio-political disruptions, demographic, lifestyle and behavioral changes such as explosive world population growth, rampant urbanization, poverty, malnutrition, ignorance, poor healthcare infrastructure, uncontrolled migration (e.g. refugees), mass transportation, international air travel, tourism and sexual promiscuity.
(c) Environmental deterioration and ecological degradation such as water pollution, adverse climatic conditions (e.g. global warming, El Nino and La Nina weather phenomena), man's encroachment into virgin jungles where unknown pathogens lurk, and deforestation facilitating contact between humans and animals, especially rodents.

Of great concern is the global resurgence of "old" diseases such as dengue, measles, malaria, cholera, meningitis, plague, pneumonia, tuberculosis, typhoid, gonorrhea, and sexually transmitted diseases. Equally alarming is the emergence of new diseases such as those that are thought to jump across species (e.g. human immunodeficiency virus or HIV; Ebola virus from primates serving as reservoir of the virus in the wild; variant Creutzfeldt-Jakob disease from "mad cow" disease; "bird flu" attributed to the avian influenza H5N1 strain which can cause multi-organ failure; severe acute respiratory syndrome (SARS)), and of diseases spread from animals (e.g. hantavirus pulmonary syndrome resulting from increased rodent population). There is also the potential risk of new retroviral infections arising from xenotransplantation, the practice of transplanting organs and tissues from animals such as pigs, to humans. More than 5 million new HIV infections occur annually, indirectly giving rise to HIV-related opportunistic infections, e.g. herpes simplex, cytomegalovirus infection, multi-drug resistant tuberculosis (MDRTB), *Pneumocystis carinii* pneumonia, candidiasis, penicilliosis (caused by *Penicillium marneffei*), toxoplasmosis, cryptosporidiosis, and isosporiasis (parasitic infection of intestines).

Anti-microbial resistance is spreading inexorably at a frightening pace, arising chiefly from the abuse by doctors, improper compliance by patients, high usage in immunocompromised patients and widespread antibiotic use in animal feed to promote livestock growth, culminating in the selection of drug-resistant strains (e.g. *Salmonella*). Methicillin-resistant *Staphylococcus aureus* (MRSA), vancomycin insensitive *S. aureus* (VISA), vancomycin-resistant enterococci (VRE), penicillin-resistant pneumococci, penicillinase-producing *Neisseria gonorrhoeae* (PPNG), MDRTB, resistant *Acinetobacter baumannii* strains, enterobacteria producing extended spectrum beta-lactamases (ESBLs), cotrimoxazole-resistant *Shigella*, and chloroquine-resistant *Plasmodium* parasites are causing enormous problems both at hospital and community levels. Also appearing with alarming rapidity are drug-resistant HIV genotypes that harbor mutations in protease and reverse transcriptase genes. To deal with the rising incidence of MDRTB, the World Health Organization is implementing DOTS (directly observed treatment, short course) as part of its tuberculosis control strategy.

Other examples of recently discovered or re-emerging microbes and related diseases include flesh-eating streptococcal bacteria; *E. coli* O157:H7 causing food poisoning outbreaks; *Helicobacter pylori* associated

Table 14.1. The emergence of significant microbial pathogens and clinical infectious diseases (1965–2000).

Viruses	Diseases
Ebola virus	Ebola hemorrhagic fever
Enterovirus 71	Hand, foot and mouth disease Encephalitis
Guanarito virus	Venezuelan hemorrhagic fever
Hantavirus	Hemorrhagic fever with renal syndrome Hantavirus pulmonary syndrome
Hendra virus (equine morbillivirus)	Encephalitis
Hepatitis A	Infectious hepatitis
Hepatitis B	Serum hepatitis Chronic hepatitis, cirrhosis Primary liver cancer
Hepatitis C	Non-A, non-B hepatitis Liver cirrhosis Primary liver cancer
Hepatitis E	Enteric-transmitted hepatitis
Human herpesvirus 6	Exanthem subitum (Roseola infantum)
Human herpesvirus 8	Kaposi's sarcoma
Human immunodeficiency viruses 1 and 2	Acquired immunodeficiency syndrome
Human papillomaviruses 16, 18 and other oncogenic types	Genital cancers
Human T-cell lymphotropic virus I	Adult T-cell leukemia and lymphoma
Human T-cell lymphotropic virus II	Hairy cell leukemia
Lassa virus	Lassa hemorrhagic fever
Marburg virus	Marburg disease
Nipah virus	Encephalitis
Parvovirus	Fifth disese Fetal death Aplastic crisis in chronic hemolytic anemia
Prion	Variant Creutzfeldt-Jakob disease
Rotavirus	Infantile diarrhea
Sabia virus	Brazilian hemorrhagic fever
SARS coronavirus	Severe acute respiratory syndrome

Table 14.1 (*Continued*)

Bacteria	Diseases
Bartonella henselae	Cat scratch disease Bacillary angiomatosis
Borrelia burgdorferi	Lyme disease
Campylobacter sp.	Diarrheal disease
Clostridium difficile	Pseudomembranous colitis Antibiotic-associated colitis
Ehrlichia chaffeensis	Ehrlichiosis
Escherichia coli O157:H7	Hemorrhagic colitis Hemolytic uremic syndrome
Helicobacter pylori	Gastritis Peptic ulcer
Legionella pneumophila	Legionnaires' disease/pneumonia
Staphylococcus aureus (toxigenic strains)	Toxic shock syndrome
Vibrio cholerae O139	Epidemic cholera

Parasites	Diseases
Cryptosporidium parvum	Enterocolitis
Cyclospora cayetanensis	Diarrheal disease
Microsporidia	Immunodeficiency-related microsporidiosis

with inflammation, ulcer and cancer of the stomach; *Chlamydia pneumoniae* and its association with coronary heart disease; *Clostridium difficile* and antibiotic-associated colitis; *Burkholderia pseudomallei* and melioidosis; *Borrelia burgdorferi* and Lyme arthritis; *Bartonella henselae* and cat scratch disease, bacillary angiomatosis; oncogenic human papillomaviruses associated with genital cancers; hepatitis B and C viruses with primary liver cancer; human herpesvirus 8 with Kaposi's sarcoma; Rift Valley fever; encephalitides associated with enterovirus 71, West Nile, Hendra and Nipah viruses; and *Cyclospora cayetanensis* protozoa with diarrheal disease. The potential spectre of biological terrorism and germ warfare (e.g. with deadly smallpox virus and anthrax spores) cannot be disregarded. New outbreaks of old enemies and of undiscovered diseases are likely to occur in the foreseeable future.

Humankind ignores the threat of these killer microbes at its own peril. It would be prudent to prepare ourselves for the worst case scenario — the appearance of a disease of high infectivity and lethality. There is thus an urgent need for politicians, doctors, and scientists to concentrate resources on fighting infectious diseases. In recognition of the importance of public awareness, surveillance of and contingency plans for emerging infections throughout the world, the World Health Organization's focus for World Health Day on 7 April 1997 was "Emerging Infectious Diseases: Global Alert, Global Response". In addition, pharmaceutical companies should continue to invest in anti-microbial drug and vaccine development, harness molecular biotechnological advances, and devise new strategies to combat microbial diseases.

14.2. Medical and Diagnostic Bacteriology

Most larger laboratories are able to provide a comprehensive diagnostic bacteriology service, though many centers offer a restricted range of virology, parasitology and mycology tests. The bulk of diagnostic clinical microbiology comprises bacteriological tests. Most bacterial pathogens are relatively easy to cultivate and manipulate under laboratory conditions. Susceptibility testing of antibacterial agents can quickly identify acquired forms of antibiotic resistance and therefore predict an increased risk of treatment failure. Moreover, some bacteria have become prominent causes of hospital-acquired infection and dictate the use of supplementary analytical methods to support hospital infection control. Paradoxically, molecular biology has been relatively late in influencing diagnostic bacteriology, and is only now beginning to alter the practices that have been laboratory routine for many years. These and other technologies present new challenges for laboratory quality control. Finally, information technology opens up a range of opportunities for making better use of diagnostic laboratory-generated data.

14.2.1. *The purpose of diagnostic bacteriology*

You may think that the phrase "diagnostic bacteriology" is self-explanatory. But it fails to recognize the complex nature of diagnostic laboratory work, which is much more than just the performance of laboratory tests. The best diagnostic laboratory can be described as a problem-solving toolbox whose components co-ordinate a process that starts with a clinical problem (i.e. a bacterial infection) and ends with a solution (i.e. a causal agent or prospective pathogen, recommendations for antibiotic therapy or guidance for infection control). Understanding what bacterial agent caused the infection in question is crucial to this process, but establishing a causal relationship can be difficult (Fig. 14.1), and the subsequent parts of the problem-solving process are the ones that make the more important contribution to the patient's eventual outcome. A problem-oriented focus is doubly important now that physicians are graduating in increasing numbers from new, problem-based medical school curricula. How effectively they integrate diagnostic bacteriology into their clinical practice will depend all the more on how effectively microbiologists can develop a "can-do" laboratory environment.

14.2.2. *General objectives and schema*

The principal product of diagnostic bacteriology is information. The process is usually initiated by a written test request and completed by a laboratory

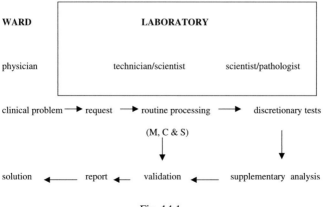

Fig. 14.1.

report. Although the sequence can theoretically be entirely electronic, the physical actions are bracketed by those critical information events. The information product or report can be judged against time, relevance, accuracy and consistency. The latter, accuracy and consistency, are key issues to be incorporated in quality control and assurance systems. Time, measured as turnaround, is a measurable often tracked by laboratory management. Relevance is more difficult to quantify and usually requires feedback from laboratory users. This is the province of the medical microbiologist and is perhaps one of the toughest tests of the laboratory's relationship with the corresponding clinical service.

14.2.3. *General methods*

Laboratory request forms often contain the phrase "microscopy, culture and sensitivity" (often shortened to "M, C and S"). These words summarize the core of traditional bacteriological testing of clinical specimens. This is shorthand for a generic range of tests that is applied to the majority of clinical microbiology specimens.

(1) **Microscopy** normally equates with Gram stain. This stain is used to assess the range of bacteria and inflammatory cells present in an inflammatory exudate. Though crude, the speed at which a Gram stain can be performed means that it is one of the truly rapid diagnostic tests available. Care must be taken not to over-interpret the data that can be obtained from a Gram stain, e.g. calling clusters of Gram-positive cocci in exudate full of neutrophil polymorphonuclear cells "*Staphylococcus aureus*" until definitive identification procedures have been performed. Other important microscopic techniques include examination of an unstained urine specimen for inflammatory cells, or of a Ziehl-Neelsen stained sputum specimen for the acid-fast bacilli characteristic of tuberculosis (Table 14.2).

(2) **Culture** is the term used to describe the biological amplification of cultivatable bacteria from clinical specimens using artificial growth media. These media may be liquid (broth), solid (due to the addition of agar, and used for their growth-supporting surface), enriched (with growth-promoting substrates), selective (with inhibitors to reduce

Table 14.2. Staining for direct microscopic detection of bacterial cells.

Method	Dye	Application	Processing time
Direct film	None	Urine deposit	circa 5 min
Gram stain	Methyl violet carbol fuchsin	Bacterial wall Shape and structure	circa 10 min
Ziehl-Neelsen Stain	Carbol fuchsin	Mycobacterial detection	circa 30 min

growth of unwanted species) or contain specific indicators to reveal growth of species of particular interest. Bacterial growth also requires the right temperature and atmospheric conditions. Many bacterial pathogens grow best at or near human body temperature (37°C). The more common atmospheric conditions used to grow bacteria are in room air, anaerobically (without oxygen), microaerophilically (low oxygen) and in the presence of carbon dioxide. The temperature and atmospheric conditions required for bacterial growth are usually provided in a purpose built incubator, but occasionally alternative systems such as gas jars or glove boxes are used. The bacteria recovered from clinical specimens in or on artificial media are said to have been "isolated" because, to be useful to the laboratory worker and requesting physician, they have first to be recovered in pure culture. This requires growth of each, individual species in single colonies (hence "isolated").

In order to identify bacteria isolated from clinical specimens a process of **identification** is followed, beginning with the simplest of test and moving onto increasingly complex, costly methods. The first identification tests performed include observation of bacterial colony morphology, alteration of the surrounding agar medium, the Gram stain (on single colonies, so no human cellular material or exudate is present), so-called benchtop tests such as coagulase, catalase and oxidase reactions and specific antibody agglutination reactions. In many cases, these tests will provide sufficient information without having to go further. As a rule of thumb, it is good practice to identify only those colonies which suggest the presence of bacteria that are capable of causing infection at the site from which the specimen was collected. Non-sterile surfaces will have a commensal flora that does not require laboratory work-up. Further identification procedures often involve a panel of substrate utilization tests, each of which will require some period of incubation to promote bacterial growth. A popular method of performing these tests efficiently is to use a multiple test panel. Results of any phenotypic identification system are subject to variation and possibly even interpretation. Additional identification procedures for difficult bacterial groups such as fastidious anaerobes, the actinomycetes, non-fermentative Gram-negative bacilli and mycobacteria may require more advanced methods including gas chromatography, nucleic acid amplification, molecular subtyping and even gene sequencing. At present these methods are beyond the means of all but the largest laboratories. But as the expense of providing these services is falling steadily, they can be expected to be in more widespread use soon.

(3) **Susceptibility testing** of bacteria isolated from a clinical specimen is one of the most important contributions that clinical microbiology has made to patient care, since the results can have considerable impact on the choice of antibiotics used to treat the patient's infection. However, it is important not to overestimate the significance of antibiotic susceptibility results. The time taken to generate a susceptibility report by conventional testing methods is too long for the attending physician to wait for the results before starting initial antibiotic therapy. The physician will therefore make a presumptive choice based on the antibiotics most suitable for a particular infection and modify this initial choice later according to laboratory results. Thus, if the problem is a wound infection thought from the Gram stain to be caused by staphylococci, the presumptive choice of antibiotic might be flucloxacillin. This would be satisfactory unless subsequent susceptibility tests show the isolate to be a methicillin-resistant *Staphylococcus aureus*, in which case the physician would probably want to substitute flucloxacillin with vancomycin.

Given the clinical importance of antibiotic susceptibility testing, it is essential that all stages of the process are strictly quality controlled. Standard operating procedures must be based on widely recognized methods. Result interpretation must be consistent and a method of dealing with indeterminate results should be agreed. An intermediate result (as opposed to sensitive or resistant) in most cases is likely to confuse requesting physicians. The most widely used methods are (a) the Kirby-Bauer disk diffusion method in which measurement is made of the diameter of an inhibition zone produced by an antibiotic-containing disk placed on agar seeded with a lawn of test bacteria; and (b) the growth of suspensions of test bacteria on agar plates containing dilutions of antibiotic (sometimes known as the "break-point" method). These and other methods must be conducted with control strains whose results are already known. Both methods have their weaknesses such as variation due to differences in the density of inoculation of the bacterial lawn in the disk diffusion method, or results close to the specific dilution chosen for the break-point method. Careful attention to results with control strains and regular review of out-of-range results (e.g. new antibiotic resistance patterns) will help minimize these sources of potential error. It should be noted that antibiotic susceptibility testing provides only a guide antibiotic therapy with the range of agents tested. Many other factors affect the outcome of antimicrobial chemotherapy and in some cases (e.g. mucoid strains of *Pseudomonas aeruginosa* and facultative intracellular bacterial pathogens such as *Listeria monocytogenes* and *Legionella pneumophila*) conventional susceptibility testing methods can be a very poor guide to clinical outcome.

In some circumstances (such as bacterial endocarditis) a more definitive susceptibility result may be required. This is usually obtained by determination of the minimum inhibitory concentration (MIC) to a given antibiotic compound, which represents the lowest concentration at which growth of the bacterial species is inhibited under standardized conditions. Several methods are in common use including broth macrodilution (the test tube version), broth microdilution (using microtiter trays), and the application of an antibiotic-impregnated plastic strip onto a bacterial lawn (E-strip). The extra work required for MIC determinations means they are conducted more selectively than conventional disk diffusion or break-point methods.

(4) **Molecular** and other bacterial **subtyping** methods are progressively displacing more traditional typing systems such as phage typing, whose historical role will prevent their complete abandonment. The typing systems now coming into wider use include gas chromatography, nucleic acid amplification-based techniques, molecular subtyping and even gene sequencing. At first glance, the choice of methods seems bewildering; an alphabet soup of acronyms. However, there are some essential criteria that can help sort out an order of priority. The ideal typing system will reproducibly subdivide or discriminate between different strains of a given species. The typing system should be easily performed at little expense and work reliably with any strain. These features are known respectively as reproducibility, discrimination, ease, cost and typability. No system fulfils all these criteria and inferences derived from one typing system may need confirming by use of a second dissimilar system. Genetic typing relies on the assumption that all strains of a given species derive from a common ancestor. All genetic variation accumulated by successive variation from the ancestral clone can be represented in terms of genetic distance. The degree of dissimilarity detected between two isolates is therefore thought to represent a measure of distinctness. The question of how much difference is necessary to call two isolates different often arises and is usually resolved by considering the number of separate genetic events (mutations) necessary to cause the differences observed. So in a DNA macrorestriction analysis (pulsed field gel electrophoresis or PFGE), a single band difference between two lanes would be regarded as insignificant, a two to four band difference borderline, and a six band difference significant.

A decade ago, molecular epidemiology was performed by relatively cumbersome methods such as plasmid gel electrophoresis (PGE), and such methods were subjected to a range of potential errors. The arrival of nucleic acid amplification techniques led to the development of systems based on polymerase chain reaction (PCR) including random amplified polymorphism detection (RAPD) and repetitive extragenic palindromic sequence analysis (REPS). These and similar methods were limited by a lack of reproducibility, and in some instances, poor typability. At present, DNA macrorestriction is widely used as a definitive molecular typing technique for medically important bacteria. This system uses infrequently cutting restriction enzymes to divide the bacterial chromosome into large fragments that are then focused in an agarose gel by a current running from hexagonal electrodes. Switching of current direction allows the migration of relatively large molecules such as DNA fragments. The bands formed by these fragments can be seen by staining the gel with ethidium bromide. The resulting ladder pattern resembles a bar code and can be scanned for computer-based analysis with interpretative software. DNA macro-restriction is time-consuming and requires a high level of laboratory skill. In larger diagnostic laboratories these methods may be used to support infection control or public health investigations. PFGE is slow and cannot be used to transfer data between laboratories efficiently. A recent development has been the introduction of an automated ribotyping device (RiboPrinter, Dupont-Qualicon, DE), which uses a restriction enzyme on an *E. coli* cDNA probe. The final result is a digital ladder or "ribotype" which can be used, stored for future reference or compared against an external digital library. In theory, the Internet should allow networking of these devices with further gains from collating multi-site information. Some bacteria are just too complex to permit conventional molecular epidemiology. Members of the CMN group can be difficult, but analysis of fatty-acid methyl ester (FAME) gas chromatograph retention times can be useful for definitive identification and typing purposes. In the event of conflicting results from other typing systems or a lack of typing data, sequencing of part of the bacteria chromosome at the 16SrRNA locus can be helpful. Gene sequencing is being increasingly used to arrive at a definitive identification of mycobacteria.

Molecular and other subtyping methods have yet to become a part of the regular repertoire of most service laboratories. Establishing advanced bacterial typing systems as a part of standard bacteriology operating procedures remains a challenge. Most molecular subtyping is performed by research laboratories and lacks the consistency expected of clinical service laboratories. Access is limited by availability. As this improves, strategies will need to be developed that triage the requests for typing. It is already apparent that not all bacterial strains that can be subtyped should be. A minimum requirement should be a prior knowledge of the genetic population structure of the species in question (how much variation can be expected in a random collection of isolates), baseline epidemiological data (person, time and place) with which to formulate a working hypothesis, and a clear line of communication/accountability (who will receive results and initiate further action). Where these basic requirements cannot be satisfied, subtyping of bacterial collections must be regarded as exploratory and will be less likely to produce epidemiologically useful results.

14.3. Bacteriological Emergencies

Many specimens are submitted to diagnostic microbiology laboratories marked "urgent", but only a small proportion will have any direct bearing on true bacteriological emergencies. Even when the patient's condition is unequivocally life-threatening, the results of any test that requires bacterial growth cannot be expected to contribute to the immediate clinical management of the patient. There is a surprisingly small collection of rapid tests which when applied to a small number of infections can help alter the course of the disease. The infections concerned are all potentially fatal and progress rapidly, necessitating prompt decisions on antibiotic choice and other therapeutic interventions.

14.3.1. *Acute bacterial meningitis*

Despite being less common than its viral counterpart, bacterial meningitis has a much higher rate of major neurological complication and of fatal outcome. The sporadic occurrence of this condition in previously healthy young adults and children underlines the potential virulence of the bacterial species that cause meningitis. Unfortunately, in centers where the condition is seen very rarely, this can lead to inappropriate and sometimes irrational requests for laboratory investigations that have little clinical justification.

The technological gains of the last decade have added little to the initial examination of cerebrospinal fluid (CSF) from patients with suspected bacterial meningitis. The most useful analyses are total and differential cell count, Gram stain on a spun deposit and biochemistry (glucose and

Table 14.3. CSF cell content and biochemistry in central nervous system infections.

CNS infection	Neutrophils (no./μL)	Lymphocytes (no./μL)	Protein (g/L)	Glucose (CSF/blood)
Normal	0	<5	<0.5	2/3
Viral meningitis	<100	10–10,000	0.5–1	2/3
Bacterial meningitis				
Not TB	>200–10,000	<100	>1	<1/2
Tuberculous	<100	50–1,000	1–5	<1/3
Brain abscess	5–100		>1	2/3

Table 14.4. Issues and exceptions in CSF analysis.

Cell count	
No detectable cells	Some cases of early meningitis
Low neutrophils, turbid CSF	Some *S. pneumoniae* meningitis
Artificially low cell count	Glass CSF container
Raised but not meningitis	Seizure, systemic viral infection, infective endocarditis
Biochemistry	
Low glucose but no bacteria	Consider mumps or lymphocytic choriomeningitis virus
Raised protein but no cells, bacteria	Consider brain abscess
Gram stain	
Neutrophilia but no bacteria	Bacteria below limit of detection in >10% cases
	H. influenzae easily missed against background of cell debris

L. monocytogenes can be mistaken for *S. pneumoniae* in CSF Gram stain

protein). When carefully performed on CSF from a previously untreated patient, a presumptive etiological diagnosis should be possible in the majority of cases. The expected results are given in Table 14.3. Important but less common exceptions from these ranges are given in Table 14.4.

As in all other areas of diagnostic bacteriology, the quality of clinical information available to the laboratory has a substantial bearing on how the results are reported. Key points to ascertain include whether the patient has a community-acquired infection or has been in hospital for some time (e.g. following a neurosurgical procedure), and whether they have recently received antibiotics that could inhibit the growth of meningitis-causing bacteria.

CSF should be cultured by inoculation of 5% horse blood agar and chocolated blood agar in 5% CO_2. It is common practice to incubate any remaining drops of CSF in glucose broth to enrich a low inoculum or resuscitate sublethally damaged bacteria. Blood cultures should be collected from any patient suspected of having bacterial meningitis to improve the diagnostic yield in view of the frequency of concurrent meningitis, and the incomplete overlap of bacterial meningitis and septicemia.

Most of the less commonly isolated bacterial causes of meningitis (*Escherichia coli, Listeria monocytogenes* and the staphylococci) can be recovered using standard bacterial isolation procedures. *Mycobacterium tuberculosis* is the exception. When tuberculous meningitis is suspected on the grounds of clinical signs or CSF biochemistry, a centrifuged deposit of CSF should be examined microscopically following acid-fast stain and inoculated into mycobacterial culture medium (e.g. Middlebrook's broth). Nucleic acid amplification can be used to obtain a quicker positive result than is possible by culture, but in the absence of a positive acid-fast stain, several days of culture-based enrichment may be necessary to yield a positive result. Few centers provide out-of-hours nucleic acid amplification for tuberculous meningitis.

Direct antigen detection of the commoner bacterial causes of meningitis immediately after receipt of CSF has been used to obtain a rapid etiological diagnosis. Latex particle and staphylococcal co-agglutination is still used in some laboratories, but is now generally out of vogue since large surveys showed that antigen detection methods add little to the clinical management of patients with suspected meningitis. Surprising though this may seem at first, it reflects the relative efficiencies of CSF Gram stain and antigen detection systems. Antigen detection loses much of its specificity when bacteria cannot be seen in the CSF by Gram stain. Moreover, the relatively poor immunogenicity of group B *Neisseria meningitidis* means that effective antigen detection reagents are more difficult to prepare against this common cause of meningococcal meningitis. Whether antigen detection reagents should be reserved for use on CSF specimens already shown to contain bacteria by Gram stain needs to be addressed by senior laboratory staff. Their decision will need to take into account the number of specimens likely to be tested during the working shelf-life of the reagent batch.

Nucleic acid amplification by PCR is coming into widespread use for the diagnosis of meningococcal and tuberculous meningitis. A general lack of 24 h access molecular diagnostic laboratories currently prevents the use of PCR-based protocols for truly rapid etiological diagnosis of bacterial meningitis. However, steadily falling costs of molecular reagents, faster thermal cycling and electrophoresis-free systems can be expected to change this in the near future. At present, the principal contribution that nucleic acid amplification makes to the diagnosis of bacterial meningitis is the detection of non-culturable or slow-growing bacterial pathogens. In the setting of meningitis, this translates into *N. meningitidis* from antibiotic treated patients and *M. tuberculosis* before conventional cultures turn positive. Both of these advances seem small gains in view of the fact that the patients will in most cases have received appropriate antibiotic therapy before an etiological diagnosis is made. However, confirmation of the diagnosis can help the attending physician narrow down on more specific antibiotic therapy, reducing the risk of toxic and other adverse effects. The falling rate of culture-positive meningococcal meningitis means that we are likely to become more reliant on nucleic acid amplification tests for *N. meningitidis*, but this should not be seen as an opportunity to ease off on attempts to culture the causal pathogen from CSF, blood or skin lesions. Not only would this be expected to reduce the overall diagnostic yield but it might also lead to missed opportunities to confirm pneumococcal and other bacterial infections. Of equal importance is the role of positive CSF culture isolates in antibiotic resistance and molecular epidemiological studies. Antibiotic susceptibility and subtyping studies can only be performed with live cultures. Clinical isolates in properly catalogued culture collections form an important resource for laboratory-based studies on antibiotic resistance trends, vaccine design and outbreak investigation. Given the importance of bacterial meningitis, these studies are performed by major reference laboratories in many countries. In smaller countries that lack the resources required to maintain a specialist reference laboratory for meningococci, pneumococci and *Haemophilus* species, collaboration with centers in neighboring countries can produce useful data on regional trends.

Given the variety of susceptibility methods used for this group of bacteria, it is essential that testing be performed under controlled conditions using a method recognized by a clinical laboratory standard or accrediting organization. Clear guidelines must be provided for difficult-to-interpret situations such as mucoid pneumococcal strains. A pathologist or infectious diseases physician should also review the susceptibilities reported against each pathogen for clinical appropriateness.

Larger clinical laboratories with an interest in molecular epidemiology may choose to perform studies on collections of bacterial CSF isolates as part of wider epidemiological investigations. These results (e.g. DNA macrorestriction by pulsed-field gel electrophoresis) must be set in the context of descriptive epidemiological information and the population structure of the bacterial species under study. For service purposes, molecular epidemiological studies will normally be reserved for investigations of case clusters.

14.3.2. *Bacterial myonecrosis*

Often known by its colloquial name, bacterial myonecrosis is another uncommon but potentially fatal infection that occurs sporadically usually in the setting of penetrating trauma, but also in patients with diabetic and other arteriopathic conditions. The infection is characterized by a rapidly progressive destruction of striated muscle in the affected limb. The speed of progression is so fast in some cases that the only prospect of halting tissue destruction is surgical debridement or even amputation. Under these circumstances, culture-based diagnostic methods have little to offer. The diagnosis is primarily clinical, but Gram stain of the thin exudate that sometimes leaks from the surface of the affected body part can be instructive providing the procedure is performed quickly, competently and communicated directly back to the requesting physician. The causal species, usually *Clostridium perfringens*, may be evident as a Gram-positive bacillus. However, it does not typically produce spores under these conditions. Although the clinical specimen is an inflammatory exudate, polymorphonuclear cells are often notable by their absence.

Confirmation of clostridial toxin production and other procedures that require culture isolates of this obligate anaerobic species have little more than a confirmatory role in the overall management of clostridial myonecrosis.

14.3.3. *Necrotizing fasciitis*

Necrotizing fasciitis is another infection that can cause rapid destruction of soft tissues. The name indicates that the principal pathological process is located in the layer overlying striated muscle, rather than the muscle itself as in clostridial myonecrosis. Necrotizing fasciitis also occurs in patients with diabetes and other arteriopathic conditions, but may also occur following minor trauma such as insect bites or in children with chickenpox. The potential for continued tissue destruction and even death after commencement of appropriate antibiotic therapy has earned the bacteria that cause this condition a reputation for "flesh eating". The specific factors responsible for setting in train the series of events that leads to unstoppable tissue destruction are still poorly understood, but clearly involve a substantial patient contribution.

Recent studies indicate that the best outcome for patients with necrotizing fasciitis are obtained when a surgical opinion is sought early in the course of the infection, when laboratory specimens are collected during a surgical exploration and when a second surgical procedure is performed to ensure physical removal of infected tissue. The most significant contribution the clinical microbiology laboratory can make is a competently performed Gram stain on a macerated tissue specimen. The commonest single bacterial cause of necrotizing fasciitis is *Streptococcus pyogenes* (group A beta-hemolytic streptococcus). But other Gram-positive cocci, enteric Gram-negative bacilli and obligate anaerobic bacteria can be involved. The principal purpose of the Gram stain is therefore to determine whether the infection is likely to be caused by a single species or is polymicrobial. Nevertheless, the requesting clinician will usually opt for presumptive antibiotic therapy based on recommendations in published series.

When necrotizing fasciitis is caused by *S. pyogenes*, the condition may be a part of the more generalized invasive streptococcal infection in which case other potentially relevant body sites should be sampled, including peripheral blood.

14.3.4. *Severe community-acquired pneumonia*

It has been recognized in recent years that the overlap in the clinical features of pneumonias caused by the major pulmonary pathogens is too great to allow a presumptive etiological diagnosis from X-ray and baseline clinical information. Community-acquired pneumonia is now managed during its initial stages on a syndromic basis, i.e. decisions are made according to how severely ill the patient happens to be, rather than from the presumed bacterial pathogen. The potential role of the clinical microbiology laboratory is reduced further by the poor quality of many respiratory specimens due to difficulty obtaining respiratory secretions uncontaminated by oropharyngeal contents, particularly in breathless or confused patients. There have been many attempts to improve the bacteriological diagnosis of pneumonia, but with little impact on early therapeutic decisions. The contribution the clinical microbiology laboratory makes to management of the patient with community-acquired pneumonia is therefore modest.

A Gram stain on a carefully collected specimen of respiratory secretions can sometimes demonstrate many Gram-positive diplococci suggestive of *Streptococcus pneumoniae*, or clusters of Gram-positive cocci consistent with *Staphylococcus aureus*. When either of these is accompanied by many polymorphonuclear neutrophils and no buccal epithelial cells, the stain results may help towards a faster etiological diagnosis. Unfortunately, this occurs too infrequently to justify the amount of work performed in Gram staining unselected respiratory secretions in many clinical laboratories.

The laboratory can make a useful contribution in the setting of severe community-acquired pneumonia by raising the possibility of Legionnaires' disease and tuberculosis. Both conditions are probably underdiagnosed. Both often require a specific request to the laboratory since investigations for the respective pathogen are rarely performed as a routine procedure.

Both can be detected by culture, though there are problems when relying entirely on this approach. In the case of Legionnaires' disease, cough is only rarely productive of sputum so there may be no specimen for culture in the first place. Even when one does reach the laboratory, culture may be negative due to the general difficulty of recovering *Legionella* species from clinical material or to prior antibiotic treatment. A rapid alternative to culture is the urinary antigen test (UAT) currently gaining wide acceptance as the most sensitive means of diagnosing Legionnaires' disease. To provide a truly rapid diagnosis, the batch test version of the UAT needs to be used. At present only *L. pneumophila* serogroup I can be detected by UAT.

As indicated above, *M. tuberculosis* is a slow growing species. Diagnosis of pulmonary tuberculosis has benefited considerably from the introduction of molecular methods. Nucleic acid amplification is widely available to rapidly confirm presumptive clinical diagnosis. To date, PCR-based and similar protocols have not replaced acid-fast stains as the principal rapid diagnostic test. However, acid-fast stains cannot establish which species of mycobacteria are present. PCR-based methods have come into their own as an effective means of demonstrating that the acid-fast bacillus present in a sputum specimen is *M. tuberculosis* or one of the other commoner mycobacteria. Nucleic acid amplification has been disappointing as means of demonstrating mycobacteria in smear-negative sputa. Molecular detection methods have not replaced culture-based techniques for detection of mycobacteria since culture isolates are still required for subsequent manipulation to determine the definitive identification, and antibiotic susceptibilities and for subtyping. These additional tests remain the province of specialist reference laboratories. Methods are currently in a state of flux. The changing epidemiology of tuberculosis has put pressure on clinical microbiologists to speed up susceptibility testing (to detect multidrug resistant strains), definitive identification (to confirm atypicals which require very different antibiotic choices) and subtyping (for epidemiological investigations into case clusters). Automated culture systems are being put into use for rapid susceptibility testing and some resistance-conferring gene loci have been identified for future PCR-based routine detection in clinical laboratories. Definitive identification can be performed rapidly by automated bacterial DNA sequencing followed by comparison with databases of known sequences. These results can be used to compare clinical isolates for epidemiological studies. However, the established molecular typing method in current use is a standardized Southern blotting method using insertion sequence 6110. Other methods available include DNA macrorestriction analysis and bacterial cell wall fatty-acid methyl ester analysis by gas chromatography.

14.3.5. *Septicemia*

Septicemia is a clinical infection caused by multiplication of micro-organisms in the bloodstream, and is usually accompanied by fever and other features of severe sepsis such as chills, sweating and confusion. In severe cases, shock, renal failure and loss of consciousness may follow. The clinical features of septicemia are common to the wide range of species that may cause the condition, though specific features localized to the skin or internal organ may suggest a bacterial etiology at the time of initial diagnosis. The generalized infective processes that constitute septicemia often follow an initial localized infection in the soft tissues or a specific organ system.

It follows that laboratory procedures used to investigate a case of septicemia form part of a wider investigative process that includes the identification of localized foci of initial infection such as the urinary tract, surgical incisions or invasive medical devices. The laboratory should expect additional clinical specimens from these sites and will gain from collating the results of all such related specimens. It is the prompt examination of these related specimens that will prove more rewarding when diagnostic results are needed urgently to assist in the immediate management of septicemic patients.

Blood culture, by comparison, is relatively slow due to its reliance on bacterial growth for a positive result. In exceptional circumstances (24 h staffing of the laboratory reception area, immediate entry of new blood cultures, automated detection of new positives, immediate Gram stain and availability of requesting physician) it is possible to return Gram stain results of positive cultures to the clinical service within 3–4 h of culture collection. From experience, patients generating very fast positive blood cultures usually have severe sepsis and cannot be left untreated for 3–4 h for the sake of a Gram stain result. However, the value of a promptly delivered preliminary blood culture result should not be underestimated. In many clinical centers, a pathologist follows up these preliminary results in a blood culture–specific ward round, using the opportunity to advise the attending physician on further bacteriology tests and antibiotic selection. The initial presumptive choice of antibiotics can often be refined at this point, even before definitive antibiotic susceptibility results are known. This contact with the clinical service may also provide the foundation for team management of patients with serious complications of infection.

Several automated blood culture systems are widely available. Bacterial growth detection is performed automatically on cultures which can remain undisturbed in the incubator cabinet until a positive has been registered. The more recent models allow entry of patient identifier data by bar code readers and interfacing with laboratory computer information systems. Many large hospital laboratories have automated blood culture detection in the last decade, mainly for logistic reasons, in particular the ability to free senior laboratory staff from the highly repetitive processes required to subculture large numbers of mainly negative blood cultures. Automation has also reduced the potential for laboratory contamination during subculture and has possibly standardized blood culture processing. However, automation has brought a number of challenges. Reliance on automated systems places a considerable strain on the laboratory in the event of computer or blood culture analyzer failure. When the equipment is running optimally, its full potential can only be realized if staff are on hand around the clock to confirm a machine positive. Very few centers have been able to justify this level of priority for blood culture processing in view of the time it takes to obtain even a fast positive.

In view of these considerations, there has been increased interest in improving the pre-analytical stage of automated blood culture processing. Studies indicate that a maximum of three blood culture sets is required per septic episode and in many cases, two will be satisfactory in view of the improved quality of modern blood culture media. Only in unusual circumstances such as truly subacute bacterial endocarditis is a further set justified on epidemiological grounds. A minimum of two types of media is required, necessitating two bottles per set. Most manufacturers include anion exchange resins to absorb out antibiotics from the patient's blood in spite of a lack of evidence that this approach significantly increases the isolation rate. Large blood volumes are preferable for inoculation of each bottle i.e. 8–10 mL, except in pediatric cases. It is common practice to review the rate of isolation of bacteria judged to be probable contaminants from the venesector's or patient's skin, in order to focus attempts to improve blood culture collection technique. This has become increasingly important in view of the rising incidence of medical device-associated infections caused by coagulase-negative staphylococci and other common skin organisms. Unless great care is taken to avoid contamination of the blood culture with skin organisms during venesection and inoculation of the culture bottle, patients may receive unnecessary and possibly toxic antibiotics.

14.3.6. *Other bacterial infections*

(1) Urinary Tract Infection
Infections of the urinary tract probably generate more clinical specimens than any other type of bacterial infection. The methods in common use are well established and have changed little in several decades of clinical microbiology. In brief, the inflammatory cell and bacterial content of a midstream specimen of urine are determined by a combination of microscopy and culture on selective agar (usually cysteine lactose electrolyte deficient or CLED agar). Pure growth of a single bacterial species at greater than or equal to 10^5 colony-forming units per mL in the presence of >500 polymorphonuclear cells per mL and in the absence of squamous epithelial cells indicate a clinically significant result. These numerical criteria were derived from studies on otherwise healthy adult women with uncomplicated bacterial cystitis and should not be regarded as absolute. There are many circumstances in which these criteria can be regarded as little more than a guide, including recurrent cystitis, pyelonephritis, urinary tract infection (UTI) in patients with neutropenia, infants or staphylococcal UTI. Nevertheless, growth of mixed bacterial species, an absence of inflammatory cells or the presence of squamous epithelial cells usually indicates a poorly collected specimen, especially if the result has been obtained a second time from a repeat specimen. Conversely, very low counts of bacteria in urine specimens have increasing

clinical significance in settings such as recurrent cystitis each time the same result is obtained. A poorly collected specimen rate of 10% or greater should lead to attempts to improve patient education and other aspects of specimen collection. Laboratory staff need to remain alert to unusual but important results that may suggest contamination at first but which have greater significance to the patient. Examples include mixed bacterial growth due to a pathological connection between the bladder and the colon (vesico-colic fistula is occasionally first suspected because of an unexpected laboratory result) or bacterial growth in a catheter specimen of urine from a patient who has catheter-related septicemia.

Probably the most significant development in this overlooked area of clinical microbiology is the introduction of nitrate and leucocyte esterase rapid tests for near patient testing. These tests have gained wide acceptance as a way of supporting clinical decisions. In clinical settings where a very low rate of positive cultures are expected, e.g. antenatal screening, these tests can be used to dispense with culture altogether providing the urine specimen is visually inspected. Surveys have found that culture of clear urine specimens negative for both nitrate and leucocyte esterase adds very little to patient care.

(2) Soft Tissue Infections

A comprehensive review of bacterial infections of the skin and underlying soft tissues is beyond the scope of this chapter. Key points in the laboratory investigation of soft tissue infections that constitute a medical emergency are covered above. The majority of soft tissue infections that generate a specimen request are in hospital patients who have suffered traumatic injury or who have undergone a surgical procedure. The work of the laboratory is thus biased towards hospital practice, though well-collected specimens from infections encountered in general practice or the outpatient clinic can provide an explanation for infections that fail to respond to first choice antibiotics or that show unusual clinical features.

Specimens collected for the diagnosis of soft tissue infections are obtained from a non-sterile surface, and may therefore be contaminated by bacteria that are either members of the resident flora or transient colonizers of an abnormal external body surface. This includes *Staphylococcus aureus*, one of the most common causes of soft tissue infection. As there is no reliable method of distinguishing colonization from invasive soft tissue infection, *S. aureus* infection is almost certainly overdiagnosed. The best that can be done to avoid overdiagnosis is not to pick colonies of *S. aureus* for further work when growth is very scanty and there is a heavy mixture of other bacterial species. It has been suggested that the quality of specimens could be improved by first cleaning exudate from surface lesions before using a fresh, sterile swab moistened with water for injection to scour the infected surface. Alternatively, a fine needle aspirate of exudate from the leading edge of a spreading inflammatory lesion can be tried. If either method is used to improve the quality of the specimen the laboratory eventually works with, fewer Gram-negative species will be obtained. Isolates are more likely to be staphylococci, streptococci or obligate anaerobic bacteria. Ward staff should be discouraged from routine collection of swabs from the surface of venous stasis ulcers and similar lesions that show no evidence of spreading or invasive infection.

Careful recording of the results of Gram stain on inflammatory exudate specimens from soft tissue infection can be very helpful in interpreting subsequent culture results. Occasionally, alternative methods will be required for detection of unusual bacterial pathogens such as *Actinomycetes* (so-called "sulphur granules" may be evident to the naked eye), which can be stained, with a modified version of the acid-fast Ziehl-Neelsen stain.

For optimal results, well-collected specimens from hospital patients will need culture on several different types of selective and non-selective agar, and incubation under aerobic and anaerobic conditions. The addition of a metronidazole disk to the surface of the anaerobic plate will help distinguish obligate from facultative anaerobes and provide presumptive susceptibility results.

The results of processing specimens from patients with soft tissue infection depend on a multi-step process that can take several days even in some monobacterial infections. Some bacteria, for instance fastidious anaerobes, can take much longer to identify and obtain definitive susceptibility results. It is therefore standard practice to issue preliminary or interim reports as soon as clinically useful information is ready for issue. Late results can then be released in a final report when they have been fully validated.

Many clinical laboratories use a selective reporting approach to antibiotic susceptibility results. It is unhelpful to the requesting physician to simply list the results of all antibiotics tested against all bacteria identified in a clinical specimen, particularly when it has been obtained from a non-sterile site. A judgement has to be made to seleet the likely pathogen, when more than one species have been isolated. Ideally, the reported choice should be agreed in advance and comply with hospital antibiotic policy and other long-term objectives. It is usual to aim to provide an element of choice for the requesting physician and to avoid reporting more than one antibiotic from a group of closely related agents (e.g. first generation cephalosporins). Antibiotic testing and reporting guidelines are published and updated regularly by organizations such as the National Council for Clinical Laboratory Standards (NCCLS). These guidelines are based on recent experience and usually reflect technological ability and clinical priorities of the organization's home country.

(3) Respiratory Infections

Laboratory diagnosis of bacterial respiratory infections poses several notable problems for the clinical laboratory. Upper respiratory tract infection is one of the most common reasons for seeking a general practitioner con- sultation and is usually a self-limiting viral infection. A small proportion may be caused by group A streptococci (*Streptococcus pyogenes*), or *Arcanobacterium haemolyticum*. The potentially fatal diphtheria (caused by *Corynebacterium diphtheriae*) is rare in this region, but may need excluding in patients with appropriate symptoms who have traveled from an endemic area. There is therefore a very low rate of return on bacteriological examination of throat swabs, despite the frequency of submission. *C. diphtheriae* infection is so rarely diagnosed in most centers that routine screening for this pathogen is not performed in the absence of a history of contact with a known case or recent travel in a diphtheria- endemic area. While many upper respiratory tract infections are self- limiting, lower respiratory infection more often leads to serious con- sequences including death. Unfortunately, respiratory secretions are amongst the most poorly collected of all clinical specimen types. Sputum has to pass through the oropharynx before it gets into a specimen pot, and this will bring it into contact with the normal bacterial flora of the upper respiratory tract. Not only will this lead to contamination with an enormous range of commensal bacterial species that make isolation of a potential bacterial pathogen challenging, it also leads to contamination of the specimen with potential respiratory bacterial pathogens such as *Streptococcus pneumoniae* and *Haemophilus influenzae*, both of which may colonize the upper respiratory tract in normal subjects. This problem is exacerbated when patients spit the contents of their buccal cavity into the specimen container, rather than coughing secretions from the lower part of their respiratory tract directly into the container. This action needs to be taught to the patient, best by a physiotherapist or equivalent. If this proves to be impossible, the patient can still wash their mouth out with sterile water for injection to reduce the bacterial load prior to specimen production. A special effort to improve sputum specimen quality should be undertaken if evidence of contamination by oral bacteria (e.g. mixed oral species, presence of buccal epithelial cells) is present in the majority of sputum specimens submitted to the laboratory.

Normally throat swabs will only be inoculated onto 5% blood agar for isolation of *S. pyogenes* (and *C. albicans*). The use of a sterile straight wire to inoculate the secondary inoculum direct into the blood agar will help demonstrate oxygen-labile hemolysin by *S. pyogenes*. In most centers, *A. haemolyticum* can be recovered on this medium if laboratory staff is aware of the possibility. Tellurite medium should be added for isolation of *C. diphtheriae* only when indicated. Mycoplasmas and *Chlamydia pneumoniae* cannot be isolated using conventional laboratory media or methods. If investigations have been requested for these bacterial species, nucleic acid amplification tests are more appropriate.

Sputum specimens are more complicated. Given the problems associated with specimen collection noted above, many centers reject specimens that appear on macroscopic examination to be mainly saliva. After Gram stain, those that contain a luxuriant mixture of bacterial types and many epithelial cells may also be rejected. The remaining specimens should be inoculated onto a minimum of 5% blood agar and hemolyzed blood agar. When the patient has been hospitalized for three or more days, MacConkey's agar should be added for recovery of Gram-negative bacteria and a dilution on homogenized sputum performed for semi-quantitative estimation (some authorities regard $>10^7$ colony forming units per mL of a Gram-negative bacillus to indicate nosocomial pneumonia). It is common practice to add a streak of *Staphylococcus aureus* to the blood agar plate to demonstrate

V factor dependent satellite growth of *H. influenzae*, and an optochin disk to the secondary inoculum on the same plate to help detect *S. pneumoniae*. *S. aureus* should also be sought in good quality specimens, particularly from patients in intensive care units who have recently been admitted following multiple trauma.

The interpretation of primary sputum culture results is a skilled task whose importance should not be underestimated. An over-enthusiasm for separating potentially pathogenic minority members of the primary culture result can lead to unnecessary identification and susceptibility determinations and in turn, misleading laboratory reports. It is common practice to comment on the presence of a heavily mixed growth or the poor quality of the specimen in order to avoid unrewarding work on poor quality bacteriological material. When the specimen has been marked "urgent" or is otherwise of greater than usual importance, the laboratory should observe the basic courtesy of informing the clinical service. This will provide an opportunity to obtain a replacement specimen quickly, and at the same time educate ward staff in the finer details of respiratory specimen collection.

Large hospitals are increasingly using flexible fibreoptic scopes to obtain lower respiratory tract specimens. These may be labeled with terms other than "sputum" such as "broncho-alveolar lavage", "BAL" or "protected brush specimen". This approach bypasses the pharynx and reduces the chance of major contamination, but the tip of the scope may still pick up large quantities of oral bacteria leading to the same contamination problem without the epithelial cells present to give the game away. Some respiratory physicians use a protected brush or BAL catheter device to overcome the contamination problem. If this has been used, any bacteria isolated are likely to be significant particularly if the patient has not yet been treated with antibiotics (a rarity). If not, interpretation of the primary culture result is largely the same as for sputum culture.

For comments about the diagnosis of pulmonary tuberculosis, see above. Many centers will not perform acid-fast stains on sputum or mycobacterial culture unless these have been requested by the patient's physician. Unfortunately, pulmonary tuberculosis does not necessarily follow the textbook descriptions, particularly in the earlier stages of infection. In communities with a high prevalence of HIV infection, the clinical features of tuberculosis are often atypical. Tuberculosis should be considered a possibility in all patients with persistent pneumonia that does not respond to conventional antibiotics, and therefore the laboratory should perform at least an acid-fast smear when the bacterial cause of a patient's pneumonia has not been determined by routine sputum bacteriology on a series of specimens submitted over a week or more. An even lower threshold of suspicion should be applied to purulent sputa that are visibly blood-stained, when the patient is a known pulmonary TB contact or in localities with a high TB or HIV prevalence. In the last of these, this is a case to be answered for screening all first sputa for acid-fast bacilli.

Specialized agar media (BCYE or ADCD) should be used on the rare occasions when a respiratory specimen has been obtained from a patient with suspected Legionnaires' disease (often a non-productive cough). Nucleic acid amplification (NAA) methods for *Legionella* nucleotide sequences are a more sensitive method of diagnosis. Although a consensus on the best means of diagnosis has not yet been reached due to substantial differences in the disease's epidemiology, the urinary antigen test is a good way to obtain a rapid diagnosis in *Legionella pneumophila* serogroup I infection. These methods are obviously much more timely than serodiagnosis which may only yield a positive result after the patient has recovered.

Mycoplasma and *Chlamydia* respiratory infections are most probably underdiagnosed. Both can be detected by nucleic acid amplification when respiratory secretions can be obtained. Unfortunately, both groups of pathogen cause a typically non-productive cough. Throat swabs can be used to obtain microbial DNA from the oropharynx for NAA-based tests. Otherwise, diagnosis is usually by serological means.

Bordetella pertussis occurs sporadically but is being recognized increasingly in older children and adults. In these groups, the full range of symptoms seen in younger children may not be present. Cough plates should be discouraged and alginate-tipped pernasal swabs used for NAA-based tests. Culture can be performed using Bordet-Gengou agar or a more modern alternative (e.g. CIN). Although this is less sensitive than NAA-based diagnosis, isolation of *B. pertussis* allows molecular epidemiological studies to be performed for disease control and vaccine design purposes.

(4) Enteric Infections

Determining the bacterial cause of an enteric infection has been likened to searching for a needle in a haystack. The problem is that most cases of enteritis result in diarrhea (frequency of defecation and/or a loose and often watery consistency of feces) irrespective of the microbial etiology. Rotavirus and other non-bacterial pathogens are common, and much diarrheal disease is non-infective. Moreover, the fecal specimen contains an abundant commensal bacterial flora. It is therefore important that the diagnostic laboratory takes a rational, planned approach to processing fecal specimens.

Microscopy plays little part in the routine handling of fecal specimens, unlike in parasitology where it is a fundamental part of the diagnostic process. It may be helpful to perform a Gram stain when either severe *Campylobacter* infection or cholera is suspected. Far more important is the macroscopic examination of fecal specimens. The consistency (liquid/solid, fluid/formed), presence of fresh blood or inflammatory exudate, or excessive mucus should all be noted if present. These, the clinical details and any travel history on the request form will help determine which bacterial species should be sought.

The presence of the commensal fecal flora dictates that selective growth media be used to inhibit all but bacteria of likely clinical significance. There is no single selective agar suitable for this task and many of the more common enteritis-causing species do not require specific antibiotic therapy. Isolation of the etiological pathogen is therefore performed in many cases for epidemiological and public health purposes. Unfortunately, this means that only a minority of cases is pursued as far as specimen collection. Surveys suggest that this figure may only be around 10% of total diarrheal cases. It is essential to make the work performed on those specimens that are sent for workup as worthwhile and cost-effective as possible.

A minimum (in many centers the normal set of media) inoculated would be a selective enrichment broth for salmonellas (e.g. selenite broth), two different media for *Salmonella* and *Shigella* (e.g. xylose lysine decarboxylase agar and deoxycholate citrate agar) and a *Campylobacter* medium (e.g. Skirrow's agar). When indicated, a sorbitol MacConkey plate should be inoculated for *E. coli* O157:H7, a selective anaerobic plate for

Clostridium difficile, thiosulphate citrate bile salts agar (TCBS) for *Vibrio cholerae*, and a selective agar for *Yersinia enterocolitica*. The indications for use of these media should be understood and consistently observed by laboratory staff as part of standard operating procedures.

Preliminary isolation requires a series of initial, benchtop tests to rule out enteric bacteria of no clinical significance. *Proteus* species are easily confused with salmonellas, but can be distinguished by performing a rapid urease test. Positives can be discarded. Agglutinating antibody preparations are used to screen *Salmonella* and *Shigella* but positive reactions should always be compared against a negative control and should be confirmed because of the ability of selective agar to suppress second species of bacteria without totally inhibiting them. Only when a suspect colony type has been picked off and subcultured onto a non-selective agar such as 5% blood agar, can one be more certain that the single colonies are in pure growth. There are over 2,000 serotypes of *Salmonella enterica*. Even major enteric reference laboratories only carry a small fraction of the antisera required to identify these. Most clinical laboratories will only keep the most basic set of antisera required to confirm *Salmonella* and leave the detailed determinations to reference laboratories. Large teaching center labs will probably want to keep the antisera for the top ten serotypes, *S. typhi* and *paratyphi*, and leave the rest to national centers who have the turnover and skills such as phase change procedures to justify the extra work. *Shigella* seroagglutination requires a smaller collection of reagents but has a different type of complexity. Again, a smaller selection of agglutinating reagents is normally stocked.

E. coli 0157:H7 and other strains responsible for hemorrhagic colitis and hemolytic uremia syndrome need to be tested for production of verotoxin. This is usually performed by either neutralization of cytotoxic effect with antitoxin antibody or by ELISA immunoassay. It should be remembered that some verotoxin positive strains are sorbitol positive on the SMAC agar developed to detect *E. coli* 0157:H7.

Confirmation of *C. difficile*-associated diarrhea requires detection of toxin in fecal specimen filtrate in addition to isolation. At present proprietary, no single immunodiagnostic method will pick up more than 90–95% cases. Reliance on any single test for *C. difficile* or its toxin may therefore result in under-diagnosis of the condition.

Preliminary and confirmatory tests for the major and other bacterial enteric pathogens are listed in Table 14.5.

Many of the bacterial enteric pathogens are important causes of community infection. They may be notifiable to the public health authority and many are either food- or water-borne infections. Typing methods are being used with increasing frequency to complement epidemiological investigations intended to identify the cause of point-source outbreaks. In the case of *Salmonella*, serotype and phage type are widely used. For other bacterial enteric pathogens, molecular typing methods such as pulsed-field gel electrophoresis or PCR-based methods are being used selectively. At present, phage typing of the commonest *Salmonella* serotypes remains the principal means of demonstrating possible links between cases. Even when molecular typing methods can provide the same level of discrimination as phage typing in a shorter time, a historical attachment to phage typing will ensure its retention as a typing method for some time to come.

(5) Genital Tract Infections

The most common bacterial infections of the genital tract are non-specific urethritis caused by *Chlamydia trachomatis* and gonorrhea. (Bacterial vaginosis is a common condition due to an overgrowth or imbalance of the bacterial flora of the female lower genital tract, but as the likely causative bacteria are difficult-to-grow anaerobic species, diagnosis is by a combination of clinical observations, non-specific tests and exclusion of other causes of vaginal discharge).

(a) Female genital tract infections

With the exception of vaginal thrush, the causes of vaginal discharge are difficult to grow in the laboratory. This is also reflected by the spread of infection mainly through intimate body contact before transmission of infection can take place. *Neisseria gonorrhoeae*, the cause of gonorrhea, is a fastidious microaerophilic Gram-negative diplococcus that requires a combination of selective agar (e.g. vancomycin-colistin-amphotericin-trimethoprim or VCAT agar) and special gaseous atmosphere for isolation. Specimens that have been collected with conventional cotton swabs or kept in transit for a long time may not allow laboratory recovery of *N. gonorrhoeae*. Chlamydial infection is even more difficult to confirm

Table 14.5. Identification of gastrointestinal bacterial pathogens.

Pathogen	Preliminary	Definitive
Campylobacter species	Motile, oxidase+, GNB Growth on selective agar at 37–42°C in 15% O_2, 5% CO_2	Species identification by cephalothin, nalidixic acid sensitivity, catalase, hippurate, H_2S and optimal growth temperature
Clostridium difficile	Spore-forming GPB Growth on selective agar at 37°C anaerobically	Assay for toxin A and B
Escherichia coli (VTEC)	Oxidase–, GNB Ferments lactose on MAC Ferments D-sorbitol	Verotoxin assay + agglutinating antisera for serotype
Salmonella enterica	GNB, oxidase–, urease– Lactose non-fermenter Agglutinates Poly-O antisera	Identification by substrate utilization agglutination of specific O and H antisera
Salmonella typhi	GNB, oxidase–, urease– Lactose non-fermenter H_2S+ on selective media Agglutinates Poly-O/Vi antisera	Identification by substrate utilization agglutination of specific O and H antisera
Shigella species	GNB, oxidase –, urease– Lactose non-fermenter	Identification by substrate utilization agglutination of specific antisera
Vibrio cholerae	Oxidase+, motile, GNB Growth on selective agar (TCBS)	Identification by substrate utilization, O/129, polymyxin sensitivity
Yersinia enterocolitica	Oxidase–, GNB Growth on selective agar	Identification by substrate utilization

since the causal agent is an obligate intracellular bacterial parasite requiring tissue culture for isolation. In both cases, the quality of specimen collection is an important determinant of successful isolation. Both these species can be detected by nucleic acid amplification tests on specimens of genital secretions, but the swabs must not have come into contact with the nutrient agar used to transport conventional cotton swabs, as this is highly inhibitory to the polymerase chain reaction. While nucleic acid amplification improves the sensitivity of *N. gonorrhoeae* and *Chlamydia* detection, it does not provide live bacterial isolates for subsequent susceptibility testing or molecular epidemiological investigations.

Other bacterial species isolated from low vaginal swabs include *Streptococcus agalactiae* (group B streptococci), *S. aureus*, *E. coli*, *H. influenzae*, *S. pneumoniae* and *N. meningitidis*. *S. agalactiae* is a minority or transient member of the commensal flora in a minority of women. It is of little consequence to the woman unless she is pregnant in which case it may require antibiotic prophylaxis if isolated shortly before delivery. *S. agalactiae* isolates should be reported in all pregnant patients, though the approach to prophylaxis will differ substantially between centers. *S. aureus* is an innocent commensal species in the majority of patients from whose lower vaginal swabs it has been isolated. Here, it probably reflects contamination of the swab by contact with the perineum where this species is quite common. *E. coli* is a common isolate in swabs from post-menopausal women, is often found concurrently with atrophic vaginitis but has no direct pathogenic role in this setting. *H. influenzae*, *S. pneumoniae* and *N. meningitidis* are occasionally noted in specimens from this location by observant laboratory staff. They are an oddity in this setting except possibly when isolated from a Bartholin's cyst.

High vaginal swabs collected with the aid of a speculum to investigate discharge are usually set up only for detection of *N. gonorrhoeae*, *Chlamydia* and yeasts. Endocervical swabs, on the other hand, are much less likely to be contaminated with the commensal vaginal flora and are therefore more suited to investigation of upper female genital tract infection. They may be used for detection of *Mycoplasma* and *Ureaplasma* infection by either culture or nucleic acid amplification. In the post-operative setting, a wider range of more conventional bacterial pathogens should be anticipated including *S. aureus* and obligate anaerobic bacteria.

Postpartum, the laboratory should be particularly vigilant for evidence of group A streptococci and clostridia.

Pelvic collections sampled during an operation or by needle aspiration should be handled as sterile site specimens, and may contain a wide range of bacteria including staphylococci, obligate anaerobes, enteric Gram-negatives, chlamydias, mycoplasmas and others.

(b) Male genital infections

The bacterial infections most commonly encountered are so-called non-specific urethritis, and gonorrhea. The former is caused by *Chlamydia* and the latter by *N. gonorrhoeae*. In view of the principal symptom being penile discharge, clinical diagnosis usually occurs faster and is more easily achieved for males than it is for females, some of whom may only be diagnosed through follow-up of their male partners. However, as in females, the success of laboratory diagnosis depends on the quality of specimen collection. The urethral swab collected in patients with non-specific urethritis should be inserted several centimeters into the distal urethral, and rotated to dislodge epithelial cells in which the bacterium resides. This should be done after a sample of discharge has been obtained for Gram stain and culture for *N. gonorrhoeae*. A Gram stain with intracellular Gram-negative diplococci is highly predictive of *N. gonorrhoeae* infection in males (much less so in females). Alternatively, a specimen of urine can be submitted for detection of both pathogens by nucleic acid amplification. Co-infection with both agents is common.

The identity of *N. gonorrhoeae* should be confirmed by a combination of oxidase reaction, superoxyl test, and either sugar utilization or co-agglutination.

Susceptibility testing of *N. gonorrhoeae* isolates is performed to detect penicillinase production and other key resistance factors that may significantly reduce the choice of therapy. Many centers publish trend data on antibiotic resistance in local *N. gonorrhoeae* isolates.

The nature of the infections caused by these bacterial species means that the results of laboratory investigations for genital tract infections should be handled with discretion and a high standard of professional confidentiality maintained at all times. Many laboratories use coding systems to ensure that these results can only be identified with a named patient by the requesting physician.

14.4. Emerging Technologies

14.4.1. *Rapid bacterial identification systems*

The requesting physician is often frustrated by the time it takes for identification procedures that rely on bacterial growth to produce a definitive result. It does not help that substrate utilization panels such as API and Microbact need an isolate in pure growth before incubation can begin. Definitive results reaching a hospital ward 48 h after specimen reception are unlikely to have much impact on acute clinical management.

The response to this problem has been varied. Some substrate identification systems have been automated and miniaturized to give a rapid, presumptive identification and antibiotic susceptibility results. These include the Vitek and Microscan systems. Results can be available on the same working day as preliminary isolation. This may appear to shorten the time to final reporting by as much as 24 h, but in reality these systems appear to have less impact on the clinical service. They must be used selectively, since definitive identification is not necessary for most bacterial isolates from non-sterile sites. Moreover, acceleration of bacterial growth can fundamentally alter microbial physiology and result in misleading results. Vitek preliminary identification of many enteric Gram-negative bacilli will often have to be confirmed by conventional substrate utilization panel, and an erroneous "*Salmonella*" result appears to be particularly common with some systems.

Another approach is to tackle the problem of identification of various bacterial species from non-sterile sites by development of specific chromogenic indicators incorporated in the primary growth medium. Chromogenic agars have yet to gain wide acceptance because of their relatively high cost but are likely to be used more widely in future.

Nucleic acid amplification has not been used as widely as originally anticipated for direct detection of bacterial pathogens. Its greatest impact has been on the diagnosis of infections caused by bacteria that are either fastidious, slow growing or easily lost in transit to the laboratory. Logistics, cost, and the problem of PCR inhibitors in particular specimen types have hindered the provision of molecular diagnostic services for out-of-hours and truly urgent specimens. Nevertheless, the new generation of thermal cycling systems such as the LightCycler and the Taqman analyser may alter this. These new systems will speed up nucleic acid amplification, reduce reliance on gel electrophoresis and enable quantitative results to be obtained. Despite these advances in laboratory technology, the complexity of genes determining antibiotic resistance and the speed with which they adapt to prevailing antibiotic use means that antibiotic susceptibility testing is likely to be determined by phenotypic methods for some time to come.

14.4.2. *Improved bacterial identification*

The identification of medically important bacteria and classification into subsidiary groups or taxa has come a long way since the original development of the Gram stain. It is often a moot point how far identification should go with a particular laboratory isolate (a suggested guide is provided in Table 14.6). Bacterial taxonomy has been assisted in recent decades by the widespread use of genomic systems with which genotypic relatedness can be explored. The use of standard names for bacteria satisfying agreed identification criteria provides a benchmark against which the accuracy of identification can be measured. Unless additional information is required for epidemiological investigations, bacteria will normally only be identified as far as species level, at best. In many cases, a genus level identification will suffice. However, species level identification of some groups of bacterial can place heavy demands on laboratory staff especially if a large range of non-standard culture-based tests are needed to follow a taxonomic key. Many laboratories will issue reports that refer to the genus or group and take the identification of these species (e.g. corynebacteria, actinomycetes or nocardias) no further. They may submit important isolates to a reference laboratory. However, there is a move to resolve some of these identification issues by selective use of newer technology. Chemotaxonomy, for instance, can be performed by analyzing fatty-acid methyl esters of the above bacteria with dedicated fine capillary column gas chromatography (MIDI System). More recent approaches are the automated ribotyping system (RiboPrinter, Dupont Qualicon) which relies on a cDNA probe directed at *Eco*RI digests of bacterial suspensions, and PCR-based gene sequencing.

Table 14.6. Guidelines on how far to extend bacterial identification procedures on bacterial isolates.

Microscopy	Colonies of potentially significant, dominant species isolated from urine, genital swabs
	Other non-sterile site swabs, sputa, feces
	Any isolates from blood cultures, CSF, other sterile site specimens
Benchtop tests	Clinically significant isolates from urine, genital swabs
	Other non-sterile site swabs, sputa, feces
	Any isolates from blood cultures, other sterile site specimens
Substrate utilization tests	Clinically significant isolates not yet identified by benchtop tests from genital swabs, other non-sterile site swabs, sputa, feces, blood cultures
	Other sterile site specimens
Subtyping procedures	Clinically significant isolates from feces, CSF
	Isolates responsible for nosocomial infections

NB: Considerable local variations will exist. More extensive identification procedures will be necessary in more serious infections e.g. urine isolates will require full characterizations to allow close comparison with blood culture isolates from a patient with urosepsis.

The use of molecular or chemotaxonomic methods for species level identification is difficult to justify on a routine basis on grounds of hardware and consumables costs. These are expected to fall as the technology comes into more widespread use. Nevertheless, specific problem isolates can be successfully identified to species level more quickly and efficiently than possible with traditional methods. There are also quality assurance issues in bacterial identification that can be resolved best by selective use of advanced microbial identification technology. The most discriminating level of bacterial identification involves the identification of reproducible subtypes of a given species. Individual subtypes may have a particular pathogenic or epidemiological significance, in which case subtype identification may be performed on a semi-routine basis. Examples of this include serotyping and phage typing of salmonellas, IS6110 typing of *M. tuberculosis*, and capsular typing of *S. pneumoniae*. The most widely used means of molecular subtyping are either PCR- of DNA macrorestriction-based (PFGE) and are generally used for epidemiological investigations, particularly in support of hospital infection control. Other methods including automated ribotyping and gene sequencing can be expected to be used with greater frequency and for a great range of applications in future. The ability to store and transmit typing data in digital form for analysis at a later date or a distant site (as with automated ribotyping) provides a particularly powerful tool for multicenter infectious disease surveillance. It is likely that outbreaks of bacterial infection will be investigated in a more timely fashion in future using molecular epidemiological methods.

As the technology settles into place, decisions will have to be taken on which methods to apply to given bacterial species in specific circumstances. In order to do this, preliminary validation of bacterial sub-typing systems must be performed, whether the system is based on phenotypic or genotypic bacterial characteristics. It is prudent to try out the system on collections of local clinical bacterial isolates as well as unrelated type culture isolates. This will give some insight into the population structure of locally-acquired bacteria, and is an essential pre-requisite to any future inferences drawn from typing studies on local isolates. The conventional parameters for comparing typing systems are typability (percentage of strains giving a subtyping result), discrimination (ability to subdivide isolates into clear-cut groups), reproducibility (ability to produce the same subtype result on consecutive occasions), and epidemiological concordance (whether clusters of indistinguishable strains within the collection are consistent with results from other subtyping systems). The term "clonal" is often used inaccurately to describe a cluster of isolates belonging to a presumed single strain after analysis with a given molecular typing system. Ribotyping examines around 1% of total bacterial genome, and DNA macrorestriction analysis around 10% of total. At present, total genome analysis is not a practical proposition for subtyping purposes. Whatever the molecular or other subtyping system used, in order to ensure a sensible use of laboratory resources, it is important that epidemiological typing addresses a properly formulated hypothesis and is performed under the same high standards of quality control as other diagnostic laboratory activities.

14.4.3. *Susceptibility testing*

Conventional antibiotic susceptibility testing methods rely on bacterial growth and therefore delay the release of critical information on clinical isolates. Susceptibility results on bacteria isolated from blood or CSF cultures typically take a further 12–24 h after primary isolation and have little impact on the outcome of severe, life-threatening sepsis. Results take even longer in the case of slow-glowing species such as *Mycobacterium*. Another problem that bedevils routine susceptibility testing is the mismatch between antibiotic resistance phenotype and genotype. Methods in common use such as breakpoint and Kirby-Bauer disk diffusion are typically reduced to "S" or "R". While many physicians will act on this kind of result as if it is an authoritative predictor of treatment outcome, the determinants of a given antibiotic resistance phenotype are obviously much more complex and only a part of the collection of factors that determine therapeutic outcome.

In some cases, the structural genes responsible for antibiotic resistance can be quickly and effectively demonstrated by nucleic acid amplification. A good example is recognition of methicillin resistance in *S. aureus* (MRSA) by probing for the *mec A* gene. This method can be used as a confirmatory test for MRSA, and in combination with rapid thermal cycling systems has the potential to significantly shorten the time to critical susceptibility test results. Unfortunately, there are far too many potential drug-bug combinations to use this approach for routine antibiotic susceptibility testing, even if the responsible gene sequences were known and stable. For the most part, genetic tests for antibiotic resistance determinants are likely to remain a research tool.

A radically different approach is to apply advanced cell biology technology for determination of antibiotic susceptibility phenotype. An example of this approach is the use of flow cytometry in susceptibility testing. Flow cytometers have become tools for the fast and accurate measurement of particles suspended in fluids. They have been used for many years in clinical haematology and can analyze data on thousands of cells in a matter of seconds. Some of the newer devices can accurately analyze particles an order of magnitude smaller than blood cells, which puts bacteria within their reach. The method that appears to provide the most effective way to measure antibiotic susceptibility is to combine viability stains (e.g. propidium iodide, which does not enter cells with a leaky cytoplasmic membrane) with cytometry and measure the viable fraction of total bacteria at a given antibiotic concentration. This method can be used to produce results on bacterial isolates in a matter of minutes. Flow cytometry has yet to be adapted for use in a routine, clinical service laboratory, but the method is promising for rapid susceptibility testing on critical bacterial isolates from blood and other sterile fluid specimens.

14.4.4. *Laboratory information systems*

The last decade has seen computerized information technology encroach on practically every corner of the diagnostic microbiology laboratory. Most larger laboratories now feed specimen request and test result data into an integrated laboratory information system, and many items of automated laboratory equipment (e.g. blood culture analyzers) are computer driven. Reliance on information technology is highlighted every time there is a central power failure, major virus alert, operating systems change and on a global scale, with the date change on 1st January, 2000. Further integration of information systems can be expected with the ultraminiaturization of nanotechnology, the fusion of computing and telecommunications, and the wider range of information portals.

   In the increasingly complex world of information technology, the ability to competently navigate the laboratory network requires progressively higher skill levels. Information specialists with competence in diagnostic bacteriology will be required in ever increasing numbers. One example is in the field of molecular epidemiology (see above) where automated ribotyping can now be networked via a modem with other centers operating the same system. Data sharing provides a powerful tool for outbreak recognition and tracking, but opens up a new world of issues that will have to be resolved if analytical results are to be applied for disease control purposes. It can be anticipated that very tight confidence limits will have to be applied on typing data when it is to be shared between centers in this way. Former methods such as subjective visual comparison ("eyeballing") of electrophoretic gels will have to give way to probabilistic comparison of gel results. These in turn will have to be integrated mathematically with other typing results, some of which will be text-based, to generate measurable levels of confidence in a proposed isolate cluster.

Another development that is well within the reach of existing information technology is in the improvement of data interfaces. Graphic user interfaces and icon-driven systems have been in widespread use for well over a decade. However, most diagnostic microbiology laboratories leave their icon-driven system behind when they enter the laboratory information system. Here the complex data that has been generated is reduced into a series of alphanumeric strings, some of which convey extremely complex concepts. Little has been done to provide the end-user, the requesting physician, with a graphic result. It is no longer difficult to trap digital images and send these as file attachments. Why not use this approach with a critical Gram stain result? And if that can be done, why not present the laboratory user with a human body image map as an entry into test results. Following the logic a little further, it should be possible to link pharmacy records of antibiotics used to those tested and depict these with temperature, leucocyte count and C-reactive protein on an in-patient timeline. Perhaps the most obvious bacteriological application of a graphic-user interface is in epidemiological cluster analysis. Infection control specialists and public health physicians have historically focused on temporal clustering to produce their epidemic curves. They have largely ignored the potential of information technology as a geographical tool. But with programs such as MapInfo, the co-location of food-borne infections such as salmonellosis and campylobacterosis would be more self-evident.

The diverse information technology developments mentioned above have an important common feature. They are all a means to obtain a second tier of information from the primary data set that comprises the laboratory's main work. It is a frequent source of frustration that the laboratory's work on a given specimen finishes with the final report. Information technology gives us an opportunity to add clinical value to that work at little additional cost.

14.5. Biotechnological Advances for Rapid, Specific and Sensitive Microbial Diagnosis

Conventional methods for the diagnosis of viral infections include tissue culture, electron microscopy, and detection of viral antigens and viral antibodies using immunological techniques. While these methods have proved useful, they suffer from certain disadvantages. Tissue culture is cumbersome and time-consuming. Certain viruses are extremely difficult, if not impossible, to propagate, e.g. hepatitis B and human papilloma viruses. The propagation of highly pathogenic viruses such as human immunodeficiency, Lassa fever, Marburg and Ebola viruses may be potentially hazardous to laboratory personnel. The interpretation of cytopathic effect is sometimes subjective and lacks type-specificity unless combined with the antibody neutralization test. Although a definitive serological diagnosis of viral infection can be achieved by observing a four-fold or greater increase in antibody titer between an acute serum sample and a convalescent sample taken about 10–14 days later, such an approach is slow. Furthermore, infected individuals (e.g. with HIV) may be negative for viral antibodies during the "window period" between viral acquisition and seroconversion.

Some viral infections may be life-threatening, e.g. chickenpox or herpes zoster in immunosuppressed patients. An increasing number of effective anti-viral agents are being introduced, which should ideally be administered early to treat certain viral infections, e.g. acyclovir for severe varicella-zoster virus infections in immunocompromised persons. In addition, the prompt recognition of viral infection allows measures to be taken to prevent further viral transmission, e.g. patient counseling, health education, vaccination of close contacts or vector control where appropriate.

Clearly there is a need for more rapid, specific and sensitive techniques to augment the conventional methods for the diagnosis of viral and other microbial infections.

Recent advances and applications of nucleic acid hybridization (NAH) with gene probes, and gene amplification by the polymerase chain reaction (PCR) have revolutionized microbial diagnosis and research in the field of infectious diseases. Being highly specific, sensitive and reproducible techniques for identifying and characterizing genes of microbial pathogens, these powerful tools are enjoying increasing applicability in the molecular diagnostics of a wide range of infectious disease agents, where large sample scale, speed and accuracy are important considerations. These improved screening methodologies allow earlier diagnosis resulting in prevention and timely intervention.

14.5.1. *Nucleic acid hybridization*

The principle of NAH is based on the remarkably similar behavior of nucleic acid molecules. Deoxyribonucleic acid (DNA) possesses a helical structure with complementary base sequences (adenine-thymine and cytosine-guanine) in the middle of the double-stranded helix, thus forming a long polymer with a net external negative charge from the phosphate groups of the sugar-phosphate backbone. Hybridization between two homologous nucleic acid strands is dependent upon the equilibrium between the attracting hydrogen bonds and the phosphate repulsions. Heat, alkali and various organic agents destabilize the attractive forces, while high salt concentrations partially neutralize the repulsive forces.

NAH involves the detection of specific target nucleic acid sequences by hybridization to a homologous gene probe. The excess probe is then washed off. In order to visualize the hybrid formation between the gene of interest and the probe, the latter must be labeled by an appropriate method.

NAH methodologies include *in situ* hybridization, filter *in situ* hybridization dot/spot/slot blot hybridization, Southern blot hybridization, sandwich hybridization and Northern blot hybridization. Depending on the size and nature of the probe, reactions for labeling probes include nick translation, end labeling, random primer labeling, RNA riboprobes, complementary DNA labeling, biotinylated bioprobes, chemiluminescence and fluorescent dyes. Non-radioactive alternatives for labeling probes are increasingly preferred in view of the hazards and relatively shorter life span of radioactive isotope labels.

14.5.2. *Polymerase chain reaction*

Despite the availability of a variety of modern amplification techniques (e.g. branched-chain DNA assay, ligase chain reaction, nucleic acid sequence-based amplification, Qβ replicase-based amplification, self-sustaining sequence replication, strand displacement amplification, transcription-based amplification), the most promising one is polymerase chain reaction (PCR) amplification which offers the important attributes of high sensitivity, specificity, speed, simplicity and sample throughput.

PCR is an *in vitro* technique for the primer-directed enzymatic amplification of specific nucleotide sequences. The exponential synthesis of millions of copies of a nucleic acid sequence of interest can be achieved in a matter of hours. Only a miniscule quantity of template, even femtogram amounts, is needed to initiate the reaction. Conventional PCR requires sequence information of the extremities flanking the target DNA segment in order that a pair of specific oligonucleotide primers can be synthesized which will hybridize to opposite strains. PCR consists of repeated thermal cycles of template denaturation, primer hybridization and primer extension. Each successive cycle thus doubles the amount of DNA synthesized in the previous cycle, theoretically resulting in the exponential accumulation of the target fragment according to the formula 2^n, where *n* equals the number of cycles. Thermostable DNA polymerases such as *Thermus aquaticus (Taq)* polymerase, are employed to catalyze DNA synthesis. Their superiority lies in their thermostability; optimum polymerization at higher temperatures with enhanced specificity; and being amenable to automation using computerized heat blocks.

14.5.3. *Molecular applications in diagnostic microbiology*

The applications of gene probes and PCR in infectious diseases have contributed immensely to a better understanding of their molecular pathology.

Rapid molecular virological assays and markers such as probes and PCR are emerging as viable diagnostic adjuncts or alternatives to slower and more cumbersome traditional methods such as tissue culture. PCR is suitable for detecting HIV during the seronegative window and perinatal periods; for latent viruses; for confirmation of herpes simplex virus, cytomegalovirus and respiratory syncytial virus infections prior to acyclovir, ganciclovir and ribavirin therapy respectively; for viral quantitation to monitor antiviral therapy; for detecting antiviral drug resistance; for hazardous laboratory pathogens e.g. Lassa fever virus; and for surveillance of insect and animal vectors. PCR is being applied to a growing list of other medically important viruses which include influenza, parainfluenza, corona, rubella, parvo B19, human papilloma, hepatitis, Epstein-Barr, human herpes 6, human T-lymphotropic, dengue, Japanese encephalitis, rabies, adeno-, entero- and rota-viruses.

Acquired immune deficiency syndrome (AIDS) represents an epidemic that threatens public health on a global scale. HIV cDNA probes and PCR have proven useful for identifying infected babies born to HIV-positive mothers, and for detecting HIV carriers who may be seronegative during the "window-period" between acquisition of infection and seroconversion.

The more rapid procedures of DNA hybridization and PCR can potentially replace cell culture and serology for the detection of potentially life-threatening pathogens such as cytomegalovirus and varicella-zoster virus in immunocompromised patients who require urgent diagnosis and treatment. PCR has already been employed for detecting herpes simplex virus DNA in cerebrospinal fluid towards the early diagnosis and therapy of herpes encephalitis.

NAH and PCR have been extensively used to detect human papillomaviruses, of which particularly genotypes 16 and 18 have been incriminated as an important co-factor in the etiology of genital cancer. Other examples of tumor viruses for which DNA probes and PCR have been employed are the Epstein-Barr virus (which is incriminated in Burkitt's lymphoma and nasopharyngeal carcinoma), and hepatitis B virus (HBV). HBV remains a global health problem with millions of chronic carriers who incur greater risks of developing chronic hepatitis, hepatic cirrhosis and primary hepatocellular carcinoma. HBV probes and PCR have been used to detect HBV DNA in serum and liver tissue specimens.

The molecular procedures of nucleic acid hybridization with gene probes and PCR technology are increasingly being applied for detecting bacterial, fungal or parasitic genomes to augment or even replace conventional cultures. PCR is ideal for identifying organisms that are dangerous, difficult or impossible to culture *in vitro*, such as *Chlamydia, Clostridium difficile, Helicobacter pylori, Legionella, Leptospira, Listeria, Mycobacterium tuberculosis, Mycobacterium leprae, Mycoplasma, Neisseria, Treponema, Yersinia, Histoplasma capsulatum, Pneumocystis*

carinii, *Leishmania*, *Plasmodium*, *Toxoplasma* and *Trypanosoma*.
Moreover, by amplifying known toxin genes, PCR can differentiate
between toxigenic and nontoxigenic strains of bacteria such as *Clostridium
difficile*.

Screening is essential to ensure that donor blood for transfusion
purposes is free from pathogenic microbes and to prevent the spread of
diseases such as cytomegalovirus, HIV infections, hepatitis B and C,
yellow fever, syphilis, malaria and Chagas' disease. Thus, a promising
area of diagnostic microbiology for the intensive application of gene probes
and PCR is in blood bank screening. A suitable technique is multiplex
PCR which involves the use of multiplex primer pairs each specific or a
particular pathogen and yielding a PCR product of diagnostic size.

The high prevalence of sexually-transmitted diseases represents an
enormous area of exploiting gene probe and PCR technology to potentially
replace slower conventional diagnostic methods.

PCR amplification of genes mediating resistance to antimicrobial agents
may serve as a useful adjunct to antimicrobial sensitivity tests and to
identify antibiotic-resistant strains of microbes.

Another potential area for the use of gene probes and PCR is
the detection of enteric pathogens in food and water samples as well
as from suspected gastrointestinal infections. Being more sensitive than
conventional cultures and immunological assays, these molecular tests may
be used to detect *Campylobacter*, *E. coli*, *Salmonella*, *Shigella*, rotaviruses,
enteroviruses, etc.

Fingerprinting of plasmid DNA isolated from pathogenic strains of
Staphylococcus species can augment diagnosis by cultures and is also
helpful for tracing the sources and spread of infection. Similarly, restriction
endonuclease patterns of DNA from microbes, such as varicella-zoster
virus, and *Mycobacterium tuberculosis*, can distinguish strains responsible
for unrelated outbreaks and even those infecting different individuals.

Methodologies complementary or related to PCR are increasingly being
exploited for improving specificity and sensitivity, as well as for molecular
epidemiological and evolutionary analyses of infectious disease pathogens.
Examples include RT-PCR for RNA viruses and viral transcripts; nested
PCR; multiplex PCR; co-amplification with human housekeeping gene
primers; *in situ* PCR; internal oligonucleotide probe hybridization;

restriction fragment length polymorphism; arbitrarily primed PCR, and direct DNA sequencing.

The absence of a diagnostic PCR product may be interpreted as a negative result or as a failure of amplification ("dropout"). To exclude the latter possibility, an additional pair of primers may be included to co-amplify a conserved human gene such as actin.

Consensus and type-specific primers are capable of detecting all four dengue virus types. Such a strategy is valuable for rapid diagnosis of dengue fever and for monitoring the presence of the virus in mosquito vectors.

Consensus or type-specific PCR primers can be employed to amplify products identified by hybridization to nucleic acid probes labeled nonisotopically e.g. using biotin/streptavidin/enzyme/substrate or chemiluminescent systems which can be adapted to a microtiter plate format.

By means of quantitative PCR, viral load (e.g. of HIV, hepatitis B or C viruses) can be assessed by determining plasma DNA or RNA levels, i.e. number of viral DNA or RNA copies per unit volume.

PCR combined with restriction fragment length polymorphism allows the fingerprinting of and differentiation between microbial strains for epidemiological purposes.

DNA cloning and/or cycle sequencing of PCR products constitute powerful tools for the analyses of hypervariable regions for studying microbial evolution and transmission pathways, and for sequencing genes encoding microbial products which are critical epitopes recognized by the host immune system, e.g. envelope proteins of viruses and the outer membrane proteins of bacteria such as *Neisseria gonorrhoeae* and *Chlamydia*.

A limitation associated with the use of PCR for diagnosis is the requirement for molecular training and instrumentation. The exquisite sensitivity of PCR is potentially also one of its greatest drawbacks in view of the danger of false positives arising from contamination with extraneous or "carryover DNA".

Notwithstanding this, contamination can be avoided by adhering to stringent pre- and post-PCR laboratory procedures and equipment to prevent cross-contamination. Suitable negative and positive controls should be valid in every batch of PCR samples for authentic results. In case of doubt, PCR of samples should be repeated. The success of routine PCR diagnosis depends on reliable internal controls, good quality assurance and DNA reference standards, and greater automation. With increasingly widespread availability and less expensive reagents, PCR and gene probe technology have found a stable niche in routine laboratory diagnosis.

14.5.4. *Other developments and advances in diagnostics*

The development of novel rapid sensitivity and specific techniques for detecting and characterizing pathogenic microorganisms has progressed with monoclonal antibody technology. Even though monoclonals are relatively less sensitive and specific than DNA methods, it is likely that both techniques will complement rather than compete with each other, in view of the generally easier and more rapid use of antibodies.

Furthermore, advances in the robust classical and IgM enzyme-linked immunosorbent assay (ELISA) as well as Western blot techniques for viral serodiagnosis are being achieved with better reagents addressing immunodominant epitopes, e.g. recombinant proteins, synthetic peptides and monoclonal antibodies. The Western or immunoblotting procedure is especially important for the confirmatory diagnosis of HIV infection.

There is currently a great demand for rapid bedside tests that are accurate and reliable in order to achieve prompt and more specific management and therapy. "Near patient" testing is ideal for obtaining almost instant diagnosis of potentially fatal or highly infectious diseases e.g. AIDS, hepatitis B, pneumococcal pneumonia, Legionnaires' disease, *H. pylori* infection, syphilis, tuberculosis, malaria which are amenable to antimicrobial therapy.

Consequently, "point-of-use" assays (e.g. latex agglutination and dipstick methods) are gaining increasing popularity for bedside and physician office testing. For example, immunochromatography involves placing the patient's specimen (e.g. urine, saliva, serum, blood) on a hand-held card, followed by the addition of a chromogenic reagent. The appearance of a single line indicates a negative result, while positivity is denoted by the presence of two lines.

With the expanding growth and importance of biotechnology in medicine and other spheres, disciplines such as computer science and biophysics can augment its development by collaborating in areas such as automation, laser and biosensor technology. Biosensors have numerous useful applications in patient care, i.e. "self-diagnostics" such as for home monitoring the concentration of chemical substances and drugs in the body (e.g. antimicrobial agents).

Diagnostic procedures are being revolutionized through miniaturization using biochip technology. Thousands or possibly millions of nucleic acid or protein/antibody probes can be embedded or printed onto chips, beads or slides made from silicon, glass or plastic, then reacted against clinical specimens, and positive signals detected by fluorescence scanners.

Biochips can perform thousands of biological reactions in seconds, thus allowing the virtually instantaneous identification of a wide range of pathogenic microbes, and facilitate prompt treatment.

Gene chips consisting of multiple DNA probes (e.g. oligonucleotides) aim to detect specific genes and even mutated genes via nucleic acid hybridization.

Protein chips are designed for recognizing specific proteins or antibodies by antigen-antibody reactions or protein-protein interactions.

Further Readings

1. Baron EJ *et al. Manual of Clinical Microbiology*. ASM, Washington, DC, 1999. (Widely regarded as the authoritative word on diagnostic microbiology. The principal emphasis is on recognition and detection of specific microbial agents of disease. Chapters are thoroughly referenced.)

2. Chow VTK. The impact of biotechnological advances in medical science: Implications for health care development. *ASEAN J. Sci. Technol. Dev.* **11**: 1–10, 1994.

3. Cleland CA, White PS, Deshpande A, Wolinsky M, Song J and Nolan JP. Development of rationally designed nucleic acid signatures for microbial pathogens. *Expert Rev. Mol. Diagn.* **4**: 303–315, 2004.

4. Inglis TJJ. *Clinical Microbiology Pocketbook*. Churchill-Livingstone, 1998. (A pocket-sized, problem-based compendium of useful

information for clinicians, scientists, infection control practitioners and public health physicians.)

5. Koneman E *et al*. *Atlas and Textbook of Diagnostic Microbiology*. Lippincott, 1997. (A comprehensive review of laboratory microbiology. that majors on the bacteriology performed in diagnostic labs. Brief accounts of clinical conditions are integrated with laboratory information. Illustrated and thoroughly referenced.)

6. Lueking A, Horn M, Eickhoff H, Bussow K, Lehrach H and Walter G. Protein microarrays for gene expression and antibody screening. *Anal. Biochem.* **270**: 103–111, 1999.

7. McFaddin JF. *Identification of Medically Important Bacteria*. Lippincott, 1999. (Widely regarded as the definitive bench-top laboratory manual on identification methods. Many flow diagrams and keys. Spiral bound for in-lab use.)

8. Searls DB. Using bioinformatics in gene and drug discovery. *Drug Discov. Today* **5**: 135–143, 2000.

9. Smith JE. *Biotechnology*, 3rd ed. Cambridge University Press, 1996.

10. Struelens MJ, Denis O and Rodriguez-Villalobos H. Microbiology of nosocomial infections: Progress and challenges. *Microbes Infect.* **6**: 1043–1048, 2004.

11. Topley and Wilson's Microbiology and Microbial Infections. *Vol. 1 Bacterial Diseases* and *Vol. 2 Systematic Bacteriology*, 1998. (These two volumes contain a comprehensive series of review chapters on medically-important bacteria and the diseases they cause. The text is available in computer-searchable, CD-ROM format.)

12. Versalovic J and Lupski JR. Molecular detection and genotyping of pathogens: More accurate and rapid answers. *Trends Microbiol.* **10**: S15–S21, 2002.

Questions for Thought...

1. Emerging and re-emerging infectious diseases represent an increasing problem worldwide. Discuss the contributory factors, and the impact of these diseases on public health. Illustrate your answer with examples.

2. Performing Gram and other staining on clinical samples may provide useful information on the etiology of an infection and may guide

preliminary antibiotic therapy. Discuss the application of staining techniques on cerebrospinal fluid, sputum and blood culture samples.

3. Discuss the role of selective culture media in the laboratory diagnosis of bacterial infections.

4. Describe the methods routinely employed to measure the sensitivity of bacteria to different antibiotics in the diagnostic microbiology laboratory. How should the results be reported?

5. How would you determine the minimum inhibitory concentration of an antibiotic for a bacterium?

6. Discuss the methods used in the laboratory diagnosis of bacterial meningitis.

7. Discuss the laboratory diagnosis of community-acquired pneumonia.

8. In what ways can the microbiology laboratory assist the clinician in the diagnosis and treatment of septicemia?

9. Discuss the laboratory diagnosis of urinary tract infections.

10. Discuss the diagnostic value of midstream urine specimens and other urine samples in establishing the etiology of urinary tract infections.

11. Discuss the techniques used in the laboratory diagnosis of bacterial respiratory infections.

12. Discuss the laboratory diagnosis and epidemiology of *Salmonella* infections.

13. Describe the principles of the different laboratory methods for the diagnosis of sexually transmitted infections. Discuss their use in a routine diagnostic laboratory, including specificity, sensitivity, cost, time and medico-legal issues.

14. Discuss laboratory information systems in a diagnostic bacteriology laboratory.

15. Discuss the advantages and disadvantages of classical (non-genetic) techniques for the laboratory diagnosis of human bacterial infections.

16. Discuss the recent technical advances for rapid diagnosis and epidemiology of infectious diseases.

17. Compare and contrast enzyme-linked immunosorbent assays with gene probe techniques for microbial diagnosis. Illustrate your answer with examples.

18. Discuss the impact of new genetic and immunological techniques on the diagnosis of microbial infections.

19. Discuss the advantages and disadvantages of nucleic acid amplification for viral diagnosis.
20. "The exquisite sensitivity of the polymerase chain reaction is both its greatest strength and its greatest weakness." Discuss this statement.
21. Discuss how advances in computerization, miniaturization, automation and signal detection are significantly enhancing the specificity, sensitivity, speed, and sample throughput of microbial diagnosis.
22. Discuss the principles, techniques and applications of molecular fingerprinting of microorganisms.
23. Discuss the techniques and applications of restriction fragment length polymorphism.
24. Discuss the application of gene probes and the polymerase chain reaction in the analysis of restriction fragment length polymorphisms.

Related World Wide Web Sites

- All the Virology on the WWW
  http://www.virology.net
- Centers for Disease Control and Prevention, USA
  http://www.cdc.gov
- Microbiology: Neal's Microbial World
  http://www.microbesite.com
- National Institute of Allergy and Infectious Diseases, USA
  http://www.niaid.nih.gov
- Program for Monitoring Emerging Diseases (ProMED)
  http://www.fas.org/promed and www.promedmail.org
- TIGR Microbial Database
  http://www.tigr.org/tdb/mdb
- World Health Organization
  http://www.who.int

Chapter 15

Microbes: Friends or Foes?

Vincent T. K. Chow[*], **Tim J. J. Inglis**[†] **and Keang Peng Song**[*]
Department of Microbiology, National University of Singapore
5 Science Drive 2, Singapore 117597
†*Division of Microbiology and Infectious Diseases*
Western Australian Centre for Pathology and Medical Research
Locked Bag 2009, Nedlands WA 6009, Australia

Robert Koch was a medical doctor and a great microbiologist who first identified *Bacillus anthracis* as the causative agent of anthrax in the late 1880's. Since then, many other bacteria were identified and later shown to be responsible for different infectious diseases. Examples of some bacteria that can cause infections include *Vibrio cholerae* (causes cholera), *Pseudomonas aeruginosa* (wound infection), *Corynebacterium diphtheriae* (diphtheria), *Clostridium perfringens* (gas gangrene), *Clostridium botulinum* (botulism), *Shigella dysenteriae* (shigellosis), and *Bordetella pertussis* (whooping cough).

From the perspective of clinical microbiology, bacteria can be divided into pathogenic and non-pathogenic groups. Pathogenic bacteria are a group of organisms causing infectious diseases, and based on the molecular definition of pathogenicity, these bacteria contain the necessary genetic information to allow them to express virulence when conditions permit. Fortunately, when compared to the huge number of bacteria known to exist, the proportion represented by pathogenic bacteria is relatively very small, even though many of them are still uncharacterized. Under most instances, pathogenic bacteria are undoubtedly viewed as one of the biggest enemies of humans. However, many people always wonder what makes bacteria pathogenic? Bacteria may be small in size, but they can certainly compete and adapt for survival. Once the host on whom they depend dies, bacteria will also lose their source of food and growth environment. Therefore, just like ants or bees found in the field, normally they will only start to attack any host when their survival is threatened. This can happen when bacteria are unintentionally brought into an environment unfamiliar to their original dwelling place. For example, during an abrasion on the skin, bacteria like *C. perfringens* (that causes gas gangrene on wounds) from the soil can be introduced into the deeper tissues underneath the skin. This can also arise when their source of food is running out, which then induces the bacteria to secrete toxins and enzymes to kill host cells so that the released material from lyzed cells can be used as nutrients. The bottom line is that bacteria may become pathogenic for their own protection and survival. It may be similar to any other biological phenomena of self-defense. Therefore, most bacteria are generally harmless, but may become pathogenic when their survival is at stake.

Although bacteria, and pathogenic bacteria in particular, have caused much misery to mankind, we cannot deny the fact that some of them have also contributed directly and indirectly to the well-being of mankind. Through advances in science and technology, many species of bacteria are exploited to produce substances of importance for both medical and industrial applications. Some of the by-products of bacteria are also being used in industry. Examples of some known uses include generation of antibiotics, production of insulin, development of vaccines, and so on. As such, although some less fortunate individuals suffer from the devastating effects caused by bacterial infections, others may enjoy the benefit and convenience offered by bacteria, either directly or indirectly. So, are bacteria our friends or enemies?

15.1. Use of Bacteria as Hosts in Molecular Cloning

Diabetes is one of the biggest killers not just in developed countries, but throughout the whole world. Patients suffering from diabetes cannot make insulin, or cannot use insulin effectively. Insulin is one of the many hormones produced by the endocrine glands of the body. It is circulated in the bloodstream and is required for the regulation and maintenance of the blood sugar level in the body by aiding in the conversion of sugar to glycogen. Approximately 10% of the patients have type I diabetes, whereby they require insulin to maintain life. The rest suffer from type II diabetes, and whose condition can be managed by a combination of diet control, exercise, using agents to lower blood sugar and insulin treatment. Regardless of which type, many diabetic patients will eventually need insulin therapy, and the standard treatment is the administration of exogenous insulin.

Insulin is synthesized by the beta cells of the islets of Langerhans in the pancreas. Only a very small amount of insulin is produced each day. When one suffers from insulin-dependent diabetes, one has to obtain insulin from external sources. Many years ago, insulin was mainly derived from the pancreatic glands of animals such as sheep, cows and pigs, or even from cadavers. Not only was the amount obtained very limited, the quality of the insulin was also inconsistent. The non-human insulin was not as effective as human insulin. Moreover, some people even developed allergy when animal insulin was used. There were many side effects associated especially with the animal sources of insulin, and it was also very expensive.

Today, diabetic patients no longer have to pay a hefty sum for diabetes treatment. The insulin they receive is also genuine human insulin, thanks to the development of genetic engineering for the mass production of recombinant insulin. Human insulin consists of two polypeptides, designated as chains A and B, which are connected by disulphide bridges. During the synthesis of insulin, proinsulin is first formed prior to cleavage to generate the active form of insulin. The recombinant human insulin currently available is prepared by cloning and expressing genes coding for proinsulin in *E. coli*. After the proinsulin is extracted and purified from the bacteria to remove other contaminating substances, the connecting peptide is enzymatically cleaved to convert the proinsulin to functional insulin.

Using *E. coli* as a host to express genes for insulin provides many advantages. Bacterial cells are generally easy to grow, and the media used to culture bacteria is relatively cheap. Moreover, being a well-characterized bacterial species, manipulation of genes in *E. coli* is relatively safe, the outcomes are predictable, and any problems encountered can be easily solved. By using fermentation techniques to grow *E. coli*, large numbers of bacteria can be cultivated at any given time. This will also translate to large amounts of protein that can be extracted and purified from the bacteria. Human insulin represents the first successful commercial product resulting from genetic engineering of mammalian genes and expressing them in bacteria. It is often referred to as a "biotech drug". After the successful introduction of insulin to the market, other human hormones like growth hormone and thyroid stimulating hormones have also been developed and marketed.

15.2. Contribution of Bacterial Enzymes in Genetic Engineering

The typical size of bacteria usually ranges from 0.5 to 2 μm. This size is at least ten times smaller than that of most eukaryotic cells, which normally ranges from 2 to 25 μm. As such, many people perceive bacteria to be small and simple organisms. However, through our understanding of the genetics and physiology of bacteria, we know that this view is inaccurate because bacteria are not only complex in their internal structural organization, they are also capable of performing many complicated tasks. This is supported by the fact that over the years, despite massive efforts by scientists all around the world to study and understand bacteria, there are still many unknowns and questions remained unanswered. Although bacteria lack nuclei and membrane-bound organelles such as mitochondria, they have almost everything that an eukaryotic cell contains. This indicates that the basic machinery necessary to keep cells alive is common for both prokaryotic and eukaryotic cells. The same ongoing events such as DNA replication, RNA transcription, gene expression and cell replication are in operation for the two cell systems. One of the main players responsible for making these events possible includes proteins.

Proteins are involved in many cellular activities of bacteria, e.g. some as enzymes to catalyze biochemical reactions of the cell, some as positive regulators for the expression of genes, and others as signal molecules for the regulation of cellular functions. Among the many known proteins, enzymes are particularly important and play a very active role in keeping the cells healthy and alive. Many cellular events and processes are driven by various types of enzymes synthesized by the bacteria. Examples of some of these enzymes include DNA polymerase, RNA polymerase, restriction enzymes and DNA ligases. Some of these enzymes are so critical that bacteria lacking the ability to synthesize them may not survive. Studies using site-directed mutagenesis and deletion mutations to construct mutant bacteria that were deficient in expressing one or more of these enzymes, demonstrated that these bacteria were weak and fragile.

Following the discovery of the structure of nucleic acids, molecular biology has become the core subject in biology and life sciences curricula in colleges. Many scientists are even prepared to switch their interests to this field. The majority of scientists who are engaged in biological research are in one way or another molecular biologists. One important tool that molecular biologists use in their research is genetic engineering, and the key players in this technology are enzymes. Enzymes such as DNA polymerases and restriction enzymes have been isolated and purified from different species of bacteria. They are used in the manipulation of genes, for example cloning, sequencing, mutagenic studies, and so on. Two of the enzymes, restriction enzymes and DNA polymerases, are described below.

15.2.1. *Restriction enzymes*

One of the many tasks that molecular biologists perform is to "cut and paste" DNA fragments into plasmid vectors, a process known as cloning. These DNA fragments are also used as templates in genetic manipulations such as gel-shift assay, DNA foot-printing analysis, probe preparation, sequencing, RNA synthesis, and so on. There are several ways to generate DNA fragments, but one easy and fast method is through the use of restriction endonucleases to digest long nucleic acids at specific locations. In fact, it was the discovery of restriction endonucleases that actually gave birth to, and laid the foundation for genetic engineering.

Restriction endonucleases were discovered in *E. coli* in the early 1970s. The first enzyme to be isolated was *Eco*RI by Robert N. Yoshimori of Boyer's laboratory in San Francisco. The nomenclature of restriction enzymes consists of several alphabets. The first capitalized alphabet is derived from the first letter of the genus of bacteria from which the enzyme is isolated. The next two alphabets of the enzyme are in small letters derived from the first two letters of the species of bacteria. There may be a fourth alphabet which is a capital letter derived from the strain designation of the bacteria. For example, the enzyme *Eco*RI is from the bacterium *Escherichia coli*. If the strain of microorganism produces just one restriction enzyme, the name ends with the Roman numeral I. If more than one enzyme is produced, they are numbered II, III, and so on.

Restriction endonucleases are a group of enzymes synthesized by bacterial cells. These enzymes usually recognize specific nucleic acid sequences about 4 to 6 base pairs long, and cleave the DNA at a very specific site. Their function is to protect bacterial cells from foreign DNA injected by invading bacteriophages. Since the discovery of *Eco*RI, many other restriction endonucleases have been discovered. Some of these are listed in Table 15.1.

One of the main *in vitro* uses of restriction endonucleases in molecular biology research is for the preparation of DNA fragments for a number of purposes. Owing to the specific activity of the restriction enzyme on the recognition and digestion of the DNA, the ends of the DNA fragment digested with the enzyme are complementary to another fragment digested with the same enzyme. These two pieces of DNA are considered to have

Table 15.1. List of restriction endonucleases discovered after the discovery of *Eco*RI.

Microorganism	Enzyme	Recognition Sequence
Bacillus amyloliquefaciens H	*Bam*HI	C*GATCC
Caryophanon latum L	*Cla*I	AT*CGAT
Diplococcus pneumoniae	*Dpn*I	GA*TC
Escherichia coli RY13	*Eco*RI	G*AATTC
Escherichia coli J62	*Eco*RV	GATAT*C
Haemophilus influenzae R$_d$	*Hind*III	A*AGCTT
Haemophilus influenzae R$_f$	*Hind*FI	G*ANTC
Moraxella bovis	*Mbo*I	*GATC
Neisseria denitrificans	*Nde*I	CA*TATG
Serratia marcescens S$_b$	*Sma*I	CCC*GGG

perfect match and therefore can be joined chemically by a process of ligation involving another enzyme called DNA ligase.

DNA ligase is another enzyme used widely in genetic engineering. The enzyme isolated from *E. coli* is a polypeptide chain with a molecular mass of 75 kDa, and its activity requires NAD^+ as a cofactor. In contrast, the ligase encoded by phage T4 is 68 kDa, and requires ATP as a cofactor and an energy source. Regardless of the source, ligase works like a "super glue" in joining two DNA fragments together, or repairing two broken strands of DNA in a double-stranded DNA fragment. Ligase was also discovered at about the same time as restriction enzymes were discovered. Both enzymes which have contributed significantly to genetic engineering are produced by *E. coli*.

15.2.2. *DNA polymerases*

DNA polymerases are required for DNA replication in both prokaryotic and eukaryotic cells. The basic and primary activity of the polymerase is to synthesize DNA, i.e. duplication of DNA, using another strand of DNA as a template. Other than DNA polymerization activity, polymerases can also function as exonucleases to cleave DNA. This cleavage activity is also an important property of polymerases where it proof-reads nucleic acid sequences, and corrects any misincorporated nucleotide when necessary. This step is critical so that the replicated DNA will contain exactly the same nucleotide sequences as the parental copy. DNA mutations arise when some of the wrongly incorporated nucleotides are not noticed and repaired accordingly. The processes of polymerization and exonuclease activities are repeated many times during DNA replication, until the DNA strands are faithfully duplicated.

DNA polymerases are used in many techniques in molecular biology research. The protruding ends of DNA fragments ("sticky" ends) can be polished to become non-protruding ("blunt" ends) by using both the polymerase and exonuclease activities of the enzyme. They can also be used to incorporate radioactive labels in DNA via a process called random primer labeling, where nucleotides containing radioactive isotopes are built into DNA strands by polymerases. DNA polymerases have also been used extensively to determine nucleic acid sequences by the Sanger or Maxam-Gilbert sequencing techniques.

A major breakthrough in molecular biology was the discovery of a way to duplicate or amplify DNA in an efficient and rapid manner *in vitro*. This procedure is well known as PCR, or the abbreviated form for polymerase chain reaction. The scientist who invented this procedure, Dr. Kary Mullis, was awarded a Nobel Prize in 1993. PCR is a rapid method that can replicate DNA to at least a million-fold within a short period of time, usually in 2–3 hours. One of the components used in PCR is DNA polymerase, but this polymerase is not an ordinary enzyme. It is an enzyme that can withstand high temperatures of as high as 100°C, and because of this property, it is called a thermostable DNA polymerase. Unlike conventional DNA polymerases, which are inactivated once they are exposed to temperatures above 70°C, this DNA polymerase used in PCR attains its optimal activity instead at high temperatures. This unique enzyme is found in bacteria that live in hot springs in places such as the Yellowstone National Park in USA. An example of this type of polymerase that is available commercially is *Taq* polymerase, isolated from an organism called *Thermus aquaticus*. Its thermostability, and the availability of sophisticated apparatus for thermal cycling have allowed the PCR technique to be performed in an automated fashion.

Since the invention and automation of PCR, the methodology has been applied to other fields such as DNA sequencing and site-directed mutagenesis. The conventional way of performing DNA sequencing is very cumbersome and laborious, taking at least a few days just to prepare the necessary components needed for the reaction. If this conventional way of sequencing were to be used for sequencing bacterial genomes, it would not be cost-effective at all because much manpower and time is involved. However, by incorporating PCR methodology to sequencing, this once laborious procedure has become almost a daily routine in many laboratories. There are now even companies offering sequencing services for only a fraction of the cost. This is attributed to the main player involved, i.e. DNA polymerase, an example of how another enzyme from bacteria has contributed to the advancement of science.

15.3. Use of Bacterial Toxins in Therapeutics

Only a small group of bacteria is known to be pathogenic, but they are sufficient to cause havoc and misery to many people. Numerous occasions in history have shown how bacteria caused epidemic infections that killed millions of people. Most of these pathogenic bacteria cause diseases by producing toxins that attack and kill host cells. The level of pathogenicity of these bacteria is usually measured by the potency of the secreted toxins to kill cells. The faster these bacteria kill cells, the more pathogenic they are. Although the ultimate effect of these toxins may be the same, e.g. death of cells, the target and pathogenic mechanism of each toxin usually differ from one another. Some toxins are milder in their cytotoxic activity, but others can be very toxic by causing instant death to cells. Therefore, toxins play a very important role in the pathogenesis of bacteria. For those toxins that are protein in nature, the best way to neutralize them is via their inactivation through the use of antitoxin antibodies. These antitoxin antibodies bind to the active sites on the surface of toxins, thereby blocking access of the toxins to host cells. In fact, this is how treatment of some infectious diseases is achieved. The primary treatment for infections like botulinum poisoning and tetanus is by the administration of antitoxin antibody to neutralize the activity of toxins. Through studies on bacterial pathogenesis, scientists can better understand how pathogenic bacteria work. Based on the knowledge gained from these studies, it was realized that toxins could in fact be harnessed as molecules with beneficial functions. It was found that there are diseases which can be treated with toxins. Research on toxicology (the study of the biochemical and physiological properties of toxins) suddenly assumed a new role in the identification of potent toxins with potential therapeutic value. So far, a few products have been developed and some are available commercially for use in medical practice. Two examples will be discussed here. The first is the use of botulinal toxin for the treatment of disabling muscle spasm. The other is the development of chimeric toxins for the killing of anaplastic cells in diseases like cancer.

Botulinal toxin is secreted by a Gram-positive, spore-forming, anaerobic bacillus called *Clostridium botulinum*. This organism is the causative agent of botulism, a highly fatal food poisoning usually contracted following the ingestion of a neurotoxin produced by the organism. *C. botulinum* can be found in soil, lake sediments, and decaying vegetation. This makes the contamination of foods such as vegetables and meats relatively easy.

C. botulinum is classified into groups A to G based on the antigenic specificity of the secreted toxins. Groups A, B and E are responsible for most cases of human disease, while the rest are found most often in animals like cattle, horses, and sheep. The toxins are neurotoxic, heat-labile, immunologically distinct, and are among the most toxic compounds known. The estimated lethal dose for humans is about 1 μg, while 1 ml of culture fluid is sufficient to kill two million mice. Botulinal toxin binds to its target at the neuron and blocks the presynaptic release of acetylcholine at the neuromuscular junction, resulting in the prevention of information relay between muscle and nerve cells. This activity of the toxin leads to flaccid muscle paralysis of patients. Some of the symptoms commonly observed in patients include double vision (diplopia) and difficulty in swallowing (dysphagia). Death results from the paralysis of respiratory muscle functions.

The introduction of botulinal toxin for medical use was initiated in the early 1980s for the experimental treatment of a painful neurological disorder called dystonia. Its use as a general therapy for dystonia was finally approved in 1989 by the Food and Drug Administration (FDA) of the US. Dystonia is a group of neurologic disorders affecting more than 300,000 people in North America. It is characterized by sustained involuntary muscle contractions and spasm of the body, which results in the development of painful movements or postures in certain parts of the body. It can affect virtually any part of the body and involve any muscle. The specific cause of the disease remains unclear. Examples of such dystonias are blepharospasm (eyelids), hemifacial spasm and hand cramps. Before the use of botulinal toxin was known, the only therapy that was given to patients suffering from the disease was through surgical destruction of the nerve endings in the area affected by the spasms. This type of surgical treatment is not only expensive and dangerous, but has a low success rate. Since dystonia involves sustained contraction of muscle, it can take advantage of the neuromuscular blocking effect of botulinal toxin to alleviate muscle spasm caused by the disease. Injecting a small amount of botulinal toxin at the affected site will neutralize the spasm caused by the disease. This was indeed proven effective. Botulinal toxin therapy provides a better alternative to surgical treatment because it is cheaper and less dangerous. However, the effect provided by botulinal toxin is only temporary, and repeated treatment of the toxin has to be administered.

The botulinal toxin used currently is the group A neurotoxin, the only one approved for medical use today. It is a large protein with a molecular mass of 150 kDa. The commercial form of the toxin is a complex with a number of other proteins resulting in a molecule that is six times the size of the original neurotoxin. Attempts are currently underway to genetically engineer botulinal toxin to make it more effective at targeting neurons so that a one-shot therapy with permanent effects will be feasible.

Although botulinal toxin is a neurotoxin that can cause death in a victim during the pathological process of the infection, it does not kill the affected cells. The person dies mainly due to respiratory failure as a result of muscle paralysis. There is a group of toxins called cytotoxins that actually kill cells through certain biochemical reactions. Examples from this group of cytotoxins include *Pseudomonas* exotoxin (produced by *Pseudomonas aeruginosa*), and pertussis toxin (produced by *Bordetella pertussis*). These toxins kill cells by inhibiting the synthesis of proteins in cells. To become functional proteins, this group of toxins usually consists of three domains when they are folded properly: (a) the receptor binding domain that binds to the receptors on host cells; (b) the translocation domain that transfers the toxin to the cytoplasm of the host cells; and (c) the enzymatic domain that carries out the catalytic function of the toxin. Toxins lacking any one of these domains will lose their ability to kill cells, and thus become inactive. Generally, the degree of specificity is determined by the receptor-binding domain of the toxin, which directs the toxin specifically to only certain cell types. Once inside the host cells, the enzymatic domain of the toxin will function to inactivate its target molecule. The specificity of the enzymatic domain is usually dependent on the target molecules present in the cell, e.g. the elongation factor and ribosome responsible for translation of proteins. Therefore, if there is a strategy to introduce the enzymatic domain into the cytoplasm of certain cells such as tumor cells in a very specific manner, this portion of the toxin will be able to kill those cells selectively. This is how the idea of the chimeric toxin was conceived.

A chimeric toxin is a toxin chemically attached to a protein molecule with high specificity for certain receptors found on the surface of eukaryotic cells, so that the action of the toxin can be directed towards those targeted cells. If this protein molecule is a portion of an antibody, it is called an immunotoxin. This type of immunotoxin can be generated by attaching the enzymatic domain of the toxin to the antibody-binding (Fab) portion of a monoclonal antibody, all achieved by genetic engineering techniques. This method of toxin targeting is potentially very useful and effective for anaplastic cells like tumor cells. The conventional ways of removing tumors include either surgical excision, chemotherapy or radiotherapy, or a combination of these. The ultimate aim of the treatment is the complete elimination of all tumor cells in the body. Despite proven success in the treatment of certain tumors, these methods are not always successful, especially when the tumor growth has already progressed to the terminal stage. The major limitation of conventional methods is the inability to achieve specific killing of only tumor cells but not normal cells. With the advent of immunotoxin technology, there appears to be hope for a better and more effective way of treating cancer patients. A group of researchers from the National Institutes of Health, US, has recently demonstrated the use of immunotoxins to destroy solid tumors. This immunotoxin was created by chemically coupling the enzymatic domain of *P. aeruginosa* exotoxin to a monoclonal antibody recognizing receptors present only on tumor cells. Results from the phase I clinical trial were quite promising, although there are still both technical and biological problems to be resolved. One of the problems is the host induction of antibodies to the recombinant toxin. This research group is currently improving on the construction of the immunotoxin, including designing a smaller recombinant immunotoxin, and the use of immunotoxin in conjunction with immunosuppressants to inhibit the development of antibodies.

15.4. Use of Biotechnology to Produce Antimicrobial and other Therapeutic Agents

Microorganisms and plasmid vectors have been manipulated by genetic engineering for use in medicine, such as cloning of genes of interest to express valuable proteins in prokaryotic (e.g. *E. coli*) or eukaryotic host cells (e.g. yeast, mammalian, human, plant).

With the advent of recombinant DNA and protein engineering technology, an ever increasing number of genetically-engineered therapeutic agents are being introduced and marketed by the pharmaceutical industry, undergoing clinical trials or have potential clinical applications.

Many important examples of these novel drugs include hormones (such as growth hormone for pituitary dwarfism; insulin for diabetes mellitus; erythropoietin for anemia or renal failure), growth factors (such as insulin-like growth factor 1 for Laron dwarfism; brain-derived neurotrophic factor), biological response modifiers, immunomodulators, immuno-stimulators, cytokines (such as interferon for hairy cell leukemia, chronic myelogenous leukemia and chronic hepatitis B and C; stem cell factor; granulocyte colony stimulating factor; granulocyte-macrophage colony stimulating factor for bone marrow failure; interleukins for certain cancers; tumor necrosis factor), hematological products (such as tissue plasminogen activator for thromboembolic lysis; coagulation factor VIII for hemophilia; albumin for shock; hemoglobin as a blood substitute), alpha-1 antitrypsin for emphysema, encephalin with analgesic properties, antibacterial lactoferrin and lysozyme, microbial vaccines (such as the hepatitis B vaccine and potential recombinant AIDS vaccines currently undergoing clinical trials), and cancer vaccines. In 1995, the world market for human growth hormone alone was worth US$1.4 billion.

Furthermore, synthetic peptides are being contemplated as therapeutic agents, e.g. vaccines. Modern biotechnology is also being exploited for the biological control of pests and vectors of human disease.

Although bacteria, yeasts and cultured cells serve as the conventional hosts for the expression of commercially important proteins, "molecular biopharming" or the use of transgenic animals and plants as living pharmaceutical factories or bioreactors to produce drugs and monoclonal antibodies on a large scale appears promising. Thus, valuable proteins can be generated in quantity in milk and from plant extracts.

The knowledge acquired from the molecular biology of important disease processes such as cancer and microbial infections including AIDS, yielding sophisticated information such as X-ray crystallography and nuclear magnetic resonance data of proteins can be processed using "virtually reality" and interactive computer graphics to permit the three-dimensional modeling and structure-function analysis of complex molecules, thus facilitating a more rational, refined, precise design and molecular engineering of effective novel drugs such as agonists, antagonists, enzyme activators and inhibitors. For example, Gleevec (STI571, imatinib mesylate) is a new signal transduction inhibitor developed against the deregulated tyrosine kinase activity of the *bcr-abl* fusion protein in chronic myeloid leukemia. Molecular mimics of the substrate of a viral enzyme that differ slightly from the natural substrate can also be created to inhibit the enzyme and interrupt viral replication.

Notwithstanding the thrust of molecular genetics in medical science, pharmaceutical companies are not neglecting the rich sources of drugs and medical products such as new antibiotics from nature. In fact, there is a renewed interest in screening for these high-value products effective against global epidemics such as AIDS, particularly in the plethora of plant and microbial flora and species of the rapidly depleting tropical rainforests of the world.

Examples of the production of antimicrobial agents from microbes are cephalosporin from *Acremonium* species, chloramphenicol from *Cephalosporium* species, penicillin and griseofulvin from *Penicillium* species, rifamycin from *Nocardia* species and streptomycin from *Streptomyces* species.

Compounds with anti-tumor activity have also been derived from microbes, e.g. actinomycin D, bleomycin, daunorubicin, mitomycin C from *Streptomyces* species, and asparaginase from *Erwinia* species. Other new drugs with anti-cancer properties include enediynes, isolated from a soil bacterial strain.

Much progress has been made in the development of antimicrobial agents, and the search for new ones is intensifying in the light of the increasing problem of antibiotic resistance and the emergence of new infectious diseases. The new antimalarial drug artemisinin extracted from the leaves of the Chinese herbal plant *qinghaosu*, shows promise in controlling multi-drug resistant and cerebral malaria currently spreading in parts of South-east Asia. Tafenoquine is a new synthetic analogue of primaquine for antimalarial prophylaxis.

Available to the physician's armamentarium is an increasing number of antiviral agents effective against herpes viruses (e.g. acyclovir, famciclovir, valaciclovir, ganciclovir, foscarnet), hepatitis B virus (e.g. interferon, lamivudine), influenza A and B viruses (e.g. neuraminidase inhibitors including oseltamivir and zanamivir), and respiratory syncytial virus (e.g. ribavirin).

The threat of the AIDS pandemic stimulated the generation of new antivirals against HIV, i.e. nucleoside analogue reverse transcriptase inhibitors (e.g. azidothymidine, dideoxyinosine, dideoxycytidine), non-nucleoside reverse transcriptase inhibitors and protease inhibitors. Highly active antiretroviral therapy (HAART) for AIDS involves multiple antiretroviral drugs prescribed in combination as a cocktail to achieve synergistic effect, to suppress viral load, and to minimize drug-resistant variants.

Novel treatment modalities for infectious diseases are being devised and tested, such as immunotherapy using anti-cytokine antibodies against septic shock, e.g. antibodies against tumor necrosis factor alpha, interleukin-1 and interleukin-6.

Microbial toxins are being harnessed for beneficial uses, e.g. botulinum toxin A for the alleviation of muscle spasms and wrinkles. Modern "magic bullets" or immunotoxins in the form of custom-made, humanized monoclonal antibodies conjugated with potent toxins which specifically target cancer cells are being studied in patients with advanced malignancies. An example is the diphtheria toxin conjugated with interleukin-2 targeted against T-lymphocytic leukemia cells that express interleukin-2 receptors.

Liposomes and biodegradable polymer particles are being developed as vehicles to deliver drugs directly to target sites in the body, thus enhancing efficacy and reducing adverse or toxic effects.

Knowledge on the formation of biofilm arising from the ability of bacteria and other microbes to adhere to and colonize surfaces has even led to the development of "non-stick" coatings for medical devices such as artificial implants.

15.5. Vaccine Development

Most pathogenic forms of bacteria owe their pathogenicity to the virulence factors they produce during pathogenic events leading to damage of host cells. Some of these virulence factors are secreted into the environment to act on surrounding or remote cells (e.g. exotoxin and hydrolytic enzymes), others are produced and deposited on the surface of the bacteria (e.g. capsules and pili). The roles of these virulence factors vary from one pathogen to another, but they are usually produced either to protect the bacteria from attack by molecules like antibiotics and antibodies, so that the bacteria can spread, or to obtain nutrients for the survival of bacteria. However, in order for these virulence factors to be successful, they must first avoid or overcome the host defense system, in particular the immune responses. The human body contains several powerful defense systems against most foreign invaders. The first line of defense starts from the skin, and extends all the way to the internal structures of the body including the Peyer's patches of the intestine. One of the immune responses is the generation of antibodies. Once the primary immune response is triggered, memory cells in the body will be established, and the subsequent encounter with the same invader will bring about speedy synthesis of the specific antibodies to inactivate or neutralize the invader. This is known as the secondary immune response, which is more efficient and rapid than the primary response. Therefore, the ideal situation for an individual is to have as many of the primary responses triggered in the body, so that the body is always ready to defend against invading pathogens. This was the concept of the smallpox vaccine introduced by Dr. Edward Jenner in the eighteenth century. A vaccine is a substance that can trigger the immune response of the body without causing harm to the host. As a result of Dr. Jenner's discovery and successful mass vaccination programs worldwide, the World Health Organization officially declared the world to be free of smallpox in 1980. Currently, vaccines are already available for several infections: poliomyelitis, diphtheria, cholera, whooping cough, tetanus, mumps, measles and rubella (Table 15.2). The use of vaccines has extended the human life span by approximately 15 years.

Vaccines can be developed in many ways. In general, to make a vaccine for a pathogen, the pathogen is first deliberately inactivated so that it is no longer harmful. This inactivated form is then injected into the body of the host. Since it is rendered not pathogenic and harmless, it will not cause damage to the host. However, although the pathogen is inactivated, some of the antigenic determinants are still retained. These remaining antigenic determinants can continue to trigger the host's immune response, resulting in the generation of antibodies against the pathogen. Consequently, the

Table 15.2. Human vaccines currently used for active immunization against microbes/diseases.

Vaccines	Dead/Inactivated	Live/Attenuated
Viral	Hepatitis A	Measles
	Hepatitis B	Mumps
	Influenza	Poliomyelitis (Sabin)
	Japanese encephalitis	Rubella
	Poliomyelitis (Salk)	Smallpox
	Rabies	Varicella
		Yellow fever
Bacterial	Anthrax	Tuberculosis (BCG)
	Cholera	
	Diphtheria	
	Haemophilus influenzae b	
	Meningococcal meningitis	
	Pertussis	
	Plague	
	Pneumococcal pneumonia	
	Tetanus	
	Typhoid	

body will be ready for any subsequent invasion of the live and active pathogens because the primary immune response has been stimulated. It is interesting to note that we can exploit the pathogen itself to boost up our own defense system, e.g. the vaccine against whopping cough is prepared using killed *B. pertussis* bacteria.

Virulence factors of bacteria such as toxins can also be targeted as a source of vaccines. The majority of bacterial toxins have conserved structures, and therefore the antibodies mounted against these toxins will be effective even if the body has not encountered the toxin for years. In other words, as long as a specific pathogen is the causative agent, the toxin produced will be more or less similar structurally. Therefore, if a toxin is targeted as a vaccine, the protective effect it can provide will be quite significant. It is thus not hard to understand why toxins are one of the best candidates for vaccine development. Before any toxin is used as a vaccine, it will have to be inactivated by treatment with formalin or heat. This is necessary to ensure that no harmful effects that may damage host cells when the toxin is introduced in the body are retained. This modified "non-toxic" toxin is called a toxoid. Most people will already have been inoculated with doses of toxoid before they reach adolescence. The most common one is the DTP vaccine which is composed of inactivated toxins of three organisms: *Corynebacterium diphtheriae* (D), *Clostridium tetani* (T) and *Bordetella pertussis* (P). Injecting toxoid in humans serves as a vaccine to protect the body from harmful toxin produced by the organism from which the toxoid is made. But when the toxoid is injected into animals, it serves to generate polyclonal or monoclonal antibodies for use in scientific research.

Another well-known vaccine is that against tuberculosis, a major cause of death worldwide. This vaccine is called BCG (which stands for bacille Calmette-Guerin), and it is prepared by using a live but avirulent strain of *Mycobacterium bovis*. It should be noted that tuberculosis is caused by *M. tuberculosis* which does not produce toxin, so the option of developing toxoid as a vaccine does not exist. The BCG srain of *M. bovis*, on the other hand, shares many common structural properties with *M. tuberculosis*, but most importantly is nonpathogenic. Therefore, it is believed that the immune response triggered against *M. bovis* will be equally effective against *M. tuberculosis*. BCG is relatively cheap and safe, and has been used to immunize children for many years.

Vaccines have also been developed based on the capsules present on the surface of bacteria. One such vaccine currently available is for protection against *Streptococcus pneumoniae*, a causative agent of pneumonia and meningitis. The vaccine against *S. pneumoniae* contains at least 20 of the commonest capsular serotypes found associated with this bacterium. Individuals who are immunized with this vaccine should be protected against the commonly encountered strains of *S. pneumoniae*. However, one major limitation is the polysaccharide nature of the vaccine. This type of vaccine is generally not effective in both infants and elderly adults, which unfortunately comprise the two main high-risk groups of *S. pneumoniae* infection among the population. At present, other vaccine targets such as surface proteins of the organism are being investigated, and it is believed that given their protein nature, such vaccines will be more effective.

15.6. Bacterial Surface Display

The outer membrane of Gram-negative bacteria contains a number of proteins such as LamB, PhoE and OmpA, and lipoproteins such as Lpp, TraT and PAL. These heterologous proteins are displayed on the surface of the bacteria. Due to their strategic location, they have been employed as tools for scientific research, especially in vaccine development. One of the limitations of vaccine development is the instability of the antigen introduced in the body. To overcome this limitation, enteric bacteria that are commensals in the intestine are employed as vehicles to deliver the vaccine antigen. Enteric bacteria are not only resident in the intestine, but are also capable of replicating. Therefore, with appropriate vaccine construction using enteric bacteria, there will be a continual supply of the antigen as long as the bacteria exist as commensals. An ideal way to develop such vaccines is to genetically engineer and fuse a subunit of toxin to a surface display protein such that the toxin can also be displayed on the surface of the bacteria.

Gram-negative bacteria such as *E. coli* and *Salmonella* species have been used for surface display of several potential vaccine candidate proteins. Preliminary results from immunization of mice with live *S. dublin* expressing flagellin with a fused cholera toxin or hepatitis B virus epitope have been quite promising. The advantages offered by this type of vaccine delivery system are related to the physiological and biochemical properties of enteric bacteria, i.e. they are part of the normal intestinal bacterial flora, easy to grow, and the cloning vectors are readily available. Moreover, no toxins or enzymes need to be extracted and purified, and no chemical inactivation of the toxin is necessary. In addition, when bacteria replicate, everything including the engineered surface display protein is simultaneously replicated.

15.7. Immunization Strategies Against Human Microbial Pathogens

Despite the global eradication of smallpox and the considerable success of immunization against poliomyelitis, infectious diseases constitute the world's major killer of children and young adults, especially HIV/AIDS, measles, diarrheal diseases, pneumonia, tuberculosis and malaria in developing countries. Launched in 2001, a multi-million dollar Global Fund for Children's Vaccines will primarily be used to vaccinate children in developing nations against hepatitis B, yellow fever and *Haemophilus influenzae* type b. Prophylactic vaccines against major infectious diseases of global public health importance are urgently required. These include malaria, leishmaniasis, schistosomiasis, tuberculosis, melioidosis, sexually-transmitted diseases, AIDS, dengue, hepatitis C, enterovirus, respiratory syncytial virus and other infections for which there are currently no effective vaccines.

Conventional microbial vaccines comprise of live attenuated vaccines, inactivated vaccines and toxoids. Although they currently contribute significantly in the protection against major infectious diseases via immunization, certain conventional vaccines suffer from drawbacks. For example, one negligible but worrisome risk of live attenuated viral vaccines is the potential generation of mutants that may possess similar or even higher virulence compared to the parental wild-type viruses, especially RNA viruses with a higher propensity for genetic variability.

Thus, much research is focusing on new strategies for vaccine development, to improve the quality of current vaccines especially against diseases for which immunization is not yet routinely available, e.g. AIDS, malaria, hepatitis C, human papillomavirus infections.

Using recombinant DNA technology, selected microbial genes can be cloned into suitable vectors for the production of immunogenic vaccines. Commercially produced in yeast or mammalian host cells capable of expressing glycosylated proteins, the hepatitis B vaccine has the potential to reduce and even eliminate the morbidity and mortality associated with chronic hepatitis, cirrhosis and primary liver cancer. *Salmonella* and lactic acid bacteria are being tested as host cells for the production of vaccines to elicit mucosal immunity. Microbial antigens are also being expressed in fruits and vegetables such as bananas, tomatoes and potatoes as "edible plant vaccines". The protective efficacy of chemically synthesized peptide vaccines (e.g. against malaria and HIV) is being evaluated.

Naked DNA vaccines (NDVs) can be engineered to encode gene(s) for microbial or viral antigen(s), e.g. envelope or core proteins. The host intracellular processing of the DNA-expressed viral protein occurs via the major histocompatibility complex (MHC) class I pathway and can evoke cell-mediated immunity. In contrast, the processing of conventional vaccines is generally through the MHC class II pathway which stimulates a predominantly antibody response.

Compared with conventional microbial vaccines, DNA vaccines possess many advantages, some of which are outlined below. NDVs are capable of inducing both humoral-mediated and cellular-mediated immune responses, with a stronger T-helper response. Antigen expression is more sustained, resulting in long duration of immunity. NDVs also allow the expression of antigens that more closely resemble native viral epitopes. In addition, DNA vaccines are safer than live vaccines and can obviate problems associated with viral vectors used in gene transfer, e.g. vacciniavirus vectors. In comparison with recombinant DNA vaccines, there is no necessity to purify the final protein product. Other practical merits which make NDVs suited to developing countries include the relatively low cost of vaccine production, and the stability of DNA versus the requirement for the "cold chain" for most conventional vaccines. A further advantage is the simple concept and route of administration of DNA vaccines, i.e. via intramuscular injection or via the "needleless" gene gun, which can influence the desired outcome of immune response.

Promising initial results have been achieved in DNA immunization against specific microbial pathogens in animal models. For example, a naked DNA vaccine based on the influenza viral core nucleoprotein has been shown to confer protection against lethal doses of heterogeneous strains in mice. Experiments have also documented the successful protection of animals by DNA immunization against challenge with various other infections, including rabies, tuberculosis and malaria. Despite theoretical concerns such as the potential induction of tolerance or autoimmunity, and the possible integration of foreign DNA into the host genome culminating in insertional mutagenesis, animal studies have so far supported the safety of DNA vaccines.

DNA immunization represents the most exciting development in vaccine technology over the past few decades. Although tests have indicated the high safety and efficacy of DNA-based vaccines, current methodologies require further optimization. The outlook on direct DNA immunizations to address a variety of medical problems is indeed very promising.

The ease and rapidity of construction of DNA vaccines coupled with advances in rapid nucleotide sequencing of microbial genes offer a clear advantage in countering the threat of newly emerging, re-emerging and drug-resistant pathogens. Moreover, in the case of highly mutable microbes e.g. influenza and human immunodeficiency viruses, DNA vaccines can be easily redesigned and tailor-made according to the predominant strains prevalent in different geographic regions.

With the advent of large-scale projects to sequence entire genomes of major microbial pathogens, there are prospects for large numbers of potential target genes to be tested as candidates for inclusion in effective DNA vaccines.

It has been demonstrated in specific instances that DNA vaccines co-expressing immunostimulatory molecules are much more efficacious than those lacking certain cytokine genes, thus justifying the need to test such "hybrid" vaccines.

In the case of typhoid, gonorrhea, chlamydial and other infections in which mucosal immunity is critical in their prevention, oral delivery of DNA vaccines can be achieved via vehicles such as specific vaccine strains of *Salmonella* and *Lactobacillus*.

While prospective prophylactic DNA vaccines should be the mainstay of research and development, there appears to be potential for the use of DNA vaccines in the immunotherapy of chronic infections e.g. hepatitis B and C, HIV infections, with the aim of enhancing cell-mediated immunity to accelerate viral clearance.

Another exciting prospect is the cloning of multiple gene cassettes of different pathogens to facilitate "single shot" protection as opposed to the current practice of multiple doses given at various time intervals. Yet another potentially successful challenge is to develop DNA vaccines to stimulate immune responses against specific tumor antigens to augment killing of cancer cells *in vivo*.

15.8. Beneficial Uses of Viral Vectors for Gene Therapy

Numerous clinical trials are underway to employ gene therapy to correct genetic diseases by supplying cells with missing or defective genes with normal functional genes, often accomplished by transferring the desired genes using vectors based on viruses such as adenoviruses, adeno-associated viruses, herpesviruses and retroviruses for gene delivery.

In September 1990, the revolutionary use of human gene therapy to treat a 4-year old girl with severe combined immunodeficiency, by infusing her with cells containing the correct adenosine deaminase gene expressing the critical missing protein with subsequent encouraging clinical improvement, marked a significant milestone in medical history.

Although many clinical trials have demonstrated that viral vectors for gene therapy are non-toxic, the unfortunate death of 18-year old Jesse Gelsinger during a Phase I trial in 2000 has forced the research community into re-evaluating public risk assessment and safety issues.

The advent of gene therapy heralds the future application of this potentially powerful technology (and its associated approaches such as gene targeting and anti-sense RNA therapy) for the treatment of hitherto incurable or debilitating maladies from genetic disorders (e.g. cystic fibrosis, hypercholesterolemia, hemophilia, sickle cell anemia and Lesch-Nyhan syndrome) to AIDS and cancer. Genetically-modified cells (e.g. tumor-infiltrating lymphocytes implanted with the tumor necrosis factor gene) are being tested as immunotherapeutic autovaccines in certain cancer patients.

Anti-sense strategy aims to block messenger RNA in order to prevent the translation of genetic information for protein expression. Anti-sense molecules are classically synthetic DNA analogues whose sequences are complementary to those of specific mRNA strands, and can thus bind to inhibit the expression of specific target genes. The anti-sense approach has been shown to interrupt viral replication in infected cells (e.g. with HIV).

The world's first Institute of Human Gene Therapy established in the University of Pennsylvania, USA in 1992 is a recognition of the impact of gene therapy in the next century, and this revolution is being hailed as a potentially dominant area of medicine as important as antibiotics and vaccines in the 20th century.

To complement this important technology, the massive Human Genome Project was launched in the US in 1990 to map and sequence the three billion pairs of bases of human DNA encoding an estimated 50,000 genes, with the aid of supercomputers for data processing. First directed by the Nobel laureate and co-discoverer of the double helical structure of DNA,

Professor James Watson, this enormous task was initially projected to take 15 years at a cost of US$3 billion. Important biomedical implications and commercial spin-offs include novel candidate genes for genetic screening, gene therapy and drug discovery.

15.9. Genomics and Proteomics of Microbial Pathogenesis

The official announcement on 26 June 2000 of the completion of the first draft of the human genome sequence opened a floodgate of important genetic information with profound implications for biomedical science and practice. Molecular biologists are harnessing the newly emerging and cutting-edge technologies of genomics and proteomics, which aim to elucidate and characterize functions of genes and proteins, with downstream applications. These projects have potential applications in novel diagnostics, new drug discovery, vaccine development and clinical trials. It is envisaged that spin-offs from these new technologies may evolve and be developed for industry, e.g. pharmaceutical and biotechnology companies.

Scientists are exploiting and developing technologies that encompass or overlap the triad of genomics, proteomics and bioinformatics. These include high-throughput DNA sequencing, gene arrays and chips, subtractive and differential gene screening, protein-protein interactions, two-dimensional gel electrophoresis, mass spectrometry, protein micro-sequencing, protein chips, transgenic and gene knock-out animals, protein identification and fingerprinting.

Microbial genomic and proteomic profiling include whole genome sequencing and proteome analysis of major causative agents of infectious diseases. The complete genomes of many bacterial species and other micro-organisms have been sequenced, and many more microbial sequencing projects are in progress.

Proteomics is facilitated in part by two-dimensional polyacrylamide gel electrophoresis which can separate complex mixtures of proteins. Interesting protein spots can be further characterized by mass spectrometry.

Bioinformatics plays a key role in mining valuable genome-wide data sources, especially for the identification of virulence factors and antibiotic resistance mechanisms, as well as for anti-microbial drug discovery.

Comparative genomic analysis can identify essential genes conserved across bacterial species but not in eukaryotes, which may lead to promising broad-spectrum antibiotics.

Comparison of gene expression levels between test (infected) and reference (uninfected) cell samples can be achieved by techniques such as differential display and microarrays, which may lead to the discovery of novel genes related to pathogenesis and antimicrobial drug targets. Microarrays or chips spotted with cDNAs permit the simultaneous expression profiling of thousands of genes. Microarray technology is useful for the analysis of cellular genetic responses to microbial infections, and for screening bacterial and other microbial isolates.

The applications of pharmacogenomics and pharmacoproteomics to study remission or resistance after antimicrobial treatment involves the profiling of successful or adverse responses to drug therapy of infections, and may lead to new diagnostic and prognostic markers of therapy.

15.10. Benefits and Implications for Healthcare Development

The extraordinary power of microbial biotechnology can be harnessed to contribute effective solutions for global health problems. Scientists are seizing the opportunities for carefully exploiting the new microbial biotechnology to meet the healthcare challenges of the future. Medical biotechnology businesses are tapping huge worldwide markets worth billions of dollars. Many countries have adopted key strategic master plans and implemented long-term as well as short-term programs to develop biotechnology relevant to regional resources and requirements, with the health care aspect constituting a strong pillar.

While the exciting advances of biotechnology in medicine augur well for a better tomorrow, many economic and ethical issues remain yet unresolved. Molecular biotechnology is expensive, and new diagnostic and therapeutic methods are expected to escalate health costs, thus bearing a negative impact on health care development of developing countries that need it most. Most vital of all, moral and ethical deliberation have not matched the progress and potential of medical biotechnology. Difficult dilemmas for biomedical ethics committees to confront include confidentiality versus abuse of an individual's genetic information, and right of access to insurance. Modern microbial biotechnology should be "environment friendly", ensuring that hazardous by-products are not indiscriminately released into the environment. Finally, given its rapid pace of progress, and if we can grapple with the core issues and implications, medical biotechnology can overcome the many challenges ahead to attain better health for all humankind.

Further Readings

1. Blau HM and Springer ML. Gene therapy — a novel form of drug delivery. *New Eng. J. Med.* **333**: 1204–1207, 1995.
2. Burkardt HJ. Standardization and quality control of PCR analyses. *Clin. Chem. Lab. Med* **38**: 87–91, 2000.
3. Clark DE and Pickett SD. Computational methods for the prediction of "drug-likeness". *Drug Discov. Today* **5**: 49–58, 2000.
4. DeFilippis V, Raggo C, Moses A and Fruh K. Functional genomics in virology and antiviral drug discovery. *Trends Biotechnol* **21**: 452–457, 2003.
5. Dertzbaugh MT. Genetically engineered vaccines: An overview. *Plasmid* **39**: 100–113, 1998.
6. Doolan DL and Hoffman SL. Multigene vaccination against malaria: A multi-stage, multi-immune response approach. *Parasitol. Today* **13**: 171–178, 1997.
7. Eisen T. Novel vaccine technologies. *Trends Biotechnol.* **15**: 483–484, 1997.
8. Ellis RW. Technologies for the design, discovery, formulation and administration of vaccines. *Vaccine* **19**: 2681–2687, 2001.
9. Galanis E, Vile R and Russell SJ. Delivery systems intended for *in vivo* gene therapy of cancer: Targeting and replication competent viral vectors. *Crit. Rev. Oncol./Hematol.* **38**: 177–192, 2001.
10. Galperin MY and Koonin EV. Searching for drug targets in microbial genomes. *Curr. Opin. Biotechnol.* **10**: 571–578, 1999.
11. Gregoriadis G. Genetic vaccines: strategies for optimization. *Pharm. Res.* **15**: 661–670, 1998.

12. Hughes TR and Shoemaker DD. DNA microarrays for expression profiling. *Curr. Opin. Chem. Biol.* **5**: 21–25, 2001.
13. Korth MJ, Kash JC, Furlong JC and Katze MG. Virus infection and the interferon response: A global view through functional genomics. *Methods Mol Med.* **116**: 37–55, 2005.
14. Lander ES *et al.* Initial sequencing and analysis of the human genome. *Nature* **409**: 860–921, 2001.
15. Lennon GG. High-throughput gene expression analysis for drug discovery. *Drug Discov. Today* **5**: 59–66, 2000.
16. Loferer H, Jacobi A, Posch A, Gauss C, Meier-Ewert S and Seizinger B. Integrated bacterial genomics for the discovery of novel antimicrobials. *Drug Discov. Today* **5**: 107–114, 2000.
17. Montgomery DL, Ulmer JB, Donnelly JJ and Liu MA. DNA vaccines. *Pharmacol. Ther.* **74**: 195–205, 1997.
18. Ogra PL, Faden H and Welliver RC. Vaccination strategies for mucosal immune responses. *Clin. Microbiol. Rev.* **14**: 430–445, 2001.
19. Pastan J and FitzGerald D. Recombinant toxins for cancer treatment. *Science* **254**: 1173–1177, 1991.
20. Ramsay AJ, Ramshaw IA and Ada GL. DNA immunization. *Immunol. Cell Biol.* **75**: 360–363, 1997.
21. Rappuoli R and Covacci A. Reverse vaccinology and genomics. *Science* **302**: 602, 2003.
22. Shearer GM and Clerici M. Vaccine strategies: Selective elicitation of cellular or humoral immunity? *Trends Biotechnol.* **15**: 106–109, 1997.
23. Smith HA and Klinman DM. The regulation of DNA vaccines. *Curr. Opin. Biotechnol.* **12**: 299–303, 2001.
24. Somia N and Verma IM. Gene therapy: Trials and tribulations. *Nat. Rev. Genet.* **1**: 91–99, 2000.
25. Stahl S and Uhlen M. Bacterial surface display: Trends and progress. *Trends Biotechnol.* **15**: 184–192, 1997.
26. Tang H, Peng T and Wong-Staal F. Novel technologies for studying virus-host interaction and discovering new drug targets for HCV and HIV. *Curr. Opin. Pharmacol.* **2**: 541–547, 2002.

27. Whelen AC and Persing DH. The role of nucleic acid amplification and detection in the clinical microbiology laboratory. *Ann. Rev. Microbiol.* **50**: 349–373, 1996.
28. Young RA. Biomedical discovery with DNA arrays. *Cell* **102**: 9–15, 2000.

Questions for Thought...

1. Using insulin as an example, discuss the advantages and disadvantages of using bacterial hosts to manufacture products from eukaryotic genes.
2. Discuss the contributions of bacterial enzymes to biotechnology.
3. With the aid of examples, discuss the role of toxins in the pathogenesis of bacterial diseases.
4. Explain the term "immunotoxin". Discuss the potential limitations of an immunotoxin when it is used as a therapeutic agent.
5. Discuss the advantages and disadvantages of using chimeric toxins to treat cancer.
6. Discuss how the growth of microorganisms can be useful to humans.
7. When preparing a toxoid vaccine, what precautions must be taken to ensure that the toxoid will be effective?
8. What are the requirements for an ideal microbial vaccine? With examples, discuss how conventional and molecular-based vaccines can fulfill these requirements.
9. Discuss the contributions of molecular biology to the development of new and improved microbial vaccines.
10. Discuss the merits of recombinant DNA vaccines for immunization against infectious diseases.
11. Discuss the molecular approaches for the development of novel vaccines against pathogenic RNA viruses. Illustrate your answer with relevant examples.
12. Discuss the use of viral vectors for human gene therapy.
13. What are the criteria for selecting an ideal inherited disease candidate for gene therapy?
14. Explain whether gene therapy can be used to cure diabetes mellitus.

15. "The tools of molecular biology have facilitated the elucidation of certain mechanisms of microbial pathogenesis." Discuss this statement, illustrating your answer with examples.
16. "Studies on microbial genes, genomics and proteomics can contribute towards a better understanding of infectious diseases." Discuss.

Related World Wide Web Sites

- DNA Vaccines
  http://www.dnavaccine.com
- Human Genome Project Information
  http://www.ornl.gov/hgmis

Part V

Microbes in Environmental Biotechnology

By the turn of the last millenium, the rapid development in various areas such as microbiology, genetics, biochemistry and chemical engineering has led to tremendous advances in the knowledge and applications in biotechnology. Previous chapters of this book describe the applications and development of biotechnology in industry, agriculture and medicine. In Part V of this book, the development of biotechnology in environmental applications will be introduced. Because there are numerous applications in this area, only selected topics will be discussed in detail with emphasis on the applications of microorganisms in the biotechnological processes for solving environmental problems.

The unique properties of microorganisms, which are not present in plants and animals, are fast growth rate, small cell size and diversified metabolic pathways. The first two properties allow microbial applications to operate in a fast mode and in a small reactor, while the third characteristic enables microorganisms to degrade or transform a large number of compounds that are harmful to our environment.

In addition, advances in genetic engineering provide further improvement and facilitate the applications of microorganisms in solving environmental problems.

Several classes of microorganisms, namely viruses, bacteria, fungi and algae are commonly used in the environmental biotechnology. Examples will be introduced to illustrate the importance of these microorganisms in environmental biotechnology.

The specific application of biotechnology to environmental management and pollution control is generally referred to as environmental biotechnology. It is used in waste and wastewater treatment as well as environmental cleanup encompassing toxic waste decontamination and the biodegradation of oils, pesticides and other organics. Microorganisms are in general more effective than multicellular organisms in environmental treatment, although higher plants are frequently used in phytoremediation and hydrophyte treatment systems. Municipal systems are usually mixed substrate, mixed culture continuous operations. Most processes are aerobic, but few are anaerobic. Some processes such as composting and anaerobic digestion are multi-stage and involve both mesophilic (15–40°C) and thermophilic (45–70°C) organisms under different operational conditions. The majority of treatment processes makes use of the indigenous populations in the wastes and provides favorable conditions for the growth and metabolic activity of these microorganisms. Industrial treatment systems often use pure strain (monoculture) of artificially screened microorganisms, while the degradation of xenobiotics may employ genetically engineered microorganisms in laboratory trials or pilot scale systems.

Chapter 16

Municipal Wastewater Treatment

Lee Man Chu and P. K. Wong

Department of Biology, The Chinese University of Hong Kong Shatin, NT, Hong Kong SAR, China

Used waters are characterized by high loading in gross organic matter as assessed by parameters such as COD and BOD, nutrients including nitrogen and phosphorus, toxic organics like hydrocarbons and inorganics such as heavy metals, as well as microorganisms such as enterobacteria (Table 16.1). The level of these parameters depends upon the source of wastewater, composition of water supply, pattern of water consumption, hydraulic flow, local habits and industrial contamination. As a consequence of urbanization and industrialization, the proper discharge of wastewater has become a problem of nearly all municipalities. Treatment is essential, and biochemical processes are usually involved in the secondary stage to remove suspended solids and organic matter after preliminary screening which removes coarse particles, and primary sedimentation which allows suspended solids to settle. The secondary treatment steps are generally considered environmental biotechnologies that make use of bioreactors for the biodegradation of organic matter and bioconversion of soluble nutrients in the wastewater. Industrial wastewater may have atypical composition and in-house treatment works may be needed prior to discharge to watercourses or further treatment in municipal treatment plants.

The discharge of municipal wastewater to open water will inevitably lead to water pollution, the degree of which depends highly on the

Table 16.1. Typical composition of municipal wastewater. Units in mg/L unless specified.

	Concentration
pH	6.5–8.0 (no unit)
Total solids	300–1,200
Suspended solids	100–350
COD	250–1,000
BOD	100–500
Total nitrogen	15–80
Ammoniacal nitrogen	10–50
Total phosphorus	5–20
Inorganic phosphorus	4–15
Total coliforms	$1 \times 10^6 - 1 \times 10^8$ (MPN/100 mL)

self-purification capacity of the receiving water. This can be typically exemplified by changes in the environmental quality (e.g. oxygen sag) and organisms in a river in response to wastewater discharge from an outfall upstream. For systems with high assimilative ability, as determined by flow rate, dilution factor, degree of aeration and water temperature, the ecosystem damage will be minimal and recovery will be rapid. The rate of degradation is also governed by the concentration of microorganisms that are either suspended in the water column or attached to plant or rock surfaces. The low biomass concentration in the natural environment is usually the limiting factor for effective purification. In bioreactors or treatment works for wastewater, the self-purification capacity is harnessed in a defined area in which compact biochemical processes are intensified by technical and engineering means, thus creating an artificially controlled environment for optimal microbial activity and subsequent biodegradation of organic matter (Fig. 16.1).

Municipal wastewater contains a mixed population of aerobic and anaerobic microorganisms that are mostly heterotrophs. The microbial composition of wastewater fluctuates widely, consisting of either pathogenic or nonpathogenic organisms, which are common microflora of the human intestine and soil. In biochemical treatment systems, the

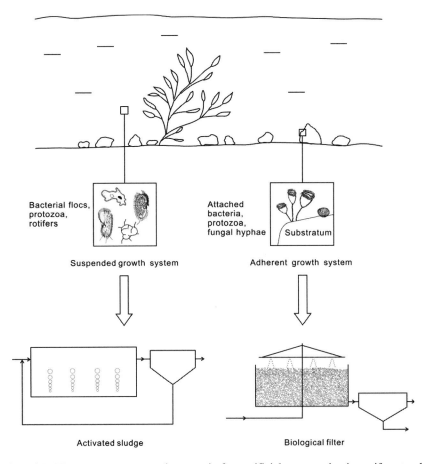

Fig. 16.1. Wastewater treatment plants are in fact artificial systems that intensify natural biological process in a defined and controlled environment so as to enhance the self-assimilative capacity of our environment (redrawn from Mudrack and Kunst 1986).

most adapted and effective microbes are selected. The selective pressure varies with the combination and intensity of various factors that can be categorized into two types. Primary selection factors, which are based on microbial physiological properties, determine whether autotrophs or heterotrophs will be selected, and include energy and carbon source as well as oxygen availability. The majority of microorganisms involved in the biodegradation of organic matter in solid wastes and wastewaters are chemoorganotrophic (Table 16.2). A small number are chemolithotrophic,

Table 16.2. All living organisms can be categorized into four major groups according to their nutritional or metabolic diversity.

Energy source	Carbon source	
	Inorganic	Organic
Solar	Photolithotrophs (autotrophs) e.g. green plants, green bacteria	Photoorganotrophs e.g. purple non-sulphur bacteria
Chemical	Chemolithotrophs e.g. nitrifiers, thiobacilli	Chemoorganotrophs (heterotrophs) e.g. animals, fungi, most bacteria

such as all nitrifying bacteria and some denitrifying bacteria, which are responsible for biological nitrogen removal. In addition, photolithotrophs such as green algae are active in shallow pond systems and surface of biological filters. Secondary factors, which are based on microbial physical properties, exert pressure on continuous flow treatment system. The electrical charge of cells determines initial attachment in fixed film reactors, and cell mass or cell density selects population based on the settling properties of microbes in the secondary sedimentation tank. These selection factors create unique microbial consortia in the various biological treatment systems. Suspended growth systems are dominated by free living organisms in dispersed colonies or cell aggregates while adherent growth systems consist mainly of surface attached organisms in stationary microbial films. Such microbial associations display different physicochemical properties, which are of fundamental significance to their ecological role and their biotechnological function in wastewater treatment. However, the two growth forms share some similarities in that they are stable synergistic microconsortia in which microbial cells are immobilized by an exopolymer matrix of high molecular weight in a three-dimensional gel-like network, enforced by divalent cations such as Ca^{2+}.

16.1. Aerobic Treatment System

Aerobic systems use dissolved oxygen (DO) in the wastewater for carbonaceous and nitrogenous oxidation. Oxygen solubility is dependent upon gas pressure, water temperature and salinity, and oxygen is usually supplied by aeration to avoid depletion by oxygen consuming processes. The transfer of oxygen at the air-water interface and its subsequent diffusion to the cellular components is affected by agitation, flow pattern, viscosity, oxygen deficit and concentration gradient. Aerobic processes are commonly used as the systems are more reliable, stable and efficient.

16.1.1. *Activated sludge process*

The activated sludge process (Fig. 16.2), which was first developed in the United Kingdom in the early 1900s is the most commonly used aerobic process for wastewater treatment plants worldwide, treating both municipal and industrial effluents, due to its operational flexibility and reliability. Activated sludge process is a suspended growth, mixed population and continuous culture system in which free living organisms are exploited to break down the biochemically oxygen demanding organic matter in the wastewater. During this process, wastewater is treated aerobically by a microbial consortium dominated by heterotrophic bacteria that are flocculated in the mixed liquor under a supply of excessive oxygen in an aeration tank 3–5 m deep which is necessary in order to form discrete clumps of microorganisms. Protozoa and rotifers that enhance floc formation and effluent clarification are also present in significant numbers. The biomass is allowed to settle out of the suspension with other flocculent solids in a secondary clarifier to form sludge, a proportion of which is

Fig. 16.2. The activated sludge system consists of a long aeration tank where the bio-oxidation of organic matter in the wastewater takes place, and a final sedimentation tank where the microbial biomass and solids are separated from the mixed liquor to produce a clarified effluent. A certain proportion of the sludge is recycled to the aeration tank as an inoculum.

Table 16.3. Bacterial population in wastewater samples from different stages in plants treating mainly domestic wastewater (adopted from Pike and Curds 1971).

	Bacterial count (number/mL)	
	Total	Viable
Primary settled wastewater	6.8×10^8	1.4×10^7
Activated sludge mixed liquor	6.6×10^9	5.6×10^7
Secondary effluent	5.2×10^7	5.7×10^5
Tertiary effluent	3.4×10^7	4.1×10^4

returned to the aeration tank as microbial inoculum to maintain a high microbial population for biodegradation (Table 16.3). The sludge contains a high concentration of active mixed micro-flora and fauna, hence the name "activated sludge". A conventional activated sludge system has a hydraulic retention time (HRT) of 3–8 hours but solid retention time (SRT) is prolonged (4–15 days) because of sludge recycle.

16.1.1.1. *Floc and mixed liquor suspended solids*

The sludge biomass or floc is the basic operational unit in the mixed liquor in the aeration tank of the activated sludge system. Under the light microscope, flocs appear as discrete clumps of flocculated slurry with irregular boundaries containing organisms that include bacteria, protozoa and metazoa in a non-living gelatinous mesh-like matrix. The cells form small aggregates to large flocs commonly of diameter ranging from 50–300 μm with a surface area of 40–150 m^2/g, the size of which is determined by cohesive strength and the degree of shear in the aeration tank. Recent size distribution studies show that flocs can vary between 1–1,000 μm; biomass in the mixed liquor is made up of flocs of size 65–260 μm. The aggregation process begins with primary particles (2.5 μm), which aggregate to form secondary particles (13 μm), which in turn form tertiary particles (125 μm). This aggregation process is aided by extracellular polymeric substances. Spherical and compact microfloc is formed when the particle diameter is less than 75 μm and specific gravity greater than 1.0. When filamentous microorganisms are predominant, the floc will be in the form of extended mat, loose (specific gravity <1.0) and irregular in shape. This macrofloc is not inducive to settlement and will result in settling problems in the secondary clarifier. Chemically, floc consists of biomass, organic matter which is mainly flocculating polymers including polysaccharides, amino polysaccharides, proteins and lipids, and inorganic material such as carbonate, phosphate, nitrate and metal hydroxides. Floc organisms usually concentrate in the slimy matrix in the peripheral region while the inner region is composed of inorganic material and inert organic matter. Cells are flocculated with polymeric matrix or polymeric bridges linked by mechanical entanglement or by electrostatic bridging. The active floc volume that contributes to aerobic degradation is a function of the floc particle diameter (Fig. 16.3). Loose irregular floc has high bulk volume but low active fraction.

Regarding biological composition, bacteria are the dominant group and over 300 strains have been isolated from activated sludge systems. The bacteria are mostly chemoorganotrophic, Gram-negative motile rods, many of which originated from human faecal materials. Common genera are *Pseudomonas, Flavobacterium, Zoogloea, Arthrobacter, Acinetobacter, Achromobacter, Alcaligenes, Cytophaga, Bacillus, Nitrosomonas* and

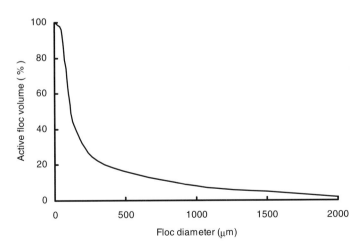

Fig. 16.3. The active floc volume (volume fraction of flocs containing oxygen) that contributes to aerobic degradation as a function of the floc diameter (Pasveer 1959).

Nitrobacter. Most of these are floc forming bacteria, which either aggregate in sludge flocs or are freely suspended in the mixed liquor. Holozoic or saprobic protozoa also play an important role in wastewater purification. Over 200 species have been identified and can be found to occur at levels as high as 50,000 per mL. Seventy percent are ciliates, such as *Vorticella* spp., *Aspidisca* spp., *Opercularia* spp. and *Epistylis* spp. They are sedentary, sessile or stalked, attaching to individual floc, crawling over the floc surface or free-swimming in the mixed liquor. They feed on bacteria and tiny suspended particles thereby regulating bacterial density and clarifying effluent. Rotifers are also present in substantial numbers; they encourage new floc formation by breaking up flocs into smaller particles, contribute to floc formation by producing undigested materials and mucus, and reduce turbidity by removing dispersed bacteria and filtering suspended particles. Fungi rarely dominate in mixed liquor except in acidic wastewater. Such a biological composition is the result of microbial selection under aerobic condition and the shear stress of aeration. Nonflocculating microbes that are unable to form settleable flocs are washed out in the final effluent. Substrates in wastewater are adsorbed to the surface and transported from the aerobic peripheral region of a floc, where carbon oxidation and nitrification take place, to the inner region. In flocs of substantial size, the inner zone may be anaerobic which allows denitrification to occur. In the aeration tank, biodegradation of organic matter by C-oxidizers is the primary goal, and the dynamic ecosystem functioning depends very much on the bacterial-protozoal relationship (Fig. 16.4).

16.1.1.2. *Biological nitrogen removal*

The biological oxidation of ammonium to nitrate by nitrifiers takes place as long as the DO content is above 2 mg/L. This is a 2-step process: (a) *Nitrosomonas* spp. converts ammonium to nitrite, and (b) *Nitrobacter* spp. converts the nitrite to nitrate. Nitrifiers are chemolithotrophic and strictly aerobic, consuming 4.2 g of oxygen for each gram of ammonium-nitrogen oxidized. They have lower specific growth rates than C-oxidizers and can therefore only establish and grow in old flocs. A sludge age of 3–5 days is required, which is longer than the solid retention time in a

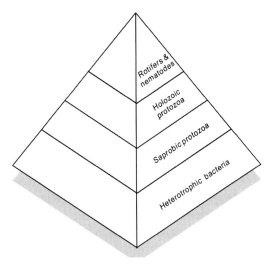

Fig. 16.4. The relatively simple bacterial-protozoal relationship in activated sludge system.

conventional activated sludge system. A two-stage activated sludge process has been developed, the first aeration tank for BOD removal and the second one for nitrification. Nitrification alone cannot remove nitrogen since nitrate is formed stochiochemically. It must be followed by denitrification in which the nitrate in the nitrified effluent is reduced to molecular nitrogen and removed as gas. Denitrifiers are facultative anaerobes, which use nitrate as an electron acceptor. They are mostly heterotrophic bacteria such as *Pseudomonas* spp., *Alcaligenes* spp. and *Achromobacter* spp., which require organic substrates. However, some are autotrophic which do not require organic substrates, such as *Micrococcus* spp. and *Thiobacillus* spp. Both types grow well at pH above 5 and DO level lower than 1 mg/L. Because of this, nitrogen removal in activated sludge process used to involve two systems (Fig. 16.5). The first system is a conventional activated sludge process with an aerobic tank for carbonaceous oxidation and nitrification and a clarifier for sedimentation. The second consists of an anoxic tank to encourage denitrification, and a clarifier for sludge settlement. As most organic matter is removed in the initial oxidation step, the concentration of organic substrate is insufficient to meet the requirement of heterotrophic denitrifiers. In such a system, an external carbon source in the form of methanol or acetate is usually added.

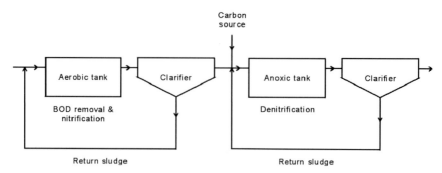

Fig. 16.5. Two-clarifier system in which the aerobic tank is for BOD oxidation and nitrification while the anoxic tank is for denitrification. An external carbon source is needed for denitrification.

To remove the requirement for carbon source addition, a one-clarifier system (also called one-sludge system) is used (Fig. 16.6). Effluent from primary treatment enters an anoxic tank where denitrification and some BOD removal take place. The mixed liquor then flows to a second aeration tank where nitrification and BOD removal are carried out. The nitrified mixed liquor will recycle back to the anoxic tank so that nitrate is converted to nitrogen by denitrification. Biological nitrogen removal is usually incorporated in modern activated sludge systems, which are no longer solely aerobic processes. There are a number of versions for such anoxic-oxic (A/O) systems. In the modified Ludzak-Ettinger process developed in South Africa, an anoxic unit, in which denitrification

Fig. 16.6. One-clarifier system with an anoxic tank for denitrification and an aerobic tank for BOD oxidation and nitrification. The nitrified mixed liquor is recycled back to the anoxic tank for denitrification.

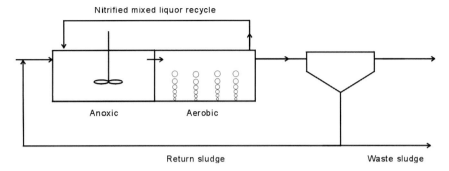

Fig. 16.7. Modified Ludzak-Ettinger process with an anoxic unit and an aerobic unit.

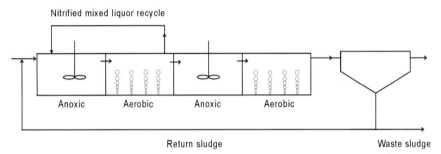

Fig. 16.8. The four-stage Bardenpho process with two anoxic units for denitrification and a second aerobic unit for the removal of nitrogen gas from denitrified mixed liquor.

takes place, precedes an aerobic unit (Fig. 16.7). In this process, effluent formed has a total nitrogen content of slightly greater than 5 mg/L. More complicated processes include the 4-stage Bardenpho process which is developed in the US. A second anoxic unit increases the total denitrification efficiency and a second aerobic unit removes nitrogen gas from denitrified mixed liquor. Total nitrogen content in the polished effluent is below 3 mg/L (Fig. 16.8).

16.1.1.3. *Biological phosphorus removal*

Phosphorus in wastewater, which cannot be removed by a conventional activated sludge system, is commonly removed by chemical precipitation.

During the past two decades, the biological removal of phosphorus has become more widely used. Enhanced biological phosphorus removal (EBPR) is the result of luxury uptake by microbial cells, which involves enhanced uptake of phosphorus during the aerobic stage and its release during the anaerobic stage. This is evident by an increase in orthophosphate concentration in the anoxic zone and a substantial reduction in the oxic zone of the activated sludge system (Fig. 16.9). Using the energy liberated from BOD oxidation, phosphorus is stored as polyphosphates in bacterial cells under aerobic conditions (Fig. 16.10). The accumulated polyphosphates are hydrolyzed and energy is released for assimilation under anaerobic conditions. Although the detailed processes are not yet entirely understood, it is postulated that that the intracellular polyphosphate storage is carried out by phosphate accumulating or removing micro-organisms (PAM or PRM) or polyphosphate microorganisms. There is a large number of nonspecific bacteria, the majority of which is *Acinetobacter calcoaceticus* var. *lwoffi*. *Aeromonas* spp., *Achromobacter* spp., *Arthrobacter* spp., *Klebsiella* spp. and *Pseudomonas* spp. also

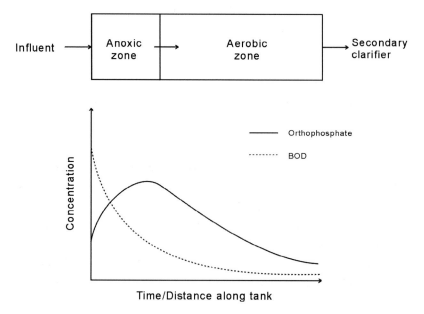

Fig. 16.9. Changes of BOD and orthophosphate in the anoxic and aerobic units of the activated sludge system.

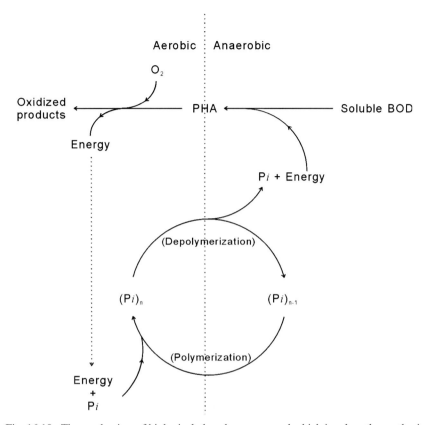

Fig. 16.10. The mechanism of biological phosphorus removal which involves the synthesis and degradation of polyphosphates and a storage product (poly-β-hydroxyalkanoates or PHA) under aerobic and anaerobic conditions.

contribute to EBPR. In the anaerobic stage, facultative bacteria convert soluble BOD to volatile fatty acids like acetate, which are then stored as poly-β-hydroxyalkanoates (PHA) such as poly-β-hydroxybutyrate (PHB), using the energy released from the depolymerization of polyphosphate; phosphorus is released from the bacterial cells. In the aerobic stage, PHB is oxidized to provide energy for growth and polyphosphate accumulation. An enhanced phosphorus uptake occurs when conditions change from anaerobic to aerobic, and phosphorus deficient cells replenish lost polyphosphate by rapid accumulation (oversurplus compensation). Growth causes an uptake of phosphorus higher than the amount released in the anaerobic stage to result in a luxury uptake, and the excess phosphorus is stored as polyphosphates.

There are a number of commercial systems for phosphorus removal. The Phoredox process is a two-stage (anaerobic-aerobic) A/O system with short solid retention time to minimize nitrification and hence denitrification. The A^2/O process (Fig. 16.11) and the Virginia Initiative Plant (VIP) process (Fig. 16.12) are three-stage (anaerobic-anoxic-aerobic) activated sludge systems. The five-stage modified Bardenpho process developed in 1978 is a three-stage process with a second anoxic stage to enhance denitrification and a second aerobic stage to ensure aerobic conditions in the final clarifier (Fig. 16.13). With the exception of the Phoredox process, all are mainstream processes and are efficient for both nitrogen and phosphorus removal.

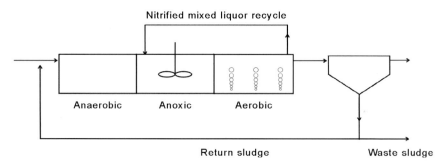

Fig. 16.11. A^2/O process consists of anaerobic, anoxic and aerobic tanks, with the recirculation of nitrified mixed liquor from the aerobic to the anoxic tank.

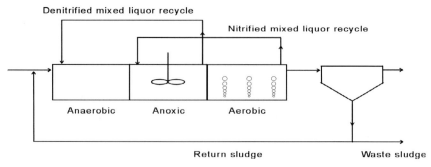

Fig. 16.12. The VIP process which is similar to the A^2/O process but with the addition of denitrified mixed liquor recycle from the anoxic to the anaerobic tank.

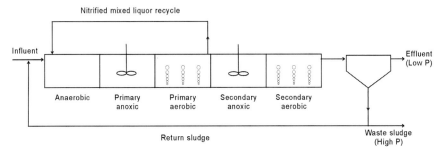

Fig. 16.13. The five-stage Bardenpho process with a second anoxic and aerobic tanks to enhance denitrification and nitrogen gas removal.

The purification efficiency of activated sludge process depends not only on the biodegradability of the organic matter and the removal of nutrients in the wastewater, but also on the separability of the suspended solids or flocs in the sedimentation process. Floc settlement separates solids from the treated wastewater to result in a clarified effluent, and allows sufficient amounts of sludge to be returned to the aeration tank in order to maintain mixed liquor suspended solids (MLSS) which is usually used as a crude measure of the biomass for routine operational control. If solids end up in the effluent, they will deteriorate its quality. Floc separability, which is affected by the settling characteristics, is governed by the physical nature of flocs. It is usually measured in terms of sludge density or sludge volume and expressed as the sludge density index (SDI) or sludge volume index (SVI) respectively. Tests are performed by determining the sludge volume of a 1 L mixed liquor sample after being settled for 30 minutes and taking into account the MLSS of the sample. The values of the two indices for sludge of good and poor settling properties are presented in Table 16.4.

16.1.1.4. *Sludge settleability*

Effluent quality depends not only on the treatment process in the aeration tank, but also on the separation of the flocs from the treated effluent by a final sedimentation tank. Settling problems result in the loss of biomass and poor effluent quality, and there are three major categories of settling problems. Floc formation problems can be resulted which are attributed to (i) dispersed growth under high BOD or high f:m ratio; (ii) formation of pin-point flocs

Table 16.4. Sludge settling characteristics. The settleability of sludge is assessed by determining the sludge volume of a 1 L mixed liquor sample after being settled for 30 minutes, and the SVI or SDI is calculated, taking into account the MLSS of the sample.

Sludge settling quality	Sludge volume index (SVI)[a]	Sludge density index (SDI)
Excellent	<80	>1.25
Moderate	80–150	0.66–1.25
Poor	>150	<0.66

[a] $SVI = \dfrac{\text{Sludge volume at 30 min (mL/L)} \times 1000}{\text{MLSS (mg/L)}}$

$SVI \times SDI = 100$

which is due to the disruption of flocs as a result of insufficient food (low f:m ratio), particularly prevalent for old flocs under extended aeration; or (iii) deflocculation which normally takes place under conditions such as low pH, low DO, the presence of toxic pollutants or severe turbulence. These result in turbid effluent. Poor settling can also be caused by density problems in the settling tank which has no aeration. Anaerobic gases such as CO_2, CH_4 and H_2S form bubbles that are trapped in the sludge and raise the sludge to surface. Nitrogen gas from denitrification under anoxic conditions can also be entrained in flocs which float. Commonly, poor settling is due to the formation of bulking sludge which interferes with the performance of many treatment works both locally and overseas. Bulking sludge has massive filamentous growth and the length of extended filament can exceed 107 μm/mL (Fig. 16.14). Filamentous bulking results in the formation of loose flocs with SVI > 150, and interferes with compaction and sludge settling. It is caused by filamentous microorganisms, predominantly filamentous bacteria which include the sheathed bacteria such as *Sphaerotilus* spp., *Microthrix* spp. and *Streptothrix* spp., and actinomycetes such as *Beggiatoa* spp. and *Nocardia* spp. Occasionally, fungi such as *Geotrichum* spp. and *Leptomitus* spp. are found especially under acidic conditions. About 30 types of filamentous microorganisms have been identified in bulking sludge so far and 10 species have been accounted to cause 90% of bulking. Of these ten species, *Sphaerotilus natans* is the most common species. Settling is poor and the settling zone is enlarged, and little clear supernatant is

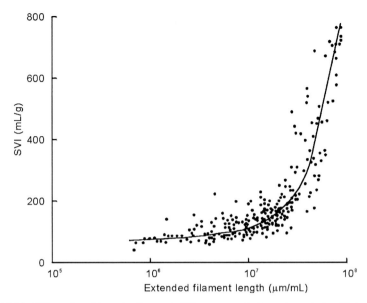

Fig. 16.14. Effect of total length of extended filament on sludge volume index (Palm *et al.* 1980).

obtained. Bulking can be caused by a variety of conditions, including high BOD, salts, H_2S and fatty acids in the influent, and low pH, DO and soluble nutrients (nitrogen and phosphorus). Filamentous microorganisms have a higher surface to volume ratio than cell aggregates, and this allows them to be more competitive under low concentrations of soluble substrates, e.g. nitrogen, phosphorus and DO, or low f:m ratio. Also, it has been empirically proven that H^+, S^{2-} and soluble organics such as phenols, organic acids and aldehydes are more toxic to flocculating cells than filamentous microorganisms. This selective toxicity explains why filamentous growth is predominant at reduced pH and high concentration of H_2S.

Apart from sludge bulking, bacterial foaming is the second most common operational problems encountered in activated sludge systems. Bacterial foaming is formed when thick brown scum occurs in the aeration tank, which is caused by excessive growth of filamentous actinomycetes such as *Nocardia amarae*, *N. pinensis*, *N. rhodochrous* and *Microthrix parvicella*. The most common species found in foams is *Nocardia amarae*, and the major genus of foam-causing agents is *Nocardia*, hence the name *Norcadia* foam. *Norcadia* is highly branched filament, which traps air bubbles and floats to the surface of aeration tank, appearing as chocolate mousse in color and texture. Foams are usually associated with high fatty material content, warm weather ($>18°C$) and long sludge age (>9 days). Foams trap MLSS which are carried out of the aeration tank when overflowed, which deteriorates effluent quality. In some cases, foam overflows onto footpaths and access ways in activated sludge treatment plants, or is dispersed by wind causing health hazards to workers in the treatment plant.

To increase treatment reliability regardless of the performance of final sedimentation, membrane-coupled activated sludge processes incorporating microfiltration have been used with success. These processes are particularly effective in case of high effluent requirement. However, the process is still not widely used due to cost considerations.

16.1.1.5. *Operational design*

Aeration is one of the key operational requirements, which ensures aerobic activities in the mixed liquor and contributes to the efficient degradation of organic matter (BOD removal). This is achieved by bottom air blowers such as porous diffusers, or surface aerators such as cone aerators and rotating brushes. Aeration serve a dual purpose: oxygenation satisfies the oxygen demand of microbes, and mixing keeps the flocs in suspension and maximizes the contact between the microbial cells and the substrates in the wastewater. Carbon oxidation takes place at an oxygen level greater than 0.5 mg O_2/L while nitrification does not occur at below 2.0 mg O_2/L. Other operational parameters which affect process efficiency include mixed liquor suspended solids, organic loading, food to microorganisms (f:m) ratio, sludge residence time and sludge recycle ratio.

There are a number of versions of the activated sludge process, such as the oxidation ditches and the deep shaft system. The deep shaft system was developed to increase the rate of oxygen transfer from the gaseous phase to the liquid phase by increasing the hydrostatic pressure in the circulating wastewater through the use of a 30–200 m deep treatment well. Such systems can achieve very good BOD removal at a very short HRT (1–2 hours). Aeration using oxygen has been used in high purity oxygen activated sludge systems for treating high strength industrial wastewaters.

Sludge is recycled to maintain microbial biomass in the aeration tank. Normally 50–90% is returned and surplus sludge is wasted. This is done in order to attain the desired sludge age. Insufficient sludge results in low MLSS, high f:m and young sludge age (<3 days) while excessive sludge will lead to high MLSS, low f:m ratio and old sludge age (>4 days). Young sludge age is usually poorly flocculated while old sludge age will result in pin-point floc, which are not inducive to good settling. The rate of sludge recycle ultimately depends upon MLSS, organic loading, HRT (a function of flow rate and reactor volume) and effluent quality.

16.1.2. *Aerobic fixed film reactors*

Another type of commonly used biological treatment process is the aerobic fixed film or biofilm reactor which utilizes surface attached organisms growing as a biological film on stationary or moving media for wastewater purification. Reactors with stationary media are exemplified by conventional biological filters, which are also called trickling or percolating filters. The first tricking filter was built in 1892 in Salford, England and is one of the earliest biological treatment systems for wastewater. This system consists of a circular or rectangular filtering tank containing bed of permeable filter media. Wastewater is sprayed from the top onto the filter bed via a rotary or reciprocating distributor for the circular or rectangular filter respectively, and trickles through the filter media which provide an extensive growth surface to support the microbial film and associated fauna necessary for wastewater purification. Most conventional filters are of the depth of 1.5–2.5 m and HRT of 0.5–1 hour. The tank is furnished with natural or forced draft ventilation with a pipe connecting the interior of the filter to the atmosphere or to a ventilation fan to provide free air movement. An underdrain collects treated wastewater from the bottom, which is then pumped to a clarifier for solid separation.

16.1.2.1. *Filter media*

The filter is filled with natural material such as mineral or synthetic material such as plastic to provide a surface for growth of organisms. Natural media is usually made up of gravel or rock of 25–100 mm in

Table 16.5. Properties of mineral and plastic media used in trickling filters.

	Surface area (m²/m³)	Void volume (%)
Mineral aggregates	50–100	50
Plastic media	100–300	>90

diameter which is good for conventional low-rate loading (<0.1 kg BOD/m³/day). Plastic media which was introduced in the 70s composed of PVC, polystyrene or polypropylene, and have a high surface area to volume ratio and high voidage. Plastic media is lightweighted and can be packed into deeper beds (8 m) for stronger influent. Plastic media can be of random pack of individual moulds of specific shape, usually in the form of rings or short tubes with external and/or internal fins and ribs to increase surface area. The other type of plastic media is modular packing either in the form of zigzag channels between vertical sheets or block of closely assembled vertical tubes with concentric tubules. Plastic media increase the surface area per unit volume by two to six times and the void volume for at least 80% (Table 16.5).

16.1.2.2. Biofilm

The percolating filter is not a physical filter but has a solid support medium on which the biofilm develops. The surface of the biofilm is not smooth but convoluted and corrugated. The biofilm is a complex community of bacteria, fungi, protozoa and other meso- and macro-invertebrates in a slimy coating on the supporting surface of the filter medium. Wastewater, as the aqueous phase, trickles through the filter bed and flows freely over and through the biofilm, which brings about purification. The biofilm is a semi-aquatic habitat with diverse life forms, and it is the basic unit of purification in biological filters (Fig. 16.15). The primary feeders are major decomposers and are responsible for the extracellular enzymatic solubilization of solids and oxidation of carbonaceous matter. They are bacterial cells and fungal mycelia growing in a microbial slime of polysaccharide matrix which forms the bulk of the biofilm. Bacteria are the largest in population, both in number and biomass and the major genera

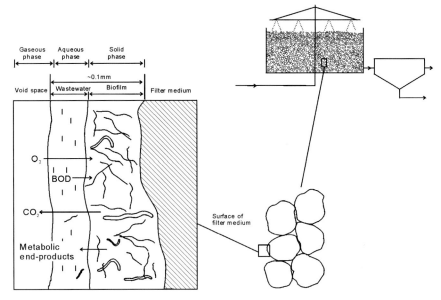

Fig. 16.15. The trickling filter works with biofilm on filter media as the basic operational unit. Wastewater percolates over and through the biofilm matrix with the transfer of substrates and end products in and out of the biofilm.

include *Zoogloea, Pseudomonas, Flavobacterium, Beggiatoa, Alcaligenes, Sphaerotilus, Nitrosomonas* and *Nitrobacter*. Fungi are also common in the aerobic regions and can flourish under low pH or high organic content. Common fungal genera are *Fusarium, Geotrichum, Penicillium, Mucor, Ascoidea, Subbaromyces* and *Sepedonium*. Algae that are present in the portion of biofilm exposed to light (usually on the upper 5 cm) contribute relatively little to wastewater treatment. Microbial composition varies seasonally with wastewater composition and filter depth. The microbial film and cells are grazed by secondary feeders which are mostly protozoa, nematodes, rotifers, annelids and insect larvae. There are about 2,000–100,000 protozoa per mL sludge. Over 200 species of protozoa have been isolated and over 50% are ciliates such as *Paramecium* spp., *Vorticella* spp. and *Opercularia* spp. Amoebae such as *Amoeba* spp. and zooflagellates such as *Trigonomonas* spp. are also present. They feed on dispersed bacterial cells and help clarify the effluent. The burrowing activities of nematodes, annelids such as epigic earthworms and larvae of insects

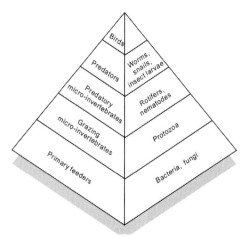

Fig. 16.16. The biological filter has a pyramid of food chain which contains several trophic levels, and is more complex than that of the activated sludge.

such as chironomid encourage nutrient and oxygen flow within biofilm. The biological filter has a food chain pyramid containing several trophic levels, and is more complex than that of the activated sludge process (Fig. 16.16). The metazoa, though not important for design purposes, also play an important role in the ecological balance needed for efficient operation.

The biological composition of biofilm is selected according to attachability as governed by the electrostatic surface adhesion of cells, degree of hydrophilicity between cells and the medium surface, surface roughness and the formation of a slimy matrix. Biofilm growth depends on waste-water strength and biofilm thickness, which affects substrate transfer and aerobic activity. Growth is unidirectional, moving out from the supporting medium to the aqueous and gaseous phases. The major factors that control biofilm composition and structure and development have been summarized in Table 16.6.

Biofilm formation is affected by factors such as hydrodynamics of the bulk fluid, nature of the substratum, species composition and nutrient availability. The traditional concept views biofilm as a virtually dual-layered structure with a base film which consists of a structured layer and a surface film which has irregular boundaries and is a transition between the base film & the bulk liquid (Fig. 16.17). However, recent

Table 16.6. Major factors that control composition, structure and development of biofilm (Wingender and Flemming 1991).

Substratum	Surface properties (e.g. roughness, hydrophobicity)
Bulk liquid (wastewater)	Physicochemical conditions (e.g. temperature, pH, salinity, ions, organic matter)
	Concentration of available organic substrates
	Hydraulic conditions (e.g. flow rate, shear stress)
Microorganisms	Morphology
	Surface properties
	Physiological activity
	Lysis
Micro- and meso-fauna	Grazing activity
Biofilm	Age

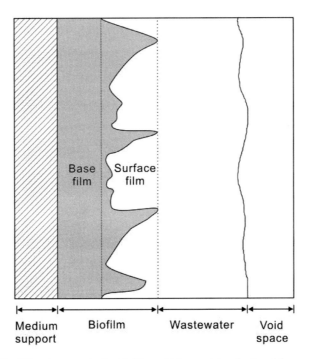

Fig. 16.17. Traditional concept of a biofilm which consists of a basal layer and a rough surface layer in intimate contact with the flowing wastewater.

experimental data suggest that biofilm is not structurally uniform and homogeneous as generally assumed, but a dynamic layer of microbial clumps adhering to and detaching from the support surface. It is a non-uniform structure of variable thickness containing microstructures such as cell clusters, cell-free pores, voids and microchannels attached to each other by an extracellular polyermic material. Spaces exist between cell clusters to form vertical voids (pores) and horizontal voids (channels). Mass transport of substrates and end-products is facilitated in the biofilm voids and channels at high flow velocities. Various biofilm conceptual models have been put forward. Among them, the penetrated water-channel mushroom-like biofilm model (1994) views biofilm as a non-uniform structure consisting of discrete cell clusters (microbial aggregates) cemented by extracellular polymeric substances; it is a porous network with voids which allow wastewater to flow through (Fig. 16.18). In the

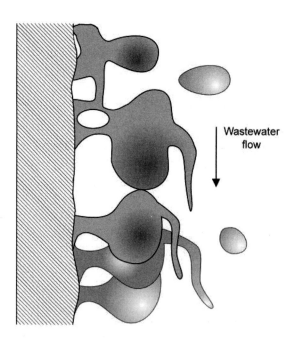

Wastewater
flow

Fig. 16.18. The penetrated water-channel mushroom-like biofilm model views biofilm as a heterogeneous and dynamic layer of microbial aggregates (microcolonies) adhering to and detaching from the support surface. The discrete clumps are surrounded by a network of interstitial voids filled with water (Redrawn from Costerton *et al.* 1994).

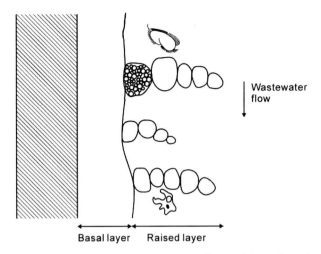

Basal layer   Raised layer

Fig. 16.19. Biofilm in the heterogeneous mosaic biofilm model consists of a basal layer and a raised layer with chains of microcolonies rising from the basal layer (Redrawn from Walker *et al.* 1995).

heterogenous mosaic biofilm model (1995), the biofilm consists of a thin basal layer of microorganisms (about 5 μm) with vertically aligned stacks of microcolonies (up to 100 μm) outgrow from the surface (Fig. 16.19). According to the dense confluent biofilm model (1991), biomass distribution within a biofilm is not uniform and the confluent biofilm has no water channels or porous structures. Fluxes are multidirectional instead of unidirectional. In addition, mass transfer in biofilm is not just diffusional because convectional transport has also been demonstrated.

16.1.2.3. *Purification efficiency*

The transport of substrates and end products take place by diffusion in the pores and channels in the biofilm. Organic substrates, nutrients and metabolic end products move between the wastewater (aqueous phase) and the biofilm (solid phase). Gaseous transfer occurs through the air-wastewater interface, but oxygen concentration in the biofilm is limited by the DO concentration in the bulk liquid phase. Oxygen is first dissolved in the wastewater and diffused through the biofilm along a concentration gradient. The total amount of oxygen transferred is governed by the surface area of the media exposed; therefore smaller gravel is better in terms of ventilation because of the greater surface area per unit volume. However, the choice of gravel size is not ultimately aimed at maximum surface area as the void space is reduced when smaller gravel is used. The thickness of aerobic layer depends on the extent of oxygen diffusion within the biofilm. The critical depth, i.e. the depth of oxygen penetration, varies from 0.05–2.00 mm, which is dependent upon the composition and density of the biofilm. The inner layer of the biofilm may be anaerobic and substrate limited. A thick film tends to have a thicker anaerobic layer than a thin film.

The biological activity of the microbial film is a function of the organic (BOD) loading, oxygen availability and temperature. BOD is removed by the active portion of the biofilm with purification efficiency being highest at optimal biofilm thickness. When the film is too thin, biodegradation is limited by low microbial biomass. For thick film, interstitial volume, which is essential to sewage flow, ventilation and microbial growth, is reduced and ventilation is hampered. In some cases, the film can be so thick that sewage flow is prevented and the filter becomes clogged, and in severe cases, ponding is caused.

The growth of the biofilm is a function of the BOD loading in the wastewater. However, gaseous and nutrient diffusion is affected when the film is thicker than 0.5 mm. When the inner layer is depleted of nutrients and becomes anaerobic, cells will undergo endogenous respiration and the film will slough off. The thickness of the biofilm is also regulated by the feeding activity of the grazing detritivores, the flushing action of the wastewater and the stripping action of detached film from the upper region. All these processes act to remove biofilm and keep the thickness at a dynamic equilibrium of 2 mm.

16.1.2.4. *Vertical stratification*

Differential growth occurs along the depth of the biological filter. At the top, higher BOD loading and a faster cycle of biofilm growth and sloughing off select microbes with short generation time (e.g. C oxidizers). Heterotrophs outgrow nitrifiers when BOD is available. Oxygen concentration in the wastewater is also limited in the upper region and this inhibits the growth of nitrifying bacteria which are strict aerobes. At the bottom, microbes of longer generation time (e.g. nitrifying bacteria) establish, with *Nitrosomonas* spp. becoming established before *Nitrobacter* spp. as ammonium is present in higher concentration in the upper zone. Nitrifying activity increases with depth and nitrification is enhanced at a greater depth when BOD < 20 mg/L. A vertical gradient is established as the pollutants are removed when the wastewater trickles downward with microbial density and bioconversion rate being highest at the top. However, nitrifiers build up only at the bottom of the tank. Chemically, BOD declines while nitrate concentration increases with depth (Fig. 16.20).

16.1.2.5. *Recirculation*

At low hydraulic and organic loadings, the conventional low-rate single-pass filtration can produce a final effluent of very high quality after solid settling. For stronger industrial wastewater, filtration towers with plastic packing are preferred. Recirculation of effluent helps, as recirculating treated effluent to the filter dilutes strong wastewater and increases the flushing action, which effectively restricts biofilm accumulation and prevents clogging of the filter. Recirculation at low flow periods can

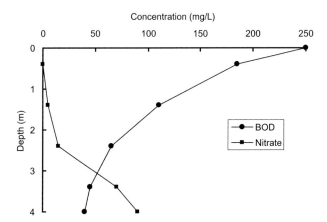

Fig. 16.20. The changes in BOD and nitrate concentration with filter depth in a trickling filter of 4 m deep. Carbon oxidation takes place in the upper region while nitrification only commences at greater depth (Mudrack and Kunst 1986).

even out diurnal fluctuations in volume, and can enhance nitrification as the recirculated effluent contains nitrate and DO at low BOD. The amount of recirculation depends on the organic strength (BOD) of the influent and effluent, and the recirculation ratio (recirculated flow to influent flow) is typically 0.1–2.0.

16.1.2.6. *Filter systems*

In the traditional single-stage process, wastewater is treated by a single filter or by filters working in parallel. This is used for a conventionally low organic loading rate of 0.1–0.5 kg BOD/m^3/day. For higher rates (>0.5 kg BOD/m^3/day), multi-pass mode is employed in which wastewater is treated progressively by filters operating in a series (Fig. 16.21). Two sets of similar filter beds can operate in series as in double filtration. For alternating double filtration, the sequence of filters used is periodically reversed to relieve overloading of the conventional filter. Partially treated effluent is applied from one filter, but the direction of flow reverses before the primary filter shows signs of ponding (7–14 days). In these situations, nitrifiers never establish and nitrification is poor. In two-stage filtration, two filters operate in series; the first one usually acts as a primary high rate filter or the secondary filter is used for nitrification.

16.1.2.7. *Nitrification and denitrification in biological filters*

Nitrification usually takes place at the lower regions of the filter with rates dependent upon temperature and the concentration of ammonium ion. Denitrification, however, takes places in the upper filter area where organic matter is abundant and biofilm is thick to allow anaerobic conditions to develop in the inner region of the film. Nitrate is supplied via recirculation to allow simultaneous denitrification.

16.1.2.8. *Rotating biological contactor*

The invention of plastic packing media has led to the development of tall tower filters (biotowers) with forced aeration for the treatment of

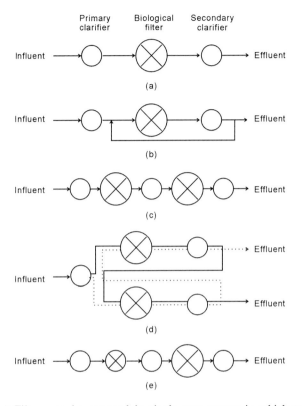

Fig. 16.21. (a) Filters can be arranged in single-pass system in which wastewater is treated by a single filter or a number of filters in parallel; (b) recirculation can take place for strong wastewater. Multi-pass systems can be used in which wastewater is treated by more than one filters in series (c)–(e); (c) double filtration consists of two sets of filters in series; (d) alternating double filtration changes sequence of filters periodically with flow direction reverses before the primary filter shows sign of ponding; (e) two-stage filtration uses two different filters in series, usually a primary high-rate filter.

high-strength wastewater. This minimizes clogging in gravel media to produce a higher quality effluent, and allows nitrification to occur.

The use of plastic media has also led to the development of the rotating biological contactor which is an aerobic fixed film reactor with horizontal flow and moving media (Fig. 16.22). It was developed in the late 1920s and was commercially available in 1965. This system is popular for small communities and industries. The moving media is a series of circular discs made of plastic, PVC or expanded polystyrene in groups, mechanically

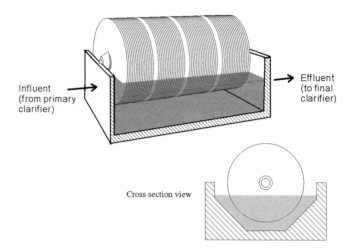

Fig. 16.22. Rotating biological contactor consists of circular discs on which biofilm develops. The discs rotate slowly in right angle to the wastewater flow.

driven around a horizontal axis. The discs are 10–20 mm thick, 2–3 m in diameter and spaced 20 mm apart. They rotate slowly (0.5–10 rpm) at right angle to the horizontal wastewater flow with 40% of the area immersed. The whole system can be covered to protect from bad weather and low temperature, and to control fly and odour nuisance. Biofilm grows on the disc surface which is alternately immersed as the discs are rotated and ventilation is achieved when the discs are exposed to the air. There is a succession along the reactor, with heterotrophic bacteria and fungi in the first few discs, protozoa on successive discs and nitrifiers towards the end of the system. Sloughing will be more frequent at the front portion of the system. Sewage flow and disc rotation ensure efficient mass transfer from liquid to film and control excessive film accumulation. This system has several advantages over percolating filters. The media is completely wetted and there is no clogging even in the case of excessive film accumulation; ventilation is good as the discs are regularly exposed to air during rotation; less space is needed (which is about 10% of that by low-rate filtration); and nuisance can be reduced when the reactor is covered. However, the system is sensitive to overloading and is more expensive to run especially in motor drive and bearing maintenance when compared to other small biological treatment plants.

16.1.2.9. *Biological aerated filters*

Though not as common as those conventionally used in municipalities, there are modified versions of the biological filter, some of which have been used for over twenty years. The activated biofilter integrates activated sludge and biological filter to provide a combined suspended and attached growth process. However, there are only a small number of full-scale operations in use due to declining interest. Technological advances led to the development of different ranges of biological aerated filters (BAF) or submerged attached growth systems. The aerated packed bed reactors use attached biomass, as in trickling filters, but the microbial film grows on smaller granular carriers (0.2–3 mm) made of mineral, glass or plastic, which are immersed in the wastewater and aerated by injection of air or pure oxygen. The small particle size provides a large specific surface area for biofilm attachment. Wastewater enters the filter either from the top (downflow filter) or the bottom (upflow filter). In the upflow system, the bed of carriers (e.g. silica sand with a diameter of 0.2–0.7 mm) can be fluidized by the wastewater flow (aerated fluidized bed reactor). An alternative is the airlift bioreactor in which the bed, which has thicker biofilm, is fluidized by the introduction of air at the bottom of the reactor. The free floating media move with the wastewater circulation in the reactor. High specific surface area (1,000–4,000 m^2/m^3) improves contact with substrate in the wastewater. High biomass density allows shorter hydraulic retention time, and clogging or ponding is no longer a problem. Microfauna with the normal species diversity and rotifers were found. These moving bed biofilm reactors are considered very efficient for treating high strength wastewater such as food industry effluents. Immobilization of nitrifying bacteria on carriers allows the development of older biofilm which is essential for nitrification. However, nitrate can only be denitrified if anaerobic condition is developed. This can be achieved using either a separated anoxic filter for denitrification before an aerobic filter (pre-denitrification), which is similar to the A/O system for activated sludge, or after BOD removal and nitrification with carbon addition (post-denitrification). Biological phosphorus removal has also been demonstrated in some of these filter systems.

16.2. Anaerobic Wastewater Treatment

Aerobic treatment processes require aeration which is energy consuming and expensive. In this regard, anaerobic plants would be an advantage and can treat strong wastewater (BOD > 500 mg/L) which is otherwise difficult to degrade aerobically. Anaerobic systems produce biogas as an end product, which can be used as a fuel. Sludge handling is also easier as about 10% of the substrate is converted to microbial biomass compared with about 50% for aerobic processes. The anaerobically digested effluent can be further processed by aerobic treatment.

Anaerobic digestion of wastewater is more or less the same as that of solid wastes or slurries in that the biological material is broken down microbially under anaerobic conditions with the generation of biogas. However, as wastewater treatment plants usually work under a high hydraulic flow rate with low (1–4%) solid content, a longer retention time is needed with conventional stirred tank reactors or very large reactors have to be built. With the development of modern digesters, anaerobic treatment is a feasible and practically better option for wastewaters of high organic loading.

16.2.1. *Microbiology of anaerobic digestion*

Anaerobic reaction is solely a bacterial process in which organic matter is degraded by a succession of mixed bacterial populations. It was previously thought to be a 2-stage process consisting of fermentation and methanogensis, but it is now clear that it has 4 stages, namely hydrolytic, acidogenic, acetogenic and methanogenic phases (Fig. 16.23). In the hydrolytic phase, the organic substrate is hydrolyzed enzymatically by facultative anaerobic bacteria. Macromolecules of high molecular weight such as carbohydrates, proteins and lipids are broken down extracellularly into soluble monomers. Cellulolytic bacteria such as *Clostridium* spp. secrete cellulolytic enzymes which convert cellulose into cellobiose and glucose; proteolytic bacteria such as *Bacteroides* spp., *Peptococcus* spp. and *Peptostreptococcus* spp. break down protein into amino acids. In the acidogenic phase, the hydrolytic products such as sugars and amino acids are acted on by acid-forming bacteria (e.g. *Lactobacillus* spp.,

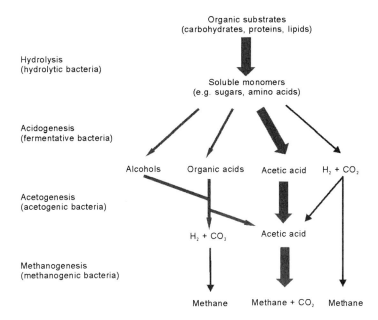

Fig. 16.23. Anaerobic digestion is a four-stage bacterial degradative process which converts polymeric substrates to methane and carbon dioxide.

Bifidobacterium spp. and *Butyrivibrio* spp.), converted to pyruvate and subsequently to acetate and carbon dioxide or alcohol and hydrogen. The fate of metabolites depends on the type of bacteria and culture conditions (e.g. redox potential). At low partial pressure of hydrogen ($<10^{-4}$ atmosphere), acetate and/or hydrogen are formed; at high partial pressure of hydrogen ($>10^{-4}$ atmosphere), ethanol or organic acid is produced. Short chain fatty acids (e.g. acetate, propionate and butyrate) are the main products, which lower the pH, and hence the name acidification phase as a synonym to acidogenic phase. Acetate is the major product of carbohydrate fermentation.

The third phase is the acetogenic phase. Alcohols and short chain fatty acids are utilized by hydrogen-producing heteroacetogens (e.g. *Enterobacter* spp., *Citrobacter* spp., *Serratia* spp. and *Syntrobacter* spp.), which produce acetate and/or hydrogen and carbon dioxide through acetogenic dehydrogenation. Hydrogen-consuming homoacetogens (e.g. *Clostridium* spp. and *Acetobacterium* spp.) use hydrogen and carbon dioxide as substrates to form acetate via acetogenic hydrogenation. Acetate and carbon dioxide formed from Phases 1 to 3 are methanogenic substrates. Acetate-removing methanogens utilize acetate and form methane and carbon dioxide through acetate decarboxylation, and this accounts for 70% of the methane formed. The remaining 30% is produced by reductive methane formation with hydrogen-removing methanogens which convert hydrogen and carbon dioxide to methane and water.

Methanogens which are strict anaerobes belong to Archaebacteria. They are morphologically diverse which are restricted to a small number of simple substrates such as acetate, formate, methanol, methylamines, carbon dioxide and hydrogen (Table 16.7). They are the terminal organisms in biogasification which produce methane as the metabolic end product. There are about 50 species. Acetate-removing methanogens include *Methanosarcina* spp., *Methanothrix* spp. and *Methanosaeta* spp. which produce methane from acetate, and *Methanococcoides* spp. and *Methanolobus* spp. which utilize methanol or methylamines. Hydrogen-removing methanogens such as *Methanomicrobium* spp., *Methanobacterium* spp., *Methanobrevibacter* spp., *Methanococcus* spp., *Methanogenium* spp. and *Methanospirillum* spp. utilize hydrogen and formate in addition to acetate.

Table 16.7. Morphological characteristics and growth substrates of different methanogenic bacteria.

Genus	Shape	Gram stain	Substrates
Acetate-removing methanogens			
Methanosarcina	Cocci in clusters	Positive	Acetate, methanol, methylamines
Methanothrix	Long rods to filaments	Negative	Acetate
Methanosaeta	Long rods to filaments	Negative	Acetate
Methanococcoides	Cocci	Negative	Methanol, methylamines
Methanolobus	Cocci in clusters	Negative	Methanol, methylamines
Hydrogen-removing methanogens			
Methanobacterium	Long rods	Positive	Hydrogen, formate
Methanomicrobium	Short rods	Negative	Hydrogen, formate
Methanococcus	Cocci	Negative	Hydrogen, formate
Methanogenium	Cocci	Negative	Hydrogen, formate
Methanospirillum	Curved rods	Negative	Hydrogen, formate

Methanogens are more demanding in environmental requirements and are sensitive to shock conditions. Oxygen, as low as 0.01 mg/L, is toxic to methanogens. Methanogens are highly sensitive to oxidized materials such as nitrate and sulphate, with an optimal redox potential at below −300 mV. Methanogens are also sensitive to pH and grow best in the narrow range of 6.5–8.0. A pH less than 6.6 is inhibitory and acidification will reduce the growth of methanogens, which further acidifies the digester content owing to an accumulation of volatile fatty acids, as the acidogenic bacteria are less sensitive to low pH. Methanogens are slow-growing with a generation time of 2–3 days at 35°C as compared with 2–3 hours for acid-formers. Methanogenesis is therefore the rate-limiting step of anaerobic decomposition.

The bacterial consortium is metabolically dependent on each other for survival and activity, and the acid-formers will be inhibited by their own end products if these are not catabolized by the methanogens. Mutualistic effects occur between different metabolic groups for the maintenance of pH and redox potential. Hydrogen is usually recognized as the controlling intermediate and an interspecies hydrogen transfer is involved. Low hydrogen concentration is maintained not just by diffusion alone, but also by intimate cell contact between acetogenic and methanogenic bacteria.

Most anaerobic digestions operate at a mesophilic temperature range of 30–38°C. As anaerobic reaction produces little free energy as heat, the digester needs energy input for heating up the content to above 35°C. Thermophilic digestion gives higher yields, allows higher loading and greater pathogen destruction, but greater heating costs and the need for insulation to reduce excessive heat loss outweigh the advantages.

16.2.2. *Anaerobic reactors*

All reactions in anaerobic digestion can take place in one reactor in single-stage system, though the use of two-stage system in which fermentation and methanogenesis are separated is not uncommon for solid digestion. Anaerobic processes can take place in open lagoons or in closed digesters. Anaerobic ponds in the stabilization pond systems are usually used as pretreatment steps for strong wastewater. In a pond of 2–5 m deep with no aeration, anaerobic conditions prevail. These are followed by facultative and maturation ponds, in series, for complete treatment. Septic tanks, which are the underground individual household systems for wastewater disposal in unsewered municipalities, are in fact low volume anaerobic systems in which solid settlement and anaerobic digestion of sludge take place in one tank.

For closed reactors, for example stirred tank digesters which are developed for the digestion of high-solid feedstocks, mixing is performed by mechanical mixers such as stirrers and screw pumps or by methane recirculation. In a conventional digestion ($0.5–1.5$ kg volatile solids/m^3/ day) of slurry in the absence of agitation, well-defined strata are formed with an upper relatively inactive scum layer, central layer of tank liquor, lower active layer of digesting solids and a bottom layer of sludge (Fig. 16.24), with a retention time of $30–60$ days. In high rate digestion ($1.5–5.0$ kg volatile solids/m^3/day), the content is homogenous due to mixing, and active digestion takes place throughout the digester content. The retention time is slightly shorter than 15 days. Digesters can be of different configurations with different height to diameter ratios (Fig. 16.25). A conical bottom retains grit and sludge for periodic removal. Egg-shaped digesters have the benefits of better mixing properties, lower energy demand for heating and smaller surface which minimizes the size of a scum blanket, and have become very popular in Asian countries such as Japan for waste-water treatment. However, all these configurations when operated as flow-through system, suffer from washout of the bacterial biomass and impractical long retention times for wastewater treatment.

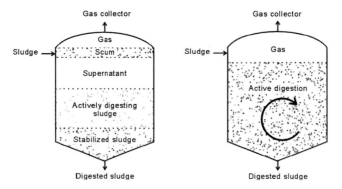

Fig. 16.24. Conventional digestion has stratified content while high-rate digestion has active digestion throughout the digester content.

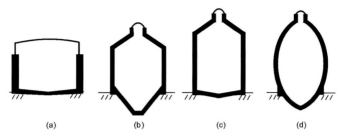

(a)   (b)   (c)   (d)

Fig. 16.25. Major shapes of anaerobic digesters most commonly used in different parts of the world. (a) Anglo-American digester; (b) classical European digester; (c) European digester; and (d) pear- or egg-shaped digester (Hobson and Wheatley 1993; Bode and Klauwer 1999).

Stirred tank digesters are not suitable for treating low strength wastewater since the hydraulic flow is great and the gas yield is very low, which would otherwise require a very large digester. The separation of solids from the effluent is another problem in that the biogas produced tends to cause the sludge to float and affects settleability. With the development of biomass-retained systems, the anaerobic treatment of wastewater becomes feasible. Contact systems are mostly used with the mean cell residence time (MCRT) far exceeding the HRT, as in aerated biological filters for aerobic treatment (Fig. 16.26). Biomass is preserved in the reactor by immobilization with the cells being trapped on a solid support in the digester or by an upward flow of the wastewater in which the biomass is suspended and prevented from being washed out. In fact, immobilized biomass containing the proper balanced anaerobic bacterial consortia is the key for success in anaerobic digestion of wastewater. MCRT is usually greater than 20 days and cells are acclimatized to the influent. Anaerobic fixed film digester in the form of anaerobic static media biofilter or anaerobic rotating disc contactor can be used. The former is actually a submerged trickling filter with packed bed of gravel, or beads of PVC or polyester of 6.0–60 mm in diameter and wastewater either flows upward or downward. Anaerobic microbial cells develop as attached film immobilized on the immersed media and as dispersed biomass in the interstitial space. Treatment performance, therefore, does not depend highly on the size and specific surface area of the media as in aerobic filters. However, when small filter media is used, there is the potential risk of

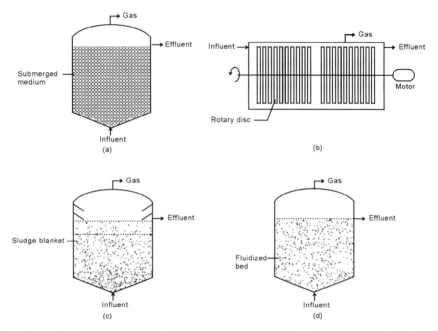

Fig. 16.26. The various anaerobic reactors commonly used for wastewater digestion. (a) Anaerobic static media biofilter; (b) anaerobic rotating disc contactor; (c) upflow anaerobic sludge blanket reactor; and (d) anaerobic fluidized bed reactor.

clogging by solids in the wastewater. The latter is similar to the aerobic form of RBC but with the discs totally submerged.

The upflow anaerobic sludge blanket (UASB) reactor is a floc-based anaerobic system. Wastewater flows upwards through the bottom sludge bed and an upper sludge blanket which consists of lighter solid fraction (flocculated sludge granules) kept in suspension by the wastewater flow. Bacterial organization and arrangement govern the granulation of sludge, and the bacterial composition, size and microstructure of the granules are dependent upon the nature of the substrate. It is generally accepted that the USBA granules, which are of 0.5–4.5 mm in size, are stratified structures. The granule core is almost exclusively composed of *Microthrix* spp., which is a key structural component in granulation. Acetoclastic methanogens and hydrogen-producing acetogens develop in the middle layer while acidogenic (fermentative) bacteria and hydrogenotrophic methanogens occupy the outer layer. Finer sludge settles back to the blanket in the upper zone while heavier sludge settles to the bottom of the digester. Mixing is done as the influent flows upward through the blanket and gas bubbles released from the digester content. The digestion tank has inclined louvres or baffles at the top, which improve sludge settleability and reduce biomass washout. Advantages of UASB granules over sludge flocs include dense compact biomass, high mechanical strength, high settleability (30–80 m/h) (which prevents washing out), balanced microbial community, closely associated syntrophic partners, high methanogenic activity, and resistance to toxic shock. There are more than 300 full-scale plants all over the world treating wastewaters mainly from food and beverage industries. A variant of the UASB reactor is the expanded granular sludge bed (EGSB) reactor or simply anaerobic expanded bed (AEB) reactor. It is a very promising version of upflow sludge bed reactors which uses granular biomass suspended at high velocities (5–10 m/h) of upflow influent to result in high hydraulic mixing, expanded granular sludge bed and better wastewater-biomass contact, and is effective in treating low strength wastewaters. At present, there are about 20 full-scale EGSB reactors in operation worldwide and the development is being considered as a sustainable integrated low cost wastewater biotechnology.

The anaerobic fluidized bed (AFB) reactor is a hybrid of fixed film and floc-based reactor in which the fine grained inert carriers, made of quartz sand, diatomaceous earth, glass beads, PVC particles or carbon granules, of size ranging 0.2–2 mm are coated with bacterial biomass. The small-sized particles provide a very large surface area and better mixing of the substrates and the bacterial cells. The upflow wastewater movement fluidizes the biomass on carrier particles (bioparticles) and the shear forces of fluidization create a thin dense biofilm with active biomass. The upflow velocity is high enough to expand the bed by typically 20–25% of the resting volume. The MCRT can be as long as 300 days at HRT of about 2 hours. This biomass-on-carrier technology has the advantages of good contact between bacterial cells and wastewater, high biomass concentration, biomass thickness under control, no potential blockage, and reduced reactor volume. It is a highly engineered system with high energy requirement for fluidization. The practical problem of controlling the attachment of biofilm to the carrier particles also limits its use. The technique is not widely applied yet with full-scale plants in the US, the UK, the Netherlands and Australia.

References

1. Bode H and Klauwer E. Advantages and disadvantages of different shapes in digester design. *Water Qual. Int.* **March/April**: 35–40, 1999.
2. Characklis WG and Marshall KC. Biofilm: A basis for an inter-disciplinary approach. In: *Biofilms*, eds. WG Characklis and KC Marshall. Wiley, New York, 1990, pp. 3–15.
3. Costerton JW, Lewandowski Z, DeBeer D, Caldwell D, Korber D and James G. Biofilms, the customized microniche. *J. Bacteriol.* **176**: 2137–2142, 1994.
4. DeBeer D, Stoodley P and Lewandowski Z. Liquid flow and mass transport in heterogeneous biofilms. *Water Res.* **30**: 2761–2765, 1996.
5. Eckenfelder WW. Trickling filter design and performance. *J. Sanit. Eng. Div. Am. Soc. Civil Eng.* **87**(SA4): 33–45, 1961.
6. Fang HHP, Chui HK and Li YY. Microbial structure and activity of UASB granules treating different wastewaters. *Water Sci. Technol.* **30**(12): 87–96, 1994.
7. Hawkes HA. The applied significance of ecological studies of aerobic processes. In: *Ecological Aspects of Used Water Treatment, Vol. 3. The Processes and Their Ecology*, eds. CR Curds and HA Hawkes. Academic Press, London, 1983, pp. 173–334.
8. Jenkins D, Richard MG and Daigger GT. *Manual on the Causes and Control of Activated Sludge Bulking and Foaming*, 2nd ed. Lewis Publishers, Boca Raton, 1993.
9. Lettinga G, Field J, van Lier J, Zeeman G and Hulshoff Pol LW. Advanced anaerobic wastewater treatment in the near future. *Water Sci. Technol.* **35**(10): 5–12, 1997.
10. Li D and Ganczarczyk J. Size distribution of activated sludge flocs. *J. Water Pollut. Control Fed.* **63**: 806–814, 1991.
11. Meaney BJ and Strickland JET. Operating experiences with submerged filters for nitrification and denitrification. *Water Sci. Technol.* **29**: 119–125, 1999.
12. Mudrack K and Kunst S. *Biology of Sewage Treatment and Water Pollution Control*. Ellis Horwood, Chichester, 1986.

13. Palm JC, Jenkins D and Parker DS. Relationship between organic loading, dissolved oxygen concentration and sludge settleability in the completely-mixed activated sludge process. *J. Water Pollut. Control Fed.* **52**: 2484–2506, 1980.

14. Parker DS and Richards T. Nitrification in trickling filters. *J. Water Pollut. Control Fed.* **59**: 679–685, 1986.

15. Pasveer A. A contribution of the development of activated sludge treatment. *J. Proc. Inst. Sewage Purif.* **4**: 436–465, 1959.

16. Pike EB. Aerobic bacteria. In: *Ecological Aspects of Used Water Treatment, Vol. 1. The Organisms and Their Ecology*, eds. CR Curds and HA Hawkes. Academic Press, London, 1975, pp. 1–64.

17. Pike EB and Curds CR. Microbial ecology of the activated sludge process. In: *Microbial Aspects of Pollution*, eds. G Sykes and FA Skinner. Academic Press, New York, 1971, pp. 123–147.

18. Pretorius WA. A conceptual basis for microbial selection in biological wastewater treatment. *Water Res.* **21**: 891–894, 1987.

19. Pujol R, Duchene P, Schetrite S and Canter JP. Bacterial foams in activated sludge plants: Characterization and situation. *Water Res.* **25**: 1379–1404, 1991.

20. Stanbridge HH. *History of Sewage Treatment in Britain. Biological Filtration.* Institute of Water Pollution Control, Maidstone, 1976.

21. Walker JT, Mackerness CW, Rogers J and Keevil CW. Heterogeneous mosaic biofilm — A haven for waterborne pathogens. In: *Microbial Biofilm*, eds. HM Lappin-Scott and JW Costerton. Cambridge University Press, Cambridge, 1995, pp. 196–204.

22. Wheatley AD. Wastewater treatment and by-product recovery. In: *Topics in Wastewater Treatment*, ed. JM Sidwick. Blackwell, Oxford, 1985, pp. 68–106.

23. Wingender J and Flemming H-C. Autoaggregation of microorganisms: Flocs and biofilms. In: *Biotechnology, Vol. 11a. Environmental Processes I. Wastewater Treatment*, ed. J Winter. Wiley-VCH, Weinheim, 1991, pp. 65–83.

Further Readings

1. Bitton G. *Wastewater Microbiology*. Wiley-Liss, New York, 1994.
2. Forster CF. *Biotechnology and Wastewater Treatment*. Cambridge University Press, Cambridge, 1985, p. 376.
3. Forster CF and Wase DA. *Environmental Biotechnology*. Ellis Horwood, Chichester, 1987, p. 453.
4. Grady CPL Jr, Daigger GT and Lim HC. *Biological Wastewater Treatment*, 2nd ed. Marcel Dekker, New York, 1999.
5. Hobson PN and Wheatley AD. *Anaerobic Digestion. Modern Theory and Practice*. Elsevier Applied Science, London, 1993, p. 269.
6. Mudrack K and Kunst S. *Biology of Sewege Treatment and Water Pollution Control*. Ellis Horwood, Chichester, 1986.
7. Scragg A. *Environmental Biotechnology*. Longman, London, 1999.
8. Wuertz S, Bishop PL and Wilderer PA. *Biofilms in Wastewater Treatment*. International Water Association, London, 2003.

Questions for Thought...

1. The aeration tank in a modern activated sludge treatment plant is not entirely aerobic. Discuss the purpose of this design.
2. What is the role of the filter media in a percolating filter? Why are plastic media superior to mineral ones in the treatment of high strength sewage?
3. Compare and contrast conventional biological filters and submerged biofilters as wastewater treatment options.
4. Discuss briefly the advantages of using activated sludge system over percolating filter for the treatment of municipal sewage.
5. Describe the operational features of an anaerobic fluidized bed reactor. What are its advantages over the other types of anaerobic reactor in high rate digestion?
6. What is meant by mesophilic anaerobic digestion? How does it compare with the thermophilic digestion in anaerobic wastewater treatment?
7. Discuss the advantages of anaerobic processes over aerobic ones for the treatment of high-strength wastewater.

Chapter 17

Industrial Wastewater Treatment

P. K. Wong and Lee Man Chu
*Department of Biology, The Chinese University of Hong Kong
Shatin, NT, Hong Kong SAR, China*

17.1. Removal of Inorganic Pollutants

17.1.1. *Removal of metal ions in industrial wastewater*

One of the most common definitions of "heavy metal" is a metal with a density of greater than 5 g/cm^3. Under this definition, 69 elements belong to this group of which 16 are synthetic and these metal ions are actually of very diverse chemical properties. Those that commonly cause serious pollution include antimony, arsenic, cadmium, chromium, cobalt, copper, iron, lead, mercury, nickel, silver, thallium, tin, vanadium and zinc. Recently, a new scheme has been provided to categorize "heavy metals". They are recognized as three classes of metals based on the electronegativity, charge and ionic radius of metal ions and equilibrium constants that describe the formation of metal ion/ligand complexes. These three categories are:

(1) Class A metals show preferences for ligands containing oxygen,
(2) Class B metals show preferences for ligands with nitrogen or sulfur, and
(3) Metals with characters between Classes A and B, which also reflect their Lewis acid properties.

Although some metal ions are essential to life (e.g. copper, iron, zinc and chromium), excessive intake of these metal ions is toxic to living organisms. The toxicity of metal ions to living organisms is well documented. Metal ions can be transported along the food chain and finally enter the human body. Toxic actions of metal ions to other living organisms also cause perturbation of ecological processes in the biogeochemical cycles. When the metal ions pass through the environment which has no natural assimilative capacity for them, the metal ion concentration in the environment will be elevated to a level that poses risk to ecosystem and human health.

Metal ions are transported to air, water and soil from various sources through diverse routes. These include natural sources released from geochemical materials and human activities such as mining, agricultural activities, industrial processes and miscellaneous activities in urban life. As a result, these metal ions are transported to the environment as mine wastes, pesticides, fertilizers, preservatives, sewage sludge and various forms of wastes produced by both metalliferous and non-metalliferous industries.

17.1.2. Metal pollution in Hong Kong

In Hong Kong, metal ions that are released into the environment come mainly from the effluent of the metal finishing industries such as electroplating industry (Fig. 17.1). The chemical composition of electroplating effluent has been determined. Table 17.1 shows the metal ion content and pH of plating wastewater discharged from sixteen electroplating factories in Hong Kong. The effluent is highly acidic or alkaline and contains toxic substances such as cyanide and metal ions. As these chemicals possess a certain degree of toxicity to human health and aquatic life, the electroplating effluent should be pretreated to minimize the impact on the receiving water bodies. Moreover, the production of metal-containing wastes is expected to increase in the future. Copper-laden waste-water production is the highest among metals now and in the future.

In Hong Kong, metal ion-containing effluent is either transported to the conventional wastewater treatment plant without pretreatment or discharged directly into the environment. The degree of pollution can be

Fig. 17.1. A plating tank in an electroplating factory in Hong Kong. Depending on the work-piece to be plated, the plating bath contains Cu^{2+}, Ni^{2+} or Zn^{2+}.

Table 17.1. Content of metal ions and cyanide, and pH of electroplating effluent discharged from 16 electroplating factories in Hong Kong (Environmental Management Division, 1986).

Parameters	Range
Chromium (Cr^{6+})	1–40 mg/L
Copper (Cu^{2+})	1–30 mg/L
Nickel (Ni^{2+})	3–365 mg/L
Zinc (Zn^{2+})	4–250 mg/L
Aluminum (Al^{3+})	10–230 mg/L
Silver (Ag^{2+})	2–3 mg/L
Cyanide (CN^-)	1–6 mg/L
pH	1.7–8.2

revealed by the elevated metal contents in the marine sediment and in shellfish that are found in the region. There is no specific process for the removal of metal ions in the wastewater treatment in Hong Kong, but large amounts of metal ions accumulate in the "activated sludge" — the major agent in biological wastewater treatment. The elevated level of metals is harmful to the microorganisms in the "activated sludge". Thus low concentrations of metal ion in untreated wastewater can significantly reduce the efficiency of biological wastewater treatment by inhibiting microorganisms in "activated sludge" process and anaerobic digestion of sludge. Cu^{2+} concentrations in excess of 1.9 mg/L inhibited anaerobic digestion in wastewater treatment plants. The threshold copper level in wastewater, defined as the concentration at which the treatment efficiency is reduced by 5%, was 1.9 mg/L. The rank of toxicity of metal ions to the "activated sludge" process and anaerobic digestion of sludge is: $Ni^{2+} > Cu^{2+} > Cr^{6+} > Zn^{2+}$.

17.1.3. *Toxicity of metal ions in the environment*

(1) Copper
Copper is a transition metal with three common oxidation states: Cu^0 (copper metal), Cu^{+1} (cuprous ion) and Cu^{2+} (cupric ion). Cu^{2+} is the most commonly occurring species that readily forms free hydrated ion in water. It prevails in the environment and is the most toxic form to living organisms among the three forms. Cu^{2+} starts to precipitate at above pH 5.5. However, in the presence of organic or inorganic ligands, Cu^{2+} forms complexes with the ligands and remains free in solution. It can be strongly complexed by electron donor groups (O-, N- and S-containing groups) in organic compounds.

Copper is an essential trace element, but it is toxic to various sensitive members of the food chains (e.g. plants, algae and human) at moderate levels and the toxicity of Cu^{2+} to microorganisms is well documented. Cu^{2+} is generally one of the most toxic metal ions to heterotrophic bacteria in the aquatic environments. The order of sensitivity to metal ions of heterotrophic microorganisms is: $Ag^+ \geq Cu^{2+} = Ni^{2+} > Ba^{2+} = Cr^{6+} = Hg^{2+} > Zn^{2+} = Na^+ = Cd^{2+}$. The toxic actions of Cu^{2+} on a living system are essentially exerted on enzymes, especially enzymes whose activities depend on sulfhydryl and amino groups, as metal ions have high affinity for ligands containing nitrogen and sulfur donors. The binding of metal ions to enzyme molecules, thus masking the catalytic sites, altering the conformations of the biomolecules, or competing with and displacing the enzyme-activating metal ions may inhibit enzymatic activities. Besides, nucleic acids and metabolites may also be targets of metal toxic actions. Cupric salts increased the frequency of non-complementary nucleotides in the synthesized DNA-double helix.

(2) Nickel

Nickel is highly toxic because of its role as an antagonist of essential metals. Nickel has two oxidation states: Ni^0 (nickel metal) and Ni^{2+} (nickel ion). Toxic effects of Ni^{2+} will exert on various kinds of organisms along the food chain. The toxicity of Ni^{2+} on microorganisms such as bacteria, algae and fungi is well documented. Besides, the toxic effects of Ni^{2+} on plants and animals have also been reported. In addition, Ni^{2+} was found to induce chromosomal aberration in mammalian cells and to suppress the immunological responses of animal cells.

(3) Zinc

Zinc, which has a molecular weight of 65.38 g/mol and a density of 7.14 g/cm^3, is a Group IIB transition metal. Zinc has only two oxidation states: Zn^0 (zinc metal) and Zn^{2+} (zinc ion). Being an amphoteric metal, Zn^{2+} exhibits a "U" shaped solubility curve with the theoretically minimum solubility at pH 10.5 and with increasing solubility at both ends. The toxicity of Zn^{2+} is considered to be mild comparing to other metal ions. This is due to the importance of Zn^{2+} to organisms in the nature where a greater need usually translates to a greater tolerance. In the terrestrial food chains, Zn^{2+} is more toxic to the plants than to the animals eating them because crop yields tends to drop before Zn^{2+} content become hazardous to the next trophic level. In aquatic food chains, the level of Zn^{2+} tends to decrease rather than increase in higher trophic levels. For example, it is shown that fish tends to die or cease reproduction before tissue accumulation of Zn^{2+} reach to a level to be toxic to humans. A generalized statement can be made: Zn^{2+} does not accumulate along the terrestrial or the aquatic food chains. It is because that certain important members of the food chains are particularly susceptible to the toxicity of

Zn^{2+} and upward transport of Zn^{2+} halts as those susceptible members deceased.

The adverse effects of Zn^{2+} toxicity on microorganisms are well documented. Growth inhibition, impaired adhesion, morphological changes and altered biochemical activities have been reported in several microorganisms.

17.1.4. *Methods of removal of metal ions from industrial effluent*

Conventional methods used for the removal of metal ions from industrial effluent can be divided into two categories. They are physico-chemical methods and biological methods.

17.1.4.1. *Physico-chemical methods*

The classical physico-chemical methods are chemical precipitation, chemical oxidation and reduction, ion exchange, electrochemical reduction, reverse osmosis, freeze crystallization, electrodialysis, cementation and starch xanthate adsorption. These methods, owing to the high operational cost or the difficulty to treat solid wastes subsequently generated, are generally expensive and ineffective especially when the metal ions are dissolved in large volumes of solutions at relatively low concentrations (around 1–100 mg/L).

Reverse osmosis (RO) is based on the application of high pressure to overcome osmotic pressure, forcing metal ion-laden wastewater to pass through the RO membrane that enclosed in vessels. Metal ion-laden waste-water will be separated by selective permeation into purified, permeate stream and concentrated stream of metal solution. The concentrated metal ion solution can be recycled into plating bath while water with sufficient purity can be reused for rinsing.

Precipitation is the most common and widely used method in wastewater treatment. Undesired metal ions in the effluent can be precipitated as insoluble metal hydroxides by adding calcium hydroxide or sodium hydroxide. They can also be precipitated as the more insoluble metal sulfides by adding sodium sulfide, sodium hydrosulfide or ferrous sulfide. The insoluble precipitates formed are then allowed to settle in the sedimentation tank by addition of coagulating agents. Finally, the sludge can be removed and the clarified effluent can be discharged to the sewerage system.

Ion exchange is a stoichiometric and reversible chemical reaction in which a metal ion (e.g. Ni^{2+}) from solution is exchanged for a similarly charged ion (e.g. H^+) attached to an immobile solid particle. The immobilization matrix can be naturally existing inorganic zeolites, synthetic resin materials or activated carbon, which are contained in a series of columns in the ion exchange system. Through the regeneration of the columns by adding an acidic solution, metal ions in the eluent of the cation exchange column may be recovered by evaporation.

17.1.4.2. Biological methods

(1) Microbial resistance to metal ions

Since the beginning of industrialization 200 years ago, there have been huge changes in the distribution of metal ions on the surface of the earth due to anthropogenic activities. Bacteria adapted to these changes and have evolved strategies to maintain low intracellular toxic metal ion concentration. Firstly, the adaptation may have innate abilities to organisms living in extreme environment. Secondly, resistance may have been easily acquired from plasmids. These metal ions resistance strategies can be acquired by a number of mechanisms:

(a) The development of active transport systems that maintain non-toxic level of toxicants within the cell. Examples of active transport systems are the *cad A* determinant from *Staphalococcus aureus* plasmid and the secondary role of arsenate detoxification of *pst* (phosphate) transport system found in various bacteria.

(b) Enzymatic mediated oxidation or reduction that converts toxicants to less toxic forms. Examples are the oxidation of As^{6+} to As^{5+} and the reduction of Hg^{2+} to Hg^0 (or Cr^{6+} to Cr^{3+}).

(c) Biosynthesis of intracellular or extracellular biological compounds that serve as traps for the removal of metal ion from solution. Siderophores refer collectively to a group of low molecular-weight iron-binding compounds secreted by various organisms. Such extra-cellular biological compounds have the ability to sequester iron and other metal ions in solution. Intracellular sulfhydryl-containing protein such as metallothionen has also been identified as intracellular trap for cadmium and copper.

(d) The binding of metal ions onto bacterial cell surfaces.

(e) The precipitation of insoluble metal salts on the bacterial cell surfaces through the actions of membrane associated sulfate reductase or through the biosynthesis of oxidizing agents like oxygen and hydrogen peroxide.

(f) Biomethylation and transport through cell membranes by diffusion-controlled processes.

Usually, metal resistant strains of the microorganisms are being used for the removal of metal ions since microorganism-based processes are more economic. For life forms, bacteria have the greatest tolerance to unusually high saline conditions and can cope with saturated solutions of salt. Moreover, they have a greater capacity to adsorb (Mechanism d) and precipitate (Mechanism e) metal ions from solution since adsorption depends mostly on diffusion and the cell surface area to volume ratio. The structural and functional simplicity and short generation time of bacteria facilitate the elucidation metal ions removal mechanism and manipulation of the genetic structure for the enhancement of metal ion accumulation ability. Bacterial cells of *Zoogloea ramigera* removed Cd^{2+} up to 40% of the microbial biomass dry weight when it was grown in medium containing about 0.345 g/L of Cd^{2+}. For instance, *Klebsiella aerogens* are capable of producing polysaccharide capsules that are known to be more tolerant to 10 mg/L Cu^{2+} than the non-capsulated strain. The details of metal ion removal by microorganisms will be discussed in the following sections.

(2) Cell wall of gram-negative bacteria

Cu^{2+}-resistant bacteria cells of *Pseudomonas putida* 5-X that can adsorb a relatively large amount of Cu^{2+} was isolated from metal contaminated sludge (Fig. 17.2). It is a Gram-negative bacterium, the cell envelope of which consists of two membrane bilayers (the outer and plasma membranes) that are chemically distinct from each other and that sandwich a thin petidoglycan (PG) layer between them. The arrangement and exposure of the various functional groups (e.g. carboxylates, phosphorylates, hydroxylates and other negatively charges sites) in cell wall produce a net electronegative charge, which allow the cell wall to interact with metal cations within their immediate surroundings.

The outer membrane of *Escherichia coli* has been studied most extensively (Fig. 17.3). It consists of an asymmetric arrangement of lipid (the outer surface of the bilayer contains mostly lipopolysaccharide (LPS), whereas the inner leaflet is mostly phosphatidylethanolamine) combined

Fig. 17.2. Transmission electron microscopic (TEM) image of cells of *Pseudomonas putida* 5-X stained with Cu^{2+}. The cells were suspended in 50 mg/L of Cu^{2+} solution for 1 hour before proceeding TEM staining. The bar is 1 μm. (Lu and Wong, unpublished results)

Fig. 17.3. Diagrammatic location of the outer membrane of a Gram-negative bacterium, *Escherichia coli*.

together with at least four major outer membrane proteins (OmpF, OmpC, OmpA and lipoprotein).

All are intrinsic proteins, except for OmpA protein which is peripherally embedded in the outer leaflet. The lipoprotein forms a chemical union between the outer membrane and the underlying PG and effectively cements the two wall layers together. The OmpF and OmpC intrinsic proteins act as porin proteins that closely associated with lipoprotein and LPS and form small, hydrophilic channels that span the bilayer. Because of the preponderance of electronegative amino acids at the channel mouth, small metal ions would be encouraged to pass. Thus, it functions as a sieve of metal ions. LPS is anionic and consists of phosphate groups. Phosphatidylethanolamine has a single polar head group and its constituent phosphate group is anionic. The same holds true with phosphatidylglycerol and cardiolipin which are found in smaller amounts within the outer membrane.

The outer membrane orients its lipid molecules in a way that the hydrophilic polar head groups align themselves along each of the bilayer's faces, while their hydrophobic acyl groups point toward the interior of the bilayer. Consequently, each hydrophilic face has reactive chemical groups available to interact with metal ions. The prime candidates for this interaction are the phosphate groups that are resident within the polar head groups of the LPS and phospholipid. Mg^{2+} and Ca^{2+} form an integral component part of the outer membrane and are required for the correct packing order of the lipid constituents. The lipid components of phospholipid or LPS constantly exhibit rotational and lateral movements, and the acyl chains of LPS that inhabit the outer leaflet are more rigid and closely packed than their phospholipid counterparts. Consequently, the packing order (and fluidity) of LPS is more dependent on the polar head groups, which requires stabilization by divalent metal ions such as Mg^{2+} and Ca^{2+}. Moreover, LPS possesses more phosphoryl groups than phospholipid, these metal ions are especially important for its retention in the outer membrane.

The PG layer is most probably a monolayer in *E. coli*. It interacts more strongly with metal ions than the outer membrane, presumably because of available carboxylate groups. This chemical reactivity of cell walls is important since they can no longer be thought of as inert, stagnant structures. They are able to interact actively with the environment. It is possible that their reactions with the counter ions of the external milieu could send a perceptible signal to the protoplast to make it aware of its surroundings. Thus, the cell wall can be considered a shroud which acts as a chemical buffer at the periphery of the cell to either collect essential ions (e.g. Mg^{2+} and Ca^{2+}) or to immobilize toxic substances (e.g. Cu^{2+} and Cd^{2+}, etc.).

(3) Metal ions and ligands

Metal ions are relatively unattracted to hard (non-polarizable) ligands compared to the attraction experienced by the metal ions, and do not form ionic bonds readily. Binding to hard ligands tends to be covalent where the metal ion and the counter ion each contribute an electron to the bond. The metal ions tend to bind to the "softer" (polarizable) ligands, such as sulfur and nitrogen. For example, Cu^{2+}, Ni^{2+} and Co^{2+} bind to ligands in the order $RS^- > RNH^- > OH^- > H_2O$. Moreover, there is variation of binding affinity within the metal ion group. For example, Cu^{2+} is attracted to the nitrogens of histidine, while Zn^{2+} has an affinity for the sulfhydryl groups of cysteine. The order of softness of metal ions is: $Cu^{2+} > Ni^{2+} > Zn^{2+} > Co^{2+}$. Biomass contains many of these softer binding sites, and therefore has greater binding affinity for metal ions. This in turn means that biomass can selectively accumulate metal ions, with a reduced loss of these ions than that of traditional ion exchangers. Ca^{2+} and Mn^{2+}, for instance, compete effectively with metal ions for electrostatic binding sites on ion exchangers.

(4) Kinetics of metal ion uptake

Metal ion taken up by microorganisms frequently exhibits two-stage kinetics (Fig. 17.4). The first stage is a very rapid, passive, reversible and metabolism-independent reaction called physical adsorption or ion exchange at the cell surface. During this rapid stage, microorganisms can adsorb metal ions onto their surfaces by many processes. This kind of adsorption is called "biosorption". The second step is slower, active and metabolism-dependent, called chemisorption with intracellular metal ion accumulation. Dead cells accumulate metal ions to the same or greater extent than living cells.

Fig. 17.4. Kinetics of metal ion removal by microbial biomass. RC is the metal ion removal capacity of the microbial cells. From 0–60 minutes, a rapid metal ion uptake phase, mainly by adsorption by microbial biomass is observed. Then from 60–120 min, slow metal ion uptake phase by intracellular accumulation is proceeded.

Biosorption is largely attributed to the presence of metal ion-coordinating ligands such as phosphoryl, sulhydryl, carboxyl, hydroxyl and amino groups on the cell surface and in membrane proteins. In addition, biosorption involved various processes such as adsorption, coordination, complexation, chelation, ion exchange, precipitation or crystallization of metal ions. These processes can occur either at the microbial cell surface or near the cells, and certain insoluble metal ions may become physically entrapped in extracellular polymers or in cavities within the cell wall. Moreover, analytic observations show those surface metal ion concentrations far exceed the expected stoichiometry of metal ion bound per reactive chemical site within the cell wall. Electron microscopy clearly establishes that metal ion precipitates within and on the cell walls are common (Fig. 17.2), and a two-step mechanism to account for this observation was invoked. Firstly, metal ions interact in a stoichiometric manner with accessible reactive cell wall groups. Then, these sites nucleate the growth of a metal ion precipitate.

Metabolism-dependent intracellular metal ion accumulation generally occurs via monovalent or divalent cation transport system and in response to a transmembrane electrochemical potential generated by plasma membrane-bound H^+-ATPases. Metabolism-dependent intracellular metal ion accumulation may result in higher levels of metal ion uptake than those attributable to adsorption processes alone. It is also sensitive to inhibition by low temperature, metabolic inhibitors, and the absence of an energy source and low light intensities in photosynthetic organisms. Moreover, this process is irreversible in many cases and requires destructive treatments for metal ion recovery. The intracellular uptake of toxic metal ions is primarily by pathways existing for the uptake of metabolically necessary metal ions.

(5) Adsorption isotherm
The adsorption of metal ions on microorganisms is a reversible phenomenon and can be represented by two adsorption isotherms — Langmuir isotherm and Freundlich isotherm.

(a) Langmuir isotherm
The Langmuir isotherm, which is a widely accepted monolayer adsorption model, is often used to describe the adsorption phenomenon of various biosorbents. This model is based on the following assumptions: (1) all the sites on the adsorbent surface are identical; (2) the energy of adsorption is constant (homogeneous); (3) there is no migration of adsorbate molecules in the adsorbent surface; (4) there is no interaction between the adsorbed molecules; (5) adsorption is only limited to a monolayer. The Langmuir model can be described as:

$$q_e = \frac{bC_e q_{max}}{(1+bC_e)} \tag{17.1}$$

in which q_e is the amount of solute adsorbed per unit dry weight of adsorbent at concentration C_e (mg metal ion/g biosorbent or mmol metal ion/g biosorbent); C_e represents the equilibrium concentration of metal ion in bulk aqueous phase after adsorption (mg/L or mmol/L); q_{max} is the theoretical maximum uptake of solute per unit dry weight of adsorbent (mg metal ion/g biosorbent or mmol metal ion/g biosorbent); b is the adsorption affinity constant related to energy of adsorption (L/mg or L/mmol). Two convenient linear forms of the Langmuir equation are:

$$\frac{C_e}{q_e} = \frac{1}{bq_{max}} + \frac{C_e}{q_{max}} \tag{17.2}$$

or

$$\frac{1}{q_e} = \frac{1}{bq_{max}} + \frac{1}{bC_e q_{max}}.$$ (17.3)

A plot of C_e/q_e against C_e (or $1/q_e$ against $1/C_e$) gives a straight line with a slope of $1/q_{max}$ and an intercept of $1/bq_{max}$ (or with slope of $1/bq_{max}$ and intercept of $1/q_{max}$). Hence the adsorption parameters, q_{max} and b, can be evaluated and used to compare the adsorption behavior quantitatively in different adsorbate-adsorbent systems.

It is worth noting that when the amount of adsorption is small such that $bC_e \ll 1$, the Langmuir equation would give a linear adsorption relationship as

$$q_e = q_{max} \, bC_e.$$ (17.4)

On the other hand, when the amount of adsorption is large such that $bC_e \gg 1$, the Langmuir equation would become

$$q_e = q_{max}.$$ (17.5)

(b) Freundlich isotherm
The Freundlich isotherm is another widely accepted model for describing the monolayer adsorption. The empirical Freundlich equation assumes that the surfaces of adsorbents are heterogeneous and therefore the adsorbents are expected to have heterogeneous energies for adsorbing adsorbates. This equation can be expressed as

$$q_e = kC_e^{1/n}$$ (17.6)

in which q_e is the amount of solute adsorbed per unit dry weight of adsorbent at concentration C_e (mg metal ion/g biosorbent or mmol metal ion/g biosorbent); C_e represents the equilibrium concentration of metal ion in bulk aqueous phase after adsorption (mg/L or mmol/L); k and n are empirical constants and indicative of adsorption capacity and intensity, respectively. To simplify the derivation of k and n, the Freundlich equation is often linearized by taking the natural logarithm on both sides of the equation as

$$\ln q_e = \ln k + 1/n \ln C_e. \qquad (17.7)$$

A plot of $\ln q_e$ against C_e would yield a straight line with a slope of $1/n$ and an intercept equal to the value of $\ln k$ for $C_e = 1$ ($\ln C_e = 0$). The magnitude of n is an indicator of system suitability, with values of $n > 1$ (or $1 < 1/n < 0$) representing favorable adsorption conditions, while the intercept $\ln k$ is related to the capacity of the adsorbent for the adsorbate.

The Freundlich equation generally agrees quite well with the Langmuir equation and experimental data over moderate ranges of concentration, C_e. However, unlike Langmuir equation, it does not become linear particularly at low concentrations but remains convex to the concentration axis.

17.1.4.3. *Methods of recovery of metal ion from metal-loaded biosorbent*

Metal ions loaded on biosorbent may be recovered by a destructive method in which the biosorbent may either be destroyed to release the adsorbed metal ions, or a non-destructive method in which the biosorbent can be regenerated for future use by treating mildly with chemicals to desorb the metal ions. The choice of methods is principally determined by the relative costs of the two recovery methods. Moreover, the metal ion uptake mechanism of the biosorbent also limits the choices. For an economically efficient biosorption operation, it is usually mandatory that the biosorbent can be conserved and regenerated. A pre-requisite for biosorbent regeneration is the non-destructive removal of metal ion from loaded biosorbent.

Many studies try to recover loaded metal ions from biosorbents. The choice of eluents for metal ion recovery and regeneration of biosorbent should be determined mainly by (1) the metal ion recovery efficiency of the eluents, (2) the cost of the eluents, and (3) the effect of the eluents on metal ion removal capacity of the biosorbent regenerated. Recovery of the loaded metal ion could be achieved by the use of an appropriate elution solution that can effectively strip the loaded metal ions from the biomass. It is desirable that the elution is complete or the irreversibly loaded metal ion could be kept to a minimum. In addition, it is desirable that the least possible damage occurs to the adsorption properties of the biomass so as to allow the reuse of the biomass in subsequent adsorption-desorption cycles.

Various kinds of eluents are used for the desorption of adsorbed metal ions from cell surfaces. Examples include eluting Cs^{2+}-laden fungal pellets with 1 M HNO_3 or 1 M $Na_2CO_3/NaHCO_3$ solutions, Au^{3+}-laden algal biomass with mercaptoethanol in 0.05 M sodium acetate (pH = 6), U-laden *Penicillium* biomass with 0.1 M EDTA (pH = 6 and 4) and sodium carbonate/sodium bicarbonate (15/5 and 30/20 g/L), Cu^{2+}-laden fungi (*Ganoderma lucidum*) with EDTA and 0.1N HCl solutions, Au^{3+}-laden immobilized cells of *Persimmon tannin* with 1 M thiourea (pH = 3) and Ni^{2+}-laden immobilized bacterial cells *Enterobacter* sp. 4-2 with 0.5 M citrate buffer (pH = 4).

17.1.4.4. *Immobilization of metal-adsorbing microbial cells*

For more rigorous industrial applications, the use of immobilized microbial cells is preferable. Immobilized microbial cells have many advantages over freely suspended cells. These include (1) better capability of repeatable reuse of the biomass, (2) easy separation of cells from the reaction mixture, (3) high biomass loading, (4) minimal clogging in continuous flow systems, (5) particle size can be controlled and high flow rates achieved with or without recirculation, (6) prevent cell wash and allow a high cell density to be maintained in a continuous reactor at any flow rate, (7) improved catalytic stability of immobilized cells, (8) provide resistance to shear for shear-sensitive cells, (9) microorganisms more inert to microbial contamination and degradation, and (10) operational stability is generally high. Moreover, immobilized microbial cells usually had good mechanical stability. As a result, reactor with immobilized microbial cells can reach a high metal loading capacity and achieve a high efficiency for metal removal.

In order to evaluate the suitability of a biosorbent (e.g. immobilized cells) for metal ion removal, it is better to select the matrix that can fulfill many of the following criteria: (1) the active immobilized biosorbent should be produced at a low cost and should be reusable; (2) particles size, shape and mechanical properties should be suitable for use in a continuous-flow system in completely mixed, packed or fluidized-bed-reactor configurations; (3) uptake and release of the metal ion should be efficient and rapid; (4) separation of the biosorbent from solution should be cheap, efficient and rapid; (5) the sorbent should be metal ion-selective; and (6) regeneration of metal ions from the sorbent should preferably be metal ion-selective and economically feasible, and the sorbent should be in a physical state that can be used. Since the use of magnetite-immobilized microbial cells have many advantages over the traditional technologies such as freely suspended cells, and possess the above criteria as a biosorbent on metal ions removal, it is an attractive method for sewage treatment.

Many studies have shown that using microorganisms (especially bacteria) that adsorbed onto magnetites (Fe_3O_4) can be applied for removal and recovery of useful substances from wastewater. Cells of *Rhodopseudomonas sphaeroides* were immobilized on magnetites to remove pesticides such as lindane, 2,4-dichchlorophenoxyacetic acid (2,4-D) and 2,4,5-trichlorophenoxyacetic acid (2,4,5-T) from wastewater. The removal of chlorinated hydrocarbons such as *p,p'*-DDT, aldin, heptachlor, dieldrin, α-hexachlorocyclohexane and α-hexachloro-cyclohexane by immobilized bacterial cells have been reported. Cells of a Mn^{2+}-oxidizing bacterial cells immobilized onto magnetites removed Mn^{2+} from water. Although there are numerous reports on the removal of metal ions by microbial biomass, mainly by biosorption, the application of biosorption using microbial biomass to remove metal ion from aqueous environment has been hindered by the difficulty of recovering metal ion laden microbial cells due to the minute size. Centrifugation and filtration are impractical due to the high cost of the operation. Immobilization of microbial cells onto appropriate matrix is the most adopted method to solve the above-mentioned separation problem. Matrices such as alginate, carrageen and even acrylamide have been used to immobilize microbial cells for adsorption of metal ions. Their drawbacks are that these immobilized cells have to be used in a packed bed reactor and the hydraulic problem reduces the applicability of the reactor to treat large volume and high flow rate solution containing metal ions. In order to solve the problem, cells of *Enterobacter* sp. 4-2 were immobilized onto magnetites and used in a fluidized bed reactor to remove Ni^{2+}. There was no hydraulic problem

Table 17.2. The removal capacity of Cu^{2+} in concentrated metal ion solution by magnetites, free and magnetite-immobilized bacterial cells of *Pseudomonas putida* 5-x (So 1992).

Adsorbent	Removal capacity (mg/g)[a]
Free bacterial cells	46.43 ± 0.72
Magnetite-immobilized cells	48.70 ± 0.25
Magnetites	0.45 ± 0.09

[a]The experimental conditions: initial concentration of Cu^{2+} before precipitation = 30 mg/L, pH for precipitation = 7.5, equilibrium time for precipitation = 40 min, acid dissolution of metal sludge = 5 mM H_2SO_4, biosorbent concentration = 50 mg/50 mL, pH for adsorption = 5.5, retention time for adsorption = 30 min and temperature = 23 ± 2°C. Data represent means ± standard deviations from triplicates.

due to the design of the reactor and the removal efficiency of the reactor was very high (Table 17.2) and suggested this technology can be used to remove metal ions from aqueous medium at a fast flow rate.

17.2. Removal of Organic Pollutants from Industrial Wastewater

Among all the wastes produced, industrial wastes demand the largest attention because they have been long regarded as being of the greatest diversity and toxicity, and are mostly recalcitrant. For many countries where the textile industry is prosperous, the industry presents a major water pollution problem since this industry not only exhausts a large quantity of clean water for processing, but also contributes a significant amount of wastewater. Figure 17.5 shows the discolored area near the wastewater outlet of a textile factory in Hong Kong. Synthetic dyes are toxic to aquatic organisms. The increasing potential hazard of synthetic dyes in textile wastewater has made them a major pollution problem that draws serious concerns.

In 1992, approximately 10,000 different synthetic dyes are used industrially, and over 700,000 tonnes of these dyes are produced annually worldwide. The loss of dyes in the effluent during the dyeing process

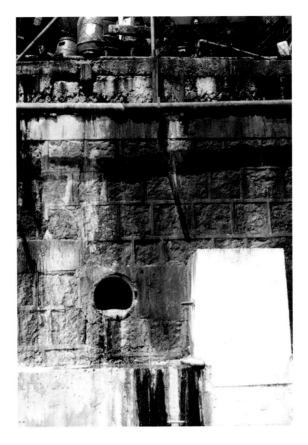

Fig. 17.5. A discolored area next to a textile factory in Hong Kong. The area was heavily polluted with wastewater discharged from the factory.

was about 10 to 15%. In other words, 84,000 tonnes or more of dyes were discharged into our environment in 1992. These hundred thousand tonnes of dye invade our environment via the air, water and soil media as synthetic dyes are recalcitrant to biodegradation and persistent in the environment. For example, Solvent Red 1, an azo dye, and Solvent Yellow 33, an quinoline dye, were transformed with half-lives of a few days and months, respectively under anaerobic condition which is favorable for dye reduction.

Obviously, the first impact of such great dye contamination is the coloring of our aquatic and terrestrial environments which are aesthetically unacceptable. However, the main concerns regarding these pollutants are the influence and toxicity of the dye-containing materials to the ecosystem and, even more importantly, to the living organisms. As one of the properties of dyes is their high color value, the treated wastewater containing non-biodegraded dyes causes the receiving water-courses to be highly colored, which would normally observable at about 1 mg/L level which is unacceptable on aesthetic grounds. The color of dye not only affects the aquatic environment aesthetically but also the growth and physiological activities of aquatic flora and fauna by reducing light penetration into the water. It has been reported that some commonly used synthetic dyes and their metabolites or derivatives were tested to be mutagenic using *Salmonella*-mammalian microsome mutagenicity test. The large number of synthetic dyes and their high potential for toxicity have made this class of compounds one of the most extensively reviewed and regulated under the Toxic Substances Control Act (TSCA) new chemical programs. The pollution problems arising from dye wastes therefore deserve more efforts to handle and deal with these wastes being discharged before into the aquatic environment.

17.2.1. *Azo dyes*

Among the many classes of synthetic dyes used in various industries, azo dyes are the largest and the most versatile, and play a prominent role in almost every type of application. Azo dyes are a group of compounds characterized by the presence of one or more azo groups (-N=N-) in association with one or more aromatic systems. Azo dyes are cost effective because of their ease and versatility of synthesis, and relatively higher tinctorial strength compare to other synthetic dyes. The versatile applications of azo dyes have initiated many studies on their adverse effects on human and the environment as azo dye-containing effluents are discharged without suitable treatment. The increase in the occurrence of intestinal cancer in highly industrialized society has been related to the increasing use of azo dyes. Some European countries have banned the application of azo dyes in the late 1990s.

17.2.2. *Treatment of industrial effluent containing azo dyes*

(1) Physico-chemical methods

The contamination of the environment with azo dyes during the 20th century has overwhelmed the natural cleaning capacities of microorganisms. Scientists had put in much effort in finding effective methods to solve the problem. More effective though less sensitive methods were developed and tested for dye removal. The feasibility of several methods was evaluated, including chlorination, ozonation, adsorption, flocculation, coagulation and Fenton's oxidation.

(2) Biological methods

(a) Activated sludge process

In the past, municipal wastewater treatment systems were mainly used for the purification of textile mill wastewater. These systems, however, depended mainly on biological activity and were mostly found to be inefficient in the removal of the more resistant, recalcitrant synthetic dyes. In the synthetic dye manufacturing industry, synthetic dyes produced have to be more and more resistant to sunlight, washing and microbial action in order to be successful in the commercial market. It is not surprising that most studies on the biodegradation of synthetic dyes produce negative result when synthetic dyes are designed to resist this type of treatment. In a study of the fate of 18 azo dyes in the "activated sludge" processes, 11 dyes were found to pass through the "activated sludge" process substantially untreated, 4 were significantly absorbed on the surplus "activated sludge" while only 3 were apparently degraded.

Some azo dyes do show a tendency to be adsorbed onto "activated sludge", but acid and reactive dyes were only slightly adsorbed. However, these studies were done on "activated sludge" plants that were not regularly treating dyeing effluent. It seems likely that in plants that continuously or regularly treat dye, the adsorption sites on microbial cells will be rapidly occupied if the dye is strongly adsorbed. Renewal of adsorption sites will be by creation of new sludge flocs, desorption of the dye or biodegradation of the dye. With normal sludge ages, the creation of new flocs is too slow to remove dyes adequately, and desorption leads to resolubilization of color. Biodegradation of dyes under the aerobic conditions in an "activated sludge" mode wastewater treatment plant is slow. Long adaptation periods are required and the enzymes produced appear to be highly specific. The net result of increased dye loads and poor dye removal on conventional wastewater treatment is an increased level of color in the receiving watercourse.

(b) Biodegradation by isolated microorganisms

Rat gut bacterial suspensions (with starved *Proteus* sp. suspension or its cell-free extract) decolorized azo dyes such as *m*-Methyl Orange, Neoprontosil, Tartrazine, Yellow 2G, Amaranth, Proceau SX, Fast Yellow, Naphthalene Fast Orange 2G and Sunset Yellow FCF. Thin-layer chromatographic study indicated that these azo dyes had been reduced

Fig. 17.6. Structures of (a) flavin adenine dinucleotide (FAD); and (b) nicotinamide adenine dinucleotide (NAD(P)). The NADP is the marked hydroxyl group (OH), is esterified with phosphate (PO_4^{3-}).

to amines. The enzyme of *Proteus* sp. that reduced Tartrazine was a flavoprotein requiring reduced forms FAD (flavin adenine dinucleotide, Fig. 17.6(a)) and NAD(P) (nicotinamide adenine dinucleotide phosphate, Fig. 17.6(b)) for reduction to take place.

A study was conducted using the established NADH-generating system incorporating in the standard assay procedure to study the azo-reductase activity of cell-free extracts of *Streptococcus faecalis* under optimal conditions. It was found that flavins (e.g. FAD) added to the preparations were initially reduced enzymatically by the crude enzymes in the presence of NADH-generating system. They rapidly and non-enzymatically reduced azo dye, Red 2G, into the primary amines. The reduced flavins were then oxidized in the process. Without $FADH_2$, NADH reduced Red 2G very slowly if at all. This azo reduction process would more likely be a double two-electron transfer via the hydrazo intermediate (R-NH-HN-R′) under anaerobic condition:

$$FADH_2 + R\text{-}N{=}N\text{-}R' \rightarrow FAD + R\text{-}NH{=}HN\text{-}R'$$
$$FADH_2 + R\text{-}NH\text{-}HN\text{-}R' \rightarrow FAD + R\text{-}NH_2 + H_2N\text{-}R'.$$

Since there was no enzyme-substrate complex formed between the bacterial azo reductase and Red 2G, it showed that FAD was the true substrate of the bacterial azo reductase, being reduced by flavoproteins, and then reduced the dyes non-enzymatically. This indirect mechanism of azo reduction would explain why both bacterial and mammalian hepatic preparations could reduce a wide range of azo dyes. They also found that under aerobic conditions, the azo dye acts as another electron shuttle for reduced flavin.

(i) Bacteria

Studies that concern the fate of azo dyes in the "activated sludge" process showed that most azo dyes were mainly and significantly removed by adsorption on the "activated sludge" flocs and only a very few appeared to be biodegraded. As early as 1937, decolorization of azo dyes in spoiled dairy products was attributed to predominant lactic acid bacteria, which turned the azo dyes into amines. Later, more studies reported that reduction of a large variety of azo dyes occurred extensively among intestinal microflora. Appreciating the versatile abilities of microorganisms in degrading and mineralizing nearly all organic matters in the nature, more studies were encouraged to determine the feasibility of using azo dye-degrading microorganisms to treat dye wastes. *Bacillus subtilis*, which was selected for having faster and stronger decolorization rate of *p*-aminoazobenzene (PAAB), an azo dye hardly degraded by activated sludge and even decreases the activity of the activated sludge, converted PAAB into aniline and *p*-phenylenediamine. *Pseudomonas cepacia* 13NA, isolated from soil of draining trenches at a dyeing factory, also degraded PAAB. The bacterium metabolized the dye better under poor nutrient conditions. *P. cepacia* 13NA also decolorized a number of synthetic dyes and their decolorization abilities were in the order Acid Orange 20 (AO20) > *p*-aminoazobenzene (PAAB) > Acid Red 88 (AR88) > Acid Orange 7 (AO7) > Acid Orange 12 (AO12) > Direct Yellow 4 (DY4) > Direct Red 28 (DR28). It appeared that when hydroxyl group was in the *ortho*-position, such as that in AO7, compared with AO20 that has the hydroxyl group at the *para*-position, decolorization was facilitated. Also, AR88 that has a naphthalene ring was decolorized faster than AO7 that has a benzene ring. There was no appreciable correlation between decolorization ability of synthetic dyes and their octane-water partition coefficient. This may indicate that the decolorization ability of synthetic dyes is not related to their permeability through bacterial cell wall, but to their molecular weight of synthetic dyes.

Growing cells, intact non-growing cells, cell-free extract and purified enzymes of *Pseudomonas* sp. S-42, which was isolated from "activated sludge", could decolorize azo dyes Diamira Brilliant Orange RR, Direct Brown M and Eriochrome Brown R under the conditions of pH 7.0 and temperature of 37°C. Oxygen inhibited the decolorizing activities of cell-free extract and purified enzymes. Studying the structure-activity relationship was difficult for intact cells due to the structure-dependent permeability of azo dyes through cell membranes. Cell-free extract of *Pseudomonas stutzeri* degraded several azo dyes and the results of the study indicated the redox potential and hydrophobicity of dyes were important to understanding the relationship between the ease of degradation of dyes and their structures. Intact cells and cell-free extract of *P. stutzeri* IAM 12097, *P. cepacia* 13NA and *Bacillus subtilis* IFO 13719 decolorized several sulfonated and non-sulfonated azo dyes. These bacteria showed very different decolorization abilities towards the dyes due to different specificity of their azo reductases, different permeabilities of their cell membranes and different structures of the dyes. Azo dye reduction is microorganism-specific and dye-specific.

Pseudomonas sp. K24 and *Pseudomonas* sp. KF46 could specifically utilize azo dyes Orange I and Orange II respectively through adaptation in chemostat culture. The azo reductases of the two bacterial strains, Orange I azo reductase and Orange II azo reductase, were found to evolve from a common precursor protein of the bacterium from which the above two bacterial strains derived. Wong and Yuen studied degradation (decolorization) of an azo dye, Methyl Red (MR), by *Klebsiella pneumoniae* RS-13 isolated from activated sludge (Fig. 17.7). The initial degradation of Methyl Red by *K. pneumoniae* RS-13 was a reductive cleavage of the azo bond and produced an aromatic acid (2-amino-benzoic acid = ABA) and a comparative toxic aromatic amine (N,N'-dimethyl-*p*-diphenylene-diamine = DMPD) (Fig. 17.8). DMPD was proven to be even more toxic than the parent dye. However with longer incubation in medium containing glucose, *K. pneumoniae* RS-13 could degrade DMPD into non-toxic intermediate, and the intermediate, like ABA, would be used by the cells for other biomolecule synthesis.

Fig. 17.7. Decolorization of Methyl Red by *Klebsiella pneumoniae* RS-13 under aerobic condition.

Methyl Red (MR)

2-Amino-benzoic acid (ABA)

N,N'-Dimethyl-p-phenylene-diamine (DMPD)

Fig. 17.8. Initial degradation of Methyl Red by *Klebsiella pneumoniae* RS-13 under aerobic condition.

(ii) Fungi

The feasibility of using fungal system to degrade synthetic compounds has been extensively studied. White-rot basidiomycetes, especially *Phanerochaete chrysosporium*, received most of the attention due to their versatile and rather non-specific aerobic lignin-degrading ability. The lignin-depending ability is initiated during secondary metabolic (idiophasic) growth, and could degrade synthetic organic compounds with structures similar to that of lignin. Lignin is an optically inactive, random phenylpropanoid polymer, which is relatively recalcitrant to biodegradation in the nature.

Many *Streptomyces* spp. (ligninocellulolytic) and *P. chrysosporium* BKM-F-1767 (ATCC 24725) (ligninolytic) degraded commercial azo dyes Acid Yellow 9, sulfanilic acid, synthesized azo dyes 1 and 2. The latter two were synthesized by conjugating guaiacol molecule (2-methoxyphenol) onto Acid Yellow 9 and sulfanilic acid via azo-linkage. Five of six *Streptomyces* spp., which were capable of assimilating vanillic acid — a compound having the same ring substitution pattern (4-hydroxy-3-methoxy) as guaiacol, could significantly decolorize the two synthesized azo dyes. These *Streptomyces* spp. could not decolorize sulfanilic acid and Acid Yellow 9. Moreover, *P. chrysosporium* could almost completely decolorized synthesized azo dyes 1 and 2 whereas it only could decolorize sulfanilic acid and Acid Yellow 9 to a limited extent. These results indicated that the degradation of sulfanilic acid and Acid Yellow 9 by *Streptomyces* spp. were rendered feasible by the bacteria which start the assimilation by attacking the guaiacol substituent first, while the linkage of guaiacol molecule to these two compounds increased their susceptibility to the degradation by *P. chrysosporium*. The newly synthesized compounds are rendered more easily biodegradable by selective linkage with readily degradable substituents into the compounds.

By comparing the degrading ability of *P. chrysosporium* BKM-1667 (ATCC24725) with that of *Streptomyces chromofuscus* A11 (ATCC55184) towards sulfanilic acid, sulfonated and non-sulfonated azo dyes, these dyes were coupled with guaiacol molecules, aromatic ring or naphthol group. *P. chrysosporium* BKM-1667 demonstrated a greater ability than *S. chromofuscus* A11 in mineralizing azo dyes. *P. chrysosporium* BKM-1667 could mineralize (completely degrade) all the sulfonated azo dyes, and the substitution pattern did not significantly influence the susceptibility of the synthetic dyes to degradation. In contrast, *S. chromofuscus* A11 was unable to mineralize aromatics with sulfo groups except those coupled with guaiacol molecule.

It seems that the lignin-like guaiacol structure attached to sulfonated azo dye was a must to initiate degradation of *S. chromofuscus* A11 and suggested that this bacterial enzymatic systems responsible for degradation may be somehow more selective to lignin-like structures than that of *P. chrysosporium* BKM-1667. A white-rot fungus, *Geotrichum candidum* CU-1, was isolated and found to be able to degrade a number of water soluble azo dyes including Procion Red MX-5B and Methyl Red under aerobic condition (Fig. 17.9). The mechanism of degradation of these azo dyes seems to be an initial oxidative cleavage of the aromatic rings of the dyes followed by further degradation of the intermediates into very "small and simple" compound(s) that could not be resolved by high performance liquid chromatography (HPLC) and gas chromatography-mass spectrometry (GC-MS) analyses. In addition, the "small and simple compounds" had no detectable toxicity towards Microtox® test (a bacterial enzymatic fluorescence test developed by Microbics Corporation).

(iii) Algae

Chlorella pyrenoidosa, C. vulgaris and *Oscillateria tenuis* degraded 31 azo dyes at 20 mg/L over 96-hour incubation. *C. vulgaris* decolorized 29 out of 31 azo dyes with the extent varying from 5 to 100%. The extent of decolorization of dyes seemed to be related to their molecular structures. Dyes with an amino or a hydroxyl group could be readily decolorized.

Fig. 17.9. Degradation of Procion Red MX-5B by *Geotrichum candidum* CU-1 under aerobic condition. (a) Complete medium with *G. candidum* CU-1; (b) complete medium with Procion Red MX-5B (50 mg/L); and (c) complete medium with Procion Red MX-5B (50 mg/L) and *G. candidum* CU-1 (part of the fungal mycelia were removed after full growth of the fungus to show the decolorization of the azo dye).

These groups also could counteract the inhibitory effect of sulfonyl group on reduction. Substitution with methyl, methyoxyl, nitro or sulfo group made dyes hard to decolorized. All three algae were found to be able to effectively degrade aniline products into simple inorganic materials. The absence of inorganic carbon and nitrogen in the medium facilitated the algae to utilize Erichrome Blue SE and Black T, azo dyes, as sole sources of carbon and nitrogen. The addition of NADH or NADPH increased the algal azo dye reduction. These observations demonstrated that NADH or NADPH could act as electron donors for the algal azo reductases. Moreover, algae that had been cultured with 10 mg/L of dye displayed higher reduction activity of that dye. This result suggested that azo reductase is inducible.

It is believed that algae serve as an oxygen source for the aerobic action of bacteria, and have a direct effect on the degradation of azo dyes.

(c) Biosorption

Almost all biological materials can be used as a biosorbent to adsorb different materials: peat, wood-shaving, biomass, starch, moss, bagasse, red mud, rice husk, biogas residue, orange peel, activated sludge, wood charcoal, coconut husk, and banana pith. It is because nearly all cell structures are basically made up of protein, fatty acid and mono/ polysaccharide. Protein constructed from a set of 20 different amino acids. Each amino acid contains at least one amino (NH_2) and one acidic group (COOH). Fatty acid composes of an acid group (COOH) and a saturated or unsaturated branched or unbranched hydrocarbon chain. Monosaccharide consists of aldo (CHO) or keto (CO) group and hydroxyl group (OH) on the carbon skeleton. Thus, biological compounds contain many potential charged binding sites for different adsorbates. Hence, the success of biosorption is dependent on the isolation and selection of the appropriate biological material.

As mentioned previously, synthetic dyes are poorly degraded in biological waste treatment systems because of their recalcitrance. The plentiful discharge of dye-containing industrial effluents into the waste-water treatment systems has initiated many studies on the fate of these dyes and the treatment efficiency of activated sludge process on these substances.

It has been reported that thirty-one water soluble and water slightly soluble azo dyes at 50 mg/L were degraded by 1,000 mg/L of sludge in 100 mL culture medium under shaking condition. The bio-degradation was limited to azo reduction and the decolorization was readily with hydroxyl or amino group, not readily with methyl group and hardly with hydrogen, chloro, methoxyl and nitro group. In addition, it was found that carboxyl group itself, similar to sulfonyl group, decreased reduction because of its hydrophobicity. The substitution of electron-withdrawing groups (SO_3H and SO_2NH_2) did not increase the reduction.

The low BOD: COD ratio (<0.1) of the majority of textile effluents leads to low biodegradability of azo dyes that are partially removed in biological treatment by adsorption onto the activated sludge flocs. This adsorption includes exchange adsorption, van der Waals forces and hydrogen bonds. The adsorption of basic dyes increased with pH and the opposite effect is observed for acidic dyes. Dye adsorption is mainly dependent on type, chemical composition, structure of dyes, structure of adsorbent surface, pH and salinity. Bacteria and fungi are the microorganisms that are always selected as biosorbents because they grow very rapidly and are easy to grow. Moreover, the surface area to volume ratio of these microorganisms is relatively higher than the other living organisms.

(i) Bacteria

Bacteria such as *Aeromonas* sp., *Rhodococcus* sp. and actinomycetes such as *Streptomyces* sp. have been reported as good biosorbents for textile dyes. The adsorption mechanisms of bacteria and actinomycetes are mainly attributed to the microbial cell wall. Many bacterial cells exert adsorption during the stationary phase in which exopolysaccharides are produced. Exopolysaccharides are primarily composed of carbohydrates such as D-glucose, D-galactose and D-mannose which are produced in the stationary phase. Moreover, exopolysaccharides are polyanionic in nature. It may explain why textile dyes, especially reactive dyes (cationic dyes), can be adsorbed by the bacterial cells. There are some exceptions, in which the bacterial cells perform better reactive dye adsorption in the earlier growth phase as in *Aeromonas* sp. These may indicate different adsorption sites and binding mechanisms between different dyes and bacterial strains. Lai *et al.* (1997) selected a bacterial strain, *Pseudomonas* sp. K-1, which was isolated from a sludge sample generated by a textile factory. The bacterium adsorbed a large amount of an azo dye, Procion Red MX-5B (Fig. 17.10) and no decolorization of the dye was observed even after incubating the dye with the bacterium on agar plate for about a week (Fig. 17.11). This result suggests that the bacterium only adsorbed the azo dye but had no degradation ability for this azo dye.

Fig. 17.10. Structure of a sulfonated water-soluble monazo dye, Procion Red MX-5B. The sulfonate groups are marked.

Fig. 17.11. Colonies of *Pseudomonas* sp. K-1 stained by Procion Red MX-5B on a minimal medium (MM) agar plate.

(ii) Fungi

Fungus such as *Myrothecium verrucaria* was isolated from soil and reported as an excellent dye-binding microorganism. Fungal biomass was as much as 100-fold more effective than "activated sludge" in the removal of dyes. The main binding materials on fungal cell wall are chitin and chitosan. Chitin is one of the most abundant organic materials in the nature. It occurs in fungal cell wall as the principal fibrillar polymer and in animal, particularly in mollusks, crustaceans and insects, as the constituent of the exoskeleton. The structure poly[β-(1-4)-2-acetamido-2-deoxy-D-glucopyranose] of chitin (Fig. 17.12(a)) is similar to cellulose (Fig. 17.13), except that the C2-hydroxyl group of cellulose is replaced by an acetamido group. In addition, chitin structure is also similar to murein in that it is the main constituent of bacterial cell wall. The principal derivative of chitin is chitosan (Fig. 17.12(b)). Chitosan is produced by alkaline deacetylation of chitin. It also occurs naturally in some fungi but the occurrence is much less widespread than that of chitin. However, in some fungal species, such as *Mucor rouxii*, the composition of chitosan is more than that of chitin.

Chitin polymer present in fungal cell wall is polycatioinc in nature due to the amino (NH) and carbonyl (CO) groups substituted to the

Fig. 17.12. Structure of (a) chitin and (b) chitosan.

Fig. 17.13. Structure of cellulose.

hydroxyl (OH^-) groups. Because of the cationic nature of chitin polymer rather than anionic nature of cellulose or exopolysaccharides in bacterial cell wall, their adsorption mechanisms are different. Using a reactive dye as an example, the adsorption sites are the electrophilic reactive group and the nucelophilic sulfate (SO_3^-) group. Under alkaline conditions, the dye molecules expose the electrophilic reactive group (positively charged group). Therefore, cellulose that exposes the nucleophilic OH^- group (negatively charged group) can adsorb the dye molecules and form covalent bond, while the chitin polymer (positively charged) will repel the dye molecules. On the contrary, under acidic conditions, since the dye molecules cannot expose the electrophilic reactive group, adsorption on cellulose surface decreases, while the chitin polymer (the amount of protonated amine (NH_3^+) groups increases under acidic condition) has more sites for sulfate groups (SO_3^-) binding. The corresponding reaction is illustrated below:

Chitin-NH_3^+ Chitin-NH_3^+

   or   + Dye-SO_3^- $\longrightarrow$ or $\longleftrightarrow$ $^-O_3$S-Dye + X$^-$

Cellulose-OH$^-$ Cellulose-OH$^-$

Therefore, chitin polymer adsorbs dye molecules more effectively under acidic condition. Chitosan is more effective in dye adsorption than chitin. Because of the high dye adsorption capacity of chitin/chitosan, they are extensively studied as a dye adsorbent. When azo dyes had high sulfonic acid substitution, poor or no adsorption of the dye by the microbial cell or cell by products would occur. This also limits the chance of aerobic biodegradation.

References

1. Lai KM. *Integration of Adsorption and Biodegradation of Azo Dyes*. M. Phil. thesis, The Chinese University of Hong Kong, Hong Kong, 1997, p. 272.
2. So CM. *Removal of Copper Ion (Cu^{2+}) from Industrial Effluent by Immobilized Microbial Cells*. M. Phil. thesis, The Chinese University of Hong Kong, Hong Kong, 1991, p. 104.
3. Yuen PY. *Microbial Degradation of Methyl Red and Its Cleavage Products*. M. Phil. thesis, The Chinese University of Hong Kong, Hong Kong, 1993, p. 230.

Questions for Thought

1. What is the definition of a heavy metal?
2. List all toxic metal ions.
3. What are the major source(s) of metal ions in aquatic environment?

4. What is the priority of treatment of toxic metal ions?
5. What are the problems associated with physical and chemical methods in removing metal ions from industrial wastewater?
6. Describe the unique features of the kinetics of biosorption of metal ion (e.g. Cu^{2+}).
7. Describe the similarities and differences between Langmuir and Fruendlich adsorption isotherms.
8. Describe the methods of immobilization of bacterial/microbial cells for metal ion removal by biosorption.
9. What are the common steps of the pathway for anaerobic degradation of azo dyes.
10. Discuss the unique features of bacteria and fungi that degrade azo dyes.
11. Define biosorption.
12. Compare the metal ion adsorption ability of chitin with that of chitosan.

Chapter 18

Municipal and Industrial Solid Waste Treatment

P. K. Wong and Lee Man Chu
Department of Biology, The Chinese University of Hong Kong
Shatin, NT, Hong Kong SAR, China

18.1. Composting

Traditionally, plant residues and animal excreta are piled up to decompose and then returned to the soil for crop production. This is composting which is the accelerated microbial decomposition of organic materials of biological origin under controlled conditions of ventilation, moisture and temperature by organisms in the wastes to form a stabilized organic product. Composting is a mixed substrate, mixed culture operation (Table 18.1). Although the process can be anaerobic, it is preferentially or predominantly aerobic in most operations, involving the oxidative breakdown of part of the carbon in the wastes. Intermediate products of carbon oxidation, secondary metabolites, microbial biomass and recalcitrant compounds form humus-like substances as the product. As aerobic decomposition is exothermic, heat is released which raises the temperature of the composting mass. Consequently, conventional composting is thermophilic and during the process, there is typical temperature change that divides composting into five stages (Fig. 18.1). There is a short latent stage, during which soil microorganisms colonize the composting mass. Microbial breakdown of the readily accessible polysaccharides, proteins and simple C substrates raises the temperature to

25–40°C in the mesophilic stage. As temperature further increases to the thermophilic stage, heat kills most mesophiles, and thermophilic bacteria and fungi take over at about 45–55°C. At a temperature above 55°C, thermophilic actinomycetes dominate, which attack hemicellulose and cellulose. At temperature greater than 60°C, spore-forming microorganisms remain viable. As readily decomposable materials are exhausted, microbial activity decreases, and the mass cools down as heat loss is much greater than heat production. Mesophilic fungi (mainly basidiomycetes) and actinomycetes re-invade and attack hemicellulose, cellulose and lignin in this cooling stage. The composting mass returns to ambient temperature, and humification and nitrification occur in the maturation stage, which may last for several weeks. Such temperature change is more apparent in

Table 18.1. Biological composition (microbial genera only) of a composting heap.

Groups		Organisms	
Decomposers	Bacteria	Fungi	Actinomycetes
Mesophiles	*Bacillus* *Pseudomonas* *Enterobacter* *Clostridium*	*Rhizopus* *Coprinus* *Polyporus* *Pleurotus* *Penicillium* *Aspergillus* *Geotrichum* *Trichoderma* *Cephalosporium*	*Nocardia*
Thermophiles	*Bacillus* *Clostridium* *Lactobacillus*	*Mucor* *Thermoascus* *Chaetomium* *Penicillium* *Aspergillus* *Sporotrichum* *Humicola*	*Stretomyces* *Thermoactinomyces* *Thermomonospora*
Consumers	Protozoa	Microinvertebrates Nematodes Mites	Macroinvertebrates Millipedes Centipedes Ants Earthworms

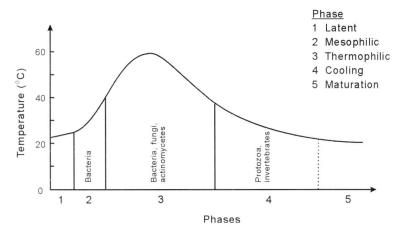

Fig. 18.1. Temperature change and biological succession in compost piles (modified from Polprasert 1996).

windrow composting (in open rows), but high temperatures (60–70°C) may not be reached in mechanical composting due to continuous agitation and thus excessive heat loss. A succession of microbial population parallels the temperature and biochemical changes, and several metabolic stages that represent different microbial populations can be observed. A dynamic community selection is acted on by a continuing alteration of environmental conditions such as temperature and substrate availability. Although composting is largely a microbial process, mesofauna such as detritivorous invertebrates (e.g. nematodes, millipedes and earthworms) and predatory invertebrates (e.g. ants, centipedes and beetles) invade the composting mass (Table 18.2), but only in the maturation stage when temperature is favorable. Color and particle size also change as a result of humification and abrasion by soil fauna.

Factors affecting decomposition will inevitably influence the rate of composting, and the physical and chemical nature of the substrate to be composted are key factors in determining the rate of the process. Particle size, which affects the surface area for microbial enzyme attack, determines the biodegradation rate. However, if the particles are too small, the structural strength and interstitial space for air circulation will be affected. Chemically, substrate assimilability as reflected by the C:N ratio is also vital in assessing the suitability of the waste as composting

Table 18.2. Organisms involved and their populations in a composting heap (Gray *et al.* 1971).

Groups	Organisms	Number/g compost
Microflora	Bacteria	10^8–10^9
	Actinomycetes	10^5–10^8
	Fungi	10^4–10^6
	Algae	$<10^4$
Macroflora	Fungi	–
Microfauna	Protozoa	10^4–10^5
Mesofauna	Detritivorous, herbivorous and predatory invertebrates	–

substrate; a C:N ratio of 20–35 is optimal for composting. For environmental conditions during composting, the optimal moisture content is 45–70%. If moisture content is too high, this will reduce aeration as water and air share the same interstices. Oxygen is essential for modern aerobic composting to avoid objectionable odors, allow compost to reach high temperature and achieve rapid decomposition. Additives can be used to improve the properties of the material to be composted. Organic or inorganic amendment can be added to adjust the C:N ratio of the wastes, sludge for high C:N wastes and sawdust for low C:N wastes. Bulking agent such as sawdust, woodchips, bark and straw can be mixed with wet and weak materials to overcome poor structural strength, lack of airspace and excessive moisture. Oxygen is usually provided by air in the interstices, which is provided by either periodic turning in windrow composting or forced aeration in mechanical composting, but too much aeration may result in excessive heat loss, which is not beneficial in terms of pathogen destruction, whereas inadequate aeration will lead to anaerobiosis and malodor.

Oxygen availability is the key factor in system design and operation (Table 18.3), and is provided by agitation or aeration. In windrow composting, agitation by turning introduce air to the interior of composting mass, expose unattached material to promote uniform decomposition and expose pathogens to heat. Periodic agitation is usually performed, either manually

Table 18.3. Different types of composting classified by the mode of aeration (modified from Haug, 1980).

Types		Modes of agitation and oxygenation
Non-reactor composting (open or roofed)		
Pile or heap	Agitated solid bed	Windrow Oxygen provided by natural diffusion and periodic turning
	Static solid bed	Aerated static pile No agitation or turning; oxygen provided by aeration system at bottom of pile
Reactor composting		
Vertical flow reactor	Moving agitated bed	Waste agitated during movement down multi-story digester; oxygen provided by forced aeration
	Moving packed bed	Waste moves downward in digester with no compartmentation; oxygen provided by forced aeration
Horizontal flow reactor	Tumbling solid bed	Rotating drum Tumbling provides dispersion; oxygen provided by forced aeration
	Agitated solid bed	Bin reactor Waste agitated; oxygen provided by forced aeration
No flow reactor	Aerated non-agitated solid bed	Container or box composting Waste usually not agitated; oxygen provided by forced aeration

or by composting machines (Fig. 18.2), the frequency of which should be adjusted to maintain good ventilation but avoid excessive heat and water loss at the surface. Turning can be more frequent in the first few days when decomposition is rapid (once or twice per day) but reduces to once every 3–7 days, depending on the rate of oxygen consumption. In aerated static pile, there is no mixing and ventilation is achieved by aeration through perforated pipes at the bottom of the pile (Fig. 18.3). Aeration can be in the form of continuous blowing (positive pressure aeration) or suction (vacuum-induced or negative pressure aeration), which differ in their pattern of heat distribution (Fig. 18.4). To obtain a more uniform heating and decomposition, alternate blowing and suction can be utilized. Composting in an open system whether roofed or not is inevitably

Fig. 18.2. Composting machine is commonly used to turn the composting piles on open ground or in roofed facility.

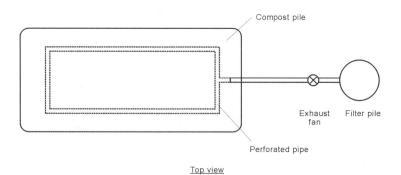

Fig. 18.3. Static solid bed composting with aeration by suction or blowing through perforated pipe at the bottom of the pile.

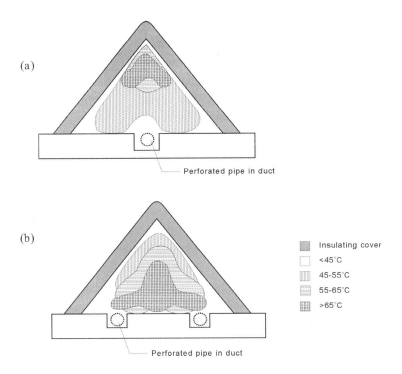

(a)

Perforated pipe in duct

(b)

Insulating cover
<45°C
45-55°C
55-65°C
>65°C

Perforated pipe in duct

Fig. 18.4. Temperature distribution in an aerated static pile with (a) blowing aeration; and (b) suction aeration (redrawn from Stentiford *et al.* 1985).

Fig. 18.5. Rotating drum is one of the most popular reactors for municipal refuse composting.

affected by ambient conditions especially temperature. To be independent of the changing environment, composting can be done in a closed system in specially constructed reactors that are designed to provide adequate aeration, temperature and moisture control. Cylindrical or rectangular reactors, either in the form of silo towers without interior floors (BAV and Triga system) or multi-story digesters with 8–10 floors (e.g. Earp-Thomas and Thermax processes) are forced aerated to provide oxygen as the waste moves down the reactor. Reactors can also be in the form of a horizontally rotating drum with or without compartments as in the Dano process (Fig. 18.5). For bin reactors which are usually roofed, forced aerated and mechanically agitated, waste being composted is moved by radial arms which have a series of rotating perforated augers which force air into the material with some mixing as in the Fairfield-Hardy process. In the Metro-Waste system, a rail-mounted agiloader passes through a long bin once per day and agitates the bin content. Agitation provides the oxygen required, which is supplemented with forced aeration through perforated floors in the bin. Recently, composting in large boxes or containers (biocontainers) has been introduced. The waste is not agitated and air is blown in for aeration. Some containers have built-in exhaust air treatment system to treat malodours produced. Container composting saves space as containers can be stacked.

Composted nightsoil, sludge and even municipal refuse are usually utilized as a slow released fertilizer or soil conditioner. For composting to be satisfactory, the product (compost) should be stabilized and cured so that its utilization poses no harms to crops and no risks to public health. The compost should be low in pathogenic microorganisms and this is entirely dependent on the temperature reached and the duration of the thermophilic stage, which is effective in pathogen destruction. This is important if composting is considered as a viable waste treatment option. Of course, the quality of compost (e.g. concentration of toxic elements) will be greatly hinged to the agronomic value of the mature product, which is ultimately dependent on the chemical composition of the waste before composting.

18.2. Bioremediation of Contaminated Surface and Subsurface Sites

Bioremediation is a collective term to express the application of biological agents (either whole cells or their components) to remove or treat toxic inorganic or organic compounds in aquatic or terrestrial environments. One of the most well studied bioremediation is the final step in the clean up of spilled oil by microbial cells with nutrient enrichment. Although there is a stringent international regulation to control the disposal of oil and petroleum into sea, accidental oil spills and leaks from tankers carrying crude oil or refined petroleum products are unavoidable. Among the bioremediation of spilled oil by microbes, the most striking example is the clean up of oil spill of Exxon Valdez at the coastal areas of Alaska in 1989. About 42,000 m^3 of oil was released from the super oil tanker and spread into the Prince William Sound off the coast of Alaska and produced a slick to cover over 100 square miles of sea. After the accident, a large portion of spilled oil was first removed by physical means such as suction, and then the remaining oil droplets were emulsified with oil dispersant(s) in order to increase the surface area of the oil droplets that facilitate the degradation chemically or biologically in water. The spilled oil was oxidized by marine bacteria; however, the sticky, recalcitrant components of oil were oxidized slowly and the residual oil spread to a very large area along the coastline. When oil spread onto the coastal areas, biological remediation became the major force to remove the spilled oil since physical and chemical treatments seemed to be impractical to remove the oil from contaminated rocks or sand on the coastal areas. Although spraying with seawater and even with chemical foam was used to wash away the oil slick from the coastal rocks and sand, most oil slick remained in the solid matrices. Enrichment by addition of nutrients was used to increase the biological activity of endogenous organisms, mainly microorganisms, to degrade the oil slick in rocks and sand in coastal area.

Other examples of cleaning up oil leakage by bioremediation were reported in 1980s. About 570,000 m^3 of oil was accidentally released from oil well into the Gulf of Mexico in 1980 and substantial amount of oil leaked from oil well into the Persian Gulf during the Iraq-Iran war in the mid-1980s. Disasters of these accidents seriously contaminated the aquatic environment of the nearby water bodies and posed extremely dangerous threat to the living organisms in these areas. Thousands of sea birds and marine mammals were killed, fishery and crustacean industries in these gulfs were seriously damaged and other ecosystems were strongly disturbed. In these occasions, microbial remediation was once again recruited to solve the pollution problem caused by oil slick. Due to the slow action of microbial degradation of oil slick, the leaked oil have to be retained by physical mean: floating booms were used to contain the oil slick and sawdust was used to help the oil sink and to give micro-organisms a good surface from which oil slick can be degraded. In addition, dispersant(s) was also used to clear the coastal areas and to help disperse floating oil droplets. However, since some dispersants are toxic to oil-degrading microorganisms, the use of dispersant was under stringent monitoring and control. Even with an excellent management of oil spill clean up strategy, the harmful effects of spilled oil can exert on the contaminated areas for several years, especially for the temperate and cold regions where microbial activity that degrades residual oil slick is low.

In order to improve the efficiency of microbial remediation of spilled oil, several aspects were investigated. As mentioned above, the addition of nitrogen and phosphate appeared to greatly enhance the oil degradation by endogenous microorganisms. On the other hand, selection of appropriate microbial strain(s) or genetic manipulation of microbial strains to have more efficient oil degradation ability have provided a significant improvement in oil degradation by microorganisms. For instance, strains of *Pseudomonas* spp. have been isolated from natural environment or genetically engineered to degrade hydrocarbon components in oil slick. A set of genes that reside on plasmid, an extra-chromosomal element, was identified in these bacteria to produce required enzymes for the degradation of hydrocarbons in oil. The trait can be easily transferred to other microorganisms since this set of genes is located in a plasmid. Genetic manipulation of this set of genes on the plasmids, enable the bacterium to degrade hydrocarbons in oil faster and more completely. Further development on the construction of a bacterial strain containing plasmids that encode genes for rapid degradation of all components in oil slick has been a main topic in this area of environmental biotechnology.

Other than the cleaning up of spilled oil, the use of microorganisms to reduce the level of toxic organic compounds such as PAHs (polycyclic aromatic hydrocarbons), PCBs (polychlorinated biphenyls) and even pesticides in the underground of abandoned agricultural land or military bases attract a great effort. There are many impressive studies on the use of natural or genetically engineered microorganisms to remove (or treat) unwanted toxic materials from the contaminated natural environment. The use of microorganisms, either as pure culture or as mixed populations, to remediate the incidents of industrial or military pollution becomes the potential key technology to deal with environmental clean up.

18.3. Microbes in Mining

During mining activity, there is a large amount of metal ions remaining in the ore. The dumping of untreated ore in the mining site will result in the leaching of metal ions into the groundwater or nearby aquatic environment, causing serious pollution problems. However, the complete extraction of remaining metal ions from the ore is impractical due to low cost effectiveness. In order to minimize the metal ion contamination problem and to recover valuable metal ions from the ore, microbial leaching has become a very attractive operation in the mining industry. Bacteria such as sulfate oxidizers can produce acidic solution for the leaching of metal ions from the low grade ore. The solubilized metal ions can be concentrated and recovered as useful metal salt or solution for commercial applications.

Most of the valuable metal ion in low grade ore are in water insoluble sulfide forms. In the presence of sulfate oxidizer, sulfide can be oxidized into sulfate and the metal sulfate becomes water soluble. Sulfate oxidizer such as *Thiobacillus ferrooxidans* can obtain energy by oxidizing either ferrous or sulfide, and produce ferric sulfate according to the following equations:

$$FeS_2 + 3.5\ O_2 + H_2 \longrightarrow FeSO_4 + H_2SO_4 \qquad (18.1)$$
$$\text{(pyrite)} \qquad\qquad\qquad \text{(ferrous sulfate)}$$

$$2\ FeSO_4 + 0.5\ O_2 + H_2SO_4 \longrightarrow Fe_2(SO_4)_3 + H_2O \qquad (18.2)$$
$$\text{(ferric sulfate)}$$

Then ferric sulfate will be available to solublize Cu^{2+} and Fe^{3+} from the CuFeS (Eq. 18.3):

$$CuFeS + 2\ Fe_2(SO_4)_3 \longrightarrow CuSO_4 + 5\ FeSO_4 + 2\ S^0 \qquad (18.3)$$
$$\text{(chalcopyrite)}$$

Ferric sulfate is a strong oxidant and is able to dissolve other metal sulfides. Leaching by ferric sulfate is an indirect reaction because it is independent of the presence of oxygen and microbial action. The rate of extraction of Eq. (18.3) depends on the concentration of ferric sulfate.

On the other hand, *T. ferrooxidans* can attach onto the mineral particles in the ore and enzymes associated with the cell membrane catalyze the oxidative attack on the crystal lattice of the metal sulfide. Oxidation of the metal sulfide such as copper sulfide proceeds as the following equations:

$$CuS + 0.5\ O_2 + 2H^+ \longrightarrow Cu^{2+} + S^0 + H_2O \qquad (18.4)$$

$$S^0 + 1.5\ O_2 + H_2O \longrightarrow H_2SO_4 \qquad (18.5)$$

Among the microorganisms that can oxidize metal ore, most are acidophilic chemolithotrophic bacteria, for instance, *Thiobacillus ferrooxidans*, *Leptospirillum ferroxidans*, and species belonging to *Sulfolobus*. *T. ferrooxidans* is a small rod, Gram-negative bacterium. It can grow in solution with pH as low as 1.5–2.5. The optimal growth temperature is 10–30°C with an upper limit of 37°C. As a chemolitotroph, *T. ferrooxidans* obtains its energy from oxidation of ferrous ion to ferric ion, and utilizes CO_2 to generate its carbon-containing macromolecules.

There are other *Thiobacillus* species such as *T. thiooxidans*, *T. acidophilus* and *T. organoparus*. These acidophilic bacteria oxidize sulfide instead of ferrous ion to obtain their energy. Usually these *Thiobacillus* species are able to co-exist with *T. ferrooxidans* and have the role of facilitating mineral leaching by *T. ferrooxidans*. *L. ferroxidans* can grow at a pH about 1.2 and more acidophilic than *T. ferrooxidans* and grow at a higher temperature (40°C). The bacterium is a highly motile, curved rod and obtains energy by oxidizing ferrous ion to ferric ion, but it is unable to oxidize sulfur compounds. *Sulfolobus*, an archaebacterium, is also involved in the mineral leaching process. The bacterium grows autotrophically at pH 1.0–3.0 and temperature of 50–90°C. It can derive energy from oxidizing either ferrous ion or sulfur compounds.

Copper ore is a smelting residual containing less than 0.5% of copper. The low grade copper ore is piled for dump leaching. Water is sprayed onto the pile and leached water is sprayed repeatedly onto the ore. After a pre-determined period, pyrite (copper sulfide) is solubilized with resultant acidic ferric sulfate solution. Continuous recirculation of the acidic ferric sulfate solution into the copper ore solubilizes the copper ion. Finally, the copper ion-rich solution is pumped into a container and precipitated by adding iron scraps. The precipitation reaction is:

$$Cu^{2+} + Fe^0 \longrightarrow Cu^0 + Fe^{2+} \tag{18.6}$$

The Fe^{2+}-rich solution remaining after copper precipitation is returned to the oxidation pond where *T. ferrooxidans* rapidly oxidizes Fe^{2+} to Fe^{3+} and forms additional H_2SO_4 through the oxidation of sulfur compound. Fe^{3+} is precipitated with NaOH to $Fe(OH)_3$. The supernatant acidic ferric sulfate solution is pumped back to the top of the ore pile.

18.4. Microbial Desulfurization of Coal

Coal is one of the major energy sources for the production of electricity. High concentration of sulfur in coal will lead to the production of SO_2 in coal combustion. As SO_2 reacts with water in mist, corrosive sulfuric acid will be formed and this leads to acid rain. The damages to our environment caused by acid rain have been well documented. Acid rain not only causes problems in the aquatic environment, but also leads to property damage in urban environment. Thus the use of coal faces increasingly stringent environmental control in industrial countries. Low-sulfur coal has been recommended in order to reduce the level of SO_2 emitted. However, the supply of such low-sulfur coal is limited. Most of the sulfur in coal is in the form of heterocyclic organic compounds. Although there are physical and chemical methods to reduce SO_2 emission before, during and after coal combustion, the reduction of sulfur content in coal is an environmental friendly approach. The operation cost of these physical and chemical methods is relatively expensive and hinders their use to reduce sulfur content in coal. Recently, microbial technology has been proven to be a good alternative for coal desulfurization.

Biodesulfurization of coal is a process that uses microorganisms to reduce the sulfur content in coal. The sulfur found in coal is either part of the molecular structure of coal (organically bound sulfur), for instance, DBT (dibenzothiophene) in coal can be used as the sole sulfur source by microorganisms; or in mineral (e.g. pyrite in inorganic form — FeS_2).

One of the simplest reactions is the oxidation of coalmine with sulfur such as CuFeS and FeS_2. Some microorganisms only remove sulfur from organic components of coal (i.e. desulfurization of DBT by *Pseudomonas* spp.), some remove sulfur from inorganic component of coal (i.e. desulfurization of pyrite by *Moraxella* spp.), while remove sulfur from both organic and inorganic components of coal (e.g. *Xanthomonas maltophila*).

For the removal of sulfur from organic components of coal, one of the most typical examples is the cleavage of DBT into inorganic sulfur and hydroxybiphenyl. Some species of *Pseudomonas* attack the sulfur in DBT without altering the carbon skeleton or affecting other valuable components of coal. Biodesulfurization mechanisms, which convert sulfur-containing components into water soluble oxidation products, could cause loss of the hydrocarbons associated with the sulfur atom in coal. These mechanisms should not be used since they will reduce the usage of coal by reducing the energy stored in the hydrocarbon-core. Recently the genes involved in the sulfur-specific DBT desulfruization were identified and the corresponding enzymes have been investigated. From the practical point of view, it has been proven that the microbial desulfurization process for DBT and its analogs is economically feasible.

A group of bacteria called sulfate bacteria can oxidize sulfur to sulfate. The acid solution can be used in mine leaching (the importance of microorganisms in mine leaching has been discussed in previous section). The bacteria can remove sulfur atom in inorganic components of coal such as pyrite. This process can minimize the problem caused by the high level of sulfur in coal and also provides an acidic solution to recover mineral resource from the ore. In addition, the use of microorganisms offers an economical and energy-saving method of desulfurization of coal.

Beside the bacteria mentioned above, many S-oxidizing bacteria oxidize the sulfur-containing compounds in coal. *Thiobacillus ferrooxidans*, *T. Thiooxidans*, *Leptospirillum ferroxidan* and *Sulfulobus* spp. have been reported to involve in the sulfur oxidation. The common drawback of using these native bacteria in biodesulfurization is their slow growth rate (i.e. slow reaction rate). In addition, most of these bacteria are oligotrophic organisms; addition of extra nutrient (including carbon and nitrogen) is required to stimulate the desulfurization rate of these bacteria in laboratory and field studies.

The idea of using acidophilic pyrite-oxidizers to clean up sulfur in coal is not well developed before 1980. The experimental and field trials were conducted in Europe and the US in 1970s–1980s. However, the large-scale application of biodesulfurization did not appear even in the large coal production countries. In early 1990, more studies were initiated on the feasibility of biodesulfurization of coal. The advantages of biodesulfurization are as listed below.

(1) Selectivity

As mentioned above, acidophilic sulfur-oxidizers attack only inorganic sulfur in coal mine, while other bacteria such as *Rhodococcus*, *Bacillus*, *Corynebacterium* and *Arthrobacter* species attack only the sulfur atom of sulfur-containing organic compounds such as dibenzothiopene (DBT) in coal mine. These bacteria do not attack the carbon core of coal. Thus, unlike physical and chemical methods, there is no significant loss of carbon or calorific content of coal.

(2) Metal removal

The production of sulfate or sulfur leads to the production of sulfuric acid. The acidic solution thus produced is able to leach out the remaining high level of metals in the coal mine. Consequently, the resultant biodesulfurized coal is low in both sulfur and metal content.

(3) Ash removal

Owing to the production of acidic solution during biodesulfurization, most metal ions in a coal mine are dissolved and mobilized. This lowers the ash content of biodesulfurized coal.

However, there are some disadvantages such as the slow reaction rate and the production of waste (including biomass and acidic metal/mineral solution) which requires further treatment after biodesulfurization. Heat production during the process is another problem that has to be dealt with. However, with appropriate design and setup, the metal ions in the acidic solution can be recovered and provide a valuable source of metals, while the bacterial biomass can be harvested and recycled back into the process to speed up biodesulfurization. The heat produced can be used directly as energy or to generate another form of useful energy — electricity. With these modifications, biodesulfurization of coal is an excellent choice to minimize pollution caused by the sulfur content in coal with the added advantage of the production of useful by-products — metal and energy during the process.

18.5. Biodegradable Plastics

Materials that are synthesized from petrochemical such as polyethylene, polyvinylchloride and polystyrene are commonly used as plastic wares in our daily life. Owing to the synthetic and recalcitrant properties of petroleum-derived plastics, the disposal of these solid wastes becomes a major problem in waste management in many modern cities. The production of photodegradable plastics becomes meaningless since many cities use landfill as the means for disposal of solid waste. Once photodegradable plastics are disposed of in a landfill and covered with soil, the photodegradable plastics cannot be exposed to sunlight and thus remain intact in landfill sites.

The alternative degradable plastics are biodegradable plastics. The first biodegradable plastic was plastic polymer blended with a biologically degradable polymeric compound such as starch. Once the biodegradable plastic was disposed of, microorganisms in the natural environment can degrade the starch molecules in the copolymer and result in the disintegration of the plastic polymer. However, the physical and chemical properties of the starch-substituted biodegradable plastics are not suitable for practical usage. In addition, a large part of non-degradable plastic polymers remains as residues that can accumulate in our environment and cause pollution problems.

The production of another types of biodegradable plastics that uses PHAs (polyhydroxyalkanoates, Fig. 18.6), which has the physical and chemical properties very similar to those of polypropylene, has been studied extensively. PHAs are microbial storage polyesters. In the presence of excessive carbon sources, many microorganisms form PHAs granules intracellularly. There are many companies engaged in the large-scale production of PHA biodegradable plastics. One of the examples is "Biopol^{TM}", which is a copolymer of P(3HB-Co-HV) [poly(3-hydroxoybutyrate-co-hydroxyvalerate)], produced by Zeneca Bio Products

$n = 1$	R = methyl	Poly(3-hydrobutyrate)
	R = ethyl	Poly(3-hydrovalerate)
$n = 2$	R = hydrogen	Poly(4-hydrobutyrate)
	R = methyl	Poly(4-hydrovalerate)

Fig. 18.6. Structures of PHAs (polyhydroxyalkanoates).

(formerly Imperial Chemicals Industries) in the UK. Later, Monsanto, a leading agricultural chemical manufacturer in the US, applied genetically engineered oilseed rape plant to produce PHA biodegradable plastic polymer. However, the properties of the PHAs produced by the plant have physical and chemical properties different from those produced by microorganisms. After that, another large chemical manufacturer in the US, Union Carbide, also produced microbial biodegradable PHA plastics called "Tone Polymer", while a chemical company in Japan, Showa Denko, produced "Bionolle" which is a chemically synthesized biodegradable PHA plastic. Among these biodegradable plastics, PHB (poly-β-hydroxybutyrate) and its copolymer, P(3HB-Co-HV) [poly(3-hydroxoybutyrate-co-hydroxyvalerate)] are the most widespread and thoroughly characterized PHAs. A number of bacteria, including *Alcaligenes* spp., *Pseudomonas* spp. and a number of filamentous genera such as *Nocardia* spp. produce these PHAs when they encounter unfavorable growing conditions.

In general, a specific microorganism, which occurs widely in nature, is utilized in a novel transgenic fermentation process to convert carbohydrates into the PHA polymers. After the fermentation, the microorganisms have accumulated 80% of their dry weight as PHA polymer. The polymer is purified by breaking open the cells and then harvesting the PHA using an aqueous based extraction method. Owing to their amorphous character, these bioploymers can be ideally handled as lattices. The bioploymer can be extracted with organic solvents and then purified to the usage form. A nonsolvent based process for recovery of PHAs from bacteria was reported. The process is to solubilize the bacterial biomass by sequential treatments of heating, protease digestion and detergent, and leave the bacterial peptidoglycan in cell wall intact for the ease of separation. Then the PHA granules will be purified by filtration and the final product is in a latex form. The estimated cost of production and extraction of PHA by this route will be US$5/kg, which is still higher than that of synthetic plastics.

PHA polymer can be compounded with safe, biodegradable additives to tailor processing and mechanical properties before being converted into plastic articles using standard techniques. Although stable in normal use, the PHAs can be biologically degraded when deposited in an active microbial environment such as compost. Certain microorganisms can metabolize and consume PHA polymers as their nutrients. Biodegradation of the PHAs under aerobic conditions yields carbon dioxide and water in the same amounts originally used during photosynthesis, completing the PHA biocycle. In order to test the biodegradability of the PHA-plastics, a series of test under laboratory and field conditions should be conducted. There are standard methods published by the American Society for Testing and Materials (ASTM) for the degradation of synthetic and biodegradable plastics under laboratory conditions. A series of microorganisms, including bacteria and fungi, has been characterized as being able to degrade PHA-plastics. Most of these microorganisms were isolated from aerobic environment. Over 10 different extracellular PHA deploymerases have been purified and characterized.

The critical factor limiting the use of PHA plastics is their high production cost. Alternatively, one can produce novel PHAs with improved physical and chemical properties that suit the specific needs of certain usages. In Hong Kong, there are several studies on the production of biodegradable PHA-plastics by microorganisms. Studies have been conducted to investigate the use of synthetic media with various amount of organic acid (e.g. butyric acid) as the sole carbon source to produce PHA with improved physical and chemical properties by selected microorganisms such as *Alcaligenes eutrophus* and *Klebsiella pneumoniae*. By varying the concentrations of nutrient supplied, the bacteria produced PHA-plastics with the ratio of HB:HV (hydroxbutyrate: hydroxvalerate) were 97:3 and 79:21. These PHA-plastics have different thermal stabilities.

These results were due to the amount and ratio of various organic acids added to control the polymeric composition of PHAs, that in turn change the physical and chemical properties of the PHA-plastics. Another approach is to produce PHA-plastics with a more economical substrate such as wastewaters from food industry and chemical industry with xenobiotic organics. There are studies in which soya wastes from a soya milk dairy and malt waste from a beer brewery were successfully used as feedstock to produce PHAs from microorganisms in activated sludge. Microorganisms, mainly bacteria, in the activated sludge fed with various amount of wastewater from food and chemical industries produced copolymer (P(3HB-Co-HV)) of PHA-plastics. Various types of PHAs were produced when the C:N ratio of the wastewater varied, with the amount of PHAs increased as the C:N ratio increased. The result was due to the wide variety of bacteria in the activated sludge that produce various types of PHA. Different bacteria have specific PHA synthetases with different substrate specificities and produce different types of PHAs. Sugars such as maltose, sucrose and fructose are present in the industrial wastewater as carbon sources for the bacteria to produce PHAs. The use of activated sludge to convert carbon substrates into PHAs not only can produce biodegradable plastics, but also serves as a wastewater treatment in which the problem of disposal of activated sludge after the wastewater treatment is solved.

Recent advances in understanding the metabolism, molecular biology and genetics of the PHA-producing bacteria and cloning of more than 20 different PHA biosynthesis genes allow the construction of various recombinant strains that are able to produce a much larger amount of biodegradable plastics with different types of PHA copolymers. These advances lead to the production biodegradable plastics with costs that are comparable to or lower than that of synthetic plastics and with greatly improved physico-chemical properties.

18.6. Detoxification of Metal Ions

18.6.1. *Demethylation and reduction of mercury*

Several hundred million tonnes of toxic mercury-containing wastes are discharged into our environment. The ionic forms of mercury such as Hg^+

and Hg^{2+} and organic mercury $(R-Hg^+)$ such as methyl mercury and phenyl mercury are extremely toxic and even mutagenic to living organisms. Many microorganisms are resistant to the toxic ionic forms of mercury. Mercury resistance found in bacteria can be categorized into two classes, narrow-spectrum resistance (resistant to inorganic mercury only) and broad-spectrum resistance (resistance to inorganic and organic mercury). Among these microorganisms, the mercury resistance of Gram-positive strains such as *Staphylococcus* spp. and *Bacillus* spp.; and Gram-negative bacteria such as *Pseudomonas* spp. and *Shigella* spp. are extensively studied. Organic mercury has been reported to be more toxic than its inorganic forms since the penetration of organic mercury into microbial cells is comparatively easier and more rapid than inorganic mercury. However, most of the discharged inorganic mercury can be rapidly and easily transformed biologically into organic forms in the environment. Therefore, the broad spectrum mercury resistant microorganisms are important to detoxify mercury-contaminated sites in our environment.

What is the difference between mercury narrow-spectrum resistance and broad-spectrum resistance in microorganisms (Fig. 18.7)? Many studies on the biochemistry and mechanisms of the mercury resistance of Gram-negative and Gram-positive bacteria reveal that the narrow-spectrum resistant bacterium has only an enzyme called mercury reductase that is

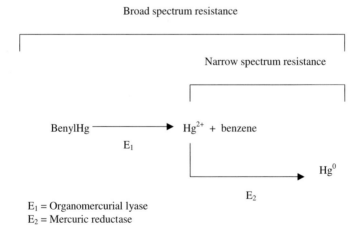

Fig. 18.7. Narrow spectrum and broad spectrum mercury resistant bacteria.

used to catalyze the reduction of Hg^{2+} to Hg^0. Further studies on the mercury resistance genes of *Pseudomonas stutzeri* indicate that a plasmid pPB confers broad-spectrum mercury resistance. The plasmid is self-transmissible. Two pPB regions, separated by 25–30 Kb and sharing homology with the transposon, Tn501 *mer* (Hg detoxification) genes. Two regions were cloned separately and each was shown to carry a cluster of functional and independently regulated *mer* genes. One cluster conferred resistance only to inorganic mercury (with at least *merA* and other mercury transporting genes such as *merT* for narrow-spectrum mercury resistance) and other gene cluster, upstream from *merA* has a novel *merB* gene encoding for organomercurial lyase.

The microorganisms with broad-spectrum mercury resistance have, in addition to the mercury reductase, an organomercurial lyase which remove the organic residue(s) of organic mercury. The function of organomercurial lyase is to remove the methyl group of methyl mercury and to produce free methyl group (CH_3) and inorganic mercury (Hg^{2+}). The inorganic mercury can be further transformed into Hg^0 by the reduction mediated by the mercury reductase. Thus the broad spectrum mercury resistant micro-organisms can grow in the presence of organic and/or inorganic mercury.

In addition, the elementary mercury (Hg^0) is volatile and most of them leave the aquatic environment and enter the atmospheric environment. Although the total mercury content of the aquatic and atmospheric environments remains unchanged, the availability of mercury to organisms in aquatic environment, mainly in ionic form, will be greatly reduced. Thus the reduction of Hg^{2+} to Hg^0 (also called mercury volatilization) is one of the most effective methods to detoxify mercury in aquatic environment.

Although many microorganisms contain mercury reductases and organomercurial lyases, only few mercury reductases, and quite a few organomercurial lyases have been purified from bacteria. Most of the properties of these enzymes are based on the purified enzymes from Gram-negative bacteria (e.g. *Pseudomonas* spp.).

18.6.1.1. *Mercury reductase*

Mercury reductases purified from the cellular membrane of Gram-negative bacteria are either dimeric or trimeric with monomers of 58,700 daltons.

The native enzymes require FAD (flavin) for reduction function and preferentially use NAD(P)H as electron donor. A reducible active-site disulfide (cys-135, cys-14) and a C-terminal pair of cysteines (cys-558, cys-559) were identified. All 4 cysteines are required for efficient Hg^{2+} reduction. These cysteine residues at its active site and the electrons donated by NAD(P)H are transferred via FAD to reduce the disulfur bond (-S-S-) of two adjacent cysteine residues at the active site; and converted them into two cysteine residues (with -SH groups). The electrons captured by the enzyme are then used to reduce Hg^{2+} to Hg^0.

18.6.1.2. *Organomercurial lyase*

The enzyme is inducible by the presence of organic mercury. It has been reported to be located in the cytoplasm of the bacterial cell. The enzyme is 19,000–20,000 daltons (19–20 KD). Its function is the cleavage of the C-Hg bond of organic mercury by protonolysis. For instance, methylmercury is detoxified by demethylation to mercuric ion by bacterial enzyme.

Other than mercury reductase and organomercurial lyase, the enzymes involved in the transportation of Hg^{2+} such as *merT* and *merP* are also important to the mercury resistance of bacteria. *MerT* encodes a transmembrane protein. The molecular biology of *merT* and *merP* proteins of Tn501 isolated from *Escherichia coli* carrying the pPB plasmid was studied. Similar to the biochemistry of *merA* protein, an active site with cys-24 and cys-25 in the first transmembrane region of *merT* protein was essential for transport of Hg^{2+} through the cytoplasmic membrane. A cys-33-ser mutation in *merP* appears to block transport of Hg^{2+} by *merT* protein, while deletion of whole *merP* only slightly reduced Hg^{2+} resistance of the bacterium and suggests that a functional *merT* protein is sufficient for Hg^{2+} transport across the cytoplasmic membrane.

18.6.2. *Reduction of chromium*

Hexavalent chromium ion (Cr^{6+}) is a common and toxic pollutant in soils and waters. Trivalent chromium ion (Cr^{3+}) is less toxic as compared with its oxidized form (i.e. Cr^{6+}). Chemical reduction of Cr^{6+} is feasible and some operations have been practised. However, the addition of reduced compounds such as NAD(P)H that are potential toxicant for living organisms render the large-scale application of chemical method to detoxify Cr^{6+}. Biological detoxification of Cr^{6+} becomes a promising approach to solve the pollution problem of toxic chromium ion. In some cases, chromate (i.e. Cr^{6+}) is used as terminal electron acceptor during anaerobic respiration by microorganisms. Once Cr^{6+} in chromate is reduced to Cr^{3+} Eq. (18.4), the toxicity of the chromium is greatly reduced. In addition, since Cr^{3+} compounds are insoluble and tend to precipitate out from the aqueous phase, the reduced chromium ion (Cr^{2+}) will not be available for uptake by microorganisms as well as other organisms in the contaminated environments.

$$Cr^{6+}_2O_7^{2-} \longrightarrow Cr^{3+}O_4^{2-} \qquad (18.7)$$

There are many cases using bioreactors with bacteria to detoxify Cr^{6+} in aqueous solution. Many bacteria such as *Bacillus* spp., *Pseudomonas* spp. and *Thiobacillus ferrooxidans* are able to reduce Cr^{6+}. Both monoculture of Cr^{6+}-reducing bacteria or bacterial consortium have been used in detoxification of Cr^{6+} in aqueous solution. Bacterial cells immobilized by appropriate matrices (e.g. DuPont Bio-Sep beads) in either packed-bed or in fixed-film (biofilm) reactors efficiently reduced Cr^{6+} in the presence of reducing compounds such as sulfite or thiosulfate. Under selected conditions, up to 200 mg/L of Cr^{6+} can be completely reduced to Cr^{3+} within 24 hours. Further study indicates that a chromate reductase is associated with the Cr^{6+}-reducing bacteria such as *Pseudomonas putida* PRS2000. Both cell suspension and cell free extract can reduce Cr^{6+}. The enzyme is a soluble protein and the crude enzyme activity is heat labile and with a *Km* of 40 μM of chromate and is NAD(P)H dependent. Neither sulfate nor nitrate affects chromate (Cr^{6+}) reduction either *in vitro* or with intact cells. In the reduction of Cr^{6+} by *Pseudomonas ambigua* G-1, an intermediate, Cr^{5+}, was identified by electron spin resonance during the enzymatic reduction of Cr^{6+}; this suggested that the chromate reductase in this bacterium reduced Cr^{6+} to Cr^{3+} with at least two reaction steps via Cr^{5+} as an intermediate.

Further Readings

1. Blazquez ML, Balklester A, Gonzales F and Mier JL. Coal bio-desulfurization: A review. *Biorecovery* **2**: 155–177, 1993.
2. Demain AL and Davies JE. *Manual of Industrial Microbiology and Biotechnology*. American Society for Microbiology, Washington, DC, 1999.
3. Evans GM and Furlong JC. *Environmental Biotechnology: Theory and Application.* John Wiley & Sons, Chichester, 2003.
4. Gray KR, Sherman K and Biddlestone AJ. A review of composting. Part I. *Process Biochem.* **6**: 32–36, 1971.
5. Haug RT. *Compost Engineering: Principles and Practice.* Technomic Publication, Lancaster, 1980.
6. Olguin EJ, Sanchez G and Hernandez E. *Environmental Biotechnology and Cleaner Bioprocesses.* Taylor & Francis, New York, 2000.
7. Polprasert C. *Organic Waste Recycling*, 2nd ed. John Wiley and Sons, Chichester, 1996.
8. Stentiford EI, Taylor PL, Leton TG and Mara DD. Forced aeration composting of domestic refuse and sewage sludge. *Water Pollu. Control* **84**: 23–32, 1985.

Questions for Thought

1. What could be done to control odor in the composting of organic wastes
2. Discuss the unique features of microorganisms that are used in bioleaching.
3. What are the differences between the biosulfurization of inorganic and organic components of coal?
4. Discuss the new developments in the production of economically feasible PHA-plastics.
5. Dicuss the difference between the narrow spectrum and board spectrum mercury resistance of microorganisms.

Annex

Regulatory Issues on Application of Natural and Genetically Engineered Microbes in Environmental Biotechnology

The rapid development of biotechnology in various areas including their applications in environment has attracted numerous legal, safety and public policy issues. The use of "unnatural" microorganisms, especially the gene-tically engineered microorganisms, in solving environmental problems such as waste treatment has lead to the intensive debate on the safety issue. The fate and risk of releasing these "unnatural" microorganisms into the environment are largely unknown. Although many studies preliminarily indicate that the problem caused by these microorganisms can be minimized by introduced a "suicide" (i.e. self-destruct) trait into the microorganisms used in environmental biotechnology. Once these microorganisms accomplished their expected roles, they will be eliminated from the environment by a specific trait specially designed and embedded in their genetic materials. An example for this approach is that the microbial strain released into the environment losses its ability in propagation into the natural environment, the released microorganisms will be eliminated until a new batch of the microorganisms are introduced into the environment. Another approach is to make these "unnatural" microorganisms such that they are removed easily from the environment by physical or chemical means. One of the common examples is the flocculant strain of microorganisms. During the operation, aeration is provided to maintain the suspension of the microbial cells in aqueous solution. At the end, microbial cells will settle to the bottom of the reactor once the aeration is discontinued and the settled microbial cells can be easily removed by physical means.

The safety of using natural and genetically engineered microbial strains in environmental biotechnological application has been debated since 1970. The public fears that the uncontrolled release of these microorganisms into our environment will disturb the balance of ecosystems in the natural environment. The imbalance will in turn cause more problems to our environment than the pollution by wastes/toxicants. There are comparatively comprehensive and stringent regulatory guidelines for the use of recombinant DNA technology and genetically engineered organisms in biotechnological applications; there are quite a few policies on the use of natural microorganisms in environmental biotechnology. In order to prepare a regulatory guideline for the use of "natural" microorganisms in environmental biotechnology, the regulatory guidelines for the release of genetically engineered microorganisms will be reviewed in order to lay the foundation for formulating the guidelines for the "natural" microorganisms.

There are several government agencies in the US that deal with the regulatory issues for the genetically engineered microorganisms. They are the National Institutes of Health (NIH), Environmental Protection Agency (EPA), Food and Drug Administration (FDA) and the United States Department of Agriculture (USDA). Some of the representative acts that have been prepared by these government agencies for the regulatory issue of genetically engineered microorganisms are Federal Insecticide, Fungicide, and Rodenticide Act (FIFRA); Toxic Substances Control Act (TSCA); National Environmental Policy Act (NEPA); Federal Plant Pest Act (FPPA) and Plant Quarantine Act (PQA).

The recombinant DNA Molecule Program Advisory Committee (RAC) of NIH established voluntary guidelines for researches involving recombinant DNA in 1976. However, the NIH did not have the statutory authority to regulate research and the guidelines, and had real impact only on the recipients of federal grants. Until the late 1970s, NIH, USDA, FDA and NSF (National Science Foundation) coordinately implemented the guidelines of NIH. However, the coordination became more difficult as the recombinant DNA technology reached the stage of commercial products. With large-scale commercial production and the potential for deliberate release into the environment, additional regulatory agencies were required. The EPA was established in 1970 to execute the authority over the activities that have potential to pollute the environment (air, water and land) in the US. At the same time, although USDA had regulations on the use of biological pesticides, each microbial pesticide was registered on an *ad hoc* basis since no consolidated policy had been issued until 1979.

In 1986, the EPA issued the FIFRA and applied TSCA to cover the genetically engineered microorganisms for a variety of uses such as environmental cleanup and industrial uses. The FIFRA also regulates microbial pesticides including viruses, bacteria, fungi and algae and derivatives of any of these microorganisms. The Act prohibits the distribution, sale and use of microbial pesticides that have not been registered with the EPA. The EPA reviews all data submitted for microbial pesticides and determines whether registration is appropriate. For registration, applicants must cite data on product composition, human health effects, environmental fate, and effects on non-target organisms. The EPA assesses this information when making decisions about safety. An experimental use permit (EUP) is required to conduct initial tests, even field studies. FIFRA has generally required an EUP even for small-scale testing of an area less than ten acres if genetically engineered microorganisms are used for the test.

The TSCA gives EPA the authority to regulate chemicals that may pose a threat to human health and the environment in research and commercial products development. The Act requires EPA to review test data within a specific period of time and determine whether the product is safe or pose a risk. TSCA was enacted to regulate organic and inorganic substances or a combination of substances, and applied to all micro-organisms produced for industrial, consumer and environmental uses with the exception of their manufacture, processing, or distribution as foods, food additives, cosmetics, medicines, or pesticides that are regulated by other acts.

The NEPA requires that all federal agencies prepare an environmental impact statement or an environmental assessment of any major federal action that might adversely affect the human environment. The impact statement must describe the predicted effect of the action, any adverse impacts, and alternatives for the proposed action. The NEPA is not a regulatory statute and it simply ensures that government agencies assess risks and environmental impact of genetically engineered microorganisms.

The FPPA and PQA provide the authority for regulating the movement of genetically engineered microorganisms that are potential plant pests into or within the US. Field tests of genetically engineered crops and plant pests require the approval of USDA under the FPPA and PQA.

Part VI

Microbes in Alternative Energy

Photosynthetic microorganisms, such as micro-algae and cyanobacteria are able to harness low-intensity solar energy and store it as latent chemical energy in the biomass, which can then be released via biochemical conversion. Various photobioreactors have been developed to allow maximum uptake and storage of solar energy. The structural and storage carbohydrates in biomass have low energy content and it is necessary to concentrate the energy content further for fuel application. Anaerobic microbial fermentation is an efficient and widely used method for such conversion processes. Useful renewable fuels produced by microorganisms include hydrocarbons, ethanol, methane and hydrogen. Biofuel cells that are able to release energy in fuel chemicals to generate electrical energy at ambient temperatures have been developed.

Chapter 19

Microorganisms and Production of Alternative Energy

Yuan Kun Lee

Department of Microbiology, National University of Singapore
5 Science Drive 2, Singapore 117597
Email: micleeyk@nus.edu.sg

The 1973 oil embargo and accompanying soaring oil prices together with the recognition that natural fuel reserves are being rapidly depleted, led to a worldwide interest in the development of alternative renewable fuels. Microbial production of fuels has the potential for helping to meet world energy demands. Living organisms assimilate and concentrate energy in their biomass and products. Hence, biomass in its various forms is an attractive alternative source of energy. Through photosynthesis, biomass collects and stores low-intensity solar energy, which can then be harvested and released via biochemical conversion. Biological processes are involved in both the harnessing of solar energy and upgrading of low energy feed stocks to biomass fuels. Useful fuels produced by microorganisms include biodiesel (produced from algal oils), ethanol, methane and hydrogen.

19.1. Harnessing Solar Energy

The sum of solar energy receives by earth per year is estimated to be 3×10^{24} J, which is about 100 times that of the proven total reserves of non-renewable energy (2.5×10^{22} J or 8×10^{11} ton coal equivalent).

This means that solar collectors covering 0.1% of earth's surface with 10% efficiency will meet all current world annual needs for energy (3×10^{20} J). Besides, the energy source is almost unlimited and the energy is renewable.

However, there are technical difficulties involved when harnessing solar energy as industrial fuel. These include the following: (i) the energy supply is variable and intermittent (solar irradiance changes from dawn to noon to dust, and total darkness in the night); (ii) the energy supply is diffuse (the highest irradiance measured at noon in the tropic and during summer in the temperate is about 4000 $\mu E/m^2/s$). Thus, it requires some form of storage system and a large area of collectors.

19.1.1. *Photosynthetic Production of Biomass*

Photosynthetic organisms (including cyanobacteria, algae, plants) use solar irradiance between 400 and 700 nm (known as the photosynthetically available radiance, PAR), this comprises about 50% of the total solar energy.

The overall photosynthetic process is represented by the following equation:

$$H_2O + CO_2 \xrightarrow{\text{sunlight}} \text{Biomass} + O_2$$

In this case, solar energy is captured by the photosynthetic apparatus, and stored as latent energy in the chemical structure of the biomass produced. The biomass can then be converted to renewable fuels, replacing fossil fuels.

To improve the productivity of cultivation, various **photobioreactors** have been developed to allow maximum uptake and storage of solar energy in photosynthetic microalgae and cyanobacteria. In general, besides having large illuminated surface area to volume ratio, these photobioreactors possess the following features:

(i) High turbulent streaming
High turbulence is required to move cells rapidly across the light path, which is perpendicular to the surface of the bioreactor (Fig. 19.1). This is to prevent the photosynthetic cells from staying too long in low light

(bottom of the culture) resulting in energy starvation, or to prevent their surface from being expose to inhibitory level of high sun light intensity. High turbulence is also necessary to facilitate mass transfer of gases (CO_2 and O_2), substrates and waste products at the surface of cell, and at gas-liquid interface.

Degree of turbulence in a flowing liquid along a pipe is proportional to the linear flow rate of the liquid, diameter of the pipe, and inversely proportional to the viscosity of the liquid. It is usually expressed as Reynolds Number, a partial turbulence is attended at a Reynolds Number

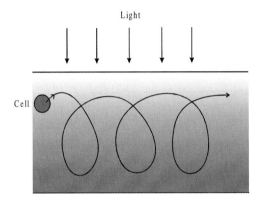

Fig. 19.1. Movement of a photosynthetic cell across the light intensity gradient in a culture illuminated from one direction.

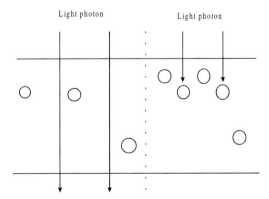

Fig. 19.2. Light penetration in a photosynthetic culture.

of 2000, and complete turbulence is achieved at a Reynolds Number of 4000.

(ii) High cell density

This is necessary to capture all light photons that impinge on the culture. A light photon would pass through the bioreactor if it is not captured by a cell while traveling through the culture (Fig. 19.2).

(iii) Light absorption

The density of photosynthetic pigments and light absorption cross section determine the absorption of light photons by a single cell in the light path. A cell of low pigment density and small light absorption cross section would allow more light to pass through and reach cells at the lower level of the light path. A cell that captures excessive quantity of light energy would lead to either dissipation of the energy (for the photosynthetic system is not able to process the captured energy sufficiently fast), or destruction of the photosynthetic components and reduce photosynthetic activity.

(iv) Shallow culture

High cell density would result in shading of cells by other cells in the light path. Discontinuous supply of light energy due to the light gradient in the

---

Box 19.1

An Enclosed α-Type Tubular Photobioreactor Developed in Singapore

The photosynthetic culture is lifted up in air-riser tubes to a receiver tank 5 m above ground, and then flows down the tubular photobioreactor to reach the opposite set of air-riser tubes. The culture was then lifted up 5 m to another receiving tank, flowing down to parallel photobioreactor tubes connected to the base of the first set of riser tubes. Because the flow of culture in the tubular bioreactor does not change direction except when lifted up in the air-lift system, a high liquid flow rate and Reynolds number, and short residence time are achieved at relatively low air supply rates. As the α-photobioreactor is placed at an angle with the horizon, a fairly constant solar irradiance is received by the culture throughout the day, allowing high biomass productivity at high culture density.

Box 19.1   *(Cont'd)*

Box 19.2

Effect of Photobioreactor Inclination on the Profile of Sunlight Measured at the Surface of the Reactor

The photon flux density incidented on the upper surface of a bioreactor panel facing the sun is $I \cdot \cos(\delta - \alpha)$, where I = the intensity of sun-light incident on a surface perpendicular to the sun rays, δ = the angle between the direction of sun and the vertical, α = the angle of inclination between the

Box 19.2   (*Cont'd*)

bioreactor and the horizontal. Thus, the photon flux density measured at the upper surface of the bioreactor panel facing the sun increased with increasing α, and reached the maximum at $\delta = \alpha$. At a fixed-angle α, the intensity of direct light measured at the bioreactor surface decreased with decreasing δ, as shown in the figure below.

Along the equatorial (like in Singapore), for a bioreactor inclined at an angle of about 45°, the photosynthetic radiance measured at the surface remains almost constant throughout the day.

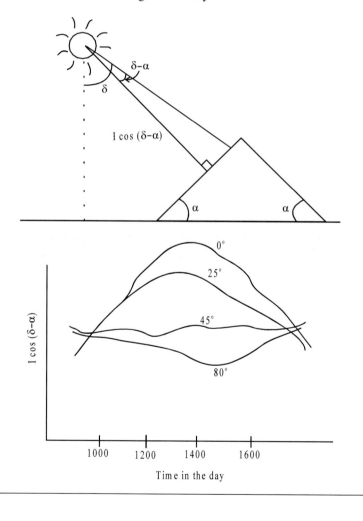

culture, would lead to enhance maintenance energy requirement and lower growth yield (see Chap. 2, Sec. 2.1.4). Shallow culture and high turbulence would allow frequent short exposure of cells to light at the surface of the bioreactor.

(v) Inclined at an appropriate angle with the horizon
Inclining a photobioreactor at an angle allows the bioreactor to receive more irradiance in the morning and afternoon, and less at noon. As such, the intensity of irradiance receives by the photosynthetic culture during the day is even out. It had been reported that higher photosynthetic conversion efficiency was achieved in this kind of set up.

In the field, photobioreactors using suitably selected microalgae or cyanobacteria achieved nearly 10% of photosynthetic conversion efficiency. This compare favorably with higher plant, which could convert <1% irradiance absorbed by its leaves to biomass energy.

19.1.2. *Photo-biological hydrogen production*

Photosynthetic apparatus, chloroplast, of some photosynthetic micro-organisms such as the green alga *Chlorella* in the presence of suitable electron acceptors is capable of producing H_2 and O_2 through direct photolysis of water.

In the system, the substrate (electron donor) is water, sunlight as the energy source is unlimited, and the product (hydrogen) can be stored

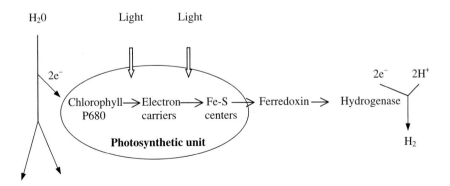

and is non-polluting. Moreover, the process is renewable, because when the energy is consumed, the substrate (water) is regenerated.

However, the process currently has no practical value, as the system is not stable (chloroplast is structurally unstable), and could function for only a few hours outside the plant cells. Artificial membrane mimicking biological photosynthetic unit is being investigated.

19.2. Conversion of Biomass Energy to Fuels

Structural and storage carbohydrates in biomass have low energy content and could not be used as fuel directly. It is necessary to concentrate the energy content further for fuel applications. The use of microorganisms to produce commercially valuable fuels depends on getting the right micro-organisms that are able to produce the desired fuel efficiently and in having an inexpensive supply of substrates available for the fermentation process. It is obviously imperative that the production of synthetic fuels does not consume more natural fuels resources than are produced. Anaerobic microbial fermentation is an efficient and widely used route for such conversion processes.

19.2.1. Alcohol (ethanol) fermentation

The microbial production of ethanol has become an important source of a valuable fuel, particularly in regions of the world that have abundant supplies of plant residues. Brazil produces and uses large amounts of ethanol as an automotive fuel. Mixing gasoline and ethanol in a 9:1 ratio is popular in some part of US. Ethanol combustion is less polluting than gasoline combustion.

Fermentation production of fuel alcohol can be accomplished through microbial conversion of low cost agricultural substrates high in starch and sugar content such as molasses (by-product of sugar industry), sugar-cane juice, maize starch, cassava starch and other fermentable carbohydrates. Though some microalgae have been reported to produce ethanol as one of the fermentation products, the production from microalgae as a finite biomass, is still being examined.

Numerous microorganisms are capable of producing ethanol, but not all are suitable for industrial processes. Yeast cultures, particularly *Saccharomyces*, have been most extensively examined because they are very efficient in converting sugars into ethanol, i.e. cost competitive, and are not as strongly inhibited by high ethanol concentrations as are other microbes.

The following equation illustrates the basic biochemical mechanism the ethanol is produced by the fermentation process.

$$C_6H_{12}O_6 \xrightarrow{\text{yeast}} 2(CH_3CH_2OH) + 2(CO_2)$$

16 kJ/g Glu   30 kJ/g ethanol

Theoretically, two moles of ethanol can be produced from one mole of glucose. On actual calculation:

    1 kg invert sugar can produce 0.484 kg ethanol (or 0.61 L)
    1 kg sucrose can produce 0.510 kg ethanol (or 0.65 L)
    1 kg starch can produce 0.530 kg ethanol (or 0.68 L)

The yeasts commonly used in industrial alcohol production include *Saccharomyces cerevisiae* (ferment glucose, fructose, maltose, maltoriose), *S. uvarum* (carlsbergensis), *S. diataticus* (ferment dextrins), *Kluyveromycesfragilis* and *K. lactus* (ferment lactose). The ethanol productivity ranges between 1–2 g/h/g cells.

Selected bacterial cultures were examined for use in ethanol production processes because of their higher temperature tolerance. However, their yield of ethanol was not as high as in yeasts fermentations. Recently, the bacterium *Zymomonas mobilis* has been selected to achieve a high productivity of 2.5–3.8 g ethanol/h/g cells.

Various processes can be employed for the fermentation production of ethanol as a fuel. The development of the fermentation processes through the years reveals the strength and weaknesses of the various culture methods as discussed in Chap. 2, Sec. 2.4.

(i) For batch fermentation, the practicable level of carbohydrate used is 16–25% (w/v), giving a final ethanol concentration of 6–12% (v/v), with an average output rate of 1.4 g ethanol/L/h. Higher carbohydrate concentrations would result in substrate inhibition or product inhibition of cell growth, as high sugar concentrations inhibit metabolism due to the increased osmotic pressure. Despite of the limitation, the advantages of batch process is that it is resistant to process abuse and can be operated by less skilled personnel.

(ii) Ethanol is relatively toxic to microorganisms, and therefore only limited concentrations of ethanol can accumulate in a fermentation process. To overcome the effect of product (ethanol) inhibition in order to achieve higher ethanol output rate, batch culture with cell recycle or continuous chemostat culture systems have been used and productivity of >10 g ethanol/L/h has been achieved. Such systems required higher operator skills, capital costs and more sophisticated control devices.

In the **batch recycle system** (Fig. 19.3), ethanol is recovered and cells are removed from the brew at the end of a fermentation cycle. As ethanol inhibition reduces the specific rate of ethanol production with resultant lower output of ethanol, higher cell density is therefore desired to maintain high output rate, and this can be achieved by recycling the cells removed from the effluent back into the fermenter in the following batch. This process allows the fermentation time to be shortened.

In a chemostat system, substrate is continuously being added to the culture system and the culture (including the product ethanol) is being harvested at the same rate, so that the volume of the culture

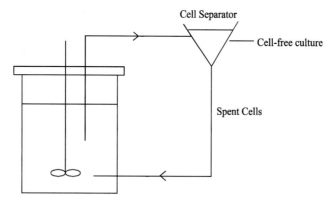

Fig. 19.3. A batch culture system with biomass recycled.

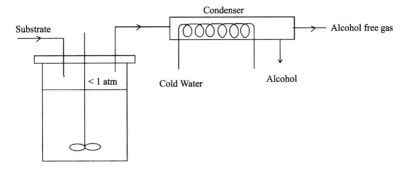

Fig. 19.4. Vacuum fermentation system for alcohol production.

remains constant and inhibitory level of ethanol does not occur to cause decreased growth rates. If the microbial population grows slower than the dilution rate due to ethanol inhibition, the culture will be diluted and "washed out" of the fermenter. High product output rate is achieved at high dilution rate (the feeding rate of medium).

(iii) Alternatively, the vacuum fermentation system has been proposed (Fig. 19.4). In this case, the fermentation is performed under reduced pressure. The ethanol is distilled off at the fermentation temperature, while concentrated substrate is fed continuously, thus the inhibitory effect of ethanol on cell growth is reduced. However, the energy balance in such a system is less than 1 (i.e. consumed more energy than being recovered in alcohol).

Alcohol fermentation is a well-established technology. However, the energy input necessary to process the feed material and product is about the same as the energy output in the form of ethanol as distillation to recover ethanol requires energy input. Besides, human food, which is relatively expensive, is often used as the feedstock; hence the cost of ethanol production by fermentation is high.

19.2.2. *Methane production by anaerobic digestion*

Methane (CH_4) is an energy-rich fuel that can be used for the generation of mechanical, electrical, and heat energy. Large amounts of methane can be produced by anaerobic decomposition of waste materials. Efficient generation of methane can be achieved by using algal biomass grown in pond cultures, sewage sludge, municipal waste, plant residue and animal waste.

In microbial production of methane, naturally occurring mixed anaerobic bacteria population is always used, and cells are retained within the digester. During the fermentation process, a large amount of organic matter is degraded, with a low yield of microbial cells, while about 90% of the energy available in the substrate is retained in the easily purified gaseous products CH_4. The end product is a mixture of methane gas and CO_2 (also called biogas).

The degradation process involves three major metabolic groups of interacting bacteria, most of which do not directly produce CH_4. Fermentative bacteria hydrolyze the degradable primary substrate polymers such as proteins, lipids, and polysaccharides and decompose to smaller molecules with the production of acetate and other saturated fatty acids, CO_2, and H_2 as major end products. The second group is the obligate H_2-producing acetogenic bacteria, which metabolize low molecule organic acids (end products of the first group) to H_2 and acetate (and sometimes CO_2). Finally, the methanogenic bacteria catabolize mainly acetate, CO_2 and H_2 to the terminal products. Methanogenic bacteria found in such mixed population include species of *Methanobacterium*, *Methanobrevibacter*, *Methanoccus*, *Methanogenum*, *Methanospirillum* and *Methanosarcina*.

The stages of the fermentation process are as follows:

(i) Stage 1: Acid-forming stage
Polymeric materials such as polysaccharides, proteins, and lipids were broken-down to form volatile organic acids (in particular acetic acid), H_2 and CO_2.

(ii) Stage 2: Methanogenesis (methane generation)
Methanogens can utilize only a small number of simple compounds as energy sources. Some utilize the H_2 produced by the other bacteria to reduce CO_2 to CH_4, and some cleave acetate to CO_2 and CH_4. The methanogens are the only organisms that are able to catabolize acetate and

hydrogen to gaseous products in the absence of light energy or exogenous electron acceptors such as oxygen, sulfate, or nitrate.

$$4H_2 + HCO_3^- + H^+ \rightarrow CH_4 + 3H_2O \qquad -32.7 \text{ kcal (free energy)}$$
$$4HCOO^- + 4H^+ \rightarrow CH_4 + 3CO_2 + 2H_2O \quad -34.7 \text{ kcal}$$
$$4CH_3OH + 4H^+ \rightarrow 3CH_4 + CO_2 + 2H_2O \quad -76.4 \text{ kcal}$$
$$CH_3COO^- + H^+ \rightarrow CH_4 + CO_2 \qquad -8.6 \text{ kcal}$$

The advantage of anaerobic production of fuel is that the energy conversion efficiency is high. A yield value ($Y_{methane/glucose}$) of 0.27 g/g is often observed, which corresponds to an energy yield (Y_{energy}) of 90%.

The disadvantage is that the digestion process is slow. Typically, for a gas consists of 60–70% methane, the production rate is 0.5 m^3/kg dry volatile solids over a resident time of 15 days.

19.3. Electricity from Biofuel Cells

Biofuel cells could release energy in fuel chemicals to generate electrical energy at ambient temperature. Fuel cells convert energy more efficiently than conventional power engines such as the internal combustion engine and produce almost no pollution. The basic set-up of the fuel cell is two electrodes placed in an electricity conducting electrolyte, separated by an ion exchange membrane. The arrangement allows the electrochemical equivalent of combustion to occur.

Example 1: To generate electricity from hydrogen gas (Fig. 19.5):

$$H_2 \xrightarrow{\text{Hydrogenase}} 2H^+ + 2e^- \quad \text{(anode)}$$

$$2H^+ + \tfrac{1}{2}O_2 + 2e^- \xrightarrow{\text{Laccase}} H_2O \quad \text{(cathode)}$$

Hydrogen gas as a fuel enters at one electrode, and oxidant, usually oxygen from the air, at the other. The anode is coated with the enzyme hydrogenase, while the cathode is coated with laccase. On the anode, hydrogen molecules split into their constituent protons and electrons. The enzymes catalyze a reduction-oxidation reaction across the membrane,

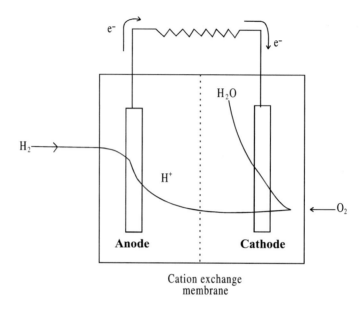

Fig. 19.5. A biofuel cell using hydrogen gas as the electron donor.

releasing energy, which pushes electrons round an external circuit. At the cathode, the protons and electrons combine with oxygen to form water. The fuel does not actually burn, therefore fuel cells do not produce pollutants associated with combustion, such as carbon oxides and oxides of nitrogen. Cells that use hydrogen generate only water as compare to fossil fuel, which produce water and carbon dioxide as waste product. This type of fuel cells is called proton exchange membrane cell (PEM).

Example 2: To generate electricity from methanol:

$$\text{Methanol} \xrightarrow{\text{Methanol dehydrogenase}} \text{Formate} + 4H^+ + 4e^- \quad \text{(anode)}$$

$$4H^+ + O_2 + 4e^- \xrightarrow{\text{Laccase}} 2H_2O \quad \text{(cathode)}$$

In this case, the fuel is methanol. The direct methanol fuel cell (DMFC) runs on a diluted mixture of about 2% methanol in water. The methanol is converted into formate on the anode. The proton then reacts with oxygen as in a PEM cell.

The metabolically active microorganisms, such as *Proteus vulgaricus* and *Anabaena variabilis* immobilized in a biofuel cell could convert energy in their substrate (glucose for the former and light for the later) into electricity. A biofuel cell in which bacteria *Proteus vulgaircus* and *Escherichia coli* were used as sulfate reduction catalysts was in operation for 5 years, thus demonstrated its long-term stability.

The disadvantage of biofuel cell is that the power output is low (1 kW at 40 mA/cm^2). Thus, it is used for specific purposes, such as small medical and military apparatuses used in the field and in space missions.

Further Readings

1. Allen RM and Bennetto HP. Microbial fuel-cells. Electricity production from carbohydrates. *Appl. Biochem. Biotechnol.* **39/40**: 27–40, 1993.
2. Dellweg H, ed. *Biotechnology Vol. 3*. Chapter 3a: Ethanol fermentation, pp. 257–386; Chapter 4: Energy from renewable resources, pp. 595–626. Weinheim, Verlag Chemie, 1983.
3. Habermann W and Pommer EH. Biological fuel cells with sulphide storage capacity. *Appl. Microbiol. Biotechnol.* **35**: 128–133, 1991.
4. Hall DO, Coombs J and Higgins IJ. Energy and biotechnology. In: *Biotechnology, Principles and Application*, Chapter 2, eds. IJ Higgins, DJ Best and J Jones. Oxford, Blackwell Scientific, 1985, pp. 24–72.
5. Laminie J and Dicks A. *Fuel Cell Systems Explained*. Wiley, New York, 2000.
6. Lee YK and Richmond A. Bioreactor technology for mass cultivation of photosynthetic microalgae. In: *Marine Biotechnology Vol. 2*, eds. M Fingerman, R Nagabhushanam and M-F Thompson. Oxford and TBH Publishing, New Delhi, 1998, pp. 271–288.
7. Tanaka K, Kashiwagi N and Ogawa T. Effects if light on the electrical output of bioelectrochemical fuel-cells containing *Anabaena variabilis* M-2: Mechanism of the post-illumination burst. *J. Chem. Technol. Biotechnol.* **42**: 235–240, 1988.

Questions for Thought

1. Using alcohol production as an example, illustrate how the principles of microbial fermentation could be applied in the development of processes for upgrading of agricultural products to liquid fuel.
2. Design a self-contained space station that is made up of space travelers and microorganisms.

Part VII

Patenting Microbial Biotechnology

Most governments see at least part of national future prosperity in the expansion of life sciences research, either by exploiting natural resources such as indigenous medicinal plants and thermophilic bacteria, or by supporting a workforce skilled enough to compete in the *de novo* design of new drugs, as well as genetically modified animals and crops. So much money will be spent on development that those who finance it will insist that wherever possible its fruits will be protected from exploitation by others, that is, patented. All scientists, including microbiologists, would do well to be aware of the prerequisites and procedures involved.[1] Put briefly, the invention must be novel (not already identifiable in practice or in the literature), it must involve an inventive step (something not obvious to those already in the field) and it must have some industrial application. In return for meeting these criteria, the invention will be protected against exploitation by others for twenty years. In regard to exactly what may be patented however, there are certain exclusions, such as devices, which offend public morality, as well as theories and discoveries. Methods of medical treatment were previously non-patentable but are now of debatable status.[2]

It is often said that scientists are the worst judges of the patentability of their work and this is generally true. They will often look upon their own advances as obvious or trivial when in fact they may be readily commercially exploitable — they may be *inventions* in the legal sense, and although the scientist may not look upon himself in that light, he is an *inventor*. Although some patents are taken out by private citizens with good ideas, inventions in the life sciences are likely to be achieved within large governmental, educational or commercial organizations. Whereas these organizations will readily recognize the scientist as an inventor, they will almost certainly be the legal owners of any patent awarded to him.

Chapter 20

Patenting Inventions in Microbiology

John Candlish

Faculty of Medicine and Health Sciences, University of Malaysia Sarawak, Kuching, East Malaysia

Email: jkcandlish@yahoo.com.sg

20.1. Territorial Nature of Patents

By law, almost universally, a resident in any country must file for a patent in that country first. The rationale is to allow the fruits of the invention to be enjoyed by the society which engenders it, and — more importantly — allow the relevant government, if it so desires, to acquire a compulsory license to exploit it. For example, if a microbiology patent is apposite to national defense — perhaps a way of halting an advancing army by subjecting them to a debilitating diarrheal illness — a government will not be slow to acquire it.

A patent granted in one country gives no protection in any other, so it is important to make applications abroad under the Patent Cooperation Treaty (PCT) as detailed below. The PCT as administered by the World Intellectual Property Organization (WIPO) is gradually making patent law more uniform worldwide. For example, the length of patent protection in the US, Australia and China has recently been lengthened to 20 years in conformity with the rest of the world.

Groupings to harmonize patent regulations within specific geographical areas are also arising. The European Patent Office (EPO) has some 20 countries as subscribers. This is not to be confused with the attempts of the European Union, with some 15 members, to identify common features within its own, strictly economically based, community. The Eurasian Patent Office deals with applications from the Russian Federation and a number of former republics of the USSR. The Association of South East Asian Nations (ASEAN) has declared itself to be desirous of a regional intellectual property framework, but this has not yet come to fruition.

The patent laws of most countries state that "any person" may make an application. That is, anybody can be an inventor. Usually the inventor cannot be a corporation or an organization. The inventor can go through the filing process on his own, or appoint another person to do so. Thus, in that sense, anybody can be a patent agent. However most countries have practicing patent attorneys whose qualifications are recognized by law and who alone are permitted to describe themselves as such. In some countries, the ordinary members of the legal profession may so describe themselves. Large universities, research institutes, and corporations often have officers who in effect act as patent agents for their employees.

20.2. Ownership and Royalties

In most, if not all, jurisdictions, an employee's inventions are regarded as property of the employer, but of course, the initiative must come from the inventor. As owner of any patent granted, the royalties from any licensing agreement accrue to the organization. However, the employer is likely to be generous in any share out, for an employee who can produce patentable inventions is obviously worth retaining. Where there are two or more inventors, they are regarded as joint owners, with "undivided shares". As a legal term, this means that they cannot claim specific proportions of the invention, say one third each if there are three of them. Rather, if one of them mortgages his share (which is possible) the lender also has an interest in the whole, not in one third.

In universities, where so much work is done by postdoctoral fellows, they are presumably to be regarded as employees, as has been indicated in a recent federal case involving the University of Chicago, in which the reasoning would probably be persuasive elsewhere.[3] However, inventions arising from graduate student projects may give rise to problems. As indicated above, there must be no public disclosure (oral or written) of the invention prior to filing (the "novelty" requirement) and this includes theses deposited in the library. Most universities then will delay emplacement in the library until the patent is filed, for it is then about to become public knowledge anyway. If the reports of projects used for the assessment of postgraduates are retained in the relevant department, one would imagine, on a reasonable interpretation, that this does not put them in the public domain. It might be as well, however, to put anyone reading commercially sensitive reports who is not already implicitly under a bar of confidentiality (such as members of the same research team) into that category. This prevents disclosure to any third party who on gaining the information innocently and publishing it, would not be liable. Thus, anybody put under a bar of confidence who discloses information rendering a patent invalid for lack of novelty could theoretically be sued, in tort, for damages; whether this would be worthwhile would depend on the circumstances.

20.3. Novelty, Disclosure and Confidentiality

It is a general proposition of the novelty requirement that a patent is not obtainable if the nature of the invention has been available in any written, oral or electronic form, anywhere in the world, before filing. (The term often used is "in the public domain"). The principle could conceivably be overturned if publication is in a really obscure form or locale. This was the burden of the decision in a keynote US case, *Gayler*[4] and also in an important UK case, *General Tire & Rubber Co v Firestone Tire & Rubber Co*[5] wherein the judge stated that novelty is negated only if the information had been accessible to the public as of right, with or without fee, anywhere in the country.

Recently, a patent for the use of Taxol in treating cancer was held to be invalid because there had been an oral disclosure at a scientific meeting attended by 500 persons.[6] This illustrates a potential pitfall for scientists, who instinctively want their work to become known as quickly as possible via learned journals, books and conferences. This can well destroy novelty. This problem is obviated to some extent in the US by way of its "first to invent" system. If there is publication between the invention itself and application (which must be not more than 12 months) the novelty requirement is not violated. Canada operates similarly. Most jurisdictions have some sort of a qualification to the strictness of the rule. In UK if the disclosure results from a display at an "international exhibition" and the applicant files notification thereof, novelty is not destroyed. However, the definition of an international exhibition is very restrictive. In Japan, there is a six month window between disclosure and filing. In Singapore, to take an example from a smaller country, if publication occurs no more than 12 months before the date of filing then it will not invalidate the patent if it was made at "an international exhibition" or in paper to a "learned society". The former seems to have the same meaning in the UK Act (which to some extent acted as a model) but the latter is broadly framed, covering a club or association anywhere in the world whose main object is the promotion of any branch of science. Even this might well become subject to judicial interpretation, however, and it might be as well to take advice before making disclosure in this manner. One of the purposes of the rather narrow 12 month window, by the way, is to prevent inventors sitting passively on potentially valuable inventions which competitors (perhaps in competing economies) might pick up.

Generally, disclosure to colleagues in the same organization will not constitute putting an invention into the public domain; they are under an implied obligation of confidentiality.

20.4. Inventive Step

All patent legislation mandates that the invention be not obvious, rather it must incorporate an "inventive step". Here the various jurisdictions introduce a legal fiction, the "hypothetical skilled artisan" or "person skilled in the art". This is a person who knows all about the technology surrounding the invention, but does not have the imagination to conceive of the inventive step in the invention claimed. If he *can* imagine it, the invention will be void for obviousness. The patent examiners, and indeed the tribunals which have to adjudicate if there is a dispute about validity, must assume the mantle of this individual.

When patent applications for biotechnological inventions fail, it is usually on the grounds of obviousness. Obviously each dispute will turn on the facts, and judges have a hard time making decisions. In the UK case of *Biogen* v *Madeva*[7] (which is famous as the first biotechnology case to go all the way to the House of Lords, the supreme court of the UK) the Biogen invention was accepted as non-obvious although the technique was well established at the time; the lack of obviousness lay in attempting a shotgun approach which disregarded the probability that introns might disrupt the expression of hepatitis B antigen in *E. coli*. Therefore, if this judicial approach is persuasive elsewhere, it means that if you try an approach to a problem which seems extremely unlikely to work, but use established techniques, and if you are lucky, then you will get a patent. Here the person skilled in the art,[8] would not have tried the shotgun technique because he did not have the imagination to believe that he might be lucky.

To sum up, you might fail to apply for a patent because being heavily involved in your field your technical advance seemed obvious to you; what the patent examiners are interested in, however, is whether it was obvious to anybody else.

20.5. Industrial Application

Some of the earlier legislation in English (the Statute of Monopolies, 1623) spoke of a patent as necessarily being a "manner of new manufacture", emphasizing the utility requirement. There is seldom any difficulty in establishing the utility of a microbial process, but the concept has caused problems in other areas of biotechnology. Many scientists and members of the public have objected to patenting genes or nucleotide sequences on the grounds that they are discoveries rather than inventions. The response has been that patents are not issued for genes in their natural state but rather on DNA sequences after they have been isolated and (usually) amplified, and even then only if the requirements of utility and novelty are met.[9]

20.6. Procedure and Form

20.6.1. *Search*

Initially, of course there has to be the realization that an invention may have been created and so a search has to be made of existing patents worldwide, to ensure that it has not been protected previously. The patent office, in the person of the "examiner" will do this check after filing, of course,[10] but it is obviously a complete waste of time and money to file when the invention already exists. The authorities, recognizing the extreme diversity of possible inventions, have devised classification systems, so that if one is a virologist, he can begin by looking at previous patents relating to viruses rather than to broom handles. There is an International Patent Classification (IPC) system,[11] devised by the WIPO based on one of 8 initial letter codes plus Arabic numerals and running to over 67,000 subdivisions. Any one invention, as might be expected, can have several different classifications. Most microbiology inventions will fall under category C. For example, the patent entitled "a vaccine-induced hepatitis B viral strain and uses thereof"[12] has six classifications beginning with C12N 15/51. The EPO generally uses this system but the United States Patent Office (USPTO) has an alternative system based on Arabic numerals, which will gather most microbiology patents under category 435. Thus "dengue virus peptides and methods"[13] has 13 categories; the first of which is 435/69.3. Where known, a published patent will list numbers under both systems.

   A patent attorney will of course do the search, but at an appropriate fee. The task is becoming much easier with modern methods of information technology. Websites with search engines, which take keywords, especially those operating Boolean logic, make the task quite feasible from one's own office. A list of sites is given in Table 20.1.

20.6.2. Drawing up the application

The document does not have to adhere to an unalterably rigid format, but must include a title, a description/specification, claims and abstract. It can also include drawings where relevant and preferred embodiments, that is forms, which the invention may eventually take. The US authorities prefer the following sequence:

Title → Abstract → Inventors → References/related patents → Claims → Description → Examples

Table 20.1. Useful websites.

Site	Comments
http://patents.uspto.gov	Front pages of all US patents from 1976
http://www.european-patent-office.org	Search of EPO, PCT and Japanese patents in facsimile
http://www.surfip.gov.sg	Singapore initiative giving access to many databases (EPO, UKPTO, WIPO, USPTO)
http://www.patent.gov.uk/dpatents/ how_prep.html	Instructions on how to frame a patent application in UK
http://qpat1.qpat.com	Front page information of all US patents since 1974
http://patent.womplex.ibm.com	IBM site — complete US patent information for last 23 years

All of these sites are free but some may need registration. Some sites offer large numbers of links to search sites, see for example: http://www-sul.stanford.edu/depts/swain/patent/patdbases.html (Stanford University) and http://www.ipmall.fplc.edu.psa/ (Franklin Pierce Law Center)

20.6.3. *Specifications*

This section of the application is intended to be read by a person having skills related to the invention. Such an addressee may be one person, but if an invention encompasses two or more disciplines, this role can be split. For example, a microbiology application may combine a genetically engineered bacterium with a chemical purification process for its product. It will state the prior art and then go on to describe the invention in terms, which allow the person skilled in the art to make it and use it. This is called "enablement".

It is often the case that a skilled person cannot perform the processes described in a microbiology invention because the microorganism is not available to him, or cannot be described, however fully, in manner which will allow reproduction of the technique unless the microorganism is cultured. There is thus a scheme, under an international convention, The Budapest Treaty of 1977, for the deposition of microorganisms in a registered depository. In Europe, this occurs at the date of publication of the application, that is, before the grant of the patent. In the US, it is at the time of the grant.[14] It is mandated by many authorities, for example the EPO.

20.6.4. *Claims*

These can be multiple and overlapping. For example, a patent for "cDNA sequence of Dengue virus serotype 1"[15] has 21 claims, the first for the isolated virus, leading on to a kit for its detection and various antibodies and antigens. The claims are often set down first when drafting a patent, and then they lead the draft of the specifications. It is therefore important that they be clear and succinct.

20.6.5. *Embodiments*

Often an invention may be conceived in several different forms, called embodiments. For example, the kit to detect the Dengue virus strain referred to above could be envisaged as using liquid reagents, freeze dried reagents, or even as the dipstick type of test. "Preferred" embodiments are often noted. The idea here is to show that the invention can be reduced to a useful, working form. If it is left at the stage of an idea or discovery, it will probably not be patentable.

20.6.6. *Filing*

This is done on prescribed forms, with the fee and the patent application attached. The costs levied by the various patent offices are now almost invariably accessible on the Internet (Table 20.1). Indeed filing on line is rapidly coming into use and its status in the selected patent office can be checked by going into these same sites.

20.6.7. *Priority*

Most countries operate a "first to file" system for priority, that is the patent is deemed to be owned from the date of the application to the relevant national patent office. Again, the US system is different in that it operates a "first to invent system" for the priority date. First to invent must be interpreted as first to reduce to practice, or in other words, show that it works. This is why properly maintained laboratory notebooks are so important in the interests of obtaining protection in the US. Should any dispute arise over priority (called "interference" in the US) the record in the notebook would be decisive.[16] Notably, an applicant from another country who files in the US has the same advantage in respect of being able to prove priority by being first to invent.

20.6.8. *International application*

The grant of a patent in your own or host country is important, but patents are highly territorial and will not prevent somebody from exploiting an invention elsewhere. Protection in the US, where your invention could be exploited by others in the face of the huge and affluent market, is obviously important, but might be equally so in Europe, possibly also Australia and Japan depending on its nature. For protection in the US — whose patent granting regime, by the way, is acknowledged by all commentators to be extremely liberal[17] — and any other country included in the PCT an international application is filed in the home patent office. There are now 109 contracting states and any number can be chosen. The local patent office will act as a post office in transmitting the application to an International Searching Authority (in effect one of the high quality national patent offices) who will prepare a list of document relevant to the prior art and send this to each of the countries included in the application. There is also the option of an international preliminary examination report, which will obviously save costs should it indicate that filing is unlikely to be successful. Only a single set of fees need be to paid in respect of an application under the PCT.

The final set of patents will however be granted separately by the countries cited in the application, if it is decided to proceed in them. This again emphasizes the enduring territorial nature of patent law.

20.6.9. *Publication*

In most countries, applications are published automatically without examination, 18 months after filing. This is also the case with PCT applications. Until recently, in the US publication only occurred at the time of the grant of the patent itself. However its Intellectual Property and Community Omnibus Reform Act (1999) harmonizes US practice with the rest of the world. Publication is important for it is only thereafter that action for infringement can be mounted. Infringement refers to the use of the invention by a person other than the inventor or the person to whom he has sold, mortgaged, or licensed it. Should this happen the patent holder has the right to challenge the infringement in court and seek damages, or alternatively an "account of profits". The law in this area is complex and replete with case precedents, so it will not be considered further here.

20.6.10. *Grant*

The actual conferment of the patent is called the "grant". If all goes well and there is no opposition, this will be between three and four years after the priority date. There are usually renewal fees payable after a certain number of years, up to 20 years of course. Some pharmaceutical patents including, for example new antibiotics from bacteria or fungi, cannot be marketed until there is extensive safety testing by way of clinical trials. There is thus provision, in some jurisdictions, for an extension of the protection period for this type of patent. An attempt at an elementary flowchart for the whole process is made in Fig. 20.1.

20.7. Recent Trends in Biotechnology and Microbiology Patents

What can be patented in the life sciences? It has been a longstanding principle that theories and discoveries cannot be patented. Thus, Crick and Watson's discovery-cum-theory of the structure of DNA, useful though it has been, was not patentable. If you discover a new thermophilic bacterium as a product of nature that will not in itself enable a patent; if however you find a way of culturing it in the laboratory the situation is entirely different. See for example "*Aspergillus* culture claim".[18] Products, processes, methods and devices, all attract protection under the patent system.

20.7.1. *Microbial products*

There is no difficulty about patenting anything produced by micro-organisms as long as the triad of novelty, inventive step and utility are met. Proteins produced spontaneously or by recombinant DNA techniques,

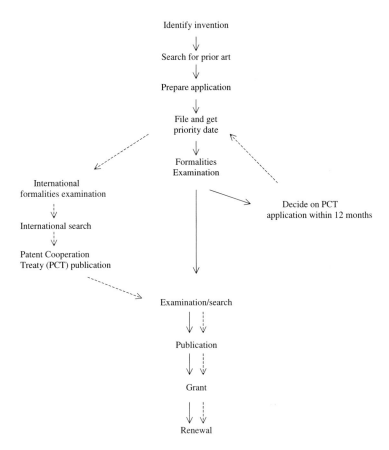

Fig. 20.1. Procedural flow diagram. This is intended as a very general scheme applicable worldwide; from filing to grant will usually take 4–5 years. PTC: Patent Cooperation Treaty.

antibiotics, flavors, and the myriad of other chemicals produced by the marvelous synthetic capacity of bacteria are all eminently patentable. See for example "The fermentation production of eicosapentaenoic acid containing oil by cultivation of heterotrophic diatoms".[19]

20.7.2. *Base and amino acid sequences*

Most people will be aware of the debate in the media over the "patenting of genes". One of the latest statements came jointly from Sir Aaron Klug

(Royal Society of London) and Bruce Alperts (National Institutes of Health) in *Nature*[20] when they stated that "The intention ... to patent DNA sequences themselves, thereby staking a claim to large numbers of human genes without necessarily having a full understanding of their functioning, strikes us as contrary to the essence of patent law ..." However it is clear that genes of known expressivity, if isolated, can be patented in most, probably all, jurisdictions. The European Patent Convention (EPC) articulates the enablement simply thus:

"Biotechnological inventions shall also be patentable if they concern (a) biological material which is isolated from its natural environment or (b) produced by means of a technical process even if it previously occurred in nature."[21]

This obviously covers nucleic acids, proteins and peptides, if they can be shown to be useful. Particularly contentious has been the status of expressed sequence tags (ESTs). Celera Genomics, Dr Craig Venter's company in the US appears initially to have been successful if the case in protecting them if it could maintain that they could be used as probes to elucidate the sequence and function of a whole gene, if only some time in the future.[22] The European Patent Office (EPO) is however opposed to the patenting of ESTs and the USPTO has recently tightened up slightly.[23] But the guidelines reiterate that no patent can be filed for a natural product occurring in nature. So a genome in itself, human, mammalian bacterial, or other cannot be protected. However, if the gene is cloned, it is then patentable like any ordinary chemical, as long as it can be demonstrated that it has "value in the real world". Notwithstanding that however, if one were to isolate a hitherto unknown microorganism from soil or a protein of unknown function from a tissue, one could claim that a use would be found for it one day.

20.7.3. *Life forms per se*

Historically there has been no difficulty in patenting lowly life forms. Louis Pasteur received a patent for a yeast strain in 1873[24] and in the landmark US case *Diamond* v *Chakraborty*,[25] an engineered bacterium was successfully protected by its inventor. Only recently, with the possibility of cloning or engineering higher organisms has a new debate erupted. The Harvard "oncomouse" was patented in US although the application was refused in several other jurisdictions, and it is still subject to opposition proceedings in Europe. Objections are partially based on the width of the claims. Thus, the patents (there are a series) seek to cover the technique of inserting the gene in any animal leading to the specter of oncocats and oncogiraffes. In Canada, the patent office refused it but on the appeal, the Federal Court allowed the grant. The Canadian Government is in the course of appealing to the Supreme Court, which has just refused it.

Human beings of course are a special case. The patent system in the US is notoriously favorable to the inventor but the 13th Amendment to the constitution forbids slavery and involuntary servitude of any human beings and so one presumes that a cloned human being would be refused as an invention although there is apparently nothing on the US patent legislation specifically blocking such applications. In any case, an application to patent any form of human being is implicitly forbidden by Article 4 of the UN Universal Declaration of Human Rights, and probably any modified germline would be refused since this could lead to a usable human form. The European Patent Convention's (EPC) exclusions policy is cogently expressed:

(1) *Processes for reproductive cloning of human beings.*
(2) *Processes for modifying the germline genetic identity of human beings.*
(3) *Use of human embryos for industrial or commercial purposes.*
(4) *Processes for modifying the genetic identity of animals … likely to cause them suffering without substantial medical benefit to man….*[26]

It is difficult to believe that in any jurisdiction with a humane basis, this will not be followed. Nevertheless, there should be minimal problems in respect of microorganisms as long as they are not environmentally dangerous — indeed they are specifically exempted from any protection in several instruments, notably Article 53 (b) of the EPC — they are too lowly to have any rights.

20.7.4. *Algorithms, methods and processes*

Historically, discoveries, theories, artistic works, methods for doing business and presentation of information (to summarize) have been excluded. Business methods are now patented in the US — (Amazon.com has received a US patent for its so called "one click" marketing system, but this is now under challenge) and in microbiology or molecular biology it is difficult to think of any new process or method of doing things, perhaps cutting down the steps or time involved in the isolation of a fermentation product, which if it works, is not useful in one way or another.

However, you may think of a method of making a bacterium produce a protein, say, in a manner distinct from that in a pre-existing patent for producing the same protein, and the owner of this might challenge. Such is the situation in the currently unresolved dispute in the US over the production of erythropoietin.[27] Amgen markets it after having cloned the gene and inserted it into Chinese hamster ovary cells; the other company Transkaryotic Therapies persuades human cells to produce it by means of a promoter.

A difficulty also arises when the process is known already but either a new starting material or new product is claimed. For example, a patented bacterium could be used to convert compound A into compound B, and either A or B is novel; obviously, A could be a component of the culture medium and B a protein. Until recently, in the US (via what is known as the *Durden* case[28]) such a maneuver was not patentable, but the law was changed.[29]

20.7.5. *Mosaicism*

Good ideas can come in many forms — one might realize that elements of a number of known processes or inventions can be abstracted, combined in some novel way, and made useful in some sense. This is known as "mosaicism" and is attacked by patent lawyers on the grounds of obviousness; however, the rejoinder must be that if assembly has been obvious then somebody else would have constructed it. Therefore, it may be that if one can put together a variety of techniques in a novel way, then the result will be patentable. In microbiology, one could think of using an unusual but established culture medium for a known bacterium, along with a known but hitherto (in that system) unused method for purifying its product. A patent application might succeed.

20.8. Potential Licensees

Even if an invention is patentable, the cost thereof may be prohibitive unless a clear commercial advantage is likely; for that reason an inventor will be at an advantage if he can identify a company or organization likely to take up a license for exploitation of the invention. It is as well to keep an eye on this when developing the invention. Before discussion with an outside party, one should get a non-disclosure agreement signed. In rare cases, especially in universities, an invention will be regarded so highly that a spin-off company might be set up with the employee taking an executive role.

Governments actively encourage inventors to take out patents. Inventions put technology into the public domain and stimulate business and further inventions. In most countries then, apart from a patent being taken over for the use of the government for perhaps defense purposes, there is a system of compulsory licensing. If an invention, having been patented, is not being developed after say three years, any person can apply for a compulsory license. Even this lag period can be waived if the invention is related to food, medicine, or treatment of disease.

References and Notes

1. As might be expected, the literature on patents in biology, including microbiology, is huge. Since the law is highly territorial, in theory there could be a separate book to explain the system in each country of the world. The scientific journals *Nature* and *Science* take a keen interest in scientific patents and keep the reader up to date. The specialist journal *Biotechnology Law Report* is especially valuable. For a bird's eye view worldwide, see Gutterman AS and Anderson BJ, *Intellectual Property in Global Markets*, Kluwer Law International, London, 1997.

2. There has been a long standing ban on patents for medical treatment in most if not all countries, in the US from 1862. Of course, the scope of "medical treatment" has to be evaluated. For a discussion, see Garris JL, The case for patenting medical procedures, *Am. J. Law Med.* **XII**: 86, 1996.

3. The case involved the claim by a postdoctoral fellow, Jenny Chou, for a share in a patent for a herpes simplex vaccine in which her former supervisor was cited as the sole inventor. The judge held that she had no standing since by law an employee's inventions are the property of the employer. See *Science*, 31 March 2000, p. 2399. For a comprehensive discussion, see Puri K, Ownership of employee's inventions: A comparative study, *12 Intellect. Prop. J.*, December 1997, p. 34.

4. The *Gayler* case is discussed in Halpern S, Nard CA and Port KL, *Fundamentals of United States Intellectual Property Law*, Kluwer Law International, The Hague, 1999, p. 201.

5. *General Tire and Rubber Co v Firestone Tire and Rubber Co, 1 Weekly Law Reports* 799, 1972.

6. *Bristol Myers Squibb v Baker Norton.* See discussion in *CIPA J.*, June 2000, p. 293.

7. *Biogen v Madeva* (1996) has been extensively discussed and analyzed. See for example Cornish WR, *Intellectual Property*, Sweet and Maxwell, London, 1999, p. 150. Although it has no binding force outside England, due to the closeness of the reasoning it is likely to be persuasive in most common law jurisdictions. In fact, the Biogen patent was revoked by the House of Lords for other reasons, including over-broadness of claim. Despite that, it is said that Sir Keith Murray, who did the work in Edinburgh in the late seventies (the first of his patents being granted in 1978), became a multimillionaire. *Biogen* also appeared in the Singapore courts in 1994 in connection with the infringement of its rights there, the patent having been registered in 1987; see *Biogen Inc v Scitech Medical Products* reported in 1 Asia Intellectual Property Reports, 1996.

8. This archaic term persists unchanged to date, being used freely for example in the *Utility Examination Guidelines for Biotechnology*, Federal Register, US Vol. 66, No. 4, 5 January 2001.

9. Federal Register, citation 8, p. 1093.

10. In some countries there is a self-assessment system, whereby, all other requirements being met, the patent will be issued without formal enquiry into novelty; it is then open to interested parties to challenge the application after publication.

11. For international classification see http://www.wipo.org/eng/general/ipip/strasbourg.htm and for US see http://www.uspto.gov/web/offices/ac/ido/oeip/taf/def/index.htm.

12. WO (World Patent) 9966047.

13. US Patent 5,824,5060.

14. For a discussion of the microbial deposit system see Cornish, citation 7, p. 234.

15. US Patent 6,017,535.

16. The US patent attorneys Ladas and Parry suggest *inter alia* that lab books be (1) permanently bound (not loose leaf); (2) gels and chromatographs be permanently mounted; (3) regularly scrutinized by a third party and signed. This last is obviously the most difficult. See http://www.ladas.com.

17. For discussions of the more liberal US system see Mahendra B, Novelty and method in drug treatment, *New Law J.*, 3 November, 2000, p. 1502; also in Cook T, Doyle C and Jabbari D, *Biopharmaceuticals, Biotechnology and The Law*, Stockton Press, New York, 1991, p. 120.

18. US Patent 4,593,005.

19. US Patent 5,567,732.

20. See *Nature* **404**: 325, 2000.

21. Article 3 of 98/44/EC: Legal protection of biotechnological inventions, 1998.

22. For discussion, see Sheraton H, Gene genius — patent rights and the human genome project, *Biotechnology*, November 1999, p. 9.

23. Federal Register, citation 8, p. 1094.

24. US Patent 141,072.

25. *Diamond v Chakrabarty* 447 US 303, 1980.

26. Art 6 of 98/44/EC.

27. See *Nature* **404**: 532, 2000 [Note added: to the surprise of many, the courts appear to have decided that Transkarystuis process infringes that of Amgen].

28. In *re Durden* 763 F.2d 1406, 226 USPQ 359 (Fed. Cir. 1985).
29. Training materials for treatment of product and process claims in the light of in *re Brouer* and in *re Ochai* and 35 U.S.C 103 (b), see http://www.uspto.gov/web/offices/pac/dapp/oppd/ppclms.htm.

Index